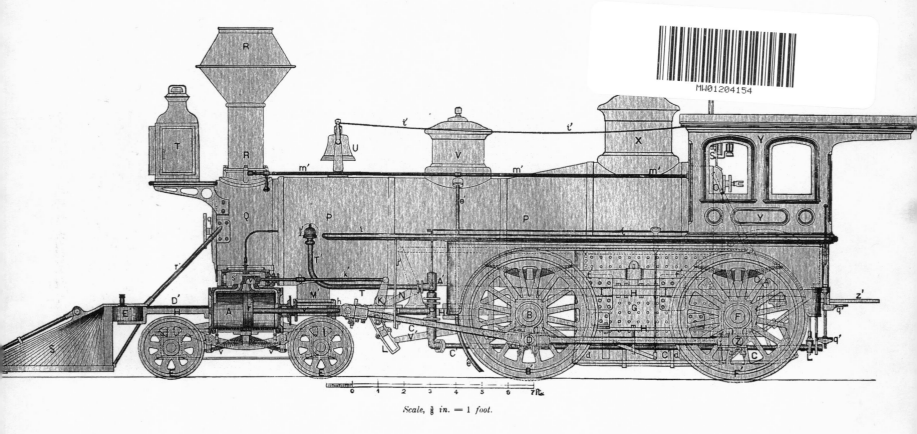

Scale, ⅜ in. = 1 foot.

AMERICAN LOCOMOTIVE

By The Grant Locomotive Works, Paterson, New Jersey.

A, A, Cylinders.
B, Main driving-axle.
C C, Main connecting-rods.
D D, Main crank-pins.
E, E, Truck-wheels.
F, Axle of trailing-wheels.
G, Fire-box.
H, H, H, Frames.
I, I, Frame-clamps.
J, J, Eccentrics.
K, K, Rockers.
L L, Links.
M, Lifting-shaft.
N, N, Lifting arms.
O O, Reverse-lever.
P P, Cylinder part of boiler.
Q, Smoke-box.
R R, Smoke stack or chimney.

w w, Sector or quadrant.
x, Blow-off cock.
e k, Reversing-rod.
y, Truck centre-pin.
z, Throttle-lever.
a′ a, Tubes.
b′ b′, Truck frame.

c′ c′, Bed-plate.
d′, Boiler brace.
e′ e′, Sand pipe.
f′ f′, Equalizing lever for driving-wheels.
g′ g′, Guide-bars or rods.
h′ h′, Receptacle for sparks.
i′ i′, Bell rope.
j′ j′, Guide yoke.
k′, Valve-stem.
l′ l′, Truck equalizing lever.
m′ m′ m′, Hand-rail.
n′, Blow-off cock in mud drum.
o′, Spring balance.
p′, Pump plunger.
q′ q′, Foot steps.
S, Pilot or cow-catcher.
T, Head-light.
U, Bell.
V, Sand-box.
W, Whistle.
X, Dome.
Y Y, Cab or house.
Z, Back or trailing-wheel crank-pin.
A′ Pump air-chamber.

B′, B′, Main driving-wheels.
C′ C′ C′, Supply-pipe.
D′ Front platform.
E′ Bumper timber.
F′ F′, Back driving-wheels.
G′ Coupling-pin.
H′ Friction-plate.
I′ Check-valve.
K′ K′, Foot-board.
L′ Lazy cock.
M′ Mud drum.
N′ N′, Driving springs.
P′ Pump.
R′ Drop-door of grate.
S′ Steam gauge.
T′ T′ Feed pipe.
U′ U′ Forward eccentric rods.
V′ V′ Backward " "
X′ Lifting-shaft spring.
Y′ Y′ Dampers.
Z′ Pushing-bar.
a a′ Tubes.
b b, Grate.
c, Fire-box door.
d d, Ash pan.

f f, Exhaust-nozzles or blast-pipes.
g, Safety-valve lever.
h h, Cross-heads.
i i, Running-board.
j, Throttle-stem.
l, Throttle-pipe.
m m, Dry pipe.
n, T-pipe.
o o, Steam-pipe.
p, Petticoat pipe.
q, Smoke-box door.
r, Piston.
s, Spark-deflector or cone.
t t, Wire-netting in stack.
u u u, Boiler-lagging.
v v, Truck-spring.

r′, Brace to smoke-box and frame.
s′ s′, Steam-chests.
t′ t′ t′, Crown-bars.
u′, Head-light lamp.
v′, Main valve.
w′, Blow-off cock handle.
x′, Bell-crank for throttle-valve.
y′, Piston-rod.
z′, Draw-bar.

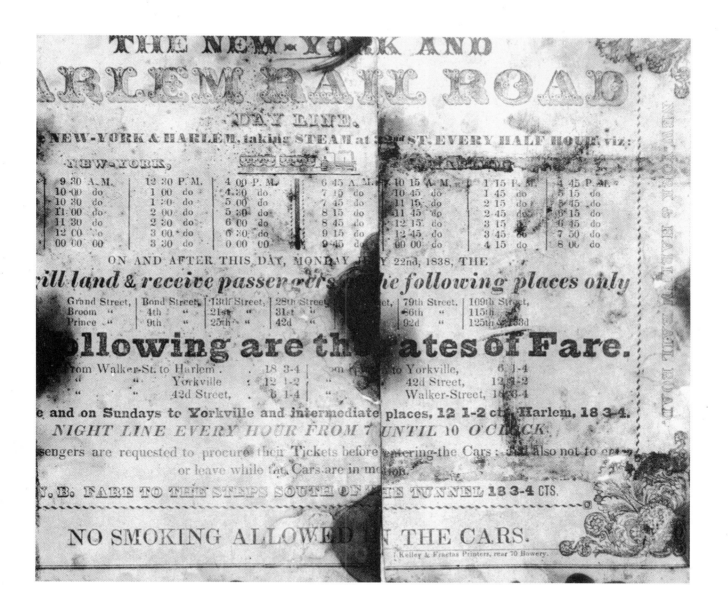

"IN FACT, THE WHOLE BUSINESS OF
RAIL-ROADS AND LOCOMOTIVES
IS NEW TO ALL THE WORLD,
AND TO THIS COMPANY IN COMMON
WITH ALL OTHERS."

Report, South Carolina Canal and Railroad
May 6, 1833

AMERICAN LOCOMOTIVES

AN ENGINEERING HISTORY, 1830–1880

REVISED AND EXPANDED EDITION

JOHN H. WHITE, JR.

THE JOHNS HOPKINS UNIVERSITY PRESS

BALTIMORE AND LONDON

© 1968, 1997 The Johns Hopkins University Press
All rights reserved. Published 1997
Printed in the United States of America on acid-free paper
06 05 04 03 02 01 00 99 98 97   5 4 3 2 1

The Johns Hopkins University Press
2715 North Charles Street
Baltimore, Maryland 21218-4319
The Johns Hopkins Press Ltd., London

ISBN 0-8018-5714-7

Library of Congress Cataloging-in-Publication Data will be found at
the end of this book.

A catalog record for this book is available from the British Library.

TO HOWARD I. CHAPELLE
1901—1975

"WHAT AN IMPOSING
SIGHT IS A LOCOMOTIVE ENGINE,
MOVING WITHOUT
EFFORT, WITH A TRAIN OF
40 OR 50 LOADED CARRIAGES, EACH
WEIGHING MORE THAN TEN
THOUSAND POUNDS!"

De Pambour, 1836

"THE LOCOMOTIVE ENGINE
MAY BE SELECTED AS
THE GRANDEST AND MOST IMPORTANT
DEVELOPMENT OF MODERN
CIVILIZATION AND HUMAN SKILL."

Weissenborn, 1871

TABLE OF CONTENTS

LIST OF ILLUSTRATIONS *xi*

PREFACE TO THE REVISED AND EXPANDED EDITION *xxi*

PREFACE *xxiii*

A NOTE ON SOURCES *xxv*

PART I
ERA OF FUNDAMENTAL LOCOMOTIVE DESIGN

1. INTRODUCTION 3

2. BRITISH IMPORTS 7

3. AMERICAN LOCOMOTIVE BUILDERS 13

   Regional Considerations *14*
   Regional Influences on Design *15*
   Manufacturing Facilities *16*
   Production *19*
   Business Aspects of Locomotive Building *23*
   Standard Design *25*
   Export of American Locomotives *27*

4. MATERIALS USED IN LOCOMOTIVE CONSTRUCTION 29

   Wood *30*
   Copper and Brass *30*
   Cast Iron *30*
   Wrought Iron *31*
   Steel *32*

5. LOCOMOTIVE TYPES AND WHEEL ARRANGEMENTS 33

   The 4–2–0 *33*
   The 4–4–0 *46*
   The 4–6–0 *57*
   The 2–6–0 *62*
   The 2–8–0 *65*
   The 0–4–0, 0–6–0, and 0–8–0 *66*

6. PERFORMANCE 71

   Train Speeds *73*
   Locomotive Power *74*
   Annual Mileage *77*
   Performance Costs *77*
   Fuel Costs per Mile *77*
   Labor Costs *79*
   Repair Costs *79*
   Lubricating Costs *79*
   Operating Cost Trends *80*

7. FUEL 83

   Coal-Burning *86*

CONTENTS

PART II
COMPONENTS

8. **BOILERS AND BOILER ACCESSORIES** 93

    Boilers 93
    Boiler Construction 97
    Boiler Tubes 99
    Boiler Lagging and Jackets 100
    Firebox Construction 102
    Coal-Burning Fireboxes 105
    Grates 108
    Grate Area and Heating Surface 110
    Smokeboxes 111
    Blast Pipes 111
     Variable Exhausts 113
    Smokestacks and Spark Arrestors 114
     The Bonnet Stack 117
     The Centrifugal Stack 120
     Perforated Cone Stacks 121
     Smokebox Spark Arrestors 123
    Feed-Water Pumps 124
    Injectors 128
    Gauge Glasses and Cocks 132
    Steam Gauges 133
    Feed-Water Heaters 137
    Superheaters 142
    Throttles 145
    Safety Valves 146

9. **RUNNING GEARS** 151

    Introduction 151
    Suspension 151
     Springs 156
    Frames 158
     Bar Frames 159
     Riveted Frames 161
     Slab-Rail Frames 162
     Frame Construction 164

    Trucks 167
     Iron-Frame Trucks 169
     Spread Trucks 172
     Safety Trucks 173
     Trailing Trucks 175
    Driving Wheels and Axles 175
     Iron Wheels 176
     Cushion Wheels 177
     Axles 178
    Tires 179
     Cast-Iron Tires 181
     Steel Tires 182
    Locomotive Brakes 184
    Rods and Crossheads 184
     Rods 184
     Crossheads 186
    Valve Gears 187
     Independent Cutoffs 189
     Variable Cutoffs 192
     Link Motion 194
     Radial Valve Gears 199
     Corliss Valve Gears 201
    Valves and Valve Ports 202
    Valve and Cylinder Lubrication 204
    Cylinders 206
     Inside Connection 208
     Compound Locomotives 209

10. **MISCELLANEOUS CONSIDERATIONS** 211

    Cowcatchers 211
    Bells and Whistles 212
    Headlights 215
    Decorative Treatment and Finish 218
    Cabs 221
    Tenders 223
    The Sandbox and Other Traction-Increasers 234

CONTENTS

PART III
REPRESENTATIVE AMERICAN LOCOMOTIVES

Stourbridge Lion, 1829   239

John Bull, 1831   248

Lancaster, 1834   269

Dunham, *ca.* 1837   280

Gowan and Marx, 1839   287

Winans' 4–4–0, 1843–49   297

Philadelphia, 1849   302

Copiapo, 1850   311

Four Fast Passenger Locomotives of the Hudson River Railroad   320

Champlain, 1849   322

Croton, 1851   328

Columbia, 1852   337

Superior, 1854   341

Susquehanna, A Winans Camel, 1854   347

Talisman, 1856–57   358

The Tyson Ten Wheeler, 1857   366

Phantom, 1857   383

Southport, 1857   392

The Flexible-Beam-Truck Locomotive, 1857   396

A Rogers Mogul of 1863   407

The Erie Mogul No. 254, 1865   416

The Rogers 4–4–0, *ca.* 1865   422

Consolidation, 1866   427

The Baldwin Ten Wheeler, 1870   437

Summary   443

PART IV
MORE REPRESENTATIVE LOCOMOTIVES

La Junta, a Cuban 4–2–2, 1843   449

The Stretch Planet, a Hinkley 4–4–0, 1845   454

Elephant, 1849: The First in the West   462

Sandusky, a Ten-Wheel Freight Engine, 1851   470

Lady Elgin and Two Lesser Sisters, 1852   476

Lisle, a Hinkley 0–4–0, 1853   483

Storey, a Western Mogul, 1869   486

Perkins Ten Wheeler, 1871   490

Some Four-Wheel Switchers of the 1870's   494

Marmora, an Eddy Clock, 1876   498

Forney Elevated Railway Locomotive, 1885   503

CONTENTS

PART V

COMMENTS AND NEW INFORMATION 507

APPENDIX SECTION

Appendix A: Biographical Sketches of Locomotive Designers and Builders 537

Appendix B: Contract for the Mohawk and Hudson Railroad's *Experiment*, 1831 547

Appendix C: Specification for the Baltimore & Ohio Railroad's 0–8–0, 1847 549

Appendix D: Specification of New York and Erie Engines, 1851 *551*

Appendix E: Specification for the Western and Atlantic Railroad's *General*, Rogers Locomotive Works, December, 1855 555

Appendix F: Parts and Weight List for the Hinkley Locomotive Works's 4–4–0, 1865 556

Appendix G: Description of a Standard Grant Locomotive Works 4–4–0, 1871 565

Appendix H: Chronology of the American Locomotive, 1795–1875 573

BIBLIOGRAPHY 577

INDEX 583

LIST OF ILLUSTRATIONS

Jacket illustration. The *Tiger*, built by M. W. Baldwin in 1856 for the Pennsylvania Railroad. This wood-burning engine was used in passenger service. (From a contemporary color lithograph.)

Frontispiece. The *Best Friend*, "the first locomotive built in the United States for actual service on a railroad," was built at the West Point Foundry for the South Carolina Railroad, and made its first excursion trip (as shown) on January 15, 1831. (From William H. Brown, *The History of the First Locomotive in America*, rev. ed., 1874.)

Figures

1. The *Baltimore* of the Baltimore and Susquehanna Railroad, built in 1837 by the Locks and Canals company. (From Thomas Norrell.) *9*
2. The *Dr. Ordway* was built in 1839 by Locks and Canals and is shown as rebuilt in 1873. (Smithsonian Photo 49,904.) *10*
3. The *Cincinnati*, built in 1835 by the Vulcan Foundry for the South Carolina Railroad. (Traced by the author from an original drawing at the City of Liverpool Museum.) *11*
4. The Norris Locomotive Works in 1855. (From *U.S. Magazine*, October, 1855.) *18*
5. Interior of the erecting shop at the Norris Locomotive Works, 1855. (From *U.S. Magazine*, October, 1855.) *18*
6. The *Brother Jonathan*, built by the West Point Foundry in 1832, is shown as rebuilt in 1833. (Smithsonian Photo 36,716–A.) *35*
7. The *Robert Fulton*, built by Stephenson in 1831 for the Mohawk and Hudson Railroad. (Traced by the author from an original drawing at the Jervis Library, Rome, N.Y.) *36*
8. The *Robert Fulton* as rebuilt and renamed the *John Bull* in 1833. (Smithsonian Photo 36,716–D.) *37*
9. An unidentified American six-wheel locomotive of about 1840. Possibly a Rogers engine. (From *Materiel Roulan . . . Chemins de fer*, Paris, 1873, Atlas, Plate LXXV.) *38*
10. The *America*, built for the Berlin-Potsdam Railway by William Norris in 1838 or 1839. (From Matthias' *Darstellung . . .*, Berlin, 1841.) *39*
11. Plan view of the *America*, 1838–39. (From Matthias' *Darstellung . . .*, Berlin, 1841.) *40*
12. Rear elevation, cross-section, and cylinder and feed-pump details of the *America*, 1838–39. (From Matthias' *Darstellung . . .*, Berlin, 1841.) *41*
13. Longitudinal cross-section of the *America*, 1838–39. (From Matthias' *Darstellung . . .*, Berlin, 1841.) *42*

LIST OF ILLUSTRATIONS

14. A Norris class "B" 4–2–0 of about 1841. (From Karl von Ghega's *Der Baltimore-Ohio Eisenbahn* . . . , Vienna, 1844 [Library of Congress].) 43
15. End elevations of the *Washington*, by Norris, ca. 1841. (From Karl von Ghega's *Der Baltimore-Ohio Eisenbahn* . . . , Vienna, 1844 [Library of Congress].) 44
16. The *Licking*, built by Rogers in 1846. (Traced from an original Rogers specifications order book.) 45
17. The Utica and Schenectady's No. 11 shown as rebuilt by David Matthew in 1844. (From *Locomotive Engineering*, September, 1893.) 46
18. The Baltimore and Ohio's No. 16, built by Norris in 1838 as a 4–2–0. (Smithsonian Photo 40,616.) 47
19. Campbell's patented eight-wheel locomotive of 1836. (From the Historical Society of Pennsylvania.) 49
20. A Rogers 4–4–0 of about 1843. (From John Weale's *Ensamples of Railway Making*, 1843.) 50
21. The *Massachusetts* and *Connecticut*, built in 1846 by Rogers. (Traced by the author.) 51
22. The *Wm. H. Watson*, built in 1847 by James Millholland in the Baltimore and Susquehanna Railroad shops. (From an original drawing in the Smithsonian Institution.) 52
23. The *Mohawk* of the Albany and Schenectady Railroad as rebuilt in 1847 or 1848 by Walter McQueen in the company shops. (After a drawing in the *Practical Mechanics Journal*, May, 1850.) 53
24. The *Tioga*, built by Baldwin in 1848 for the New York and Erie Railway. (Traced from an original drawing by the author.) 54
25. The New York and Erie's No. 13, built in 1848 by Swinburne. (Chaney Photo 24,277.) 54
26. A Baldwin 4–4–0 of about 1848. (From an original engraving.) 55
27. The *Allegheny*, built in 1850 by Baldwin for the Pennsylvania Railroad. (From the Pennsylvania Railroad.) 55
28. The *New Jersey* illustrates the modern 4–4–0 first constructed by Rogers in 1852. (From C. L. Winey.) 56
29. The *James S. Corry*, built by Baldwin in 1859 for the South Carolina Railroad. (Smithsonian Photo 46,562–C.) 60
30. The Philadelphia and Reading's No. 405, built in the company shops in 1875. (From Alois Feyrer's *Der Lokomotivbau*, Vienna, 1877.) 61
31. The *Cumberland*, built by Baldwin for the Pennsylvania Railroad in 1852 after a Millholland design. (Traced from an original drawing by the author.) 63
32. The *Cumberland*, remodeled in 1855 as a 4–6–0, is shown here as rebuilt in about 1863. 63
33. The Baltimore and Ohio's fast passenger Mogul No. 600, built in 1875 at the company shops. (From Alois Feyrer's *Der Lokomotivbau*, Vienna, 1877.) 64
34. An eight-wheel connected locomotive of the Baltimore and Ohio Railroad dating from about 1848. (From J. S. Bell's *The Early Motive Power of the Baltimore and Ohio Railroad*, 1912.) 67
35. The Baltimore and Ohio's *Hero*, built by Thatcher Perkins in 1848 at the company shops. (Smithsonian Photo 47,995.) 68
36. Diagram table showing the development of the American locomotive from 1835 to 1875. 76
37. Horizontal coal-burning boilers developed by Ross Winans from 1844 to 1857. 106
38. Griggs's firebrick arch and diamond smokestack for coal-burning locomotives. (From Thomas T. Taber.) 107
39. Millholland's coal-burning passenger locomotive the *Hiawatha*, built in 1859 at the Reading shops. (From Alexander L. Holley's *American and European Railway Practice*, 1861, Plate 53.) 109
40. Composite drawing showing many different styles of spark arrestors devised before 1860. (From *The Engineer*, 1862.) 115
41. Typical bonnet stack for coal- and wood-burning engines. (From Douglas Galton's *Report to the Lords of the Committee of the Privy Council* . . ., 1857–58, Plate 18.) 117
42. The earliest known style of bonnet stack is shown in this drawing of an 1833 Camden and Amboy locomotive. (From *Master Mechanics Report*, 1884). 118
43. Wood-burning spark arrestors showing early development of the bonnet stack. (From F. A. Ritter von Gerstner's *Die innern communication der vereinigten Staaten von North-America*, 1842–43, Plate 14.) 119

LIST OF ILLUSTRATIONS

44. Radley and Hunter centrifugal spark arrestor patented in 1850. (From *Lowe's Railway Guide*, 1861.) *122*
45. Modified Radley and Hunter spark arrestor with a section removed to show interior arrangement. (From the Baldwin-Lima-Hamilton Corp.) *123*
46. Worthington feed-water-pump locomotives. (From *American Engineer*, 1857.) *127*
47. Injector introduced in the U.S. by William Sellers & Co. in 1860. (From Alexander L. Holley's *American and European Railway Practice*, 1861, Plate 63.) *129*
48. Drawing of injectors based on an illustration in a Sellers circular of December, 1862. *130*
49. An early injector on the Pennsylvania Railroad's No. 214. (Chaney Photo 17,484.) *131*
50. Back-head view of a Pennsylvania Railroad engine of about 1850, possibly the *Wyoming*. (Chaney Photo 10,858.) *134*
51. Back-head view of a Grant Locomotive Works eight wheeler of 1871. (From the Grant catalog, 1871.) *135*
52. Bourdon steam gauge advertising circular issued in 1852 by E. H. Ashcroft. (From the Historical Society of Pennsylvania.) *136*
53. Hudson's exhaust-steam feed-water heater of 1859. (From Alexander L. Holley's *American and European Railway Practice*, 1861, Plate 46.) *139*
54. Ebbert's exhaust-steam-smokestack feed-water heater used by the Chicago and North Western Railway in the mid-1850's. (From *Railroad Advocate*, 1857.) *140*
55. The Mine Hill Railroad's No. 30, built in 1860 by Baldwin. *141*
56. Richardson automatic safety valve introduced in 1867. (From M. N. Forney's *Catechism of the Locomotive*, 1879.) *148*
57. Andrew Eastwick's patent of 1837 was devised to improve the suspension of the 4–4–0 locomotive. (From the National Archives.) *152*
58. Harrison's 1838 patent for locomotive equalizers. (From the National Archives.) *153*
59. Norris and Knight's 1843 patent for locomotive suspensions. (From the National Archives.) *154*
60. The *Virginia*, built for the Winchester and Potomac Railroad by Norris in 1842. (From Karl von Ghega's *Die Baltimore-Ohio Eisenbahn* . . ., Vienna, 1844 [Library of Congress].) *155*
61. A Baldwin flexible-beam-truck locomotive of 1844 with a Bissell air spring. (From a Baldwin advertising card sent to S. M. Felton in 1844.) *157*
62. A Baldwin 4–2–0 of about 1840 showing an outside iron frame. (Smithsonian Photo 26,799–C.) *160*
63. Detail of the slab-rail frame on the El Paso and South Western Railroad's No. 1, built by the New York Western Railroad's No. 1, built by the New York Locomotive Works in 1857. (Drawn by the author.) *162*
64. Detail of the frame on the Boston and Albany Railroad's No. 39, built in 1876 by Wilson Eddy. (Drawn by the author.) *163*
65. The *Massachusetts*, built by Hinkley for the Philadelphia and Reading Railroad in June, 1849. (From Emil Reuter's *American Locomotives*, 1849.) *165*
66. The New York and Harlem Railroad's *Amenia*, built by Rogers in 1850. (Smithsonian Photo 45,947–D.) *166*
67. Longitudinal cross-section of the *Davy Crockett* built by Robert Stephenson & Co. in 1833 for the Saratoga and Schenectady Railroad. (From a lithograph by J. H. Bufford of Boston, dated about 1833.) *169*
68. The inside-connected locomotive No. 201 of the Baltimore and Ohio Railroad, built in 1853 by Norris. (From a drawing made by J. S. Bell in about 1865.) *170*
69. Live-spring truck used on the Baltimore and Ohio Railroad's No. 201 and on other 4–4–0's of the so-called Dutch Wagon class. (After an original drawing dated September 29, 1853.) *171*
70. Norris' independent cutoff, introduced in about 1845. (Drawn by Keith Buchanan.) *191*
71. The Cuyahoga variable cutoff valve gear devised in 1849 by Ethan Rogers of the Cuyahoga Steam Furnace Co. (Drawn by Keith Buchanan.) *193*
72. The Baldwin-Cuyahoga variable cutoff valve gear introduced in 1852 and produced by Baldwin as late as 1860. (Traced by the author.) *195*
73. The Morris and Essex Railroad's *Essex* showing the crosshead-drivers valve gear devised by Seth Boyden in 1837–38. (From *Railroad Gazette*, 1902, after an

LIST OF ILLUSTRATIONS

original drawing.) *200*

74. The Erie Railway's No. 122, rebuilt in 1868 or 1869 by the Shepard Iron Works, is thought to have been the first compound locomotive built in America. *209*
75. Combination bell and handrail stand introduced by the Mason Machine Works, 1864. (From C. E. Fisher.) *214*
76. The square-case headlight. (From Bowles Railroad Supply Catalog, 1860.) *216*
77. The round-case headlight. (From Bowles Railroad Supply Catalog, 1860. *217*
78. The New Hampshire Central's *Reindeer,* a product of John Souther, 1854. (From Thomas Norrell.) *222*
79. Drawing of the tender for the *West Point,* built by the West Point Foundry in April, 1831, for the South Carolina Railroad. (From *Railroad Gazette,* May 25, 1883.) *224*
80. A Planet-class locomotive built by Robert Stephenson & Co., shown as it appeared on the Bangor and Piscataquis Railroad. (Smithsonian Photo 22,707–M.) *225*
81. Diagram drawings of tenders, 1844–45, redrawn from sketches in Baldwin's specification books. (Data courtesy De Golyer Memorial Library, Dallas, Texas.) *227*
82. The six-wheel tender accompanying the Württemberg State Railway's *Necker* was built by Baldwin in May, 1845. (From the *Eisenbahn Zeitung,* October 11, 1845.) *228*
83. The Philadelphia and Columbia Railroad's *Tioga,* built by Norris in 1848. (From Thomas Norrell.) *229*
84. A six-wheel tender built at the shops of the Saratoga and Schenectady Railroad by E. S. Norris in about 1849–51 after a design developed earlier in Philadelphia. (From Frederick Moné's *A Treatise on American Engineering,* 1854.) *230*
85. A standard eight-wheel tender built by Baldwin in about 1860 for the Cleveland and Pittsburgh Railroad. *231*
86. Engraving of the *Stourbridge Lion* prepared by James Renwick from the original locomotive. (From *The Steam Engine,* 1830.) *243*
87. The *Agenoria* was built by Foster, Rastrick & Co. on the design of the *Stourbridge Lion.* (Courtesy of the Science Museum, London.) *244*
88. Reconstruction drawing of the *Stourbridge Lion* prepared by the Delaware and Hudson Railroad in 1932. (From the Delaware and Hudson Railroad.) *245*
89. End elevations and cross-section of the Delaware and Hudson's reconstruction drawing of the *Stourbridge Lion.* (From the Delaware and Hudson Railroad.) *246*
90. Drawing of the *Stourbridge Lion's* tender as reconstructed by the Delaware and Hudson Railroad in 1932. (From the Delaware and Hudson Railroad.) *247*
91. This reconstruction shows the *John Bull* as built by Robert Stephenson in June, 1831. (Drawn by the author.) *255*
92. Photograph of the *John Bull* as it was modified and rebuilt by the Camden and Amboy Railroad, believed to have been taken in about 1865. *256*
93. A drawing of the *John Bull* published by the *American Railway Review* on February 20, 1862. *256*
94. The *North America,* a sister engine of the *John Bull,* was built in 1835 by the Camden and Amboy Railroad. (Drawing from the *American Railway Review,* May 8, 1862.) *256*
95. A drawing of the *Maryland,* built in 1832 by Robert Stephenson & Co. for the Newcastle and Frenchtown Railroad. (After an original drawing at the Stephenson Works of English Electric.) *257*
96. The boiler drawing believed to have been used in the construction of the *John Bull.* (After an original drawing at the Stephenson Works of English Electric.) *258*
97. General arrangement of the *John Bull* as it exists today. (From the Pennsylvania Railroad.) *259*
98. Arrangement of the valve gear, cylinder, and throttle of the *John Bull.* (From the Pennsylvania Railroad.) *260*
99. Detail of the outside and inside frames of the *John Bull.* (From the Pennsylvania Railroad.) *261*
100. Detail of the leading wheels and axle of the *John Bull.* (From the Pennsylvania Railroad.) *262*
101. Detail of a wooden driving wheel believed to be from the *John Bull* as originally constructed. (After the original wheel in the Smithsonian Institution.) *263*
102. Detail of the present cast-iron driving wheel of the

LIST OF ILLUSTRATIONS

John Bull. (From the Pennsylvania Railroad.) 264

103. Arrangement of the pilot of the *John Bull.* (From the Pennsylvania Railroad.) 265
104. General arrangement drawing of the present four-wheel tender of the *John Bull.* (From the Smithsonian Institution.) 266
104–A. General arrangement drawing showing left elevation and cross-section of the four-wheel tender of the *John Bull.* (From the Smithsonian Institution.) 267
105. The original wooden frame and valve gear of the *John Bull.* (Traced by the author.) 268
106. Drawing traced from an original labeled the "Lancaster" but possibly the drawing for a 4–2–0 of about 1836–38. 273
107. The Utica and Schenectady Railroad's No. 1, built by Baldwin in 1836. (Smithsonian Photo 36,716–G.) 274
108. The Philadelphia and Columbia Railroad's *Martin Van Buren,* built by Baldwin in 1839. (Smithsonian Photo 26,799–B.) 275
109. An advertising lithograph issued by Baldwin in about 1855. (From C. L. Winey.) 276
110. A reconstruction of the *Lancaster* dating from 1834. (Drawn by the author.) 277
111. A reconstruction of the *Lancaster* showing details of the running gear. (Drawn by the author.) 278
112. A reconstruction of the *Lancaster* showing front elevation, cross-section, rear elevation, and a detail of the crank boss. (Drawn by the author.) 279
113. H. R. Dunham's foundry as shown by an 1837 lithograph. (From the New York Historical Society.) 282
114. Front and side elevation of Dunham's locomotive of about 1837. (From P. R. Hodge's *The Steam Engine,* 1840.) 283
115. General plan and details of Dunham's 1837 locomotive. (From P. R. Hodge's *The Steam Engine,* 1840.) 284
116. Cylinder details of Dunham's 1837 locomotive. (From P. R. Hodge's *The Steam Engine,* 1840.) 285
117. Boiler and pedestal detail of Dunham's 1837 locomotive. (From P. R. Hodge's *The Steam Engine,* 1840.) 286
118. Portrait of Joseph Harrison, Jr. (1810–1874). 287
119. An Eastwick and Harrison advertisement from the 1839 Philadelphia directory. 290
120. The *Ontalaunee,* built for the Philadelphia and Reading Railroad by the Newcastle Manufacturing Co. in 1843. (Photo from Thomas Norrell.) 290
121. A French engraving of the *Gowan and Marx* published in 1843. (From *Annales des Points et Chanssees's,* 1843.) 291
122. A side elevation of the *Gowan and Marx.* (From Karl von Ghega's *Die Baltimore-Ohio Eisenbahn . . .,* 1844.) 292
123. A reconstruction drawing of the *Gowan and Marx* showing side and front elevations. 293
124. A reconstruction drawing of the *Gowan and Marx* showing details of the engine. 294
125. A reconstruction drawing of the style of tender used on the *Gowan and Marx.* 295
126. The Peoples Railway's No. 3, now exhibited by the Franklin Institute. 296
127. A reconstruction drawing of the *Juno* showing the engine as originally constructed for the Baltimore and Ohio Railroad by Ross Winans in 1848. (From a drawing by C. B. Chaney.) 298
128. A reconstruction drawing of the *Juno* showing the engine as rebuilt by the Baltimore and Ohio Railroad in 1856. (From a drawing by C. B. Chaney.) 298
129. Detail drawing of a 4–4–0 built by Ross Winans in about 1845. (Traced from an original drawing in the Smithsonian Institution.) 299
130. Detail drawing of an eight-wheel tender built by Ross Winans in about 1845. (Traced from an original drawing in the Smithsonian Institution.) 300
130–A. Drawing of an eight-wheel tender built by Ross Winans in about 1845. (Traced from an original drawing in the Smithsonian Institution.) 301
131. Norris advertisement. (From the *American Railroad Journal,* 1845.) 303
132. Portrait of James Millholland (1812–1875). 304
133. The *Philadelphia,* shown as remodeled by Millholland in 1848 or 1849. (From Emil Reuter's *American Locomotives,* 1849 [Library of Congress].) 306
134. Longitudinal cross-section of the *Philadelphia.* (From Emil Reuter's *American Locomotives,* 1849 [Library of Congress].) 307

LIST OF ILLUSTRATIONS

135. Cross-section and elevation of the *Philadelphia*. (From Emil Reuter's *American Locomotives*, 1849 [Library of Congress].) 308
136. Details of the *Philadelphia*. (From Emil Reuter's *American Locomotives*, 1849 [Library of Congress].) 309
137. A Norris six-coupled freight locomotive of about 1855. (Smithsonian Photo 41,770.) 310
138. The *Copiapo*, built by Norris Brothers in 1850 for the Copiapo Railway, Chile. 315
139. The *Copiapo* as presently exhibited at the School of Mines, Copiapo, Chile. (Courtesy of the Chilean State Railway.) 315
140. The *Copiapo* on exhibit in Copiapo, Chile. (Courtesy of the Chilean State Railway.) 316
141. The *Beaver*, built by Norris Brothers in 1849 for the Pittsburgh, Fort Wayne and Chicago Railroad. 316
142. Side elevation and plan view of the running gear of the *Copiapo*, 1850. (Drawn by the author.) 317
143. Frame, truck, and valve-gear details of the *Copiapo*, 1850. (Drawn by the author.) 318
144. End elevation and cross-section of the *Copiapo*, 1850. (Drawn by the author.) 319
145. The *Gasconade*, built by Taunton in 1853 for the Pacific Railroad, Missouri. (Photo from the National Museum of Transport, St. Louis, Mo.) 324
146. The *Champlain*, built by Taunton in 1849 for the Hudson River Railroad. (Drawing from the *Practical Mechanics Journal*, III, 1850.) 325
147. A detail of the *Champlain's* frame and valve gear. (From the *Practical Mechanics Journal*, IV, 1851.) 326
148. Two cross-section elevations of the *Champlain* showing crank axle, driving wheel, and yoke. (From the *Practical Mechanics Journal*, IV, 1851.) 326
149. Patent drawing of the Asa Whitney truck wheel dated March 19, 1850 (No. 7202). 327
150. Portrait of Walter McQueen (1817–1893). 328
151. The *Essex* of the Great Western Railway of Canada, built in 1853 by the Lowell Shops. 331
152. A schematic drawing of the *Croton* showing the principal parts of a typical mid-nineteenth-century inside-connected locomotive. 332
153. The *Croton*, built by the Lowell Shops in 1851 for the Hudson River Railroad. (From Frederick Moné's *An Outline of Mechanical Engineering*, 1851.) 333
154. A longitudinal section of the *Croton*. (From Frederick Moné's *An Outline of Mechanical Engineering*, 1851.) 334
155. Front and rear elevations with cross-section drawings of the *Croton*. (From Frederick Moné's *An Outline of Mechanical Engineering*, 1851.) 335
156. The six-wheel tender of the *Croton*. (From Frederick Moné's *An Outline of Mechanical Engineering*, 1851.) 336
157. The *Columbia*, built in 1852 by the Lowell Machine Shop for the Hudson River Railroad. (From Frederick Moné's *A Treatise on American Engineering*, 1854). 338
158. Longitudinal section and details of the *Columbia*. (Drawing from Frederick Moné's *A Treatise on American Engineering*, 1854.) 339
159. End elevations and cross-section of the *Columbia*. (Drawing from Frederick Moné's *A Treatise on American Engineering*, 1854.) 340
160. The New York Locomotive Works; an advertisement appearing in the Jersey City directory of 1854–55. (From the Library of Congress.) 342
161. The *Superior*, built by Breeze, Kneeland & Co., was delivered to the Hudson River Railroad in March, 1854. (Drawing from Frederick Moné's *A Treatise on American Engineering*, 1854.) 343
162. Details of the *Superior*, 1854. (Drawing from Frederick Moné's *A Treatise on American Engineering*, 1854.) 344
163. Details of the *Superior*, 1854. (Drawing from Frederick Moné's *A Treatise on American Engineering*, 1854.) 345
164. Details of the *Superior*, 1854. (Drawing from Frederick Moné's *A Treatise on American Engineering*, 1854.) 346
165. Engraving of the Ross Winans Locomotive Works in Baltimore. (From *Rambles in the Path of the Steam Horse*, 1855.) 348
166. The *C. E. Detmold* of the Cumberland and Pennsylvania Railroad, built by Winans in 1859. (From

LIST OF ILLUSTRATIONS

W. R. Hicks.) *349*

167. The *Tuscarora* of the Huntington and Broad Top Mountain Railroad, a medium-furnace Camel built in about 1859. *350*
168. The 199, the last Camel purchased by the Baltimore and Ohio Railroad (in 1863), was built in 1860. (From a photo taken by R. K. McMurray at the Mt. Clare Shops in 1863.) *351*
169. The *Susquehanna*, built in 1854 for the Philadelphia and Reading Railroad, shown as it appeared in 1859 after slight modifications. (From A. Bendel's *Aufsätze Eisenbahnwesen in Nord-Amerika*, Berlin, 1862.) *355*
169–A. Detailed views of the *Susquehanna*'s valve motion. (From A. Bendel's *Aufsätze Eisenbahnwesen in Nord-Amerika*, Berlin, 1862.) *356*
170. The last Camel being scrapped at the Mt. Clare Shops, Baltimore, in 1898. (Smithsonian Photo 49,654.) *357*
171. The *J. H. Devereux* of the U.S. Military Railroad, built in March, 1862, by the New Jersey Locomotive and Machine Company (From the Library of Congress.) *360*
172. The *Talisman*, built in about 1857 by the New Jersey Locomotive and Machine Company. (From Gustavus Weissenborn's *American Engineering*, 1857 and 1861.) *361*
173. Details of the *Talisman*, 1857. (From Gustavus Weissenborn's *American Engineering*, 1857 and 1861.) *362*
174. Details of the *Talisman*, 1857. (From Gustavus Weissenborn's *American Engineering*, 1857 and 1861.) *363*
175. Details of the *Talisman*, 1857. (From Gustavus Weissenborn's *American Engineering*, 1857 and 1861.) *364*
176. Tender of the *Talisman*, 1857. (From Gustavus Weissenborn's *American Engineering*, 1857 and 1861.) *365*
177. A Tyson Ten Wheeler with the Artists Excursion Train on the Baltimore and Ohio Railroad west of Piedmont, Va., in June, 1858. *368*
178. The Baltimore and Ohio Railroad's No. 230 at Keyser, W. Va., in 1862. *369*
179. The 227, built by Denmead in December, 1857, shown as rebuilt. *370*
180. The Tyson Ten Wheeler No. 225, built by Denmead in August, 1857. (From Douglas Galton's *Supplement to the Report on the Railways of the United States*, 1858.) *371*
181. Detail of Tyson's ten-wheel boiler. (From Douglas Galton's *Supplement to the Report on the Railways of the United States*, 1858.) *372*
182. Detail of the frame for the Tyson Ten Wheeler. (From Douglas Galton's *Supplement to the Report on the Railways of the United States*, 1858.) *373*
183. Cylinder saddle and variable exhaust for the Tyson Ten Wheeler. (From Douglas Galton's *Supplement to the Report on the Railways of the United States*, 1858.) *374*
184. Truck detail for the Tyson Ten Wheeler. (From Douglas Galton's *Supplement to the Report on the Railways of the United States*, 1858. *375*
185. Valve-gear detail for the Tyson Ten Wheeler. (From Douglas Galton's *Supplement to the Report on the Railways of the United States*, 1858.) *376*
186. Cab and rear-frame detail for the Tyson Ten Wheeler. (From Douglas Galton's *Supplement to the Report on the Railways of the United States*, 1858.) *377*
187. Throttle and safety-valve detail for the Tyson Ten Wheeler. (From Douglas Galton's *Supplement to the Report on the Railways of the United States*, 1858.) *378*
188. Driving-wheel detail for the Tyson Ten Wheeler. (From Douglas Galton's *Supplement to the Report on the Railways of the United States*, 1858.) *379*
189. Side elevation of the tender for the Tyson Ten Wheeler. (From Douglas Galton's *Supplement to the Report on the Railways of the United States*, 1858.) *380*
190. Plan view of tender tank and frame for the Tyson Ten Wheeler. (From Douglas Galton's *Supplement to the Report on the Railways of the United States*, 1858.) *381*
190–A. Tender truck of the Tyson Ten Wheeler. (From Douglas Galton's *Supplement to the Report on the Railways of the United States*, 1858.) *382*
191. Portrait of William Mason (1808–1883). (From John Livingston's *Portraits of Eminent Americans*, 1853.) *383*
192. The Baltimore and Ohio Railroad's No. 232, built by

LIST OF ILLUSTRATIONS

Mason in June, 1857. 386
193. Elevation of the *Phantom*, built by Mason in 1857 for the Toledo and Illinois Railroad. (From Daniel Clark and Zerah Colburn's *Recent Practice in the Locomotive Engine*, 1860.) 387
193–A. Longitudinal section and plan view of the *Phantom*. (From Daniel Clark and Zerah Colburn's *Recent Practice in the Locomotive Engine*, 1860.) 388
194. End elevation and sections of the *Phantom*. (From Daniel Clark and Zerah Colburn's *Recent Practice in the Locomotive Engine*, 1860.) 389
195. Side elevation of the *Phantom's tender*. (From Daniel Clark and Zerah Colburn's *Recent Practice in the Locomotive Engine*, 1860.) 390
196. Tender built by Mason in May, 1862, for the U.S. Military Railroad. 390
197. A Mason tender of 1856 or 1857 built by Mason for the Baltimore and Ohio Railroad and sketched in 1863 by J. S. Bell. (Redrawn by the author.) 391
198. The *Southport*, built in 1857 by Danforth, Cooke & Co. for the Delaware, Lackawanna and Western Railroad. (From Daniel Clark and Zerah Colburn's *Recent Practice in the Locomotive Engine*, 1860.) 393
198–A. Longitudinal section of the *Southport*. (From Daniel Clark and Zerah Colburn's *Recent Practice in the Locomotive Engine*, 1860.) 394
198–B. End view of the *Southport*. (From Daniel Clark and Zerah Colburn's *Recent Practice in the Locomotive Engine*, 1860.) 395
199. The *John C. Calhoun*, one of the first flexible-beam-truck locomotives built by Baldwin, 1843. (After an original drawing traced by John Stine.) 400
200. A flexible-beam-truck locomotive of 1844. 401
201. A flexible-beam-truck locomotive of about 1845. (From Max Mayer's *Esslinger Lokomotiven: Wagen und Bergbahnen*, Berlin, 1924.) 401
202. A flexible-beam-truck locomotive of about 1848. 401
203. A flexible-beam-truck locomotive of 1857. (From Daniel Clark and Zerah Colburn's *Recent Practice in the Locomotive Engine*, 1860.) 402
204. Front and end elevations of a flexible-beam-truck locomotive of 1857. (From Daniel Clark and Zerah Colburn's *Recent Practice in the Locomotive Engine*, 1860.) 403
205. Tender drawings for a flexible-beam-truck locomotive of 1857. (From Daniel Clark and Zerah Colburn's *Recent Practice in the Locomotive Engine*, 1860.) 404
206. *The Northumberland* of the Pennsylvania Railroad, shown as rebuilt in 1864. 405
207. A flexible-beam truck of about 1855–60. (From Baldwin Locomotive Works, *Record of Recent Construction*, No. 25, 1901.) 405
208. The *Santiago De Cuba*, built by Baldwin in May, 1866. (Photo from H. L. Broadbelt.) 406
209. Builder's photograph of the New Jersey Railroad and Transportation Company's No. 36, built by Rogers in October, 1863. 409
210. The New Jersey Railroad and Transportation Company's No. 36 in service. 410
211. Reconstruction drawing of the New Jersey Railroad and Transportation Company's No. 36, side elevation. 411
212. Reconstruction drawing of the New Jersey Railroad and Transportation Company's No. 36, details. 412
213. Reconstruction drawing of New Jersey Railroad and Transportation Company's No. 36, details. 413
214. Reconstruction drawing of the New Jersey Railroad and Transportation Company's No. 36, details. 414
215. The standard ten-wheel tender of the New Jersey Railroad and Transportation Company. (New drawing traced from Gustavus Weissenborn's *American Locomotive Engineering and Railway Mechanism*, 1871.) 415
216. Mogul freight locomotive No. 254, built in 1865 for the Erie Railway by Danforth, Cooke & Co. (Photo from Walter Lucas.) 418
217. Side elevation of a Mogul freight locomotive built for the Erie Railway by Danforth, Cooke & Co. (From the Railway and Locomotive Historical Society.) 419
218. Longitudinal section and plan view of a Mogul freight locomotive built by Danforth, Cooke & Co. for the Erie Railway. (From *Engineering*, May 11, 1866.) 420
219. Front elevation and cross-section of a Mogul freight locomotive built by Danforth, Cooke & Co. for the

LIST OF ILLUSTRATIONS

Erie Railway. (From *Engineering*, May 11, 1866.) *421*

220. The New Jersey Railroad and Transportation Company's No. 8, built by Rogers in 1865. *423*

221. A coal-burning passenger locomotive built by Rogers in about 1865. (From *Locomotive Engineering*, 1864 and 1871.) *424*

222. A coal-burning passenger locomotive built by Rogers in about 1865. (From *Locomotive Engineering*, 1864 and 1871.) *425*

223. Details of a coal-burning passenger locomotive built by Rogers in about 1865. (From *Locomotive Engineering*, 1864 and 1871.) *426*

224. Portrait of Alexander Mitchell (1832–1908). *427*

225. The *Consolidation*, built by the Baldwin Locomotive Works in 1866 for the Lehigh and Mahanoy Railroad. (Photo from the Pennsylvania Railroad.) *431*

226. The *Consolidation*. (Photo from H. L. Broadbelt.) *432*

227. Reconstruction drawing of the *Consolidation*, 1866. (Drawn by the author.) *433*

228. Reconstruction drawing of the *Consolidation* showing frame and boiler details. (Drawn by the author.) *434*

229. Reconstruction drawing of the *Consolidation* showing truck and wheel details. (Drawn by the author.) *435*

230. Reconstruction drawing of the *Consolidation*'s tender. (Drawn by the author.) *436*

231. A Baldwin ten-wheel freight locomotive of 1871. (From *Manual for Railroad Engineers*, 1878.) *439*

232. A soft-coal-burning, class 24½D locomotive built by Baldwin for the Evansville, Hendersonville and Nashville Railroad in May, 1870. *440*

233. An anthracite-coal-burning, class 24½D locomotive built by Baldwin for the Danville, Hazleton and Wilkes Barre Railroad in November, 1870. (From the original erecting drawing.) *441*

234. An anthracite-coal-burning, class 24½D locomotive built by Baldwin for the Danville, Hazleton and Wilkes Barre Railroad in November, 1870. (From the original erecting drawing.) *442*

235. *La Junta*, 1843, built by Thomas Rogers in Paterson, New Jersey, for service in Cuba. (Wayne Weiss photo.) *451*

236. *La Junta*, drawings made from measurements taken from the original machine while it was on display at Lenin Park, Havana, Cuba, in 1983. (Drawn by the author.) *452*

237. More drawings of *La Junta* and its tender, now preserved in Havana, Cuba. (Drawn by the author.) *453*

238. A lithograph of an inside-cylinder 4–4–0 issued by Hinkley and Drury, locomotive builders of Boston, in about 1845. (Collection of the late C. L. Winey.) *457*

239. A longitudinal cross-section representing a Hinkley locomotive of the same design as shown in the previous illustration. (Drawn by the author.) *458*

240. A reconstructed front elevation of a Hinkley 4–4–0 of about 1845. (Drawn by the author.) *459*

241. A reconstructed rear elevation and a detail of the crank axle of a Hinkley 4–4–0 of about 1845. (Drawn by the author.) *459*

242. The Fitchburg Railroad's *Brattleboro*, 1844, an outside-connected Hinkley eight wheeler with Hinkley's standard six-wheel tender. (Railway and Locomotive Historical Society photo.) *460*

243. A reconstruction of a Hinkley six-wheel tender of about 1845. (Drawn by the author.) *461*

244. The *Elephant*, the first locomotive in California, built by John Souther of Boston in 1849. (Drawn in 1993 by the author.) *465*

245. The *Roanoke*, built in 1854 by John Souther for the Virginia and Tennessee Railroad. (Norfolk and Western Railway photo.) *466*

246. A reconstructed side elevation of the *Elephant*. (Drawn by the author.) *467*

247. A reconstruction of the *Elephant*. (Drawn by the author.) *468*

248. An eight-wheel tender of the type used by the *Elephant*. (Reconstructed drawing by the author.) *469*

249. The *Sandusky*, built by the Portland Company in 1851 for the Mad River and Lake Erie Railroad. (Traced by the author from the original in the collection of the Maine Historical Society.) *472*

250. The *Wilmore*, built in 1856 by Smith and Perkins of Alexandria, Virginia, for the Pennsylvania Railroad. (*Locomotive Engineering*, December, 1900.) *473*

251. M. W. Baldwin produced his first ten wheelers in 1852 on a plan very similar to the one shown here. (Copied by the author from Fig. 109.) *474*

LIST OF ILLUSTRATIONS

252. A Baldwin 4–6–0 built in 1855 for the Pennsylvania Railroad. (Smithsonian Institution, Chaney Neg. 3759.) *475*
253. The *Lady Elgin*, built by the Portland Company in 1852 for service on the first railroad in what is now Ontario, Canada. (Traced by the author from an original drawing in the Maine Historical Society.) *478*
254. The *Lady Elgin* as it appeared late in its service life on the Grand Trunk Railway. (Canadian National Railway photo.) *479*
255. A copy of a drawing from which the Portland Company built two locomotives, the *Danville*, and the *Consuelo*, in 1852 for service in Canada. (Traced by the author from an original drawing in the Maine Historical Society.) *480*
256. The *Forest State*, built by the Portland Company. (Smithsonian Neg. 59238.) *481*
257. The tender floor plan for the *Danville* and the *Consuelo*. (Traced by the author from an original drawing in the Maine Historical Society.) *482*
258. The *Lisle*, built in 1853 for the Syracuse and Binghamton Railroad. (Traced by the author from an original at the Boston Public Library.) *485*
259. The Virginia and Truckee Railroad's No. 3, the *Storey*, built in 1869 by the Union Iron Works of San Francisco. (*Engineering*, March 10, 1871.) *488*
260. A cross-section engraving of the *Storey*. (*Engineering*, March 10, 1871.) *489*
261. The Louisville and Nashville Railroad built several ten-wheel freight engines at its Louisville shops to the design of Thatcher Perkins in 1871. (*Engineer*, February 16, 1872.) *492*
262. Engravings showing end and cross-section drawings of Perkins ten wheelers. (*Engineer*, February 16, 1872.) *493*
263. The *Suffolk*, built by Rhode Island in 1871, a prototypical four-wheel switcher of its day. (Collection of Thomas Norrell.) *496*
264. The Burlington Lines' No. 335, produced in 1879. (*National Car Builder*, February, 1880.) *497*
265. The *Marmora*, produced at the Boston and Albany Railroad's Springfield, Massachusetts, shops in 1876. (Drawn by Richard K. Anderson.) *501*
266. The *Barnes*, built in 1877, a near duplicate of the *Marmora*. (Private collection.) *502*
267. The Brooklyn Elevated Railway's first set of engines were copied from those serving on the New York Elevated lines. Rhode Island built these engines in 1885. (*American Machinist*, February 28, 1885.) *505*
268. End elevations of the Brooklyn Elevated's first group of locomotives. (*American Machinist*, February 28, 1885.) *506*
269. The *South Carolina*, one of Allen's double enders, built in 1832 by the West Point Foundry for the South Carolina Railroad. (*American Railroad Journal*, February–March, 1890.) *511*
270. The Chicago and North Western's *Missouri*, 1865, the first of a series of standard 4–4–0's. (*Railway Master Mechanic*, June, 1900.) *515*
271. The Chicago and North Western's *Crawford*, 1867, a Baldwin version of Cushing's standard 4–4–0. (Baldwin Locomotive Works photo.) *515*
272. The *Yates*, built in 1848 by Rogers for the New York and Erie Railway. (Z. Colburn, *Locomotive Engineering*, 1872.) *519*
273. Drawings for a 66-inch-gauge 4–4–0, prepared at the Portland Company in around 1855, showing beautifully detailed cross-sections through the firebox. (Traced from an original drawing in the Maine Historical Society, Portland, Maine, by Robert J. White.) *522*
274. The Cheshire Railroad's *New Hampshire*, built in 1847 by the Hinkley Locomotive Works. (Smithsonian Neg. 96-1521.) *525*
275. A light eight wheeler for the Cleveland and Toledo Railroad produced by the Amoskeag Manufacturing Company in 1853. (Norris Pope Collection, courtesy of Stanford University Press.) *526*
276. Interior hardware of an oil-burning headlamp from Adams and Westlake catalog of 1887. (Smithsonian Neg. 87-3887.) *527*
277. Preserved tender exhibited with the 1846 locomotive *Lion* at the Maine State Museum, Augusta, Maine. (Drawn by the author.) *530*
278. A large iron-framed tender built in 1884, representative of the larger, 3,000-gallon-capacity tenders coming into favor because of larger freight locomotives. (*National Car Builder*, November, 1884.) *532*

Preface to the Revised and Expanded Edition

The steam locomotive, from its beginning, was seen as a noble machine. Few other mechanisms have exercised such a compelling hold on the public's imagination. Nor has mankind had a more faithful friend and servant. It could be so human and gentle, and at the same time, so powerful and fleet. Locomotives were named for presidents, governors, railroad officials, Greek and Roman gods, and creatures of the forest. They were painted like peacocks, with highlights of gilt and polished brass.

Here was a machine that literally had fire in its belly. It roared down the track, a big clattering thing with a noisy bell and whistle. It was heavy and ponderous but cocksure: it rolled on, self-reliant and unafraid. Its only bad habits were drinking and smoking. It is no wonder, then, that thousands have been seduced by the charms and power of the steam locomotive. I am but one of the many who have attempted to pay homage to this marvelous device that cannot speak for itself. The enormous literature about steam locomotives is a testament to the lure of this fascinating mechanism. I feel fortunate to have witnessed the steam locomotive's last years in commercial service. Indeed, I can remember when steam locomotives were commonplace and diesels the exception. At that time, roundhouses were crammed with hot, smoldering engines awaiting the call to take out their next train. Yards were bustling places filled with busy switchers puffing to and fro with cuts of cars. The main lines were a succession of trains propelled by engines that moved over the landscape with columns of steam and smoke bellowing high above their stacks. These were everyday scenes, so commonplace that they went unnoticed by the majority of citizens. As a historian I was, of course, drawn into the subject to discover more about the origin of the species. And so began the investigations that led to this book.

I must acknowledge that this book does not answer all questions regarding early American locomotives. If only it did, I should be a very happy author. There is much more work to be done. My time for such laborious tasks is nearly over, so I can only hope that historians of the future will produce a more complete history.

PREFACE TO THE REVISED AND EXPANDED EDITION

I have added 92 pages and forty-five new illustrations to this expanded reprint. This was not done to increase the scope of the book, which remains largely confined to American locomotives dating from before 1880. The new material fills in certain voids in the old text and allows me to print many drawings never before published. The preparation of the new material (Parts IV and V) was made easier by the work of many scholars and collectors over the past thirty years. There has been a dramatic expansion of railroad historic literature during that time. I will not attempt to list these works here, But I would direct readers seeking more information to a series of locomotive histories published by Alvin B. Staufer. William D. Edson's locomotive rosters of major U.S. railroads and locomotive builders have appeared in separate publications or within the pages of the journal *Railroad History*. They provide much-needed details on the American engine fleet. Thomas T. Taber's *Guide to Railroad Historical Resources* is an exhaustive four-volume survey listing repositories for records on every railroad in the United States and Canada.

The help of a number of individuals was essential to the production of this edition. When it came to finding obscure pieces of data and illustrations, no one proved better than my longtime colleague John Stine. Richard K. Anderson and my brother, Robert J. White, produced two of the more intricate drawings for this book. Wendell Huffman generously shared his detailed knowledge of early western railways with me. Others who helped include the late Gerald M. Best, Stephen E. Drew, Albert S. Eggerton, Jr., Norris Pope, Thomas T. Taber III, and Jim Wilke. Over the years, Edward T. Francis's reading of early newspapers has provided me with many fascinating and obscure news items on our premier locomotives. The manuscript was typed by Mary E. Braunagel and Edwin Wynn.

I am also indebted to many officials at Miami University, in Oxford, Ohio—especially Karl R. Mattox, Charlotte N. Goldy, and Judith A. Sessions—for their support, which aided in the completion of this project. Similar and no less sincere thanks must go to the staff at the Johns Hopkins University press, especially Miriam Kleiger, senior manuscript editor, and Valerie A. Dolan, assistant design and production manager. My final thanks must go to Henry Tom, executive editor at the press, who first suggested that we should move ahead with this reprint.

PREFACE

The pioneer period of the American locomotive ended in about 1855. The machine had developed from a squat little boiler-on-wheels in 1830 to an elegant, well-proportioned mechanism. The pattern created by mid-century was so satisfactory that it underwent little change in the following twenty-five years. Much has already been published on this subject, but none of these accounts has, in my opinion, adequately treated the engineering history of the early American locomotive. The limited documentation makes this undertaking no simple task and it is difficult to offer more than conjecture or qualified guesses on many points.

The present study was written for those already acquainted with locomotive history and construction. It is hoped that the specialist will find new insights and materials not readily available elsewhere.

The general reader and those seeking an elementary treatise on the workings of the engine, the properties of steam, or a glossary of terms should consult the numerous books already published on the subject. These are available in most city or university libraries. I would recommend M. N. Forney's *Catechism of the Locomotive* for a sound introduction to the subject.

I have chosen to confine my remarks to the early period, with 1880 as the cutoff date. The second major phase of locomotive development, 1880–1910, when the machine underwent a dramatic growth in size, must be left for another time or another author. The final or modern era of development has been well covered by Alfred Bruce in his classic work *The Steam Locomotive in America*.

This book has three major sections. The first deals with the operating conditions that influenced the principal design concepts, the makers of locomotives, and the materials used. Part II covers the development of components. The third section is a series of case histories of representative American locomotives. Emphasis is on standard designs and practice.

PREFACE

After seven years of preparation I find myself indebted to many persons and institutions. Howard I. Chapelle, former Curator of the Division of Transportation and now Senior Historian in the Department of Science and Technology, Smithsonian Institution, suggested this study, and unfailing interest and prodding are largely responsible for its completion. Many of the illustrations and sources used are from Thomas Norrell, who has proved himself not only a mentor but also a friend.

Harry Eddy and John G. McLoed of the Bureau of Railway Economics (Association of American Railroads), and Jack Goodwin and Charles Berger of the Smithsonian Library were unsparing in their efforts to secure many obscure publications. The Library of Congress, The National Archives, and the U. S. Patent Office furnished many illustrations.

Among the libraries outside the Washington area which cooperated by supplying material or data were: the Jervis Library, Rome, New York; The American Antiquarian Society, Worcester, Massachusetts; the Historical Society of Pennsylvania, Philadelphia, Pennsylvania; and the City of Liverpool Museums, Liverpool, England.

I must also thank Charles E. Fisher, President of the Railway and Locomotive Historical Society, Professor S. R. Wood, Everett De Golyer of the De Golyer Foundation Library, and Thomas T. Taber of the Railroadians. Keith Buchanan prepared and contributed two excellent valve-gear drawings. Robert M. Vogel and Monte Calvert, both of the Smithsonian staff, offered useful comments on the manuscript. Mary S. Chroniger, Terry Lynn Parrish, and Mary E. Braunagel deserve praise for their care in typing. John Gallman, Penny James, and Gerard Valerio of The Johns Hopkins Press offered freely of their skills in the final preparation of this study.

A NOTE ON SOURCES

The bibliography annotates the major published works consulted, but I feel a few additional comments at the beginning of the text will more clearly define the problems and limitations encountered during the preparation of a work on early American technology.

Only a few technical studies of the locomotive were published before 1870. The works of De Pambour, Colburn, and Reuter are valuable, though incomplete, records of early locomotive practice. Weissenborn's excellent *American Locomotive Engineering* (1871), with its profusion of detailed working drawings, presents a remarkable picture of American locomotive design in the late 1860's and early 1870's.

By 1870 the technical press began regular publication of mechanical drawings. Therefore, a relative abundance of material can be easily found on locomotive design after this date. Secondly, a fine collection of these drawings was brought together as *Recent Locomotives* (1883) by the *Railroad Gazette*. This was followed by a second expanded edition in 1886 and in turn by a similar updated folio entitled *Modern Locomotives* (1897). Other works followed, the most familiar and valuable being the *Locomotive Cyclopedias,* a publication that in many editions presents a fine survey of locomotive engineering to the end of steam. The works dating before 1870 are extremely rare, many being known only to a few specialists. One function of the present work is to reprint the drawings from these early works and make their existence more widely known. A bibliography appended to this publication lists these works completely.

I would prefer to use original drawings exclusively but, as already indicated, remarkably few working drawings have survived. Unlike sailing ships, where vast numbers of antique drawings have been retained as public or military records by government agencies, locomotives have been the concern of private corporations in this country. Railroad companies and locomotive manufacturers were quick to destroy obsolete records. Few were saved by museums or collectors, and those that have survived are often fragmentary. Whole periods are barely represented; the 1840's is a particularly barren time where only

A NOTE ON SOURCES

a few incomplete drawings are known to exist. Therefore, it is difficult to make many meaningful comparisons or to comment precisely on what was standard practice. The actual arrangement and the often subtle changes in locomotive design can only be understood from complete erecting drawings. Line cuts, showing only a side elevation, and photographs give only a *picture* that does not permit a full understanding of the design. Therefore, since we are largely concerned with the development of design and construction, complete working drawings have been used wherever possible. In a few cases reconstruction drawings have been prepared. These are, of course, conjectural but may help the reader better understand the original machine. In all cases the illustrations consulted have been reproduced or cited close to the reconstruction.

It may be noticed that what might appear to be an obvious source for good information—surviving locomotives—has been largely ignored. This was done intentionally, for surviving nineteenth-century locomotives in this country are not authentic representatives of locomotives of their respective periods. Most have been rebuilt to the point that little remains of the original mechanism. Inept restorations have further destroyed the value of these antiques as a reliable record of past practice. The *John Bull,* discussed later in this work, is a case in point. Its numerous rebuildings during 30 years of service leaves only the boiler and a few incidental components. Even these original parts have been altered. The famous *Pioneer* preserved by the Chicago and North Western Railway is another example of a much rebuilt machine, although it has not suffered as much alteration as the *John Bull*. While it is the sole surviving 4–2–0, the first distinctive American wheel arrangement, it is no longer representative of this class of engine as built in 1836. Another *Pioneer,* built in 1851 for the Cumberland Valley Railroad, appears to have survived in a reasonably unaltered state. However, its unusual wheel arrangement (2–2–2 T) is so untypical of American practice that it cannot be regarded as a representative locomotive of the 1850's. It is not possible to determine how much of the original mechanism has survived because no illustrations dated before the 1870's are known to exist. All of this is not to say the surviving relics should be discarded as worthless but rather that for the purposes of this study their inclusion would confuse rather than clarify design trends.

ERA OF FUNDAMENTAL LOCOMOTIVE DESIGN

I

INTRODUCTION

1.

The development and improvement of the locomotive engine was governed as much by economics and traffic as it was by technical breakthroughs or inspired insights of designers. The combination of light traffic and insufficient capital compelled early United States railroads to devise a cheap plan of construction. Light bridges and rails, steep grades, and sharp curves were the result of economy-minded railroad builders. Speed was impossible on such roads, even if desired. Yet the steep grades and sharp curves called for powerful engines.

The commercial railway originated in Britain during the 1820's. Accordingly, late in that decade the earliest railroad enterprises in the United States sent engineers to study the pioneer British lines. The Delaware and Hudson Canal Company sent Horatio Allen; the Baltimore and Ohio Railroad sent George W. Whistler, Jonathan Knight, and Ross Winans. A few years earlier (1825) William Strickland had made a study of British practice for the Pennsylvania Society for the Promotion of Internal Improvement. The reports of these men strongly influenced the construction details of the earliest United States roads. However, it was soon discovered that the British railroad was not practical for our use.

The British railways were models of a civil engineering enterprise, with carefully graded roadbeds, substantial tracks, and grand viaducts and tunnels to overcome natural obstacles. Easy grades and generous curves were the rule. Since capital was plentiful, distances short, and traffic density high, the British could afford to build splendid railways. In this country it was precisely the reverse; though rich in unexploited natural wealth, we had little ready cash. Distances were great, the population was scattered, and, accordingly, traffic density was light. While the British spent $179,000 per mile, partly owing to the high cost of land, few American roads between 1830 and 1860 spent more than from $20,000 to $30,000 per mile.[1] By 1850 Britain was spending $1,000 million for 5,000 miles of railway while the United States spent only $265 million for

---

[1] These cost figures are from Tanner, Gillespie, and Lardner. They represent the total cost of the road, bridges, rail, and equipment. There were exceptions, of course. The Philadelphia and Reading was carefully graded and used 45# T rail. But most roads were quickly and cheaply built only to be rebuilt in later years at enormous cost. The operating costs of United States roads (before 1880) were very high when compared to those of Britain.

6,500 miles. The economy of American roads was effected by building along the natural contours of the land. Lines were built around and over hills, thus avoiding tunnels or expensive cuts and fills. Cheap, light wooden trestles were used instead of substantial stone viaducts. Our first tracks were copied after the stone blocks and chair rails of the British. The Philadelphia and Columbia was one of several pioneer United States roads built on this plan at a cost of $10,179 per mile. Later, it and other roads abandoned this form of track for cheaper but less permanent wood stringer and strap rail at a cost of only $5,604 per mile.[2] Little or no ballast was used. In the winter the ground would freeze, thus making the tracks a rigid unyielding pavement. With the spring rains the roadbed became a soggy mass of mud, providing uncertain support for the tracks. Sharp curves were common, a 1,000-foot radius being quite ordinary. Many roads had 600-foot curves, and the Beaver Meadow Railroad had one as sharp as 250 feet. Grades were also severe; an incline of 30 to 50 feet per mile was not unusual. The Baltimore and Ohio's seventeen-mile grade (near Piedmont, West Virginia) of 116 feet per mile is an extreme example of the steep inclines intended to challenge locomotive power.

Captain Douglas Galton summarized the essence of American railroad construction in the following paragraph from his 1857 report on the subject:[3]

In a new country, where time is not so valuable, high speeds are not required, but a cheap and certain means of locomotion for all classes, and of transporting the produce of the country to a market, is a first necessity. A railway is the best instrument for satisfying this want. Any saving in the cost per mile of a railway adds to the means available for extension; and, in a rapidly developing new country, capital is dear. Hence a rough and ready cheap railway, although it entails increased cost for maintenance, is preferable to a more finished and expensive line. These considerations have influenced the construction of American railways, and the system which has grown up under them is well adapted to the wants of the country.

[2] H. S. Tanner, *A Description of the Canals and Railroads of the United States* (New York, 1840), p. 117.
[3] *Report to the Lords of the Committee of the Privy Council for Trade and Foreign Plantations, on the Railways of the United States* (2 vols.; London, 1857–58), Vol. 1, p. 26.

Accordingly, the requirements for early locomotives in this country were, in order of importance: flexibility, simplicity, power, low initial cost, and ease of maintenance. Fuel economy was decidedly a secondary consideration. Moreover, like any useful tool of civilization, the locomotive did not develop in a vacuum isolated from practical considerations; it was meant for work.

Nineteenth-century American locomotive building was distinguished by conservatism and a steadfast resistance to the acceptance of novel or "new-fangled" designs. This conservatism was essentially an intelligent rejection of many foolish reforms and patents eagerly promoted by impractical or even fraudulent inventors. Although thousands of patents were issued, it would be difficult to account for more than a dozen or so that significantly altered or improved locomotive design. The first concern of early railroad mechanics was the development of a dependable locomotive that would not break down when sent out on the road. Such a machine was perfected by the early 1830's. It would consume more fuel than theory permitted but it wouldn't forever be in shop for repairs. Most attempts to improve fuel efficiency, such as feed-water heaters, called for a more complex mechanism. This in turn meant more maintenance and a greater first cost which readily offset any operating economies. Most railway officials regarded extra accessories as more parts to go wrong and more road failures. This sentiment was expressed by the president of a small Pennsylvania railroad, who, while speaking of the Sellers grade-climbing locomotive, said: "We are all practical men on this road, and don't believe in thy gimcracks."[4]

The understandable tendency to retain successful designs should be regarded as an intelligent conservatism. However, late in their lives certain prominent master mechanics (such as Griggs, who championed the inside-connected engine through the 1860's) exhibited a perverse allegiance to worn-out designs. Basically, however, I believe that genuine improvements were recognized and freely adopted as they slowly evolved to meet the more demanding operating conditions that developed late

[4] *Journal of the Franklin Institute*, January, 1868, p. 15.

in the nineteenth century. Broadly speaking it was possible (before 1900) to increase locomotive capacity satisfactorily simply by increasing boiler and cylinder size or by raising the steam pressure. This simple, straight-line development reached its limits in about 1900. From that time forward more complex alterations such as super heaters, boosters, and stokers were needed to increase capacity and maintain roadway and weight limitations.

Lest any false impressions be made, it should be emphasized that while no revolution in basic locomotive design was tolerated, the nineteenth century was alive with experiment. Both master mechanics and builders brought forward numerous imaginative machines. Baldwin's geared locomotive (1840), Winans' high speed *Carroll of Carrollton* (1849), Allen's articulated double enders (1832–33), and G. A. Nicolls' twin boiler *Novelty* (1847) are but a few examples of "splendid failures." Later in the century Strong, Holman, and Fairlie demonstrated their ideas with varying success. It should also be noted that the equalizing lever and leading truck, both quickly adopted in orthodox practice, were in their beginnings experimental.

The proper credit of priority or invention is one of the greatest problems in preparing any history of technology. Who first invented the truck or link motion or a host of other valuable improvements so necessary to the locomotive's development is a natural question. While such a question cannot always be definitely answered, it surely cannot be entirely avoided. Innovations were rarely reported in the contemporary press during the early years of the nineteenth century. Only a few railroad publications were in existence and they concerned themselves primarily with financial or corporate news. The patent records cover freak and imaginary designs but rarely include the practical reforms that constituted the real advances in design. Railroad mechanics of the last century almost to the man were practical shopmen not given to literary pursuits or lengthy published reminiscences. Few papers, letters, or documents of the great locomotive designers have been preserved. True, much of the early correspondence of M. W. Baldwin is preserved by the Historical Society of Pennsylvania, but these are business papers and rarely comment on technical matters. Furthermore, many claims for invention were made years later, often in careless second-hand accounts, the main purpose of which seems to be the telling of a good story. The railroad periodicals of the 1890's and early 1900's are full of such accounts.

Surely some invention was the result of inspiration or sudden imaginative flashes, but more often it appears to have been the result of experience and obvious changes suggested by operating difficulties. Since identical difficulties were experienced by railway mechanical officers, it is not surprising that more than one intelligent mechanic arrived at an identical solution independently. The spread truck and level cylinder, a basic reform, was apparently devised at about the same time by several designers. The history of science is replete with similar instances.

Along the same line revolutionary ideas were often advanced, then shelved, only to be revived years later, often without knowledge of the previous invention. Although the poppet throttle valve was developed in the 1830's by E. A. G. Young, it did not attain any notable success until Millholland began to use it on the Reading in the late 1840's. Possibly Millholland knew of Young's design, but it is also conceivable that, being an ingenious engineer, he designed the throttle valve without knowledge of Young's work.

No inordinate stress has been placed on the assignment of inventions or "firsts," even though priority is a natural and important issue in any history of technology. Claims and counter-claims can be found for nearly every significant innovation, and it is a tedious and often inclusive task to sort through such statements. It appears far more important to establish when a mechanical improvement was *accepted* in regular practice.

More perplexing than the inconsistencies of chronology is the almost total lack of statistical information for the period before 1880. Such basic information as the number of locomotives in service is not available until the 1870's. Wheel arrangements in service, the number of steel boilers, or any number of other figures that might point out significant design trends are not available. Rough estimates can be manufactured by assembling scraps of data but these can never be more than educated projections. Even the recorded weights of locomotives are in question if they are given in tons rather than pounds. Long (2,240 pounds) and short (2,000 pounds) tons were not always so identified. The state of New York used the short ton, but the statement that other states did the same "to a greater or less extent" does little to assure our confidence in the actual weight specified in many nineteenth-century reports.[5]

While locomotive design was primarily dictated by operating conditions and was altered principally to meet changes in these conditions, other obvious factors contributed to its development. Mechanics and designers could not help but learn from, and improve upon, the work of their predecessors. Creative engineers made individual contributions through the gift of intellect. The advances in the iron and machine tool industries also obviously permitted improvement in the size, efficiency, and dependability of the locomotive. The development of a specialized locomotive building industry and the rise of the professional engineer made their contributions. Even so, we must return to our original premise that increased traffic and the demand for faster service—in short, a revolution in operating conditions—created a challenge and a need for improved locomotives.

---

[5] D. K. Clark and Z. Colburn, *Recent Practice in the Locomotive Engine* (London, 1860), p. 52.

BRITISH IMPORTS

2.

It is not surprising that the first locomotives in this country were imported from Britain, but it is surprising how quickly they were found unsuitable for American roads and how quickly our rural nation launched its own locomotive establishments. This is not to say that we are not enormously indebted to the British builders. It was the British who perfected the basic design of the locomotive and introduced the separate firebox, multitubular boiler, direct connection to the wheels, blast pipe, and other fundamental features that remained with the steam locomotive to the end of its production. American improvements—mainly in running gears—were hardly as fundamental or as far-reaching as the work of the British designers who had perfected their basic design by 1830.

The four locomotives imported in 1829 by the Delaware and Hudson Canal Company were the first locomotives in North America. These engines represented the British colliery engines, with their single-flue boilers, and differed little from the primitive engines built some fifteen years earlier by Stephenson, Hedley, and Hackworth. The next machines imported, beginning in 1831, were built on the reformed British plan. The famous *John Bull* of the Camden and Amboy Railroad, *Herald* of the Baltimore and Susquehanna, *Whistler* of the Boston and Providence, were among the next locomotives to enter this country from Britain. More were imported in succeeding years as new railroads were built, with a maximum shipment in 1835 of about twenty-six locomotives. Imports fell rapidly after 1837, owing to the financial panic of that year, with only about eight arriving in 1838. In 1841 the Philadelphia and Reading received the *Gem*, which is generally considered the last locomotive to have been imported from Britain. In all, between 1829 and 1841, about one hundred twenty locomotives were imported. Assuming that nearly all were still in use in 1841, they would constitute about 25 percent of the locomotives in service in the United States at that time.

More than one-third of these engines were built by Robert Stephenson at Newcastle upon Tyne. The Vulcan Foundry (Charles Tayleur and Co.) of Warrington, closely associated with Stephenson, followed his design in building locomotives.

Edward Bury, Braithwaite, Rothwell and Hick, and Foster, Rastrick and Company were among other builders who supplied smaller numbers of engines to the United States.

We have already described the differences between the British and American railways; since cheap railroads were being built in this country by the early 1830's, it is apparent that the rigid British engines were found unsatisfactory for our use. The *Robert Fulton* (Fig. 7), built by Stephenson in 1831 for the Mohawk and Hudson Railroad, illustrates the ineptness of the British design for United States service. The short wheel base, (only 51 inches) and the lack of leading wheels caused the engine to rock and derail easily on rough track. Generally, its limited capacity, the consequence of low steam pressure and a small boiler and cylinders, prevented it from hauling a paying load over the steep grades and rough track of American railroads. The *Robert Fulton* and most other early British imports were stock designs modified in no way for service on United States roads.

The following quotations from two reports of the Pennsylvania Canal Commissioners are examples of the bitter and not always objective criticism aimed at the British locomotives:

These engines were not obtained from England . . . (as has been generally supposed) with the view of getting *better* engines than could be procured in this country, but simply because locomotives could not be manufactured here . . . fast enough to meet the wants of the road.[1]

Two of them, viz: the "Fire Fly" and "Red Rover," both British engines [built in 1833 by Tayleur], have been recently sold [to Camden and Woodbury], and it would have been a saving to the Commonwealth had they been given away for nothing the first day they were placed on the track.[2]

Even so, the British imports did many years of good service. The Camden and Amboy's *John Bull*, for example, was used from 1831 to 1866. The *Herald* of the Baltimore and Susquehanna was another long-lived British engine. Built in 1831 by Stephenson as an 0-4-0, it was rebuilt as a 4-2-0 in 1832.

Again rebuilt in 1846, as a six-wheel geared engine, it ran until 1857.

The British imports prompted several famous United States builders to enter locomotive production. Baldwin's experience in assembling the *Delaware* of the Newcastle and Frenchtown Railroad and his inspection of the *John Bull* provided much useful background for the building of his first locomotive, *Old Ironsides*. This engine was an exact copy of Stephenson's "Planet" class. Similarly, Thomas Rogers was introduced to locomotive work in 1835 when he assembled a Stephenson-built engine, the *McNeill*, for the Paterson and Hudson River Railroad.

The Locks and Canals company (later the Lowell Machine Shop) gained experience in locomotive construction by assembling two British locomotives for the Boston and Lowell Railroad in 1831 or 1832.[3] The company's first engine, built in 1835, was an exact copy of a Stephenson "Planet," a class of engine which had already demonstrated its ineptness on United States roads. The Locks and Canals was not dissuaded by this but continued to build engines of the same pattern until 1840. As late as 1839 an officer of the firm said to Robert Stephenson: "We have followed your pattern thinking it the best we have seen, either from England or built here."[4] The company somehow managed to pick up orders from roads desperate for new power. The *Baltimore* (Fig. 1) is an example of the standard Locks and Canals locomotive as it was built originally. The *Dr. Ordway* (Fig. 2) was identical to the *Baltimore* but is shown after its rebuilding in 1873 by the Portland Locomotive Works for the Johns River Railroad. This engine was built in October, 1839, as the *Hampshire* of the Western Railroad of Massachusetts. It was sold to the Cheshire Railroad in 1848 and renamed the *Rough and Ready*.

The position of American railroad managers on the importa-

[1] *Report of the Canal Commissioners of Pennsylvania*, 1835, p. 33.
[2] *Report of the Canal Commissioners of Pennsylvania*, 1836–37, p. 36.
[3] The full title of this company is the Proprietors of Locks and Canals on the Merrimack River. Mainly a water power company, it engaged in locomotive manufacture from 1835 to 1864.
[4] Letter of P. T. Jackson to Robert Stephenson, February 9, 1839, in *Railway and Locomotive Historical Society Bulletin*, No. 4 (1923), p. 45.

Fig. 1. The Baltimore of the Baltimore and Susquehanna Railroad was built in 1837 by the Locks and Canals company. It was closely patterned on the Planet class originated by Robert Stephenson in 1830.

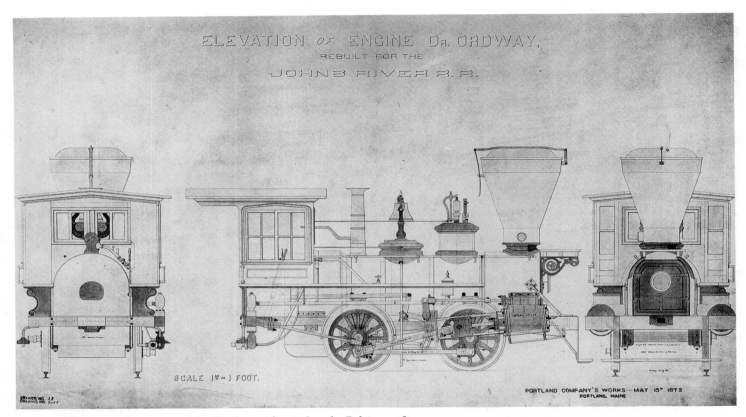

Fig. 2. The Dr. Ordway *was originally almost identical to the Baltimore. It was built in 1839 by the Locks and Canals company and is shown here as rebuilt in 1873 by the Portland Company.*

ERA OF FUNDAMENTAL DESIGN

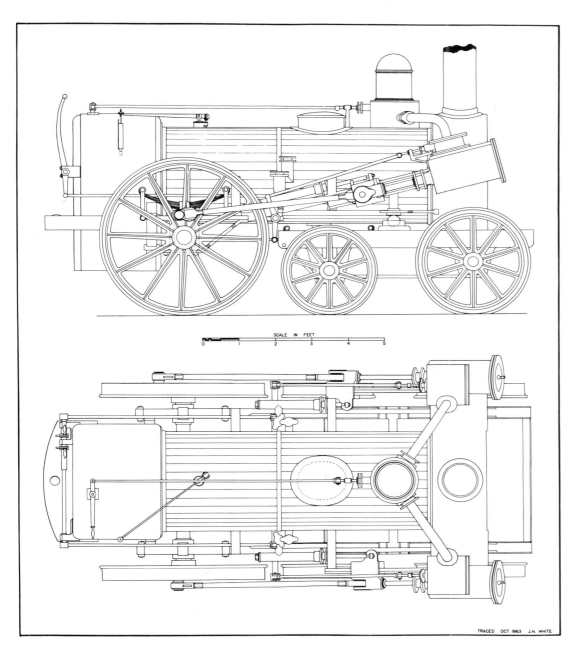

Fig. 3. The Cincinnati *was built in 1835 by the Vulcan Foundry for the South Carolina Railroad. It represents British efforts to keep United States customers by building truck locomotives. The cylinders measured 10 in. by 16 in.; the driving wheels are 54 in. in diameter. The engine was scrapped in 1850.*

tion of British locomotives was unsettled before 1835. The Baltimore and Ohio Railroad, fearing that local mechanics would not supply satisfactory locomotives, petitioned Congress in 1828 to remove the duty from imported iron products in order to encourage the procurement of foreign locomotives and rail.[5] The management of the road reversed its opinion three years later, however, when advertising its contest specifically for locomotives of "American manufacture." The 1832 Baltimore and Ohio annual report reinforced this sentiment by noting that despite the "glowing and flattering description of the efficiency of British locomotives," the road was determined to encourage the production of such machines by native talent.[6] The Baltimore and Ohio proved remarkably consistent in this policy by not buying British locomotives at a time when nearly every American road was tempted to do so.

The Newcastle and Frenchtown Railroad did not share the confidence of the Baltimore and Ohio in native suppliers to produce good engines. Its difficulty with Baldwin in assembling the *Delaware* and the poor performance of Colonel Long's locomotive apparently destroyed confidence in American mechanics. In a memorial of 1833 or 1834 the Newcastle and Frenchtown asked Congress to remove all duties on British locomotives.[7] Such action was thought to be justified because of inadequate domestic workshops and the alleged inferiority of American mechanics. The memorial was not honored but it did result in an outraged rebuttal in the pages of the *American Railroad Journal,* which strongly defended the capacity of native machinists to produce first-class railway engines.[8]

The failure to remove the tariff on British locomotives worked to the advantage of American builders. The duty amounted to about $1,000 or 25 per cent of the engine's selling price.[9] This figure is confirmed by the charges paid in 1831 by the Mohawk and Hudson Railroad for its first imported locomotive.[10] Transportation fees of $100 to $250 per engine added another burden to the importation of locomotives.[11]

British builders, aware of the shortcomings of their standard design for American customers, began to build engines to United States specifications in order to hold that market. The first of these machines was the *Davy Crockett* built by Stephenson in 1833 for the Saratoga and Schenectady Railroad. It was a 4–2–0 designed by John B. Jervis, who had perfected this plan for locomotives a year earlier.[12] Stephenson and Tayleur built other 4–2–0's for the South Carolina Railroad (see the *Cincinnati,* Fig. 3), for Camden and Woodbury, and for the Mohawk and Hudson Railroad. In all, the British built only about twenty truck locomotives for use in this country. Despite attempts to please their American customers, the cause of the British builders was doomed. The shipping and import duties added materially to the first cost. This cost, the rise of United States builders, and the desire to "buy at home" combined to close the American market to foreign makers.

While locomotives no longer crossed the Atlantic, the British enjoyed a profitable trade with United States builders well into the nineteenth century in the supplying of high-quality boiler plates, tires, and other forgings.

---

[5] *Memorial of the Baltimore & Ohio Railroad . . . ,* Petition to U.S. Senate, 20th Cong., 1st Sess., March 17, 1828.

[6] *Baltimore and Ohio Annual Report,* 1832, p. 109.

[7] Minute Books, Newcastle and Frenchtown Railroad, December 13, 1833, as transcribed by C. B. Chaney in about 1930 from original manuscripts owned at that time by C. L. Winey.

[8] *American Railroad Journal,* June 20, 1835, p. 371.

[9] MS record books of Robert Stephenson & Co., 1833, in the files of the English Electric Co., Ltd.

[10] F. W. Stevens, *Beginnings of the New York Central Railroad* (New York, 1926), p. 46.

[11] MS record books of Robert Stephenson & Co., 1833.

[12] An unidentified drawing, possibly of the *Davy Crockett,* appears on p. 305 of J. G. H. Warren's *A Century of Locomotive Building by Robert Stephenson & Co., 1823–1923* (Newcastle, England, 1923). It may well be one of the drawings that Jervis stated he had sent to Stephenson for a six-wheel engine (see Fig. 67). *American Railroad Journal,* May 11, 1833.

American Locomotive Builders

3.

The early emergence of a strong locomotive building industry directly influenced the design and character of railway engines in the United States. Domestic builders produced a flexible, cheap machine well suited to severe local operating conditions. Commercial builders controlled locomotive design more directly than did similar British firms, for few American railroads attempted to build their own engines; most of the big designers and the significant design reforms are associated with commercial builders, not with railroad shops. British locomotive development depended upon locomotive superintendents like Daniel Gooch and company shops like Swindon, whereas American locomotive development was achieved by such business houses as Baldwin, Rogers, and Norris. Because of the direct influence of commercial builders on mechanical development, a brief survey of the characteristics of this industry is necessary.

The locomotive industry began early in this country and developed with amazing speed at a time when many railroads were little more than paper organizations. The first locomotive built in this country was produced by the West Point Foundry Association of New York City in 1830 for the South Carolina Railroad. This engine, the *Best Friend*, was completed before many British engines had been imported. The British engines were imported not because of the absence of home industry but because American builders could not satisfy the sudden and large demand for locomotives.

Many small foundries and machine shops, such as Dunham, Sellers, and West Point, were attracted to locomotive work by the great demand for engines occasioned by the rapid rise of the railway in this country during the 1830's. All of these shops—counting stationary engines, boilers, and other heavy iron fabrications among their regular products—were equipped to build locomotive engines. In fact, until the 1870's most domestic builders did not specialize in building locomotives but were often engaged in several lines of work. As early as 1840 about ten locomotive builders were in existence. By mid-century there were forty firms engaged in this industry; the Panic of 1857 had closed over half of the shops. Never again were there more than twenty firms at any one time actively engaged in locomotive building. In all, between 1830 and 1950 there were about 150 establishments engaged in the manufacture of locomotives.[1] Most of these concerns built locomotives for a short time only and, accordingly, their production was limited.

Like other American iron manufactures, the locomotive building industry experienced rapid growth in the nineteenth century. In the first years, many small inefficient producers satisfied a growing market. After 1860 more adequately equipped plants, professional engineers and managers, and the general uplift in all parts of iron technology led to the manufacture of better locomotives by a few specialized makers. By the end of the century the industry was dominated by fewer than six major builders.

REGIONAL CONSIDERATIONS

The strongest commercial builders were concentrated in the Middle Atlantic States with Philadelphia and Paterson as the nineteenth-century centers of the American locomotive industry. The largest firms were Norris, Baldwin, and Rogers. The 1860 census estimated that three-fourths of all the locomotives were produced in these two cities. In the early 1850's Rogers overtook Norris and Baldwin and continued its lead for nearly ten years. Rogers was the most progressive builder in the country at that time and championed such design reforms as wagon-top boilers, spread trucks, level cylinders, and link motion. Baldwin regained its leadership in the 1860's and maintained that position until the end of the century. Several small builders, led by Schenectady, founded the American Locomotive Company in 1901 in an almost desperate attempt to challenge Baldwin's overwhelming dominance of the industry.

The New England builders were next in importance to the Philadelphia and Paterson manufacturers. Starting somewhat later than the Middle Atlantic builders, the first New England builder was the Mill Dam Foundry of Boston. This firm built only two engines in 1834. The Locks and Canals machine shop of Lowell, Massachusetts, was the next firm to take on locomotive building. Despite an early beginning in 1835, its total production in some thirty years of building was less than 150 engines. In the late 1840's other small shops entered the field; among these were Taunton, Portland, Lawrence, and Souther. The Boston Locomotive Works, operated by Holmes Hinkley, largest of the New England concerns, enjoyed a production nearly equal to the great Paterson and Philadelphia builders up until about 1855. Unfortunately, the Hinkley shop, like the other northern shops, adopted the inside-connected engine which rapidly fell from favor after the early 1850's. The New England shops soon acquired a bad reputation, not only for obsolete design, but for cheap and indifferent workmanship.[2] They never fully recovered from this decline despite the efforts of William Mason and other gifted New England mechanics to reverse the trend.

The New England shops followed a peculiar shop labor system, known as "job hands." The job hands were foremen of various departments of the locomotive works who were hired on contract to manufacture or assemble specific parts of the locomotive. The system is analogous to our present-day subcontractor. The job hand plan as it was practiced in the early 1850's was described by Colburn in *Engineering* (January 3, 1862):

The "job hands," who contracted for the work, worked themselves, and employed each from one to thirty workmen, we "finding everything," shop-room, tools, power, oil, steel, &c.; except files, emery, emery cloth, &c., for bench work, which were found by the job hands themselves. But, which may appear surprising, the men employed by the job hands were mostly engaged by the day. The ordinary wages of a good journeyman were from 40s. to 50s. a week, ordinary hands being paid from 30s. to 40s., and apprentices

---

[1] See *Railway and Locomotive Historical Society Bulletin*, No. 58, for a list of U.S. locomotive builders. A similar list will also be found in E. P. Alexander's *Iron Horses* (New York, 1941).

[2] *Engineer* (Philadelphia), August 30, 1860, p. 20; hereafter cited as *Engineer* (Phila.). Other reports on the poor design of New England engines are given in the *American Railroad Journal* during the 1850's.

14s. The hours were ten a day, viz., from 7 a.m. till noon, and from 1 to 6 p.m. Of course, under a system whereby fixed prices were paid for the several parts of all engines, the small locomotives cost hardly less than the largest—the cost of labour being the same for both.

A more detailed idea of the work done by job hands is recorded in the notebooks of P. I. Perrin, foreman of the Taunton Locomotive Company.[3] Among the many entries made between 1847 and about 1885, the following show the variety of work performed:

1847

29 October
Luther Anthony
    Contracts to make the pistons & packing, springs Bolts & nuts for the Same, fit and key the Same to the Rod the whole of which shall be completely finished and the rods turned & draw filed the whole to be done in a workmanlike manner for the Sum of 32 dollars for each Locomotive Engine.

. . .

    Also to Finish 2 Crossheads complete fit & key the same to the Piston Rods as above for the Sum of 15 dollars to each Engine.

. . .

1848

6 January
Wm. F. Hathaway
    One Set of Connecting Rods for Locomotive Engine viz
2 Inside rods
2 Outside d°
to be fitted with Brass boxes keys & Bolts & Nuts as per Draught in a workmanlike manner for the Sum of 50 dollars.

Chas. Sylvester
    For turning and finishing 6 Bells for Locomotives with arm for Rope & Brass yoke for the Same to be all fitted finished complete at 6.50 Each.

$30.00

As railroads spread south and west, local foundries were attracted to locomotive work. Cincinnati, which built its first locomotive in 1846, was the largest nineteenth-century center of locomotive building in the West. Its three builders produced about five hundred engines between 1846 and 1868. Other shops opened in Chicago, Louisville, St. Louis, and Cleveland but their life and production was limited; most of these plants were closed by the Panic of 1857. Virginia promised to be a center of locomotive building for a few years; factories engaged in such work were located in Alexandria, Richmond, and Petersburg. The largest of these was Richmond's Tredegar Iron Works, which built upwards of one hundred locomotives between about 1850 and 1860.[4] Aside from a handful of engines built in Charleston in the 1830's, the deep South did not enjoy much in the way of a locomotive industry. In the far West a few engines were built in San Francisco after 1860 by the Union Iron Works and the Vulcan Iron Works. The high freighting fees required to bring in locomotives from the East should have encouraged a strong western shop, but the Union Works is known to have produced only thirty machines from 1865 to 1882.[5] The locomotive builders of the Midwest, the South and the far West were, even in their collective production, small and unimportant. Most built only a few machines for local use. It was not until after 1915, when the Lima Locomotive Works began production of road engines, that the eastern builders recognized that formidable western competition was to be reckoned with.

REGIONAL INFLUENCES ON DESIGN

Until the late 1850's there was a tendency for builders to adhere somewhat to regional designs. The New England builders favored the inside-connected engines and built almost exclusively on this design from 1845 to 1855. These machines were sold to railroads in all areas of the country; hence the type cannot be associated only with the northeastern states. Further-

[3] The Perrin notebooks are in the collections of the Old Colony Historical Society, Taunton, Massachusetts.

[4] Kathleen Bruce, *Virginia Iron Manufacturers in the Slave Era* (New York, 1931), pp. 281–85.
[5] *Railway and Locomotive Historical Society Bulletin*, No. 68 (1946), pp. 40–49.

more, many builders outside this area built this type of engine; Rogers and Baldwin may have preferred outside-cylindered engines but they built to the wishes of their customers. Inside-connected engines were built even in the West; the Detroit Locomotive Works built a sizable number for the Michigan Central Railroad.

The Bury boiler was favored by builders in the Middle Atlantic States and was virtually a trademark of all engines built in this region from the early 1830's to about 1855. It appears to have been used rarely by New England builders and almost never by western builders because the design was obsolete when most of these firms opened.

It is even more difficult to make much of an argument for distinctive regional locomotive designs after 1855. By this time basic arrangement and the details of design were so universally agreed upon that it was difficult to distinguish a Philadelphia-made engine from a machine built in Portland, Maine; the only identifying mark was the maker's name plate or the style of dome cover. Even in the first years of locomotive building the basic plan of the machine was agreed upon by commercial makers in all regions of the United States. The horizontal fire tube boiler, leading truck, and direct connection to the wheels were designs employed universally. Maverick designs—and these were mainly alternates to details rather than conceptual changes in the basic plan—were usually the product of an independent master mechanic, such as Wilson Eddy or James Millholland, and not the result of geographic location.

MANUFACTURING FACILITIES

Except for the largest builders, most early locomotive shops were poorly equipped, understaffed, and undercapitalized. The engineering departments of such firms were equally small; they included only a superintendent, who acted as chief engineer and designer, and one or two assisting draftsmen. Such a small concern, having neither staff nor funds, could not afford to experiment but instead concentrated on the production of a simple engine for the market.

The *American Railroad Journal* for August 25, 1853, estimated that a locomotive shop capable of producing three engines a month could be built and equipped for slightly less than $42,000. Another $10,000 was required for material and immediate cash needs. Funds or connections for financing in the amount of $100,000 were also necessary. This relatively modest capital requirement was largely responsible for the organization of numerous locomotive shops during the nineteenth century. Proprietors of machine shops and foundries were tempted by the prospect of profitable local markets to add locomotives to their existing line of products.

An example of an average-sized locomotive producer was the Portland Company of Portland, Maine. Founded in 1846 by Septimus Norris and prominent local citizens, it was opened to supply locomotives for the projected railway intended to connect Montreal and Portland.[6] The first engine was built in 1848. A capacity of sixty locomotives per year was claimed in 1855 but the company's production averaged about ten engines per year up until 1860. Production continued intermittently until 1907; a total of 628 locomotives was built. Capitalization was $180,000, and the plant employed 365 men. The company was located on the harbor and consisted of the following major shops plus several smaller auxiliary buildings:

Machine Shop	63' × 225'
Foundry	63' × 175'
Car Shop	50' × 200'
Blacksmith Shop	60' × 200'
Boiler Shop	50' × 125'

Unfortunately, no record exists of the number and style of machine tools at the Portland Company. A contemporary estimate of the tools required for a locomotive shop of similar ca-

[6] E. T. Freedley, *Leading Pursuits and Leading Men* (Philadelphia, 1856), p. 314; see also *American Railroad Journal*, November 14, 1846, p. 685.

pacity is available and will serve to illustrate this point.[7] The list of tools follows:

- 30 Lathes in size from 19″ to 82″ swing
- 1 Boring lathe, 48″ swing
- 8 Planers, 6′ to 16′ long
- 3 Polishing lathes, no size specified
- 10 Upright drills, no size specified
- 2 Bolt cutters, no size specified
- 2 Boilers, 42″ in diameter
- 1 Stationary engine, 14″ × 42″ cylinders
- ? Power saws
- 1 Power shear
- 1 Trip hammer

plus drills, dies, taps, bench vises, pulleys, shafting, etc.

The facilities of the Norris Locomotive Works in Philadelphia offer an interesting contrast to the Portland Company. Fortunately, an elaborate description of the Norris plant was published in 1855 when the firm was recognized as the largest builder of locomotives in this country.[8] The Norris factory evolved from a small shop in a converted stable in 1834 to a mighty industrial works located in central Philadelphia by 1855. The plant was located on Spring Garden Street near Broad and occupied nearly two acres. The ten buildings, many of them three-story brick structures, and the machinery and stock were valued at one million dollars.

The shops and their over-all floor size are listed below:

Steam Hammer Shop	80′ × 104′
Truck Shop	68′ × 100′
Tank Shop	30′ × 100′
Boiler Shop	80′ × 100′
Erecting Shop	130′ × 179′ (2 buildings)
Engine Room and Stables	200′ × 254′
Blacksmith	116′ × 153′
Finishing	153′ × 166′
Foundry	70′ × 103′
Office and Drafting	size not given

[7] *American Railroad Journal*, August 25, 1853, pp. 549–50.
[8] *United States Magazine*, October, 1855, pp. 151–67.

The erecting or assembly shop is the three-story double building to the left of the center of Fig. 4. Twenty-four tracks were served by a combination transfer-turntable located in the open central court between the two structures.

The procedure for building a locomotive was well organized. After an order was cleared by the front office the engineering department prepared a lengthy specification on printed forms. Pertinent portions of these forms were sent to the foundry, hammer shop, and the other shops involved so that the needed parts would be made. When finished, the pieces were brought to the erecting shop for assemblage in the following sequence:

—First, the boiler is placed and accurately leveled—the frame is set and fastened—the braces are set and fastened—the check and whistle stands are riveted fast—set the cylinder—then the rock arms and pedestals—the boiler flues put in—the driving boxes are put up, and laid out for boring—the center pin set—the driver and the guides are set—pumps put up—next, the reversing shafts—the foot-plate put on and fastened—the valve rod and valves—the steam pipes—the throttle valve—the dome—the safety valve—the feed and supply pipes—the wheels—the eccentric hooks—the connecting rods—the valves set—and the whole of the working part, being now about together, steam is raised, and after "giving her a good blowing out," to get rid of dirt, etc., the cylinder heads are screwed on and the working of the new engine thoroughly tested. At these times the senior proprietor is always present.

When the assemblage was completed, the machine was moved across the transfer table to the opposite building for finishing and painting. An engine could be assembled by a fourteen-man crew in fifteen days. Another two weeks were required for preparing the parts for assembly; this included casting, forging, and machining. Thus, total fabricating time was about thirty days. It might be noted that in 1880 Baldwin required four to five weeks for assembly, not including the preliminary work needed for the preparation of the parts.[9]

Six to seven hundred men were employed in 1855, but times were dull; as many as eleven hundred had been employed two years earlier. A yearly capacity of one hundred and

[9] U.S., Bureau of the Census, *Tenth Census: 1880. Report on the Manufactures of the United States*, II, 57.

Fig. 4. The Norris Locomotive Works in 1855, at the time the largest manufacturer of railway engines.

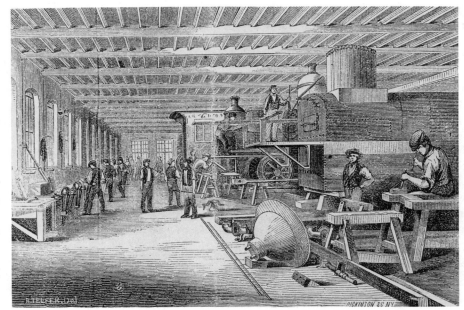

Fig. 5. Interior of the erecting shop at the Norris Locomotive Works, 1855.

fifty locomotives was claimed. It is doubtful if this production was achieved in any one year, but one hundred were built in 1853. Few shops equaled the Norris Works in physical plant or production until after 1870, thus the Portland Company described earlier presents a more typical picture of an early nineteenth-century locomotive building factory.

PRODUCTION

While no production figures for the early years are available, a few estimates can be determined from the total number of engines in service and the capacity of the different builders. In 1840, total United States production was undoubtedly less than 100 locomotives. Ten years later about 400 were produced annually. The 1860 census states that only 470 locomotives were constructed; this figure reflects the dull times that followed the Panic of 1857. Production increased rapidly through the next years as railroad mileage expanded. By 1880, 1,045 locomotives had been built.[10] The production of few builders averaged better than 50 per cent of their claimed capacity. In 1856, for example, Hinkley claimed a capacity of 80 to 100 engines per year but built only 33.[11] In 1854 the forty active shops were credited with the huge capacity of 1,200 engines a year.[12] The figure was undoubtedly optimistic and the projected production of from 600 to 800 locomotives per year was inflated, even for a boom year. The Panic of 1857 closed many of the smaller shops but the industry continued to be plagued by overcapacity. In 1877, John Cooke, a Paterson builder, reported that the sixteen active builders had a combined production capacity of 2,254.[13] Cooke estimated that 780 engines wore out each year, leaving an excess of some 1,500 units to be absorbed by new roads, expanding traffic, or export. The most disquieting note was that only 78 new engines were ordered during the first four months of 1877.

If times were occasionally bad, at least the commercial builders did not face any formidable competition from railroad shops. Relatively few railroads attempted to produce their own locomotives in this country. Incidentally, some of the larger roads did build engines in their repair shops, and much of the rebuilding amounted to new construction, but total production was small. With a few notable exceptions, repairs and maintenance were the main functions of these facilities.

The Baltimore and Ohio Railroad began locomotive construction at its Mt. Clare shops (near Baltimore) in 1832. The facilities were operated under a contract arrangement, however, very much like that of an independent builder; they were returned to the railroad in 1839 when the last contractor, Ross Winans, moved to a new plant. Mt. Clare was largely a repair shop after this date.

The Pennsylvania Railroad's Altoona shops did not undertake locomotive construction until 1866 and were not substantial producers until after 1875.[14] By 1880 a total of 500 machines had been completed; 1,000 had been produced by 1886. Altoona must be viewed as a notable exception to the general situation.

The production of such notable master mechanics as George S. Griggs, Wilson Eddy, and James Millholland was rather modest. Millholland built fewer than one hundred machines during his twenty-eight-year career as a railroad master mechanic, Griggs finished only about thirty, and Eddy just over one hundred.

In 1855 the maximum production of railroad shops was estimated to be one hundred engines.[15] Twenty years later total

---

[10] *Historical Statistics of the United States* (Washington, D.C., 1957), p. 416, lists locomotive production from 1880 to 1957.
[11] Freedley, *Leading Pursuits,* p. 306.
[12] *Railroad Advocate* commented on locomotive production in its issues of November 11, 1854, October 20, 1855, and June 6, 1857.
[13] *Railroad Gazette,* May 18, 1877, p. 220.
[14] Manuscript register of locomotives built at Altoona Shops, prepared by J. D. Lovell in about 1946. (Copy in the Pennsylvania Railroad Library, Philadelphia.)
[15] *Railroad Advocate,* October 20, 1855, p. 1.

capacity of such shops was estimated at two hundred and fifty, but the figure on actual production was not offered.[16] By the late 1880's the figure topped three hundred.[17]

The following tables provide more information on locomotive production. The general figures presented are estimates; this is particularly true for the number of engines in service. The number of engines built by the builders named is more positive and is based on surviving register lists or known construction numbers.

[16] *Railroad Gazette*, May 18, 1877, p. 220.
[17] *National Car and Locomotive Builder*, January, 1888, p. 12.

TABLE 1: TOTAL PRODUCTION OF MAJOR AMERICAN LOCOMOTIVE BUILDERS AND TOTAL NUMBER IN SERVICE

	Baldwin	Norris	Rogers	Hinkley	Schenectady	Total No. of Locomotives in Service
Year of Production						
1835	20	15	0	0	0	175
1840	159	100	26	0	0	475[a]
1845	241	350	76	56	0	1,000
1850	410	420	242	291	0	3,000
1855	677	800	637	581	139	5,000[b]
1860	988	1,000	979	660	234	8,500[c]
1865	1,444	1,250	1,320	750	407	9,500
1870	2,328	0	1,818	950	680	12,500
1875	3,813	0	2,432	1,250	994	15,567[d]
1880	5,430	0	2,676	1,320	1,315	17,949[e]
1885	7,726	0	3,601	1,600	2,037	25,662[e]
1890	11,489	0	4,436	0	3,267	31,812[e]
1895	14,615	0	5,102	0	4,390	35,699[e]
1900	17,315	0	5,654	0	5,700	37,663[e]
Total No. in Service	(1832–1900)	(1834–67)	(1837–1900)	(1844–89)	(1851–1900)	
	17,315	1,400	5,654	1,820	5,700	

[a] Estimate based on U.S. Congressional Report, 1838, lists 345 locomotives; Von Gerstner accounted for 463 in 1839; Ghega accounted for 524 in 1842; *Railway and Locomotive Historical Society Bulletin*, No. 101, lists a total of 589 for the years 1828–40.
[b] *American Railroad Journal*, 1853, estimated 3,500 locomotives; *Railroad Record*, 1855, estimated 6,000. Both figures are based on track mileage. However, *State Engineers Report* (New York), 1856, gives only 4,115 locomotives for the U.S.
[c] *American Railway Review*, 1861, estimated that there were 9,000 locomotives in the U.S.
[d] *Poor's Manual of Railroads*, 1876, stated this figure.
[e] *Historical Statistics of the United States*, 1957, notes figures given for 1880–1900. These differ slightly from those given by *Poor's Manual of Railroads*.

TABLE 2: ESTIMATED YEARLY PRODUCTION OF AMERICAN LOCOMOTIVE BUILDERS[a]

Year	Baldwin	Norris	Rogers	Hinkley	Schenectady	All Other Builders	Total for All Builders
1835	14	10	0	0	0	10	35
1840	9	15	7	0	0	50	85
1845	27	50	14	26	0	75	200
1850	37	20	43	35	0	200	350
1855	47	75	82	29	32	250	500
1860	83	40	88	15	31	200	470[b]
1865	115	40	95	20	46	250	600
1870	280	0	145	40	76	300	850
1875	130	0	42	70	21	300	550
1880	517	0	125	15	115	650	1,405[c]
1885	242	0	73	45	103	350	800
1890	946	0	35	0	313	1,000	2,300
1895	401	0	80	0	140	480	1,101
1900	1,217	0	170	0	415	1,350	3,153

[a] Yearly production for major builders based on builders' lists or representative construction numbers.
[b] Total production given by 1860 census.
[c] Total production for 1880–1900, *Historical Statistics of the United States, 1957.*

Cost was an important consideration for the capital-poor American roads, and builders, aware of this requirement, managed to produce a sound but remarkably cheap locomotive throughout the last century. Locomotive prices for a standard 4–4–0 were constant between 1840 and 1900, $8,000 being a good average figure for the entire period. This general price prevailed, despite rising labor and material costs and the increasing size and complexity of the machine, which the following table illustrates:

Year	Name or type	Maker	Weight	Cost of Locomotive and Tender
1828	*Stourbridge Lion*	Foster, Rastrick	7 tons	$2,914
1828	*Pride of Newcastle*	Stephenson	7 tons	3,663
1830	*Best Friend*	West Point	4½ tons	4,000
1831	*De Witt Clinton*	West Point	3½ tons	3,200
1831	*John Bull*	Stephenson	10 tons	4,000
1832	*Experiment*	West Point	7 tons	4,600
1834	*Lancaster*	Baldwin	7½ tons	5,580
1836	4–2–0	Baldwin	10 tons	6,700

Year	Name or Type	Maker	Weight	Cost of Locomotive and Tender
1843	4–4–0	Rogers	11 tons	$ 8,000
1848	4–4–0	Souther	20 tons	7,300
1855	4–4–0	Lawrence	25 tons	8,825
1860	4–4–0	Baldwin	24 tons	9,725
1860	0–8–0	Baldwin	27 tons	11,331
1863	2–6–0	Rogers	36½ tons	12,500
1866	2–8–0	Baldwin	45 tons	19,950
1870	4–4–0		35 tons	10,000
1875	4–4–0		35 tons	7,000
1880	4–4–0		35 tons	15,000
1885	4–4–0	Baldwin	40 tons	6,695
1885	2–8–0	Baldwin	46 tons	7,888
1905	4–4–0	Baldwin	51 tons	9,410
1905	2–8–0	Baldwin	91 tons	14,500

Zerah Colburn, who was actively involved in the locomotive trade, contended that few customers paid the actual asking price.[18] This is understandable, considering the large number of builders and the lively competition for orders. Roads having little cash, however, were in no position to bargain. They undoubtedly paid the asking price in addition to interest and the discount charges for payment in stock. Patent fees were another source of additional charges. Ross Winans added $750 to the basic $9,000 charge for a Camel locomotive. Rogers charged the Paterson and Hudson a $50 fee for a patent stack in 1848 on the locomotive *Passaic*.[19] Naturally, the price of individual locomotives varied with the addition or deletion of "extras." Again, using the *Passaic* as an example, we find extra charges of $100 for a "house" (cab) and $290 for a copper (rather than an iron) firebox. It should also be noted that broad-gauge engines, being larger, cost more than standard-gauge machines.

In 1854 the Erie paid $10,500 for such an engine, or roughly $2,500 more than the price of a standard-gauge engine of the same wheel arrangement.

British builders could not understand how American makers produced engines of a capacity equal to their own at a savings of from $2,500 to $4,000.[20] American engines were more ornamental, having much expensive brass work, and American labor and material costs were higher. Although material costs were high, United States builders made extensive use of cast iron instead of expensive wrought iron. They made wheels, rocker bearing brackets, and even crank axles and valve gear links from the cheaper metal. However, British overhead costs were high; more elaborate shops and machine tools and larger engineering departments accounted for greater production costs. Then, too, the British built a more finished engine and machined many non-running surfaces. These factors may not

[18] *Engineering,* January 3, 1862, p. 5.
[19] Walter A. Lucas, *From the Hills to the Hudson* (New York, 1944), p. 294.

[20] A comparison of British and American locomotive manufacture and costs was reported in *The Engineer,* January 3, 1862, and in the *American Railway Review,* January 16, 1862.

fully account for lower American prices, but it is not difficult to understand why we held our own locomotive market and rapidly moved into world-wide locomotive trade.

The effect of financial panics on locomotive prices is not difficult to assess, although production was curtailed more drastically than price. The table above does indicate a definite fluctuation in prices according to the times. Clark states that a locomotive that sold for $10,000 before the 1873 panic could be purchased for only $5,000 a few years later.[21] The same writer notes that the boom between 1878–82 drove prices up to $11,000 or $13,000.

The Civil War, a national emergency far overshadowing any financial panic, had the most dramatic effect on locomotive prices in the nineteenth century. A standard eight wheeler costing $8,000 at the beginning of the war brought $25,000 in 1864.[22] This was a temporary situation brought about not only by the increased demands of regular railroads (the result of inflated wartime traffic) but also by the requirements of the United States Military Railroads. Many roads benefited from the tremendous wartime demands by unloading their obsolete power at a handsome profit. Locomotives were in such short supply that some roads considered placing orders with British builders during this period.

BUSINESS ASPECTS OF LOCOMOTIVE BUILDING

Locomotive building was at best a speculative venture. Profits were high, but the market fluctuated widely and credits were risky. The histories of most builders were chronicles of financial crisis and near-bankruptcy. The basic problem was a chronic lack of capital. Most railroads exhausted their funds in acquiring rights of way and in building the line. When the road opened it had no cash for equipment. Locomotive builders either extended long credits or saw the orders go to a rival.

The early years of M. W. Baldwin are a case in point.[23] Encouraged by large orders, Baldwin built a new and much enlarged shop in 1835 and increased his employment from about 30 to 300 men. The suspension of payments and the declining orders for new engines following the Panic of 1837 resulted in Baldwin's failure. His creditors, realizing they could collect only a quarter of his debt, extended their notes for another three years, but hard times continued and it took more than twice this amount of time for Baldwin to meet his obligations. During these years several partners were taken in in an effort to save the firm. The threat of foreclosure and a sheriff's sale was ever present, but Baldwin weathered the storm that closed many less fortunate firms.

According to the scant evidence available, credit was extended cautiously in the early years of locomotive development. In 1831 the West Point Foundry required 80 per cent payment when the locomotive was completed and ready for shipment.[24] Three years later Baldwin wanted half payment when the boiler and cylinders were completed, with full payment on delivery.[25] In the late 1830's Rogers sometimes deferred payment for nine or even twelve months but he often required a substantial down payment or payment in full if a railroad's credit was thought to be unsound.[26] Rogers lost an order for several engines to Norris, who not only underbid his competitor but

---

[21] Victor S. Clark, *A History of Manufacturers in the United States* (New York, 1929), p. 338.
[22] James L. Ringwalt, *Development of Transportation Systems of the United States* (Philadelphia, 1888), p. 210.
[23] *Pennsylvania Magazine of History*, October, 1966, pp. 423–44.
[24] See Appendix A for contract.
[25] See the *Lancaster* section of Part III (p. 269) for the contract of the Baldwin and the Philadelphia and Trenton railroads, June 2, 1834.
[26] Typed copy of portions of Rogers order book (1838–46) is in the U.S. National Museum. Location of the manuscript is unknown.

also agreed to accept three thousand dollars in Erie stock for each engine.[27] A more liberal policy was evident some ten years later when Rogers agreed to accept half payment in cash thirty days after delivery and the balance in bonds of the purchasing railroad six to twelve months later.[28] Other builders eager for business were even more liberal. One builder accepted western land.[29] Terms of credit ranged up to two years, and substantial quantities of railroad securities were accepted in payment but often at a heavy discount.[30] The ability to absorb paper in lieu of cash was more often the basis for success in locomotive building than was the production of a technically advanced product. Because of their location in large financial centers, the eastern builders were in a better position to negotiate securities. Rogers' chief partner, Morris Ketchum, was an influential New York banker who undoubtedly was helpful in arranging financing for new engines and the profitable unloading of railroad securities. It might be added that, while a director of the Illinois Central Railroad, Ketchum funneled most of the locomotives ordered by that road to Rogers.

The acceptance of used locomotives as partial payment on a new machine was another measure used to obtain orders. Baldwin accepted his first locomotive, *Old Ironsides,* from the Philadelphia, Germantown, and Norristown Railroad as partial payment on a new machine in 1846. Acceptance of the Reading's *Gowan and Marx* in 1860 is another instance of a similar transaction. Other builders are known to have rebuilt and sold old engines, and, presumably, many of these machines were acquired as trade-ins.

The imprudent extension of credit proved the undoing of many firms that were not supported by adequate capital to withstand the occasional financial upset of the last century. Seth Wilmarth was engaged by the Erie Railway to build a large number of engines. Payment was promised in cash, but a financial flurry in 1854 caused the railroad to renege; payment in stocks was offered instead.[31] Wilmarth's creditors and suppliers demanded cash, thereby forcing him to suspend production.

William Swinburne of Paterson was only one of about twenty locomotive concerns closed by the Panic of 1857.[32] The Chicago, Alton and St. Louis Railroad owed Swinburne $60,000 on a group of new engines. After a long delay partial payment was made in the depleted paper of the company and Swinburne was forced to accept a 25–50 per cent discount. Breese, Kneeland and Company of Jersey City was closed during the same panic by the "iniquitous conduct of certain western railroad managers who were buying engines on credit while they knew their companies were hopelessly insolvent."[33]

The dangers and uncertainties of locomotive building have been outlined, but it should be noted that not all was loss and ruin; many a fortune was made by these enterprises. Baldwin regained a sound footing by 1845 and was to achieve an annual income of more than two hundred thousand dollars by 1864.[34] Richard Norris accumulated a fortune of one and a half million dollars.[35]

A rewarding profit could be expected on each locomotive. In 1838 the Locks and Canals machine shop realized a profit of $2,200 on a $7,000 engine.[36] Between 1848 and 1851 John Souther, a small builder of Boston, sold engines for an average price of $7,302. The average cost was about $5,225, thus leaving a profit of $1,077.[37] In 1853 the *American Railroad Journal* estimated profit per locomotive at $1,500.[38]

---

[27] E. H. Mott, *Between Ocean and the Lakes* (New York, 1901), p. 391.
[28] Lucas, *From the Hills,* p. 299.
[29] Lands owned by the Lawrence Manufacturing Co. were sold off with the assets of the company. *American Railway Times,* February 8, 1862.
[30] *Railroad Advocate,* June 6, 1857, p. 4.
[31] *Railroad Gazette,* October, 1907, p. 382.
[32] L. R. Trumbull, *A History of Industrial Paterson* (Paterson, N.J., 1882), pp. 146–47.
[33] *Engineer* (Phila.), November 1, 1860, p. 93.
[34] *Pennsylvania Magazine of History,* October, 1966, p. 443, note 103.
[35] *Railroad Gazette,* June 20, 1874, p. 237.
[36] George S. Gibb, *The Saco-Lowell Shops* (Cambridge, Mass., 1950), p. 95.
[37] *Engineering,* January 3, 1862, p. 5.
[38] *American Railroad Journal,* August 25, 1853, p. 549.

ERA OF FUNDAMENTAL DESIGN

STANDARD DESIGN

The commercial builder of locomotives had an obvious interest in a standard design; it meant less engineering. When a good plan had been conceived of and tested it was pointless to cast it aside for another. Design is not only taxing creative work; it is also time consuming and costly. But more important, standardization means small inventories. If only a few basic styles are produced, the number of patterns, repair parts, and the related storage and overhead costs are considerably reduced. The complexity of a business operation is simplified, and this was important to the early builders who operated small enterprises, often as single proprietors.

As early as 1833 the American Steam Carriage Company of Philadelphia proposed to build standard locomotives in three sizes.[39] This policy was continued by William Norris after he assumed full control of the above firm. In a published catalog of 1841 Norris offered four engine sizes, Extra A and A to C. These machines ranged in weight from 15 to 7½ tons and all were of the 4–2–0 wheel arrangement. By 1845, Norris advertisements appearing in the *American Railroad Journal* claimed that six classes or sizes of engines were in production, but in truth only two wheel arrangements were offered.

Baldwin also decided at an early date to adhere to standard designs as far as possible. In a letter of June, 1836, while refusing to produce a special order, he stated: "I make but one kind of machine and am persuaded from actual test that they are the best."[40] Baldwin was speaking here of the 4–2–0, a successful, practical design that he was ready to manufacture in quantity. His catalog of 1840 offered three sizes of this wheel arrangement in weights of 10, 11½, and 13 tons. Two years later Baldwin adopted a letter designation, A to E, for the different styles of engines built. A was reserved for special order, high-speed machines and was rarely used. The remaining classifications showed the number of driving-wheel axles; B was for a single driving-wheel axle engine, C for two driving axles, D for three axles, and E for four axles. Minor subclasses were established for varying weights and cylinder sizes. By 1870, twenty standard classes were offered, but again the variety is not as great as might be assumed, for only six wheel arrangements were involved.[41]

Beyond the early establishment of production classes, standard design is evident from a comparison of advertising lithographs issued by the major builders. A series of lithographs issued by one builder to show several classes of machines always reveals the employment of standard fittings. Bell stands, dome covers, feed-water pumps, and other minor parts will be identical.[42] Major assemblies, such as boilers, cylinders, and valve gears, were undoubtedly identical for different wheel arrangements of the same weight. Where this was not possible a characteristic leading design was followed. But in general it was possible to build a 4–2–0, a 4–4–0, or even a 0–6–0 from the same boiler and fittings by substituting a different frame and combination of wheels and rods. Obvious savings in assembly time, inventory, and engineering were realized from this practice.

Enough illustrative evidence is available to solidly support the concept of standard design for parts, but no claim is made that these parts were literally interchangeable. A considerable amount of hand fitting was undoubtedly required. However, interchangeable parts were in production at least as early as the 1860's. Baldwin had begun production of interchangeable repair parts by 1865 and may have employed gauges and templates ten years earlier.[43] The Grant Works built their engine

---

[39] *Proposals of the American Steam Carriage Company* (Philadelphia, November, 1833), p. 4.

[40] Letter of M. W. Baldwin to a Mr. Hoffman, June 17, 1836, in the files of the Historical Society of Pennsylvania.

[41] *Railroad Gazette,* November 19, 1870, pp. 173–74.

[42] The use of standard fittings is illustrated by the Baldwin lithograph reproduced as Fig. 109, Part III, *Lancaster* section; see also Figs. 61 and 200, Part III, Flexible-Beam section.

[43] Baldwin Locomotive Works, *Record of Recent Construction, No. 60* (Philadelphia, 1907), p. 28; C. T. Parry is credited with introducing gauges and templates for interchangeable use after he became superintendent of the Baldwin Works in 1854.

to a rigid system of templates and jigs. According to a report made in 1867, "every piece . . . [was] executed by gages"[44] as early as 1863 or 1864.

The building of engines for stock by commercial makers is another indication that standard designs were followed. Such machines were built to the maker's rather than the buyer's plan. Many builders are known to have followed this practice. Hinkley built six engines in 1842 in the hope that a buyer could be found.[45] The same maker took to stock building during the dull times following the Panic of 1873.[46] During the 1850's Taunton built engines on speculation and leased them to local railroads until a buyer could be found.[47] The Lawrence Manufacturing Company, encouraged by large orders and good profits, decided to build forty-two standard engines for stock.[48] About one-half of the lot was completed when the market collapsed, and the builder experienced considerable trouble in selling the machines; the entire lot was never completed.

As already mentioned, the vast majority of locomotives built in this country were produced by commercial makers. Hence these firms were more directly involved with the standardization of design than were the purchasing railroads. This worked to the disadvantage of the railroads because the individual makers, beyond the agreement on general arrangement, built according to their own plans. Most roads bought from several builders and would in the course of a few years have a variety of engines in service. The consolidation of small roads only aggravated the situation. The Cleveland, Columbus, Cincinnati and St. Louis Railroad, an amalgamation of a dozen or more short lines, owned 463 locomotives of 114 different patterns.[49] In 1874 the Erie had 469 engines of 83 varying styles.[50] The resultant potpourri of repair parts made the acquisition of more uniform classes of engines imperative.

Wilson Eddy attempted to standardize the locomotives of the Boston and Albany Railroad by stocking the road with engines built at the company's Springfield, Massachusetts, repair shops.[51] Two classes of 4–4–0's were designed by Eddy, one for freight, the other for passengers. The freight engines had smaller driving wheels and larger cylinders but were otherwise copies of the passenger engines. The first of these machines was built in 1852; by 1881, 136 had been constructed. It should be noted, however, that while a general leading design was followed there was considerable variation in size among these machines. A marked increase in weight was evident; the first machine weighed 20 tons, the last more than 40 tons.[52] It should also be noted that the Boston and Albany remained dependent upon outside builders for most of its locomotives, all of which were built to the contractor's design, and did not at any one time operate exclusively with the standard engines built by Eddy.

The Pennsylvania Railroad embarked on a comprehensive program to standardize its motive power in 1867. From this date forward all engines built at the company shops or purchased from outside contractors would be built to the standard design of the railroad. The program was pushed vigorously and by 1873 more than 40 per cent of the road's engines were of the new standard design.[53] Three years later 56.6 per cent were standard, and by 1900 the entire system was powered by uniform locomotives.[54] Three classes were established initially, and efforts were made to keep the variety of designs rigidly limited; by 1893, however, twenty-five classes were in existence. The number of interchangeable parts was large; in some

---

[44] *American Railroad Journal*, February 22, 1867, p. 150.
[45] Freedley, *Leading Pursuits*, p. 305.
[46] *Railroad Gazette*, April 27, 1877, p. 192.
[47] *Railway and Locomotive Historical Society Bulletin*, No. 46, p. 48.
[48] Annual Report of the Lawrence Manufacturing Company, 1854; the published report is on file at the Essex Institute, Salem, Massachusetts.
[49] *National Car and Locomotive Builder*, September, 1891, p. 131.
[50] American Society of Civil Engineers, *Transactions*, Vol. 11 (September, 1882), p. 292.
[51] *Railway and Locomotive Historical Society Bulletin*, No. 22, pp. 9–39.
[52] *Ibid.*
[53] G. H. Burgess and M. C. Kennedy, *Centennial History of the Pennsylvania Railroad Company* (Philadelphia, 1949), p. 718.
[54] *Ibid.*

cases boilers were used for several classes. Thus, the economy and convenience of the system were never in doubt. Only the largest roads could institute such measures, however; smaller lines had neither the engineering staff nor the capital to acquire their own wants independent of the standard product of the commercial builder. In general the routine product of the builder's shop satisfied the needs of most purchasers. Stock designs were modified and an occasional special-order machine was built, but the degree and consistency of the standard product cannot be denied.

EXPORT OF AMERICAN LOCOMOTIVES

The importation of British locomotives to the United States is a well-established fact, but the early exportation of American-made locomotives is a much neglected episode of our industrial history. The regular production of locomotives had hardly begun before we were shipping locomotives to Europe and elsewhere.

The first locomotive exported from the United States was undoubtedly the *Columbus,* sent to the Leipzig and Dresden Railroad by Ross Winans in 1836 or 1837.[55] Another Winans engine was sent to Russia in about 1843.[56] Norris sent the *Philadelphia* to Austria in 1838. This was followed by the shipment of 100 Norris engines to Italy, France, England, and Austria between 1838 and 1845.[57]

Norris was so eager to promote European trade that a nineteen-page prospectus in English, German, and French was issued in March, 1838. Following the business collapse brought on by the 1837 panic William Norris toured Europe in search of new orders. He was successful in obtaining substantial orders and decided to open a plant in Vienna. A small shop was opened in 1844 and a few engines were produced, but the venture failed and Norris returned to this country in 1848.[58]

One American firm, however, was successful in establishing itself in the Old World. At the invitation of the Russian government, Eastwick and Harrison of Philadelphia moved their works to St. Petersburg in 1843 to manufacture locomotives for the Moscow to St. Petersburg Railway then under construction.[59] George Washington Whistler had been hired earlier as chief engineer of the projected line and he undoubtedly promoted the selection of American contractors. Eastwick and Harrison, joined by Thomas Winans, built several hundred locomotives for Russian railways between 1843 and 1862. The engines were built on the American plan and set the general pattern for future Russian motive power.

Baldwin entered the export trade in 1838 when several locomotives were sent to Cuba. Three years later two engines were sent to Austria. These were followed in 1845 by three flexible beam locomotives for the Royal Württemberg Railway (Fig. 82). While Baldwin's European trade never matched Norris', he too was interested in developing more overseas business. A

[55] *American Railroad Journal,* December 24, 1836, p. 805, gives notice of the order by the Leipzig and Dresden Railway.
[56] Joseph Harrison, *The Iron Worker and King Solomon* (Philadelphia, 1869), p. 50.
[57] *Railway and Locomotive Historical Society Bulletin,* No. 79, is devoted entirely to Norris; special attention is given to the engines exported to Europe.

[58] The exact history of William Norris' Vienna shop is uncertain. The facts I have are from Thomas Norrell's unpublished history of the Norris Locomotive Works.
[59] Joseph Harrison, *The Locomotive and Philadelphia's Share in Its Early Improvement* (Philadelphia, 1872), pp. 52–53.

handsome lithograph with a multilingual caption, issued in about 1840, failed to bring in the orders it was intended to attract.

Rogers never entered the European market but he did seek such business. The lack of activity during the early 1840's prompted him, as it had Norris, to make a personal visit to Europe in search of more orders.[60] Reportedly, some buyers were interested but not on terms Rogers would accept.

European trade, at a standstill by the mid-1840's and destined not to resume for another fifty years, had been limited to about 120 engines. As railways spread in South America, Cuba, and Canada, a brisk export trade developed, particularly after 1870. Between 1871 and 1894 nearly 2,900 locomotives were sent overseas by American builders.[61]

[60] *American Railroad Journal,* May 3, 1856, p. 280.

[61] C. M. Depew (ed.), *One Hundred Years of American Commerce* (New York, 1895), Vol. 2, p. 342. The export figure does not include locomotives built for Mexico or Canada.

4. MATERIALS USED IN LOCOMOTIVE CONSTRUCTION

This topic is treated more extensively in Part II, but a brief survey of materials now will provide convenient introduction to a subject that is otherwise scattered throughout the text.

The experience gained from the construction of stationary steam engines, water wheels, and other mechanical works had empirically established over many generations a reserve of information on the characteristics of engineering materials available for locomotive construction. Wood was the traditional building material, but in general it was found to be too bulky for the construction of machinery. Iron was well established as the principal material for steam engine construction by the time the locomotive was introduced as a commercial product.

The variety of cheap materials available to the builder was limited to cast iron, wrought iron, and copper and its brass and bronze alloys. Steel was available but was too costly before 1860 to warrant more than occasional use. This limited selection of metals was resourcefully employed by early designers to produce a serviceable, practical locomotive. Ease of manufacture often prompted the selection of seemingly unsuitable materials for a given purpose; cast iron, for example, was often used where steel or wrought iron would have appeared to be a more logical choice of material. Cost was also an obvious factor in the selection of materials. A less obvious factor, but one that unquestionably existed, was the capricious and often inept selection of materials by mechanics seeking a novel solution to an old design problem.

One of the most perplexing problems faced by the early locomotive-builder was the lack of uniformity in the irons available. Scientific testing of metals had begun, but very few standards had been established. Even if such standards were recommended, as were the Franklin Institute's requirements for boiler iron, issued in the mid-1830's, there was no machinery to enforce the rules.[1] The reputation of the iron-maker and his distributors was the only ready gauge of a metal's quality. The tensile strength of iron varied greatly from one furnace to the next, and the wary engine-builder bought selectively if he wanted to produce a first-rate product. The judgment and skill of the ironmaster determined the soundness of the product; quality control and the scientific production of iron and its alloys were developed too late to benefit the pioneer locomotive manufacturer.

[1] *Technology and Culture,* Winter, 1966, pp. 13–14.

WOOD

Wood played a minor role in locomotive construction. In the early and mid-1830's it was used for locomotive frames but it was entirely superseded by iron frames within a few years. Pilot beams, however, the foremost cross-brace of the frame, continued to be made of wood until almost 1900. The cab, the cowcatcher, and the boiler and cylinder lagging were made of wood, but these were relatively minor parts of the locomotive. By 1880 it was estimated that only 7 per cent of the locomotive was made from this material.[2] Tender frames continued to be made of wood until late in the nineteenth century and this assembly remained the only substantial wooden fabrication associated with the locomotive.

COPPER AND BRASS

The importance of copper and brass in locomotive construction steadily declined as the machine itself developed and grew in size. At first copper was largely employed in locomotives; before 1855 the firebox, the tubes, and much of the auxiliary piping were made of copper. A 20-ton locomotive of 1850 would contain about 1½ to 2 tons of copper. The metal was ductile, homogeneous, and easy to work. It was a good conductor of heat and was therefore appropriate for firebox walls and boiler tubes. It was also weak and expensive. It became weaker as heated so that very heavy sheets were required for safety in boiler work.

At 30 cents per pound it cost five to six times as much as wrought iron.[3] Much of its first cost could be recovered as scrap, but the saving in first cost made wrought iron an attractive substitute. It should also be noted that the introduction of coal-burning contributed to the abandonment of copper fireboxes and tubes because of the erosion of this soft material by fly ash.

Brass was the favored material for small fittings, valves, gauges, and cocks; it was a necessary material for bells and whistles. In higher class engines brass was used for feed-water pumps in place of cast iron, which was cheaper. Decorative jackets for cylinders, sandboxes, and steam domes were made of polished sheet brass. Number and builder's plates were made from this material. During the years of heavy decoration (*ca.* 1850–60) about five hundred pounds of brass embellished the locomotive. Bearings or wheel journals were constructed of various brass or bronze alloys; the same materials were used for side-rod bearings.

CAST IRON

Cast iron, a favored material of early machine-builders, possessed many desirable qualities. A complex shape could be made in quantity by casting; this process was far easier and cheaper than attempting to fabricate assemblies by welding or forging together many small pieces. Cast iron was also easy to machine. The material was cheap—only 3 cents a pound in 1851—thereby prompting its employment by cost-conscious manufacturers.[4]

The metal was not malleable and this brittle quality restricted its use to parts not subject to tension or impact stresses although it was occasionally used in such conditions. This was possible because the best grades of cast iron possessed a tensile

[2] Bureau of the Census, *Tenth Census*, II, 46.
[3] A. L. Holley, *American and European Railway Practice* (New York, 1861), p. 20. Zerah Colburn notes the price of copper as 30 cents per pound in his *The Locomotive Engine* (Boston, 1851), p. 157.

[4] Colburn, *The Locomotive Engine*, p. 157.

strength of 35,000 pounds or more.[5] Ordinary gray iron has a tensile strength of only half this amount and is of course suitable only for more restricted service. Gray iron was used for valve handles, dome bases, and other parts not subject to stress. Top-quality iron was used for every part of the locomotive, as will be shown.

If massive enough, cast-iron parts can be made very strong, and such design practice is acceptable and is often less costly than alternative materials used in the production of stationary machinery. But for locomotives, size and weight are more important considerations, and it was necessary to keep the overall bulk of the machine within the clearance and weight limitations of the line. When cast-iron parts were increased in size for strength beyond a reasonable limit, it was necessary to redesign the part for wrought iron or steel.

Certain parts of the locomotive would logically be made of cast iron. Because of internal steam and exhaust passages the cylinders and valve boxes logically required casting. The cheapness of cast iron, and its porous character, which makes it easy to lubricate, lent weight to its employment in the construction of cylinders. Journal boxes, smokebox fronts, feed-water pumps, bell stands, and cab decks are other parts that fit into a similarly plausible argument for the use of cast iron.

The employment of cast iron for frame pedestals, driving and truck wheels, tires, and even crank axles seems entirely preposterous, yet it was done with notable success. The one feature of American locomotive construction that forever amazed British engineers was our use of chilled cast-iron wheels. Their technical journals and engineering proceedings repeatedly commented on this unconventional use of cast iron. Good service and low first cost fully justified the American mechanic's adherence to the cast-iron wheel. Tires as well as truck wheels and driving-wheel centers were made of cast iron.

After about 1850, the role of cast iron began to decline in American locomotive construction. Cast-iron frame pedestals and valve rocker arms had been replaced by wrought iron by the mid-1850's. Cast-iron tires were made obsolete in the 1860's and 1870's by the introduction of steel tires. However, as late as 1880 it was reported that cast iron constituted one-third of the material in American locomotives.[6]

WROUGHT IRON

The vibration of the locomotive when in motion subjects many of its parts to constant flexing and twisting. Wrought iron, a malleable, strong metal, was capable of withstanding the stresses produced by the locomotive's undulations. In fact, wrought iron was used in greater quantities than any other single material in locomotive construction. The frame, boiler, axles, rods, valve gear, and other major components were fabricated from this material.

If carefully manufactured, wrought iron was entirely adequate for the work required of it. Having a tensile strength of about 50,000–60,000 pounds, it was safe for boilers of from 100 to 150 pounds per square inch and for axles and connecting rods subjected not only to road shocks but to the working forces of cylinders.[7] Unfortunately, the methods used in manufacturing a wrought-iron boiler plate and large pieces, such as axles, invited trouble. Large pieces were usually made as faggoted iron. In this process wrought-iron rods, old boiler sheets, and bits of scrap are piled, heated, and rolled or hammered into a single piece. If the heat is right and care is taken not to mix in foreign scrap, such as cast iron or steel, a sound sheet or bar of wrought iron will result. Too often, however, a poor weld,

---

[5] Lionel S. Marks, *Mechanical Engineers Handbook* (New York, 1930), p. 545. Nineteenth-century figures agree with the modern estimates of tensile strength; Douglas Galton, in his 1878 report on American railways, notes that Lake Superior charcoal iron used for car wheels had a tensile strength of 35,000 pounds.

[6] Bureau of the Census, *Tenth Census,* II, 46.

[7] Clark and Colburn, *Recent Practice,* pp. 1–2. Marks's handbook gives a slightly lower figure for wrought iron.

or a pocket of cinders hidden under the surface of the metal, was not discovered until the piece failed in service. These defects were often unintentional, but the scandalous practice of British rail makers who purposely rolled in cinders to produce cheap rails for export is well known and casts doubt on the reliability of other wrought-iron producers in this period.[8] Yet the best faggoted iron produced contained enough cinder, often nearly 1 per cent, to prevent complete homogeneity. Before the introduction of cheap steel, however, wrought iron was the strongest metal available for locomotive construction. Its shortcomings were recognized but tolerated out of necessity.

STEEL

Steel was far too expensive for large-scale use in machinery before the advent of the Bessemer process in 1856. Prior to that discovery, true steel was made by the crucible process, a slow, costly method in which the impurities were burned out of wrought iron at a moderate heat. Bessemer or "cast steel" was an iron alloy produced by a quick burning off of impurities and a simultaneous combination with manganese. The resultant metal was homogeneous, malleable, and strong. Unlike wrought iron, sheets and large bars or shapes could be readily produced from a single ingot and did not need to be fabricated by faggoting. Thus, a uniform, sound piece was more likely to result.

It would seem likely that steel would have been rapidly adopted for locomotive construction, but this was not the case. Indeed, the early Bessemer steels were cheaper than crucible steel but they were still more expensive than wrought iron. The early steels contained too much carbon and were too hard and brittle for boiler construction.[9] As the price fell and softer alloys were manufactured, steel gradually became the accepted material for boiler construction but not, surprisingly, until about 1890. Similarly, steel was slow in being adapted for axles, although it would seem a natural material for this purpose. The poor performance of the first steel axles prejudiced railway mechanics against their use until late in the nineteenth century.[10]

In contrast, steel was rapidly adapted in place of wrought iron for driving-wheel tires and firebox sheets. By 1870 both of these components were invariably made of steel. The high mileage of steel tires—anywhere from three to five times that of wrought iron—explains the conversion.[11] A similarly spectacular increase in the life span of the firebox accounts for another conversion to steel, yet the present author is puzzled by the reluctance of builders to adopt steel for axles or for the entire boiler shell. No substantial explanation has been found other than simple resistance to innovation. In instances where steel's superiority was so substantial that it could not be ignored, the new metal was adopted, but where this was not the case mechanics preferred to use wrought iron because its "quality is pretty well known and its uniformity can be depended upon."[12] The continued preference for wrought-iron axles over steel axles was explained in this way as late as 1904.[13]

In general, however, it can be stated that steel played an increasingly important role in locomotive construction after 1870. It began to replace wrought iron in most major assemblies. By 1880 many parts formerly made as wrought iron forgings were now cast steel.[14] Within the next twenty years steel castings began to displace cast iron for wheel centers, cylinders, and other major components; eventually, almost the entire machine was made from this material.

[8] Philadelphia, Wilmington and Baltimore Railroad, *Annual Report*, 1858, p. 4.
[9] William M. Barr, *Steam Boilers* (Indianapolis, Ind., 1880), p. 30.
[10] *Railroad Gazette*, December 13, 1878, p. 599; see also the section in this volume on axles.
[11] Documentation and more detail on material used for firebox construction are given in the section on that subject in this volume.
[12] G. Weissenborn, *American Locomotive Engineering and Railway Mechanism* (New York, 1871), p. 175.
[13] J. G. A. Meyer, *Modern Locomotive Construction* (New York, 1904), p. 214.
[14] Bureau of the Census, *Tenth Census*, II, 46.

LOCOMOTIVE TYPES AND WHEEL ARRANGEMENTS

5.

Locomotives have always been classified into general types by number of wheels. More precision was given to this broad designation when the Whyte system came into use early in the twentieth century.[1] The ambiguities of such terms as "six-wheel engine" (which could mean a locomotive with a four-wheel leading truck and two driving wheels or a machine with six driving wheels) were thus defined by the Whyte symbols 4–2–0 and 0–6–0. In this system the first numeral represents the number of leading wheels; the second figure the number of driving wheels, and the final figure the number of trailing wheels.

Locomotives capable of general service were the most popular type with American railroads of the nineteenth century. Three wheel arrangements, 4–2–0's, 4–4–0's, and 4–6–0's, were particularly favored during this period. The 4–2–0 was the first general service engine but it soon proved too light and was replaced by the 4–4–0, by far the most popular wheel arrangement in this country. The 4–6–0, while popular, was never as predominant nor did it succeed in replacing the 4–4–0. The following discussion will briefly treat the more popular wheel arrangements and the reasons for their success or failure.

THE 4–2–0

The six-wheel engine or "Jervis type" enjoyed a brief but intense popularity in the United States.[2] It performed well in mixed service and was our first national type, a distinctive American locomotive. The great years for the 4–2–0 were between 1835 and 1842 when it was built almost to the exclusion of any other wheel arrangement. In 1840 nearly two-thirds of the locomotives in this country were of this type.

The *Experiment* (later renamed the *Brother Jonathan*) of the Mohawk and Hudson Railroad was the first 4–2–0 constructed. It was built by the West Point Foundry after a design prepared by John B. Jervis in the fall of 1831. The boiler and

---

[1] The numeral symbol system devised by Fredric M. Whyte of the New York Central Railroad was reported in the *American Engineer and Railroad Journal*, December, 1900, p. 374. It was adopted by the American Locomotive Company in 1903 and became the accepted system within a few years.

[2] Walter Lucas suggested that the 4–2–0 be named the "Jervis type," recognizing John B. Jervis as the originator of this wheel arrangement.

the valve gears were copied directly from the standard Stephenson engine of the period, but the running gear was a radical departure from the usual British design. A four-wheel leading truck with side bearings carried the front of the engine; a single driving wheel set behind the firebox supported the rear weight. Jervis observed that the rigid, short wheel base English locomotive was not successful on twisting, uneven American tracks. His solution was the *Experiment,* which had a swiveling leading truck and a long wheel base. The machine was delivered in August, 1832, and proved an immediate success as far as the running gear was concerned. Jervis' triumph was somewhat deflated, however, by the failure of the engine to make steam. The long (5 feet long), shallow firebox intended for anthracite coal-burning was a failure. This firebox is shown in the commonly reproduced line drawing of the *Experiment*.[3] Wood was tried, but even this failed, and finally in the winter of 1833 the engine was rebuilt with a new wood-burning firebox.[4] A fine contemporary drawing showing the engine as it was rebuilt in 1833 has survived and is reproduced here as Fig. 6. Jervis, proud of the excellent tracking performance of the 4–2–0, wrote several letters to the *American Railroad Journal* in 1833.[5] These letters, substantiated by a similar report in the same journal by E. L. Miller, focused the attention of other railroad mechanics on Jervis' scheme. Baldwin, encouraged by Miller, visited the Mohawk and Hudson Railroad to inspect the Jervis truck engines; favorably impressed, he built his next locomotive, the *E. L. Miller,* on the Jervis plan. Other builders were quick to adopt the 4–2–0, and within three years it was in common use. In line with the promoters of many other basic locomotive reforms Jervis did not patent the arrangement and it was freely given over to the industry.

Not only were the new engines in production almost exclusively 4–2–0's, but a certain number of 0–4–0's were rebuilt as six-wheel engines. The rebuilding of the Mohawk and Hudson's *Robert Fulton* (Fig. 7), of the Stephenson "Sampson" class, was a particularly difficult conversion. Because it had inside cylinders it was necessary to use bell cranks to make the connection to the rear driving wheels.[6] The machine, renamed the *John Bull* after its rebuilding in 1833 by Asa Whitney at the Mohawk and Hudson Shops, is shown in Fig. 8.

Despite its excellent tracking abilities, many roads soon found that the 4–2–0, which had only one set of driving wheels, lacked tractive power. E. L. Miller devised a scheme to increase traction by adding part of the tender's weight to the rear of the locomotive frame. Miller obtained a patent for his invention in June, 1834. It was claimed that 1¼ tons of the tender's weight could be shifted to the locomotive. This device was used on many 4–2–0's built by Baldwin; Miller received a patent fee of $100 per engine.

A second plan for increasing the 4–2–0's tractive power was patented on May 22, 1835, by Charles and George E. Sellers of Philadelphia. In this plan the weight of the train when ascending a grade caused the engine to tip up, thus increasing the distribution of the locomotive weight on the drivers. This scheme was used on two engines built by Sellers for the Philadelphia and Columbia Railroad and may have been used by other builders as well. The classic solution to the problem was Norris' scheme of placing the drivers in front of the firebox. This placed more weight on the driving wheels but it also produced a less steady running engine by shortening the wheel base. Later, Baldwin, Griggs, and Winans devised schemes for throwing more weight on single-axle locomotives.

Sandboxes were reported to be in use as early as 1836 and they undoubtedly were adopted to improve the traction of slippery 4–2–0's. E. Tolles's patent model (1841) for a locomotive sander shows the device attached to a 4–2–0.

These several attempts to increase the capacity of the six-wheel engine were not entirely successful. By the early 1840's

---

[3] The line drawing of the *Experiment* first reproduced in the *Railroad Gazette,* February, 1872, p. 47, was based on an original drawing presently preserved by the Jervis Library, Rome, New York.
[4] H. M. Flint, *Railroads of the United States* (Philadelphia, 1868), p. 152.
[5] *American Railroad Journal,* January 5, May 11, and July 27, 1833.

[6] Jervis described the rebuilt *Robert Fulton* in the *American Railroad Journal,* July 27, 1833, pp. 468–69.

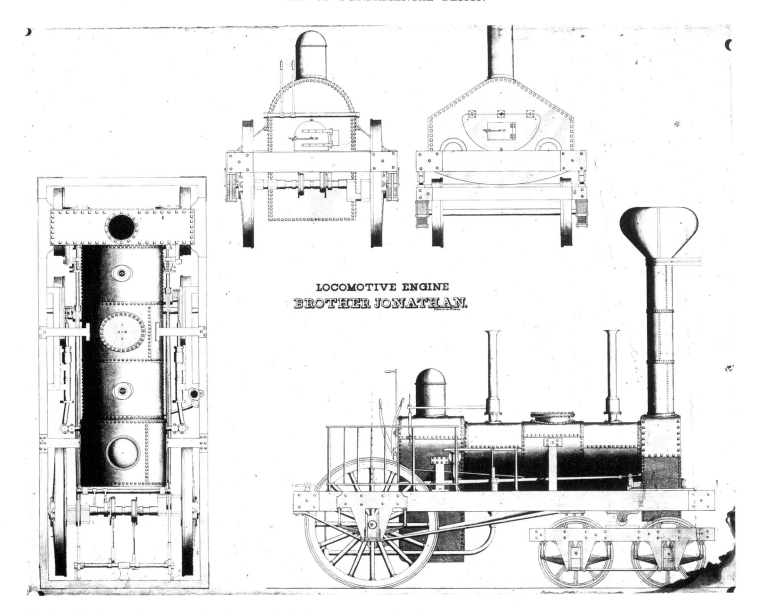

Fig. 6. The Brother Jonathan, *originally named the* Experiment, *was the first 4–2–0 built in the United States. Assembled by the West Point Foundry in 1832, it is shown as rebuilt in 1833. The cylinders measured 9½ in. by 16 in., the driving wheels 60 in. in diameter. The engine weighed 7 tons.*

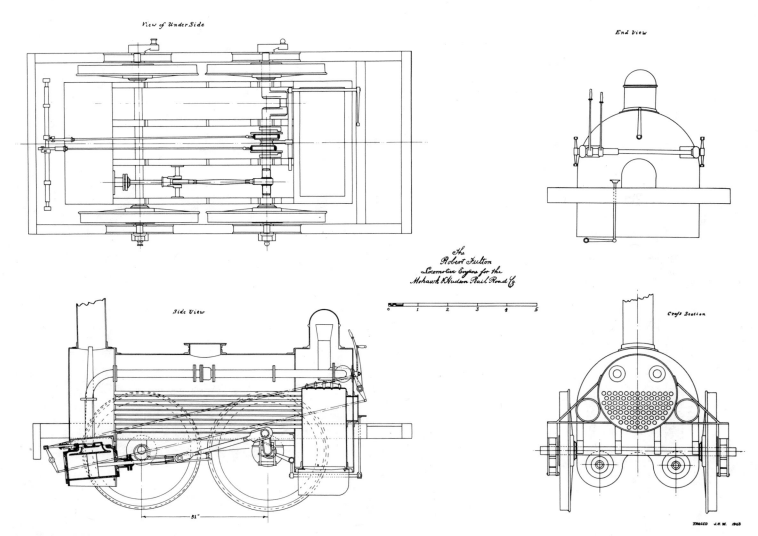

Fig. 7. *The Robert Fulton, built in 1831 by Stephenson for the Mohawk and Hudson Railroad, was a typical rigid British locomotive. Found to be entirely unsatisfactory for light American tracks, it was rebuilt as a 4–2–0. Its cylinders measured 10 in. by 14 in., the driving wheels 48 in. in diameter.*

ERA OF FUNDAMENTAL DESIGN

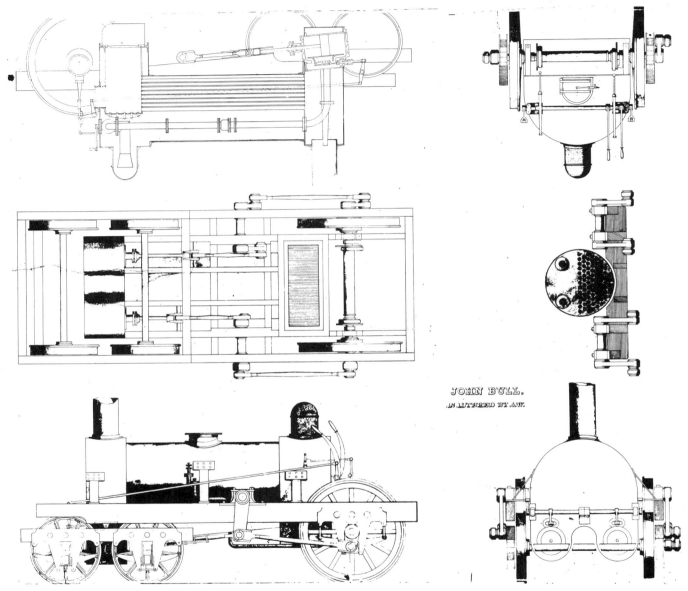

Fig. 8. This drawing shows the Robert Fulton *as rebuilt and renamed the* John Bull *in* 1833.

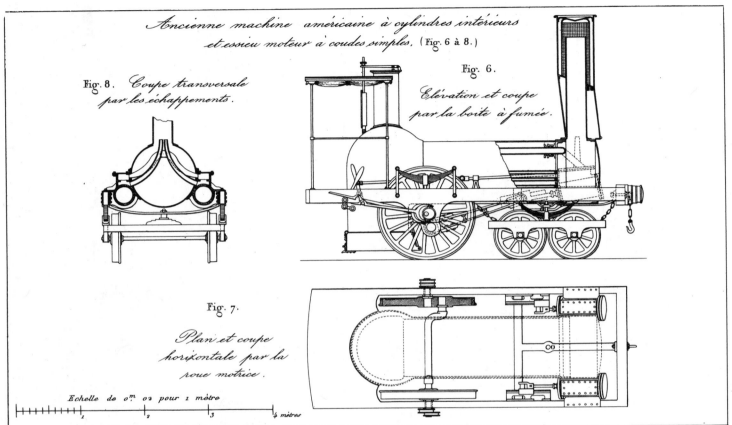

Fig. 9. An unidentified American six-wheel locomotive of about 1840, possibly a Rogers engine.

ERA OF FUNDAMENTAL DESIGN

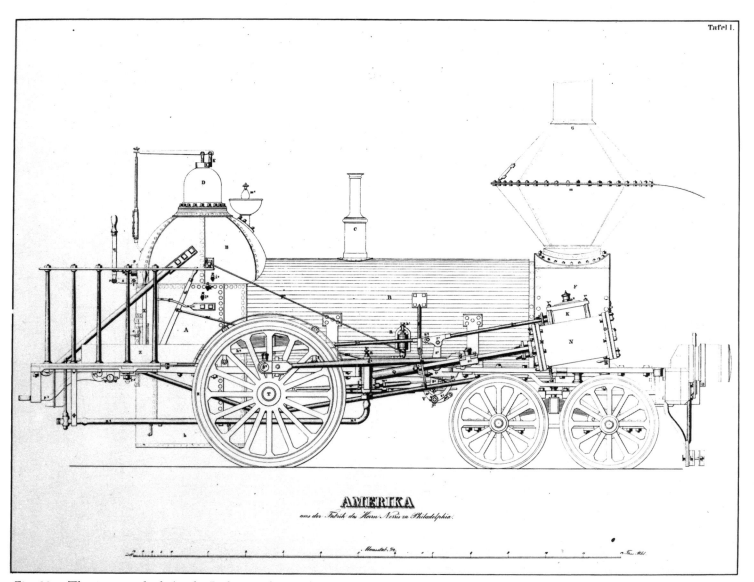

Fig. 10. The America, built for the Berlin-Potsdam Railway by William Norris in 1838 or 1839. Cylinders measured 10½ in. by 18 in., wheels 48 in. in diameter. This was a standard engine designated as "Class B" by the maker.

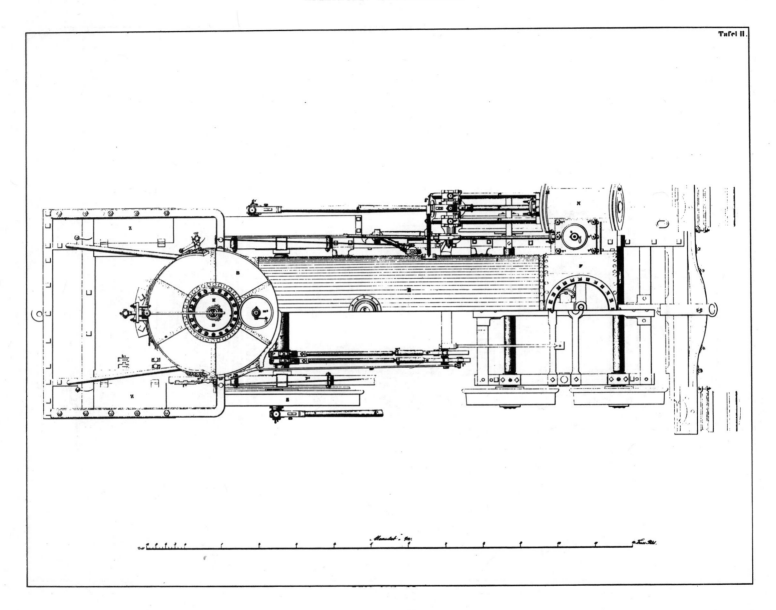

Fig. 11. Plan view of the America, 1838–39.

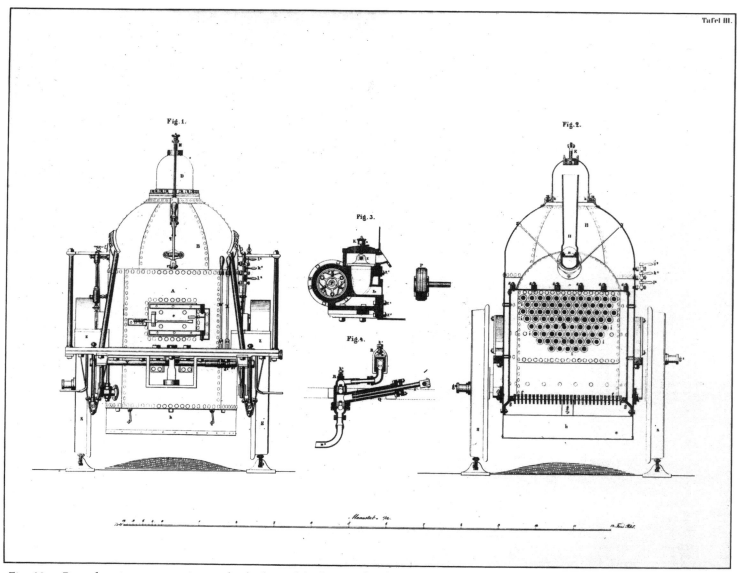

Fig. 12. *Rear elevation, cross-section, and cylinder and feed-pump details of the* America, *1838–39. The boiler contained seventy-eight copper tubes 1⅞ in. in diameter.*

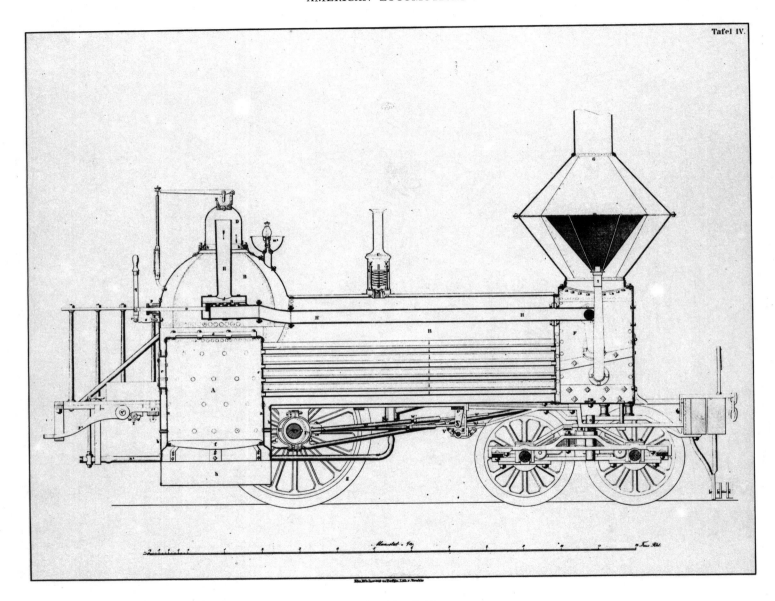

Fig. 13. *Longitudinal cross-section of the America, 1838–39. Note the Shultz spark arrestor, lock up safety valve, and iron-frame truck.*

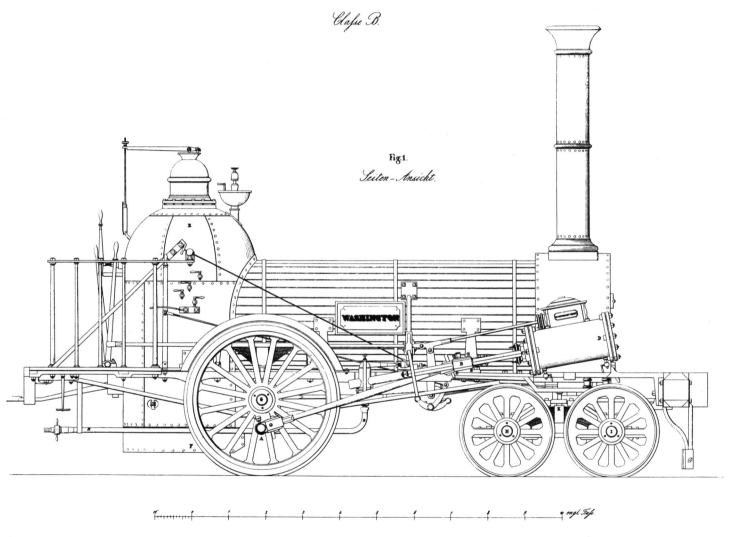

Fig. 14. *A Norris "Class B" 4–2–0 of about 1841. This machine is an excellent example of the straightforward and well-proportioned design perfected by early U.S. builders.*

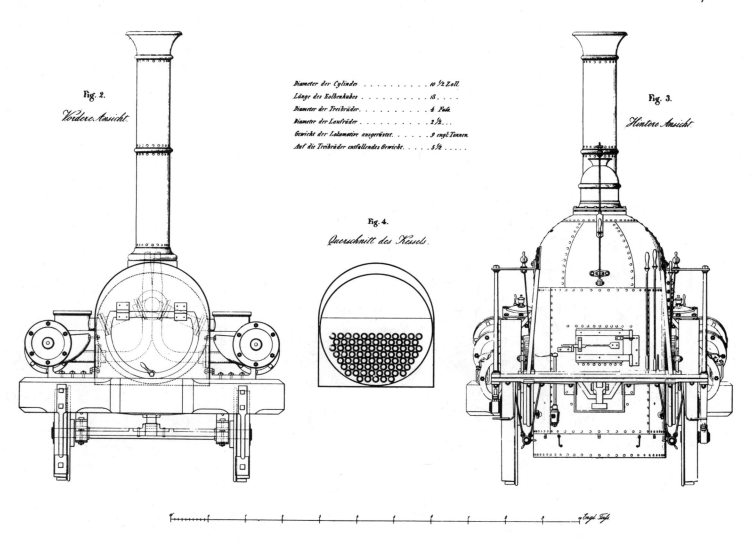

Fig. 15. End elevations of the Washington, *by* Norris, dated about 1841.
See Fig. 14 for the side elevation.

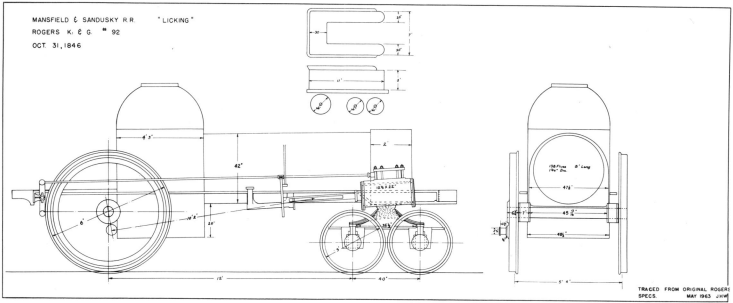

Fig. 16. The Licking, *built in 1846 by Rogers, was possibly the largest six-wheel truck engine constructed in the U.S. The small diagram shows the locomotive's tender.*

the seven to ten ton 4–2–0 was found to be too small to move a paying load. Its production was accordingly curtailed in favor of the more powerful 4–4–0. This occurred despite the loyalties of such conservative builders as Baldwin, who as late as 1845 refused to build any 4–4–0's. Rogers, for example, built only 4–2–0's between 1837 and 1840, yet between 1840 and 1845 only five out of a total production of 56 engines were 4–2–0's. Limited production of the 4–2–0 did of course continue. In 1846 Rogers built a giant six wheeler, the *Licking* (Fig. 16), for the Mansfield and Sandusky Railroad. This machine, probably the largest 4–2–0 ever built, carried 110 pounds of steam and pulled a 380-ton train up a grade of 16 feet per mile.[7] The Baldwin works continued to solicit orders for 4–2–0's in its advertisements of the 1850's and it built a six wheeler, the *Rockport*, as late as 1868.

While some roads, including the Baltimore and Ohio and the New York Central, continued to operate 4–2–0's in light service through the late 1840's, most roads had found them totally unsuited for any kind of work years earlier. Accordingly, a number of 4–2–0's were rebuilt as eight wheelers. In 1845 Walter McQueen converted the Mohawk and Hudson's *John Bull* (Fig. 8) into a 4–4–0 and renamed the engine *Rochester*.[8] The next year McQueen similarly rebuilt the venerable *Brother Jonathan* (Fig. 6), adding larger 12-inch by 18-inch cylinders and replacing the two 60-inch wheels with four 54-inch drivers. David Matthew of the Utica and Schenectady

[7] The performance of the *Licking* was reported in the *American Railroad Journal*, June 10, 1848, p. 371.

[8] Clark and Colburn, *Recent Practice*, p. 49.

Fig. 17. The Utica and Schenectady's No. 11 shown as rebuilt by David Matthew in 1844. Originally thought to be a Baldwin 4–2–0.

rebuilt one or more old Baldwin six wheelers as 4–4–0's (Fig. 17). A final example of such activity is shown by the conversion of the Baltimore and Ohio's No. 16, originally a Norris 4–2–0 similar to Figs. 10 and 15, into an eight wheeler in 1848 (Fig. 18). This engine performed satisfactorily and was in service as late as the mid-1860's.

THE 4–4–0

The 4–4–0, variously called the American type or the standard eight-wheel engine, was the most popular wheel arrangement in nineteenth-century America.[9] It succeeded because it met every requirement of early United States railroads. It was well suited to all service, including passenger, freight, and switching. It was flexible, having three-point suspension and a leading truck, and it operated well on uneven tracks. It was simple, having relatively few parts, which made it easy to repair. It was low in first cost, and it was relatively powerful because of its four connected driving wheels. Finally, it was the national engine, a machine without peer in this country, because it answered every need.

The 4–4–0 was originated by Henry R. Campbell, a native of Philadelphia and long-time associate of M. W. Baldwin, while he was chief engineer of the Philadelphia, Germantown, and Norristown Railway. Campbell secured a patent on February 5, 1836, and began work on his engine a month later. The machine was built in Philadelphia at James Brook's shop. When finished in May, 1837, it was a giant for the times as indicated by the following dimensions:[10]

Cylinders	14″ by 16″
Wheels	54″ in diameter
Weight	12 tons (long tons)
Heating surface	723 sq. ft.
Steam pressure	90 lbs.

It was estimated that at about 15 miles per hour the engine could pull a 450-ton train on level ground. This represented a 63 per cent gain in tractive force over the standard Baldwin 4–2–0. Campbell had succeeded in producing a machine

[9] The class name "American" was suggested for the 4–4–0 in the *Railroad Gazette*, April 27, 1872, p. 183. This is the earliest recorded instance of the term known to the author.

[10] *American Railroad Journal*, July 30, 1836, pp. 465–66; see also the *Railway and Locomotive Historical Society Bulletin*, No. 35, for P. T. Warner's excellent general history of the 4–4–0 locomotive.

ERA OF FUNDAMENTAL DESIGN

Fig. 18. The Baltimore and Ohio's No. 16, built in 1838 by Norris as a
4–2–0 (see Fig. 15). Shown here as rebuilt in 1848.

with increased power over the six wheeler, but otherwise the engine was not successful. Its suspension was too rigid and it was prone to derail.[11]

A contemporary engraving of Campbell's eight-wheel engine was printed in the *American Railroad Journal* for July 30, 1836. This cut was undoubtedly reproduced from a large engraving issued at the time by Campbell to advertise his patent (see Fig. 19).

A limited amount of evidence has been uncovered indicating that Campbell sought to enforce his patent. He threatened to prosecute Garrett and Eastwick for infringement in 1837.[12] Several years later the minute books of the Philadelphia and Reading Railroad refer to a dispute with H. R. Campbell for the unlicensed use of his patent for several new eight-wheel engines built by the Locks and Canals machine shop.[13] Settlement was made for forty shares of Reading stock. Baldwin was reported to have purchased the right to use Campbell's patent in 1845.[14] The extent of Campbell's success in capitalizing on his patent is unknown, but even the collection of a trifling fee for each engine built on this plan would have resulted in a fortune for the patentee, considering the popularity of the 4–4–0 during the life of the patent.

At about the same time that Campbell was testing his first engine, two other Philadelphians, Eastwick and Harrison, built an eight-wheel engine named the *Hercules*. Completed early in 1837 for the Beaver Meadow Railroad, it was surely the second 4–4–0 built. Little is known of its mechanical particulars except that it weighed 15 tons. It had a flexible running gear and, unlike Campbell's engine, could adapt itself to uneven tracks. The *Hercules* was fitted with a separate "truck frame," after a design of Eastwick (see Fig. 57), which equalized the two driving axles. Eastwick's equalizer was not entirely satisfactory and was succeeded a year later by Harrison's equalizing lever. The equalizing lever, which allowed three-point suspension, was possibly the most important American contribution to locomotive design. More will be said on this subject in the section on suspension.

Eastwick's and Harrison's perfection of the 4–4–0 did not immediately bring a general adoption of this style of engine. For the next two years they were the only builders to offer 4–4–0's. Norris, probably the next builder to produce eight wheelers, built his first in 1839. Rogers, Locks and Canals, and Newcastle followed, building their first 4–4–0's in 1840.[15] Even so, production was limited; only about twenty 4–4–0's were in service by 1840.

After 1840 the production of 4–4–0's increased sharply. The eight wheeler had proved itself a practical road engine with greater capacity than the 4–2–0 and was readily accepted as the new national type. Even Baldwin, a dedicated advocate of the 4–2–0, was finally forced in 1845 to build eight wheelers since few roads were purchasing six-wheel engines.

The typical 4–4–0 of the early 1840s was a compact little machine of short wheel base which rarely weighed more than 12 tons. Such a machine is illustrated by the Rogers engine apparently built for the Utica and Syracuse Railway, in Fig. 20. Another example is the *Gowan and Marx* discussed at some length later in this study. Nearly all of these early 4–4–0's were characterized by the connection of the main rod to the rear driving wheels.

The 4–4–0 took on a more familiar appearance in the mid-1840's when the lengthened boiler resulted in the wider sepa-

---

[11] Campbell's first 4–4–0 was later sold to the Long Island Railroad, according to a note in C. B. Stuart's *Civil and Military Engineers* (New York, 1871), p. 330. This is confirmed in part by a note in the *Railway and Locomotive Historical Society Bulletin*, No. 10 (p. 12), which stated that a 4–4–0 named *Chichester* was acquired by the Long Island from H. R. Campbell in 1842. It is also claimed that the above locomotive was a Baldwin 4–2–0 rebuilt by Campbell.

[12] *American Railroad Journal*, August 26, 1837, p. 534.

[13] Minutes of the Philadelphia and Reading Railroad, January 21, 1843, in the files of the Reading Co., Philadelphia, Pa.

[14] *The History of the Baldwin Locomotive Works* (Philadelphia, 1923), p. 41.

[15] M. N. Forney in his history of the Rogers Locomotive Works *Locomotives and Locomotive Building* (orig. pub. New York, 1886; 2nd ed., Berkeley, Calif., 1964), p. 17, incorrectly states that Rogers built his first 4–4–0 in 1844. This error has been repeated many times. Rogers built nearly twenty 4–4–0's between 1840 and 1844. The *Oneida* of the Syracuse and Utica Railroad, completed in June, 1840, was probably his first.

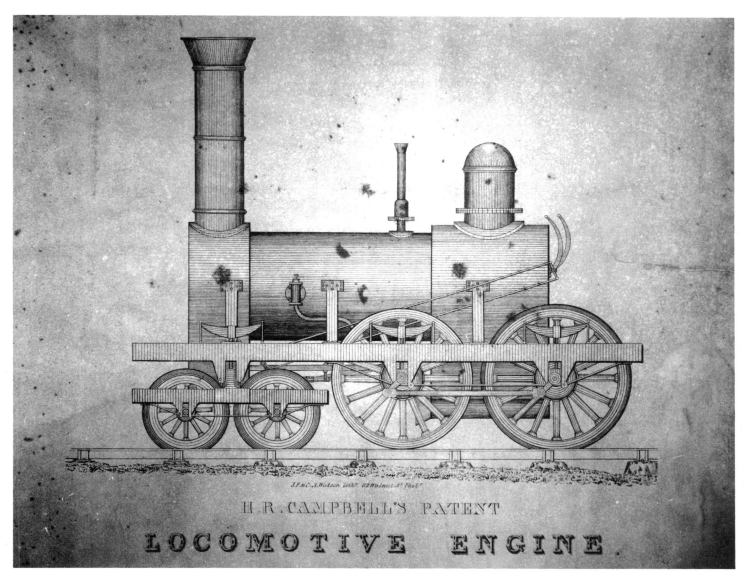

Fig. 19. Campbell's patented eight-wheel locomotive. A machine similar to the one in this illustration was built in 1836 or 1837; it was the first 4–4–0 constructed.

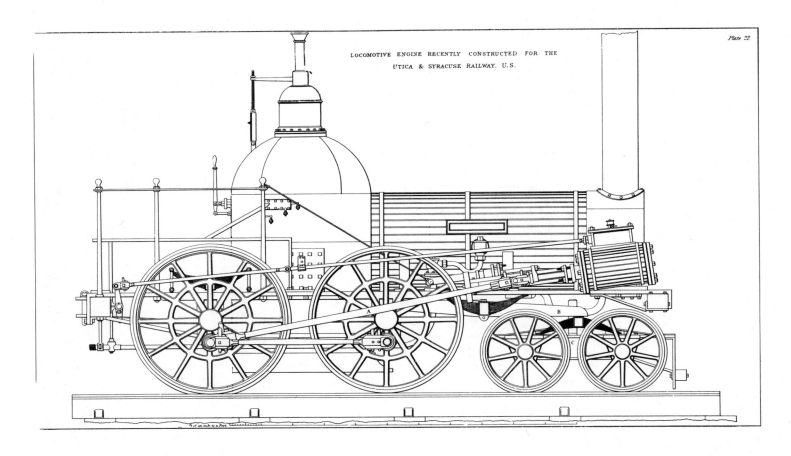

Fig. 20. A Roger's 4–4–0 of about 1843. Wheels, 60 in.; cylinders, 12 in. by 18 in.; weight, 11 tons.

ration of the truck and drivers. The cylinders were now so far from the rear drivers that the main rod was commonly attached to the front drivers. Many significant changes in general arrangement occurred during the mid-1840's; it is unfortunate that no complete working drawings for any 4–4–0 of this period are known to exist. There are, however, several fragmentary drawings for eight wheelers of this period that are worthy of consideration.

The earliest of these drawings, reproduced in Fig. 21, is for the *Massachusetts* and the *Connecticut*. Built in October, 1846, by Rogers, these engines are typical 4–4–0's of the period. The drawing traced from an original Rogers order book is re-

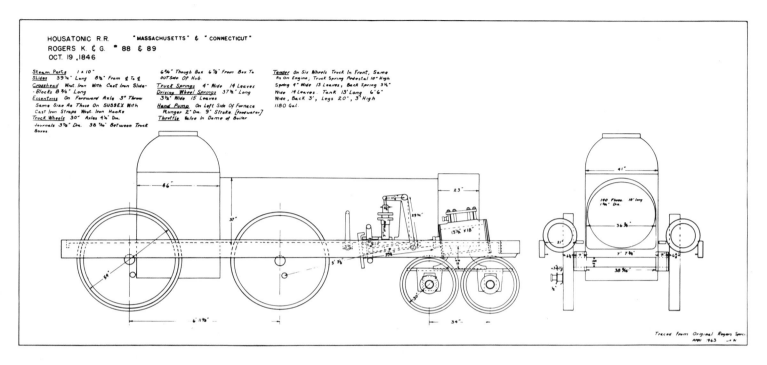

Fig. 21. The Massachusetts *and the* Connecticut, *built in 1846 by Rogers. Drawing and data copied from an original Rogers order specification. Weight of each, 18 tons.*

grettably sketchy. The other three drawings of 4–4–0's of the 1840's, Figs. 22–24, while more complete, show more atypical designs. The *Wm. H. Watson* (Fig. 22) was an unusually large locomotive for its time, weighing nearly 27 tons, and embodied several unusual, though not freakish, design features. The engine was built in the Baltimore shops of the Baltimore and Susquehanna Railroad to the design of James Millholland. Completed in March, 1847, it was used for heavy freight service. The inside cylinders were not unusual, but the massive *cast-iron* crank axle was a feature decidedly peculiar to Millholland. The cylinders measured 18 inches by 18 inches; here again the fact that the stroke was equal to the diameter of the cylinder was at variance with usual practice. The truck was also unusual since the frame was outside the wheels and *below* the axles. The wooden outside frame was rather antique for the period.

The *Mohawk* (see Fig. 23), while more conventional than the *Watson*, was also quite a large machine even for the late 1840's. It weighed 22 tons and had 15-inch by 25-inch cylinders and 60-inch wheels. The *Practical Mechanics Journal* for January, 1850, noted that the engine was rebuilt by Walter McQueen in the Albany and Schenectady Railroad shops. The rebuilding apparently took place in 1847 or 1848, but no evidence is available on the original builder nor on the exact

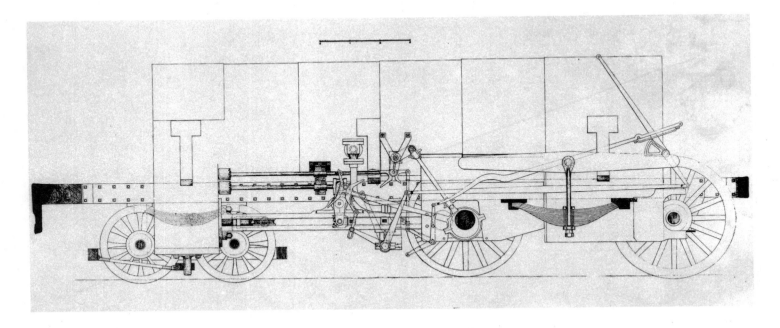

Fig. 22. The Wm. H. Watson, *built in 1847 by James Millholland in the Baltimore and Susquehanna Railroad shops.*

nature of McQueen's alterations. The *Mohawk* had one remarkable design feature, the cylinder saddle, which generally is not credited until the early 1850's.

The final 4–4–0 of this group is the *Tioga* (Fig. 24) built by Baldwin for the New York and Erie in December, 1848. It was a six-foot-gauge engine and was therefore large for the period. It weighed slightly over 28 tons and had cylinders 16½ inches by 20 inches with 72-inch drivers. Except for its size it is probably a fair representative of Baldwin's standard practice of the late 1840's. The outside frame (note the heavy cast-iron pedestals) was not used much in this country. The outside cranks (see Fig. 25) were necessary because of the outside frame. Other mechanisms worthy of note are the double throttle valves, the forward steam dome, and the separate cutoff. Baldwin built eight engines on this plan for the Erie Railway between December, 1848, and June, 1849. They were among the last half-crank engines built by Baldwin.

Generally, the development of the 4–4–0 between 1840 and 1850 was simply an enlargement of the machine as it was first introduced. Boilers were set low, conforming to a near panic about high centers of gravity, short wheel base trucks prevailed, and the Bury firebox and hook-motion valve gears were standard. All of this was changed in the early 1850's. A "modern" machine emerged that revolutionized the industry. It combined several notable features such as spread leading trucks, Stephenson link motion, and the wagon-top boiler. Zerah Colburn, writing in 1860, believed that Thomas Rogers deserved credit for the combination, though not for the inven-

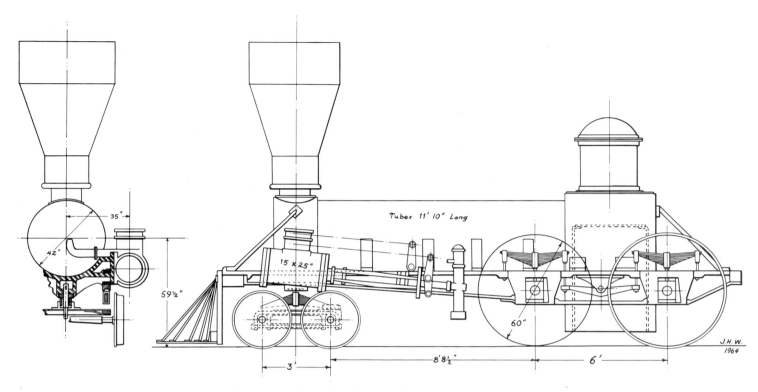

Fig. 23. The Mohawk *of the Albany and Schenectady Railroad as rebuilt by Walter McQueen in the company shops in 1847 or 1848.*

tion, of these features: "Thomas Rogers . . . may be fairly said to have done more for the modern American locomotive than any of his contemporaries. In America the standard wood-burning engine of today stands precisely where he had brought it in 1852, and where he left it at his death in 1856."[16] While not wishing to detract deserved praise from Rogers, it should be noted that William S. Hudson became superintendent of the Rogers works precisely when the first modern engines were built, in 1852. In addition, in 1852 William Swinburne, also from Paterson and formerly with Rogers, built a locomotive named *America* that combined all the reforms credited to Rogers. The question, then, is who was responsible for these reforms—the proprietor, the superintendent, or some obscure draftsman? Credit properly given or not, Rogers' new style of engine was rapidly adopted so that by 1855 every progressive American builder was producing engines on the "Rogers pattern." This pattern was little changed, except by an increase in size, for the next 30 years. The *New Jersey* (Fig. 28) clearly illustrates the "modern" style of eight wheeler introduced by Rogers in 1852. The lithograph from which this illustration is taken is undated. Rogers is known to have built four engines named *New Jersey* between 1852 and 1854. However, none of these had wheels as large as those indicated by the lithograph.

[16] Clark and Colburn, *Recent Practice,* p. 48. In its issue of October 20, 1855, the *Railroad Advocate* credited Rogers with the most advanced design.

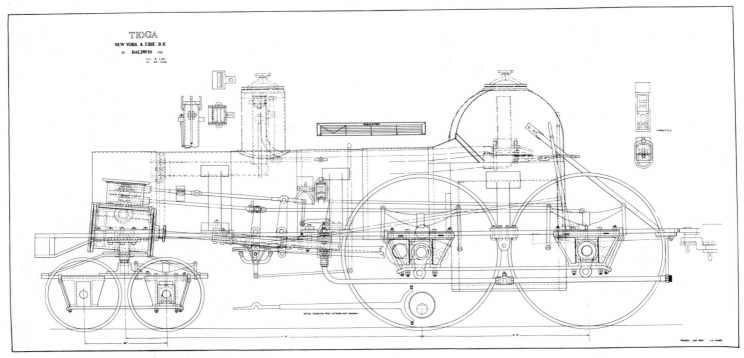

Fig. 24. *The Tioga, built by Baldwin for the New York and Erie Railway in 1848. Six-foot gauge.*

Fig. 25. *The New York and Erie's No. 13, built in 1848 by Swinburne. Shown as rebuilt with a spread truck, link motion, and other minor modifications. Note the similarity of this engine and the Tioga.*

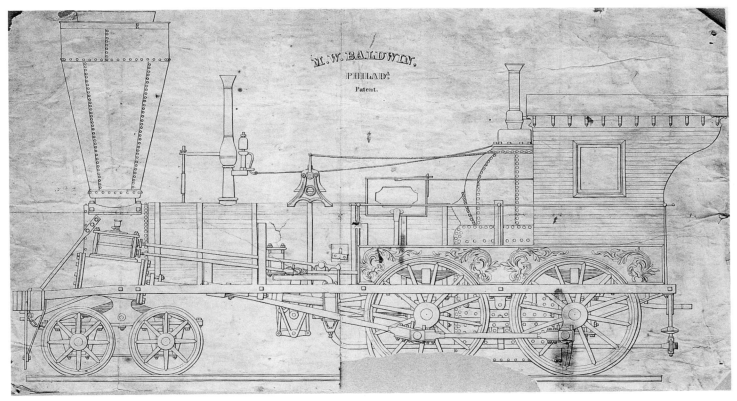

Fig. 26. A Baldwin 4-4-0 of about 1848.

Fig. 27. The Allegheny, built by Baldwin for the Pennsylvania Railroad in 1850.

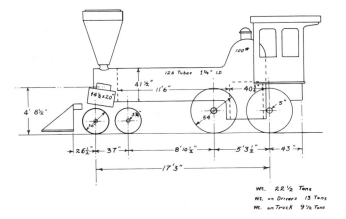

Fig. 28. *The New Jersey illustrates the modern 4–4–0 first constructed by Rogers in 1852. Note the wide spread truck, wagon-top boiler, and link motion.*

The American type continued unrivaled as a general service engine through succeeding decades. Angus Sinclair claims it reached its peak of popularity in 1870 when 85 per cent of the engines in service were of this type.[17] Sinclair's estimate might be somewhat exaggerated, yet in 1872 4–4–0's accounted for 60 per cent of the Baldwin works's production. The Master Mechanics Association concluded after comparing all wheel arrangements that "the eight wheel engines do the same work at a much less cost."[18] In the same report Wilson Eddy, master mechanic of the Boston and Albany, attacked the "Mogul" and "Consolidation," saying, "They run much harder, and with much more friction and more wear and tear of the track than a common eight wheel engine."[19] Eddy supported his contention in 1876 by testing one of his remarkable eight wheelers known as "Eddy Clocks" against a Rhode Island Mogul named the *Brown*.[20] Except for wheel arrangement the two engines were nearly identical in size, but the *Adirondack*, Eddy's engine, had a firebox 6½ inches wider than its rival because of Eddy's clever use of a slab-rail frame. The results of the test were a decided triumph for the *Adirondack* which consumed $190 less fuel while doing the same work as the *Brown*. Eddy's point was that a good American could do anything a Mogul could, and do it cheaper! The actual point was that in the 1870's more roads were finding the American too small for freight service.

The decline of the American type was rapid by the mid-1880's. This is most noticeable in the decreasing number of new 4–4–0's built. In 1884, 60 per cent of the new engines purchased were 4–4–0's.[21] Two years later only half of new construction was of this type, and by 1891 just over 14 percent was of the American type.[22] The eight wheeler maintained its popularity for passenger work into the early 1890's until heavy wooden vestibule coaches and faster schedules proved too much for its limited capacity. By 1900 the 4–4–0 was an obsolete design. It continued to be built in very limited numbers as late as 1945 when Baldwin turned out a tiny 33-ton 4–4–0 for the United of Yucatan Railways.

It would be difficult to exaggerate the importance of this single class of locomotive to the nineteenth-century American railway. No other style of general purpose locomotive enjoyed a greater popularity, and few proved as useful or satisfactory in performing the work they were required to do.

THE 4–6–0

The ten wheeler was the second most popular wheel arrangement in the nineteenth century. It did not seriously rival the eight wheeler, however, nor was it built in large numbers until after 1860, some thirteen years after its introduction.

The *Chesapeake,* completed in March, 1847, by Norris for the Philadelphia and Reading Railroad, is commonly cited as the first 4–6–0 constructed in the United States. Credit for the design is regularly given to Septimus Norris, but George E. Sellers claims that the honor belongs to John Brandt.[23] Sellers states that Brandt prepared the design when he was with the Erie Railway but he does not give a positive date; presumably the plans were made before 1847. Brandt was employed by the Erie between 1842 and 1851. The Erie's management was not impressed with the design and would not permit its construction. The design was then shown to Baldwin and Norris. Baldwin would have none of it, but Norris seemed impressed. James Millholland saw Brandt's plan and ordered Norris to build a locomotive after it for the Reading—supposedly the *Chesapeake*.

At this point Sellers' story loses validity. First, Millholland did

---

[17] Sinclair, *Locomotive Engine,* p. 636.
[18] *American Railway Master Mechanics Association Annual Report,* 1872, p. 171; hereafter cited as *Master Mechanics Report*.
[19] *Ibid.,* p. 130.   [20] *Ibid.,* p. 177.

[21] *National Car and Locomotive Builder,* March, 1884, p. 30.
[22] *Ibid.,* January, 1887, p. 12, and January, 1892, p. 10.
[23] G. E. Sellers, "Early Engineering Reminiscences," *American Machinist,* September 12, 1885, p. 5.

not join the Reading until 1848, a year after the *Chesapeake* entered service. Second, Sellers gave the engine's name as the *Susquehanna*; writing forty years after the fact, Sellers apparently mistook the Erie's first ten wheeler, the *Susquehanna*, for the *Chesapeake*. The accuracy of the essence of his tale (that Brandt originated the 4-6-0) cannot be positively established at this late date. Generally, Sellers' recollections were remarkably precise, but in this instance he was confused on several important details.

The often-reported "patent" obtained by Septimus Norris in 1846 is another questionable story surrounding the building of the earliest ten wheelers.[24] No mention of such a patent can be found in the Patent Office records for the years 1846 to 1853. It is possible that this idea originated from an advertisement issued at about this time by Norris, indeed showing a 4-6-0, but in fact announcing a patented axle box. This illustration is commonly used to represent the *Chesapeake*.

Septimus Norris did, on September 26, 1854, obtain a patent (No. 11733) for locomotive running gears. The drawing showed a 4-6-0, but the specification was uncommonly vague, claiming only improvements regarding locomotives "not having more than one pair of flanged driving wheels" combined with a truck, but it did *not* specifically mention the 4-6-0 wheel arrangement. This evasive claim was made with the knowledge that ten wheelers had commonly been manufactured by all builders before 1854 and that, in fact, the idea was not patentable. The indefinite wording of the patent would not sustain Norris' claim and he was obliged to amend the patent in 1858 and 1859, making it a clear statement on the ten-wheel arrangement. In later years advertisements appeared in the technical press threatening suit against anyone who built ten wheelers without securing a license from the patentee.[25]

These matters aside, the *Chesapeake* was, as already mentioned, one of the first machines built on this plan and it is worthy of further study.

It was a large engine for the time, as will be noted from the data listed in the following table:

Tried in shops March 15, 1847.
Placed on road March 19, 1847.
Fire-box, 37½ inches square inside. Height from grate to crown, 50 inches. Upper surface of grate bars, 17 inches from rail. 2½ inches water space on throat and 2 inches around sides, which will extend down to form ashpan.
Cylinders, 14½" × 22"; 6 feet 3 in. center to center; flanged drivers, six, 46 inches diameter; axles, 5 inches diameter.
Sand-box with copper pipes 2 inches diameter leading in front of drivers.
Roof extending from front of dome to extreme hind of platform, with wrought-iron columns.
Whistle on engine.
Bumper after plan of company.
Smoke-pipe after plan of company, 14 ft. 3 in. from rail.
Tender held 2 cords wood and 2,000 gallons water.
Platform on each side of engine.
Duty: To haul 100 loaded coal cars weighing (without engine tender) 710 tons of 2,240 lbs., with the wood as used for fuel on the road.
Weight of engine with wood and water, 44,074 lbs.
Truck, 4 wheeled, 26" diameter; axles, 4" diameter.
Tender, 8 wheeled, 30" diameter; axles, 4" diameter.
133 tubes, 2 inches diameter, 12 feet long.
Pumps under cylinders worked by arm from piston rod.
Wrought-iron rock arms, bearings inside of frame.
Frames straight from end to end, pedestals fitted with wedges and vibrating boxes.[26]

After two months service the Philadelphia and Reading reported to Norris in May, 1847, that "the *Chesapeake* possesses to a greater extent than any other engine, . . . the combined qualities of efficiency and ease to the rail and bridges."[27] It was further reported she had 14¼ tons (long tons) of her total 19¼ tons on the drivers and easily hauled her "*allotted* load of 384 tons of coal." Yet this glowing testimonial did not commit

---

[24] In *The History of the Baldwin Locomotive Works*, p. 47, it is stated that Septimus Norris obtained a patent for a 4-6-0 in 1846.
[25] *American Railroad Journal*, January 5, 1861, p. 21, carries an advertisement of Octavius J. Norris which specifically mentions Septimus Norris' ten-wheel patent.

[26] Specifications for the *Chesapeake*, consisting of data reportedly from the original records of the Norris works, were printed in *Locomotive Engineer*, August, 1889.
[27] *American Railroad Journal*, November 27, 1847, p. 754.

the Reading to a wholesale purchase of 4–6–0's; in fact, the road was slow to acquire engines of this type and had no other ten wheelers on the road until after 1863.

Hinkley delivered his first ten wheeler, almost to the day (March 18, 1847), at the same time that Norris completed the *Chesapeake*. This engine, named the *New Hampshire*, was built for the Boston and Maine Railroad.[28] A sister engine, the *Maine*, was completed in June of the same year. The Erie's *Susquehanna*, built in June, 1848, was another early 4–6–0 and was probably the first such machine manufactured by Rogers.[29] Other builders gradually began to offer 4–6–0's as the demand grew. Baldwin, possibly not yet adjusted to the adoption of 4–4–0's in favor of his beloved 4–2–0, did not build any 4–6–0's until 1852. A fine example of an early Baldwin ten wheeler is the *James S. Corry*, shown in Fig. 29. This machine was completed in December, 1859, and is representative of a heavy, small-wheeled ten wheeler designed for freight service. The following data are from the original order books now at the De Golyer Foundation Library:

"Jas. S. Corry"   South Carolina Railroad.
M. W. Baldwin & Co. December 1859
Construction No. 905

23 ton D   5 ft. track   November 1859
Boiler   Waist 41" diameter 15 ft. long
Firebox   45 × 36½ × 58" deep
Cylinders   15½ × 22"
Link Motion   Solid Links
Tubes of iron with copper on ends (outside)
Wheels   50" diameter
Tires   wrought iron 2" thick, 2 pr. flange, 1 pr. plain, ¾" play
Journals   5¼ × 5½" diameter
Pumps—of iron
Feed Pipes—of iron
Throttle Valve   in smoke box; copper steam pipes
Smoke pipe [Stack] Radley & Hunter
Cow Catcher   flat bar (Peaks plan)

[28] Colburn describes the *New Hampshire* in *The Locomotive Engine*, pp. 112–13.
[29] The *Susquehanna* is described in the *American Railroad Journal*, September 2, 1848, p. 564.

Piston Follower & *Heads*   ⅞" thick
Steam gauge   Allens
Outside Frames   4 × 1
Stub key   Screw by set screws under straps
Safety Valves   of brass, seat for do. of brass
Names   "Wm. M. Stockton" & "Jas. S. Corry"
Lamps   Charged extra
Tender   With bracket frames two wrought iron trucks. Brakes on all the wheels. Tank 17' long; 7' 6" wide; 42" deep to contain 1800 gallons.

The earliest engines of this type were built with small wheels and were intended for freight service only. By the early 1850's the ten wheeler had been recognized as a useful general service engine. The Pennsylvania and the Baltimore and Ohio railroads were earlier users of 4–6–0's and found them well suited for fast freight and heavy passenger trains. The Erie Railway, which at first rejected Brandt's ten-wheel engine, was an enthusiastic convert by 1853, having 28 such machines on their roster.

After 1860 the ten wheeler came into its own as more roads found it serviceable for heavier work. By 1870 it was serious competition for the 4–4–0. Yet, at the very time of its ascent as the standard engine, the dogma of the general service locomotive was under attack by railroad officials who advocated specialized motive power for each class of service. This new philosophy gained favor as the century wore on, with the 4–6–0's primary assignment being passenger traffic. It continued to be built for such work until about 1910. After that time it was found to be too light for regular, main line service but continued to be built for secondary or branch line work.

In closing, it might be well to repeat the several complaints commonly voiced against the ten wheeler. These account in large part for its failure to succeed the eight wheeler. The most common criticism was that it derailed easily. Too little weight on the truck was the usual reason cited. Whereas the 4–4–0 had too much weight on the truck, often one-third of the total engine weight, the 4–6–0 had only about one-fourth of its tonnage on the truck. In reality the ten wheeler was a better

Fig. 29. The James S. Corry, *built by Baldwin in 1859 for the South Carolina Railroad. Scene, the Baldwin works. Weight, 23 tons; cylinders, 15½ in. by 22 in.; drivers 50 in.*

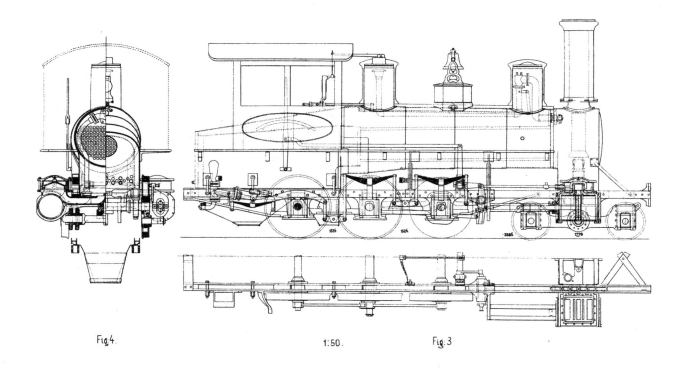

Photolitographie von C. Haack in Wien.

Druck v. Eduard Sieger, Wien.

Fig. 30. The Philadelphia and Reading's No. 405, built in the company shops in 1875. This engine was of the Gunboat class and differed little from the first Gunboats designed by Millholland in 1863. Cylinders, 18 in. by 24 in.; wheels, 54 in.; weight, 45 tons.

balanced engine than the 4-4-0; the true cause of its tracking problems can more likely be found in its suspension. The six drivers were equalized by four levers. The truck was not connected to the suspension of the driving axles; in short, there were five points of suspension. This was in contrast to the ideal three-point suspension of the 4-4-0, a locomotive celebrated for its ability to stay on the roughest track.

A final complaint against the 4-6-0 was that it was more complex, with its extra set of drivers, connecting rods, axle boxes, and springs, than the standard eight wheeler. The implication was that it had more parts but could do no more work. This biased criticism was leveled at nearly every competitor of the 4-4-0 by many prejudiced motive-power officials.

THE 2-6-0

The Mogul, or 2-6-0, was best suited for heavy, fast freight service. This wheel arrangement was an improvement on the ordinary six-wheel connected engine because of the addition of a two-wheel leading truck for better speed and safety in merchandise traffic.

Pseudo 2-6-0's were built as early as 1852. They should not be confused with the true Mogul type, for the leading wheels were rigidly mounted to the locomotive's frame and not carried by a truck. Millholland built the first engine with this wheel arrangement, the *Pawnee*, in 1852 for the Philadelphia and Reading. The design was successful for drag service. Other engines of the same pattern were built, not only for the Reading, but also for other roads (including the Pennsylvania Railroad and the Delaware, Lackawanna and Western) by Smith and Perkins, Norris, and Baldwin. In all only about twenty-five or thirty locomotives were built on this design. One of these machines, the *Cumberland*, is illustrated in Fig. 31. It copied Millholland's design exactly, except for the wagon-top boiler and the inclined cylinders. Notice that the leading wheels are fixed to the main frame behind the cylinders. The leading wheels were probably fitted with a certain amount of lateral "slop" or play to reduce flange wear and improve the guiding qualities of the engine. The Pennsylvania found the *Cumberland*'s front end overloaded and after only three years of service replaced the single pair of guide wheels with a swiveling four-wheel truck as shown in Fig. 32.

The rigid 2-6-0 enjoyed brief popularity confined to a few roads and on the whole was limited to slow freight work. Its leading wheels offered no startling improvement in operation that would warrant its construction in favor of ordinary 0-6-0's and 0-8-0's.

For the next several years the 2-6-0 was primarily inactive until Levi Bissell patented the swiveling two-wheel leading truck in 1858. The several 2-6-0's built late in 1860 by Baldwin for the Louisville and Nashville Railroad are the earliest known 2-6-0's built with a swiveling truck. The New Jersey Locomotive and Machine Company followed in 1861 with a similar but better-proportioned machine, the *Passaic*, for the Central Railroad of New Jersey. The next year the Erie placed the first large order for 2-6-0's. By the middle 1860's the Erie was moving a sizable portion of its freight traffic with a modest fleet of 2-6-0's. The Rogers works built its first 2-6-0 (occasionally referred to as the first built in the United States) in 1863 for the New Jersey Railroad and Transportation Company.

These early 2-6-0's, while remarkably advanced over the pseudo Moguls of the 1850's, were not entirely successful. They were easy on the tracks but they were prone to derail easily. Many mechanics credited this failure to too little weight on the leading wheels. The real cause, as with the ten wheeler, was defective suspension. The leading truck was not equalized with the driving wheels; when running on uneven tracks it would at times be carrying too little weight to keep it on the rails. William S. Hudson, superintendent of the Rogers works, was the first of several mechanics to perfect a practical plan for equalizing the truck and driving axles so that they

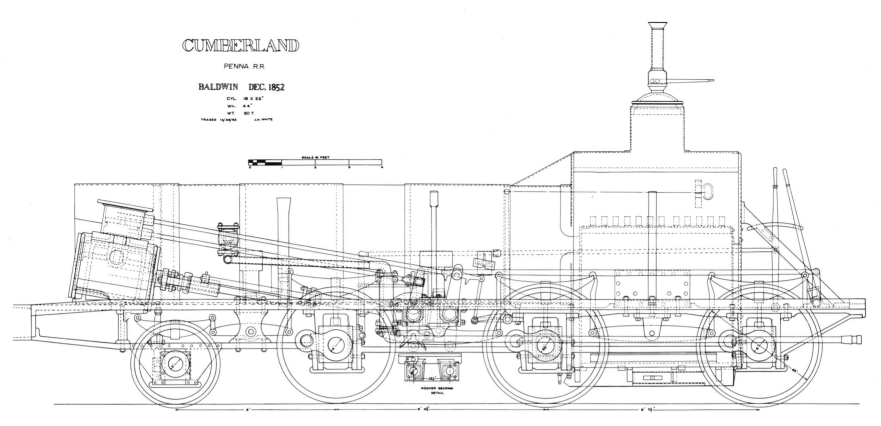

Fig. 31. *The Cumberland was built by Baldwin in 1852 for the Pennsylvania Railroad after a Millholland design. This is not a true Mogul, because the leading wheels are rigid.*

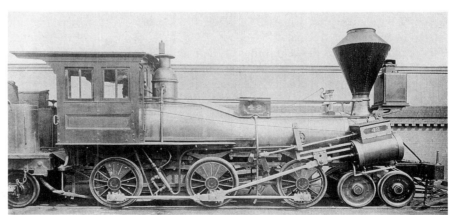

Fig. 32. *The Cumberland was remodeled in 1855 as a 4–6–0. It is shown here as rebuilt in about 1863 with a new cab, crosshead, injector, link motion, and stack.*

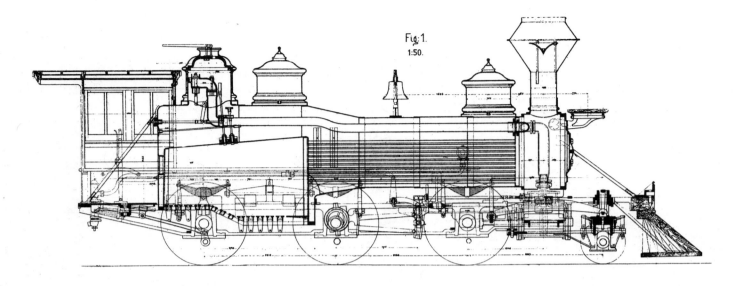

Fig. 33. The Baltimore and Ohio's fast passenger Mogul No. 600, built in 1875 at the company shops.

would work in harmony when passing over rough tracks. Hudson achieved the desirable three-point suspension, making the Mogul the first wheel arrangement, in this respect, competitive with the 4–4–0. Hudson's plan, patented in 1864, is discussed in more detail in the section on trucks.

Hudson's equalizer marked the beginnings of the true Mogul as well as the success of the 2–6–0. The New Jersey Railroad and Transportation Company's No. 39, completed in 1865 by Rogers, was probably the first Mogul so equipped. Baldwin built its first 2–6–0 equipped with the Hudson equalizer in 1867. Between 1868 and 1870 the company built only three engines of this type a year, but in 1875 Baldwin built twenty-six, thus attesting to the growing popularity of the Mogul.

The term "Mogul" was popularly used to describe any large freight locomotive, irrespective of its wheel arrangements, but in the trade it meant simply the 2–6–0. The earliest generic use of the term was found in the Master Mechanics Association Report for 1872. Baldwin described 2–6–0's as "Moguls" in an advertisement of 1871–72.[30] However, it is possible that the Central Railroad of New Jersey's *Mogul*, built by Taunton in 1866, was the origin of the class name. This was true in the case of the *Consolidation,* the first 2–8–0.

Before leaving the discussion of the 2–6–0, mention should be made of the giant Mogul built by the Baltimore and Ohio for fast passenger traffic. This machine, the No. 600, is familiar to many readers because of the handsome engraving published by the *Railroad Gazette*, November 17, 1876, but a better mechanical appreciation is rendered by the longitudinal cross section shown here as Fig. 33. Designed by J. C. Davis, then master of machinery, it was built at the company's Mt. Clare shops in 1875. It weighed 45 tons, had 19-inch by 26-inch

---

[30] *Poor's Manual of Railroads,* 1871–1872; advertising pages were not numbered.

cylinders and 60-inch drivers, and was meant to pull a 117-ton passenger train over the Baltimore and Ohio's steeply graded Third Division between Keyser and Grafton, West Virginia. The 600's ultimate failure in this service is illustrative of the general failure of the Mogul as a passenger locomotive in this country.

The Mogul, after Hudson's improvement, showed great promise of becoming a new national type, at least for freight service. It was a simple, serviceable machine that stayed on the rails and bettered the 4–4–0's tractive power by nearly 50 per cent. Yet the more powerful 2–8–0, introduced in 1866, replaced the 2–6–0 before it was firmly established. Moguls were built for domestic service in this country as late as the 1920's, but they were not at any time widely favored.

THE 2–8–0

The 2–8–0, better known as the Consolidation type, did not become an important wheel arrangement in the United States until 1875. It was introduced in the late 1860's for slow, heavy, pusher service. In a few years its merits as a road engine had become obvious and it was built in large numbers. Indeed, it became the most popular type of freight locomotive in the United States and was built in greater quantities than any other single wheel arrangement. Bruce estimates that 33,000 were constructed between 1866 and 1950.[31]

John P. Laird, then master mechanic of the Pennsylvania Railroad, probably built the first 2–8–0. Actually, it was a rather makeshift affair and was a true Consolidation only in a limited sense. Laird rebuilt an old eight-wheel flexible-beam locomotive, the *Bedford*, in 1864 or 1865. The flexible-beam truck was eliminated. The front and rear driving wheels were set far apart from the two middle drivers. A square water tank running the full length of the boiler was added. A two-wheel leading truck, probably of Laird's peculiar design, was another basic addition.[32] In all, the *Bedford* was something of a hybrid and had little direct influence on the subsequent development of the 2–8–0. It is mentioned here because of its priority, not its excellence.

There is no question that Alexander Mitchell deserves full credit for the Consolidation locomotive. He designed the namesake of this class in late 1865 and incorporated all the elements necessary for the 2–8–0's success in that plan. Mitchell did not patent the idea but lived to see it adopted by every major railroad in the United States. A more complete history of the *Consolidation* is given in the section devoted to that machine later in this volume.

Only a few roads purchased Consolidations in the first years after its introduction; it was believed to be suitable only for heavy, slow service. The Baldwin works, at first the sole producer of this style of engine, understood the initial prejudice against the Consolidation. In 1867 it delivered seven engines; between 1866, when the first 2–8–0 was built, and 1873, only 30 were constructed.[33] The Baltimore and Ohio, whose heavy coal trains were well suited to this wheel arrangement, did not purchase any 2–8–0's until 1873. However, the road immediately recognized the merits of the 2–8–0 and by 1885 had 180 in service.

The day of the Consolidation suddenly brightened after 1875. The next year the Pennsylvania adopted it as its standard freight locomotive and reported moving 80 or 90 car trains (1,000 tons) at 14 miles per hour with such engines. The Erie Railway in the same year began to replace its 4–4–0's with Consolidations for freight service. They found that the heavier machines moved trains of twice the weight and reduced expenses from 0.958 cents to 0.526 cents per ton mile.[34] The road

---

[31] A. W. Bruce, *The Steam Locomotive in America* (New York, 1952), p. 287. It should be noted that few 2–8–0's were built for main-line service after 1920.

[32] See the section on trucks for data on Laird's truck design (p. 174).

[33] *Baldwin Locomotive Works Illustrated Catalogue*, 1873, p. 43.

[34] Ringwalt, *Transportation Systems*, p. 318.

masters at their annual meeting of 1878 concluded that 55 Consolidations would do the work of 100 4–4–0's.

The Consolidation furthered the trend toward heavier power which did not end until giant 600-ton Mallets were built in the 1940's. Engines of larger capacity reduced the number of locomotives and crews required to move a given amount of tonnage. They reduced repair parts inventories and permitted standardization and all the other attendant economies that are possible when fewer units are required.

By 1880 the Consolidation had largely relegated the once standard 4–4–0 to passenger traffic and was without a serious rival for freight service.

THE 0–4–0, 0–6–0, AND 0–8–0

The four-wheel connected engine is the most elementary wheel arrangement possible. Its history as a road engine is extremely brief and accordingly will be treated briefly. Several of the British imports, notably Stephenson's "Sampson" class, a few early American products such as the *West Point* and *De Witt Clinton*, and the Baltimore and Ohio's Grasshoppers were all intended for road service. Their limited power and poor tracking soon reduced the 0–4–0 to switching service, however, after which it played a minor role in locomotive development.

What would have been the first American 0–6–0 was ordered from Stephenson by the Baltimore and Ohio in 1829, but the machine was not delivered, because it was lost at sea.[35] The *Nonpareil*, built by the Beaver Meadow Railroad in 1837 or 1838, was probably the first 0–6–0 constructed in this country. Baldwin's first flexible-beam truck engines, introduced in 1842, were 0–6–0's. They were probably the most numerous machines of this wheel arrangement in the United States during the early years of the 0–6–0's development. Between 1842 and about 1864 some 200 were manufactured. Norris also produced six-wheel connected engines; they built a number of 0–6–0's from the early 1840's through the 1850's. The *Philadelphia* (Fig. 133) is an example of a Norris six wheeler. A number of powerful six-wheel freight engines were built between 1854 and 1856 by Rogers for the Buffalo and Erie, the Buffalo and State Line, and the Lackawanna railroads. A lithograph was issued of one of these engines, the *Volcano*. This engine weighed 23 tons and had 16-inch by 22-inch cylinders and 54-inch drivers. The New Jersey Locomotive and Machine Works and Danforth-Cooke also built engines (mainly for the Lackawanna) on the same plan during the mid-1850's. All 0–6–0's were intended for slow freight traffic and were a minor wheel type on early American railroads. Before 1870 it is questionable if more than 300 were constructed in this country. During the 1860's and 1870's they were entirely replaced by 2–6–0's or 2–8–0's for road service and after that date were used only for switching.

The first eight-wheel connected engine was, like the 0–6–0, built at an early date. The Camden and Amboy Railroad built an eight-wheel engine, the *Monster*, between 1835 and 1838.[36] An awkward machine, it seemed to favor marine rather than railway engineering practices. It was not truly an 0–8–0, for spur gears rather than connecting rods were used to couple the second and third driving axle. The machine was indeed a monster; it weighed about 18 tons and had cylinders 18 inches by 30 inches. Four heavier engines were built on this design in 1852 and 1854 for the Camden and Amboy by the Trenton Locomotive Works. Another was built by the Camden and Amboy in 1852, thus making a total of six "monsters."

Ross Winans designed the next 0–8–0 in this country after the original *Monster*. The *American Railroad Journal* of March 15, 1841, describes this engine as weighing 19⅓ tons with

---

[35] C. F. D. Marshall, *A History of the Railway Locomotive Engine Down to the End of the Year 1831* (London, 1953), p. 145.

[36] In his report of 1838 J. Knight described the *Monster* as nearly completed.

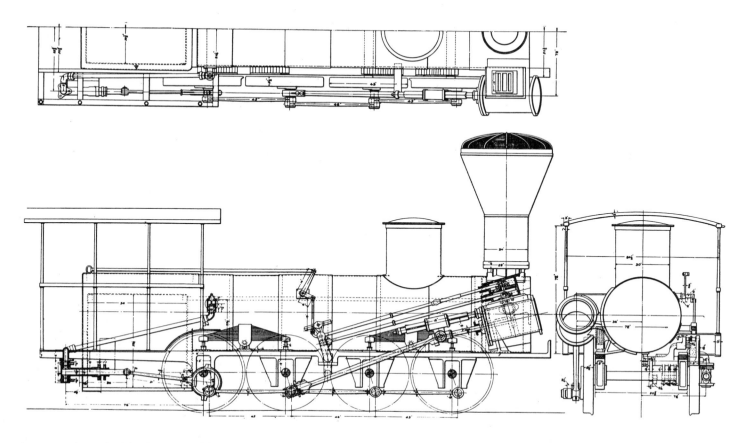

Fig. 34. *An eight-wheel connected locomotive of the Baltimore and Ohio Railroad. An exact identification cannot be made (it may be only a design study), but its date of about 1848 is certain.*

cylinders 14¼ inches by 24 inches. The machine was apparently new at the time and was undergoing tests on the Baltimore and Ohio. In September, 1841, the Western Railroad (Massachusetts) purchased this or an identical engine from Winans and named it the *Maryland*. The *Maryland* had a vertical boiler and spur gear drive and was in general arrangement an enlargement of Winans' earlier "Crab" engines. Winans and Baldwin each built three more eight-wheel engines of this design in 1841–42. The Western Railroad found them unsatisfactory and all were retired by about 1850.

The Baltimore and Ohio found the 0–8–0 well suited to its heavy coal traffic and became the largest single user of eight-wheel connected engines in the nineteenth century. Between 1844 and 1846 the road received twelve locomotives from Winans, which, except for horizontal boilers, were identical to those of the Western Railroad. These curious machines became known as "Mud Diggers." A thirteenth "Mud Digger" was built in the company's shop in 1847. In 1847 Winans materially improved his eight-wheel engines by eliminating the gear drive and making a direct connection to the driving

Fig. 35. The Baltimore and Ohio's Hero, *built by Thatcher Perkins at the company shops in 1848. This photograph was taken in 1867. The engine was rebuilt in 1857, but the exact alterations made are not known.*

wheels. Four machines were built on this reformed plan for the Reading in 1847.

The following year the Baltimore and Ohio purchased Winans' first "Camel" locomotive. This machine was the primogenitor of the most famous class of 0–8–0's built in this country. Designed for coal-burning and heavy traffic, some two or three hundred were used successfully by several roads, despite several design defects. The Camel's history is more fully discussed on page 347.

In addition to Winans products, the Baltimore and Ohio built and purchased 0–8–0's from other builders. An example of one of the company-built eight wheelers is the unidentified machine shown in Fig. 34. Note the similarity of this drawing to the *Hero* (Fig. 35), designed by Thatcher Perkins in 1848. In 1865 Perkins designed an improved 0–8–0 that was intended to supplement the Baltimore and Ohio's aging fleet of Camels. Twenty-four Perkins 0–8–0's were built. They represent the final example of this wheel arrangement designed for road service.

Aside from the Baltimore and Ohio and Winans, Baldwin was the only other major figure associated with the 0–8–0. In 1846 he built his first 0–8–0, which, like the 0–6–0, was a flexible-beam truck engine. Subsequently, Baldwin built about three hundred 0–8–0's, ending production of this design in 1866.

At best, six or seven hundred eight-wheel connected engines were constructed in this country before 1870. Their use was restricted to a few roads and their part in the early development of American locomotives was accordingly minor.

PERFORMANCE

6.

The small American locomotive of the nineteenth century registered a remarkable performance record in spite of many operating difficulties. We have already mentioned poor tracks, but note should also be made of inadequate maintenance shops, untreated water, and a lack of competent enginemen. Against all of these handicaps, small, 20-ton locomotives regularly hauled 200-ton trains, year after year, often serving for twenty years or more.

Specific examples of performance are given whenever possible in the sections dealing with representative locomotives, but general examples and some conclusions will be presented here. It is difficult to make sweeping statements on early train operations because so few reports of *typical* operations have survived. Extraordinary and unusual performances were reported almost without fail and unhappily constitute the bulk of presently available data.

During the first decade of railway operations in this country (1830–40), short, light trains were used. Freight trains rarely consisted of more than fifteen loaded cars and weighed, not including locomotive and tender, about 75 tons. A passenger train in this period probably averaged six four-wheel cars and fully loaded might have weighed a total of about 20 tons. The above figures are offered as an average based on contemporary annual railroad reports and the *American Railroad Journal*. These probable averages are tempered by the following examples of spectacular performances. One of the Baltimore and Ohio's early "Grasshoppers" (probably the *Atlantic*) was reported on August 16, 1832, to have pulled a six-car passenger train with ninety passengers 82 miles at an average speed of slightly over 13 miles per hour.[1] The light, 7½-ton engine pulled its train, estimated at 30 tons, up grades varying between 32 and 57 feet per mile. The *Atlantic*'s tractive force was calculated at 1570 pounds, or 63 indicated horsepower, at 15 miles per hour.[2] Another Baltimore and Ohio "Grasshopper," the *Arabian*, pulled a 113-ton load on level ground at nearly 12 miles per hour.

The unprecedented work of Norris' *George Washington* on

---

[1] Clark and Colburn, *Recent Practice*, p. 49.
[2] *Baltimore and Ohio Annual Report*, 1832.

the Philadelphia and Columbia Railroad in July, 1836, was one of the most celebrated performances on record. The engine, a 4–2–0 weighing 7½ tons, was credited with pulling nearly a 16-ton load up the Belmont Incline at a speed of 15.5 miles per hour. This section of railroad, operated as an inclined plane, and thought to be impractical for regular locomotion, was 2,800 feet long and had an impressive grade of 369 feet per mile.[3] The much-publicized performance of the *Washington* won Norris an enviable if undeserved reputation as well as many orders for his locomotives. Another notable performance for this early period was made on the New Jersey Railroad and Transportation Company's line by a Rogers' 4–2–0, probably the *Uncle Sam*. This machine hauled a twenty-four-car train, said to weigh 120–180 tons, between Jersey City and New Brunswick. At one point the engine hauled the train over a grade of 26 feet per mile at 24.5 miles per hour.[4]

Many reports are available on the operation of Reading's locomotives because of the road's heavy trains and the advanced nature of its operations. The Reading ran exceptionally long coal trains from the first years of its existence. In February, 1840, the famous *Gowan and Marx,* an 11-ton 4–4–0, pulled a 423-ton train of 101 four-wheel cars at an average speed of 9.8 miles per hour. Only a year later the *Hichens and Harrison* pulled a train weighing 481½ tons. By the mid-1840's the Reading was regularly operating 500- and 600-ton trains and in one instance moved a 150-car coal train weighing 1,180 tons. This enormous load was moved by a six-wheel freight locomotive, the *Ontario*.[5] In the late 1850's the Reading's average coal trains were said to weigh 708 tons.[6] Yet the Reading was exceptional, and its trains weighed two to three times the national average. The New York *State Engineers Report on Railroads* is one of the few available sources of early operating statistics.[7] The following table shows average train weights for all railroads in the state of New York. Locomotive and tender are not included, but passengers and baggage are.

1855	Passenger train (Passengers and Cars)	62.5 tons
1859	" " " " "	77.5 tons
1865	" " " " "	ca. 95 tons
1868	" " " " "	ca. 96 tons
1855	Freight train (Freight and Cars)	267 tons
1859	" " " " "	212.6 tons
1865	" " " " "	226.5 tons
1868	" " " " "	252.5 tons

In 1875 the Pennsylvania Railroad reported that its passenger trains averaged 128 tons in weight (not including the weight of the cars). This lack of growth in train size explains in part the stability of locomotive design between about 1855 and 1875. On the whole there was relatively little change in size or design during these years. Except for the introduction of new wheel arrangements (notably the 2–6–0 and 2–8–0) small, light 4–4–0's remained the standard engine into the 1870's. Its continued dominance must be credited largely to a lack of growth in train size or speed, and the reluctance of old-guard master mechanics to adopt any other wheel arrangement. Stimulated by expanding industrial production, train weights climbed during the 1880's and by the early 1890's several major railroads were reporting average freight train weights of 553 tons.[8] Only after the expansion of operating conditions did heavier and more complex locomotives come into their own and displace the old eight wheeler. The point is very simple; operating conditions directly influenced locomotive design.

---

[3] Colburn and other experts question the *Washington*'s performance, noting that weight on drivers and steam pressure would not produce the needed tractive force for the reported performance.
[4] *American Railroad Journal*, December 15, 1839, pp. 366–67.
[5] Colburn, Z., *The Locomotive Engine* (Boston, 1851), p. 167.
[6] Z. Colburn and A. L. Holley, *The Permanent Way and Coal-Burning Locomotives* (New York, 1858), p. 117.

[7] Even the New York State Reports are questionable. Colburn claims that the train weights given in these reports are estimates only.
[8] *American Railroad Journal*, January, 1892, p. 5.

ERA OF FUNDAMENTAL DESIGN

TRAIN SPEEDS

Compared to train weight, speed was a decidedly secondary consideration to railway managers in the nineteenth century. Nicholas Wood, the eminent British railway authority, wrote in 1825 that talk of railway travel at 20, 16, or even 12 miles per hour was "nonsense" and, being entirely visionary, would do nothing but harm to the railway movement.[9] Six years later John B. Jervis shared Wood's opinion and stated: "The expectations of the public have been so much excited in reference to rapid travelling (and that must be by locomotive steam power) that they will not be satisfied with moderate speed, say 10 to 12 miles per hour; they must have 15 as a regular business."[10] This conservative stand on speed was not some perverse desire to frustrate the public. Slow speeds were favored because operating costs were directly linked to the rate of travel. A report of the South Carolina Railroad noted: "Mr. Stephenson, in reply to the enquiry made by the President of the Boston and Lowel [sic] RailRoad Company, viz. What do you consider the economical rate of speed at which Locomotives should travel? States that they should not exceed eight miles per hour with freight cars, nor sixteen (16) miles per hour with passengers, the latter speed yielded to, not from considerations of economy or durability, but solely to gratify the public in their wishes for rapid travelling."[11]

The Chicago and Rock Island Railroad effected major economies, a necessary adjustment to bad times after the 1857 panic, by reducing speeds. Its report of 1859 stated that locomotive running costs had been cut down from 30 cents per mile in 1857 to 21 cents, "the saving arising from diminished speed, both of freight and passenger trains." According to Jervis a train could operate at 20 miles per hour for half the cost of running at 30 miles per hour.[12] Costs were directly related to speed to the extent that few roads felt compelled to drastically increase their operating costs merely to satisfy a public whim. This whim could not be ignored, however, when a competing line was opened that offered better service. Under this stimulus passenger train speed showed some improvement during the 1870's, but apparently competition was not severe enough until the 1890's to have any material effect on the upgrading of schedules. It should also be noted that faster speed not only boosted fuel cost but meant more wear and tear on rolling stock, track, and bridges. It also increased the likelihood of accidents, which meant loss of equipment and increasingly heavy claims for deaths and injuries. In addition a more rapid rate of travel obviously called for heavier rail, stronger bridges, and a straighter line, all involving a larger financial outlay than the capital-poor American railroads could afford in their early years.

Locomotives proved their capacity for high speed from the beginnings of the public railway. Stephenson's *Northumbrian* averaged 36 miles per hour in a desperate run to save William Huskisson, a member of Parliament, when he was seriously injured at the Liverpool and Manchester Railway's opening in 1830. David Matthew claims he ran the *Experiment* at better than 60 miles per hour on the Mohawk and Hudson in 1832. Other early fast runs could be mentioned, but they are incidental to standard operating practice. Ghega noted in 1840 that Baltimore and Ohio passenger trains regularly ran at 18–20 miles per hour and, including stops, averaged only 15 miles per hour. Reporting at the same time, Von Gerstner noted that the average on all United States roads was about 15 miles per hour.

Moderate gains were made in succeeding years. By the mid-1850's New York passenger trains were averaging 24 miles per hour; however, this included the express trains of the Hudson

---

[9] Nicholas Wood, *A Practical Treatise on Railroads* (London, 1825), p. 290.
[10] *Railway and Locomotive Historical Society Bulletin*, No. 55 (May, 1941), p. 10.
[11] South Carolina Canal and Railroad Co., *Report of the Committee on Cars . . .* , 1833, p. 11.

[12] J. B. Jervis, *Railway Property* (New York, 1861), p. 156.

River Railroad and should be viewed as higher than the probable national average. *Appletons Railway Guide* for June, 1864, showed American express trains averaging 32 miles per hour, regular passenger trains, 26 miles per hour, and branch line trains, 17 miles per hour.[13] One cause of low average speed was frequent station stops. There was only one continuous run of 40 miles in this country in 1867, and most trains stopped every few miles.[14] The mere passage of years, however, did not guarantee improved service; the railroads of Massachusetts between 1851 and 1857 reported a slight decline in average speed and a 50 percent increase in the cost of service. As late as 1887 the New York Central's fastest train averaged just over 40 miles per hour; this represented almost no improvement on the schedule offered some 35 years earlier on the same run by the Hudson River Railroad.[15] The Pennsylvania Railroad express between Jersey City and Washington averaged 45 miles per hour, while its fast train between Jersey City and Pittsburgh averaged 39.2 miles per hour. No road in this country had a scheduled *running speed* of over 53 miles per hour, and the average, of course, was far less.

If improvement in passenger train speed was conservative, the increase of freight train speed was negligible. The goal of the railroad, however, was cheap transportation, and because speed so directly affected costs, slow freight schedules were accepted without question during the nineteenth century. In fact, freight trains were commonly referred to as "burden" or "tonnage," both terms connoting slow and ponderous traffic. According to Colburn, "The policy of American railway managers . . . was maximum loads at slow speeds, involving a minimum resistance per ton, and correspondingly a minimum working expenditure per ton."[16]

The New York railroads ran their freight trains at an average of 12 miles per hour in the mid-1850's. During the same period, the Reading's heavy coal train had a road speed of 11.5 miles per hour; the road's average speed was far slower. As late as 1876 the United States Bureau of Statistics stated: "The rate of speed for freight trains yielding the maximum profit is about ten miles an hour." It can be assumed with some certainty that, throughout the nineteenth century, average freight train speeds rarely exceeded this rate.

LOCOMOTIVE POWER

A single, comprehensive measurement of locomotive capacity has never been formulated, because of the diverse factors governing locomotive power. Horsepower and tractive force are often given as positive measures of locomotive capacity, yet they are largely calculated or "rated" measures that do not encompass many practical but highly variable conditions affecting the actual ability of railway engines to move trains over the line. Locomotive hauling capacity is directly affected by track conditions, gradients, curve radii, weather conditions, and the working condition of the locomotive itself and the cars it pulls.

Rough, uneven tracks and sharp curves resist the passage of trains and require more power to traverse them; icy or greasy rails cause drivers to slip; and steep grades obviously call for more power. Strong head winds, the friction of the locomotive cylinders, valve gears, and wheel bearings, and the lubrication or lack thereof also reduce the engine's hauling capacity. These factors naturally vary from one railroad to the next and from one season or day to the next.

Because the locomotive is a vehicle, the power produced by its boiler and cylinders is only partially delivered to the rails by the driving wheels. Thus, adhesion—the friction between the rail and the driving wheel tire attributable to the weight on

[13] Z. Colburn, *Locomotive Engineering and the Mechanism of Railways* (London and Glasgow, 1872), p. 80.
[14] *Engineering*, July 26, 1867, p. 64.
[15] *Railroad Gazette*, February 11, 1887.

[16] Institution of Civil Engineers, *Proceedings* (London, 1869), Vol. 28, p. 360.

the drivers—materially affects the pulling power of the locomotive. Under ordinary operating conditions, adhesion is generally estimated to represent one-fifth of the weight on the driving wheels. Thus, an engine with 20 tons on the drivers would be expected to have an adhesion weight of 4 tons. Yet this important consideration is not reflected in horsepower or traction effort formulas.

Horsepower was established as a standard measure of power by James Watt in the late eighteenth century and continues as a satisfactory measure for stationary engines. It has only rarely been applied in railroad circles as a measure of locomotive power. Tractive effort has been the standard measure of locomotive power since about 1900. It was presented in De Pambour's 1836 locomotive handbook and has been repeated in nearly every railway text on the subject since that time. The formula reads

$$T = \frac{C^2 \times S \times P}{D}$$

where T = tractive force in pounds, C = diameter of cylinders in inches, S = stroke of cylinders in inches, P = effective steam pressure (80–85 per cent of boiler pressure), and D = diameter of driving wheels in inches. Tractive force indicates the starting power of the locomotive but it does not, of course, show how well the engine can be expected to perform under all of the variable conditions mentioned at the beginning of this discussion. It was intended to provide only a convenient bench mark figure for comparisons.

Tractive force, though understood, was rarely used to designate locomotive power in the nineteenth century. The railway literature of the period other than text books rarely alludes to this standard. Annual reports, locomotive registers, builders records, and other sources known to the author do not mention tractive force. The only criterion regularly employed was that of rated hauling capacity which was measured by the tonnage a given locomotive could pull on level ground or up a specific grade.

Locomotive builders' catalogs contained a table for each class of engine that showed its rated tonnage on level ground and on gradients. The published catalogs of Baldwin, Rogers, and Dickson for the years 1873–85 are examples of this practice. Here was a direct, practical measurement that readily addressed itself to the operative mechanic. By referring to the table a railway superintendent could immediately determine what size engine would be required to surmount the heaviest grade on his line for a given train weight. The figure thus obtained was no more comprehensive than tractive force but it was more readily understandable and provided a quick reckoning for motive-power requirements.

The growth of locomotive power and size during the years encompassed by this study is summarized in the diagram chart, Fig. 36. It can be seen that an increase in cylinder size, heating surface, steam pressure, and weight on the drivers improved the locomotive's power with the passing years. Every effort was made to list representative machines of ordinary size for the period; extreme examples were avoided. For this reason the 4–4–0 was chosen to represent the years 1865 and 1875, even though far heavier locomotives, such as the 2–8–0 and the 2–6–0, were in service. The latter machines were relatively limited in number compared to the 4–4–0 and their use would exaggerate the collective growth in engine size for the years designated. From the information available there appears to have been only a modest growth in size and power between about 1855 and 1880. The 4–4–0 was regarded as adequate for most needs during these years; however, some roads required heavier, more powerful locomotives for coal traffic. The development of these machines is given in the section on wheel arrangement.

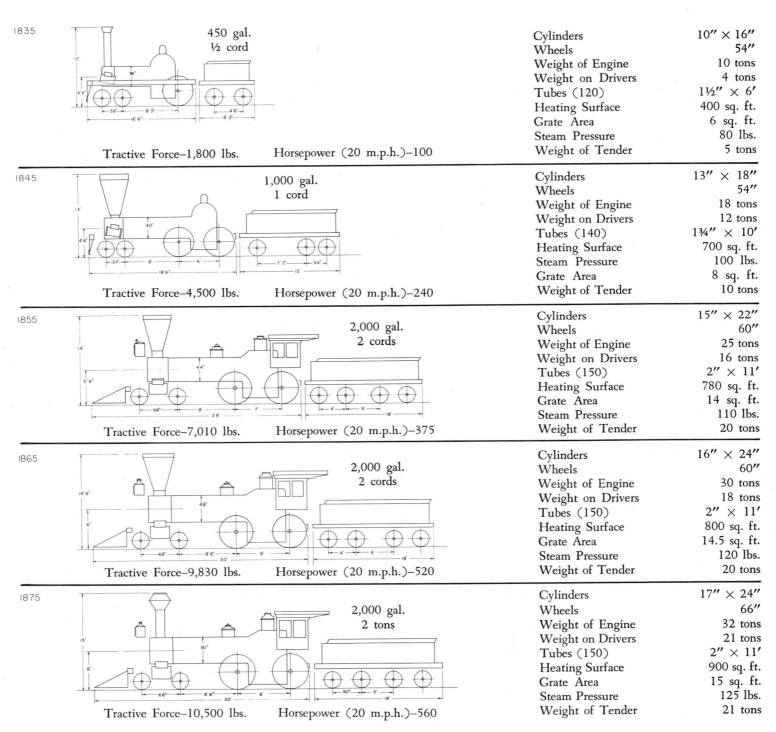

Fig. 36. *Diagram table showing development of the American locomotive from 1835 to 1875.*

ANNUAL MILEAGE

A final consideration of locomotive performance is annual mileage. Here again, considering track and shop facilities, early locomotives turned in a commendable record. As early as 1837 the Boston and Lowell Railroad reported that each of its engines averaged over 21,000 miles a year. This figure was not unusual (as shown by other annual reports) and was an ordinary average for American locomotives until the 1870's. Because of improved tracks, shops, and operating techniques annual mileage climbed rapidly during the 1870's. In 1877 the Pennsylvania Railroad reported that its engines averaged from 48,000 to 72,000 miles per year. The New York Central and Hudson River Railroad in the same year stated that its engines averaged nearly 40,000 miles. Some locomotives achieved exceptional mileage records. One of these, the Illinois Central's No. 23 (built by Rogers in 1853) traveled more than one million miles over a period of thirty years, or some 32,000 miles per year.

Thus, typical operations before 1875 were characterized by light, short trains, low speeds, and high annual mileage.

PERFORMANCE COSTS

The three major items in locomotive running expenses are fuel, wages, and repairs. Other, more minor expenses include water, oil, and cleaning. Depreciation and interest, while major expenses, are difficult to calculate and are beyond the intended scope of this work. As early as 1835 John B. Jervis recognized the importance of these costs and criticized the Philadelphia and Columbia Railroad for neglecting to include them in its operating costs.[17] The railroad stated that its locomotives operated at 19 cents per mile; Jervis noted that water and repairs ($3.00 per day) and interest and depreciation ($9.00 per day) increased *total* operating costs to 36 cents per mile. Jervis' criticism was not ignored; the Baltimore and Ohio and several other roads began to include depreciation or "renewal" costs in their annual reports at an early date.

FUEL COSTS PER MILE

Fuel was, as a rule, the largest single item of expense, and most roads included some statements on this expense in their earliest annual reports. These reports, a vast array of data, permit very few meaningful comparisons. One road, for example, might report splendid mileage while operating only light trains over easy grades at low speeds. Another road, moving heavy trains over steep grades or offering faster schedules, would naturally be doing more work and thus would consume more fuel per mile. This point was aptly made by the *Railroad Advocate* when noting the costs of three different trains on a 100-mile run:[18]

	One Car, 50 passengers	Four Cars, 200 passengers	Merchandise, 100 tons
Wages of Engineer	$2 50	$2 50	$3 33
Wages of Fireman	1 25	1 25	1 67
Wages of Conductor	2 00	2 00	2 00
Wages of Baggage Master and Brakemen		1 50	4 00

---

[17] *Railway and Locomotive Historical Society Bulletin*, No. 55, pp. 34–35.
[18] *Railroad Advocate*, May 3, 1856.

	One Car, 50 passengers	Four Cars, 200 passengers	Merchandise, 100 tons
Wages of Brakemen for freight			5 00
Fuel	12 00	16 00	30 00
Oil, Waste & Water	2 00	2 25	3 50
Repairs of Engine	3 00	5 00	8 00
Repairs and refitting Cars	2 00	8 00	10 00
Sundries and extra	5 00	6 00	8 00
Total	$31.25	$47.00	$71.50
Wages of Stat'n and Sw'chmen & gen'l Expenses not included above	8 00	10 00	25 00
Expenses of Roadway and Renewal of Track	10 00	15 00	25 00
	$49.25	$72.00	$121.50

For these reasons it is difficult to determine with much precision the rise or decline in operating expenses between 1830 and 1870 because operating conditions changed materially during these years. The geographic variations in fuel cost are another consideration and make a comparison of fuel costs all the more uncertain. Nevertheless, a few examples of fuel costs are given.

Knight's Report of 1838 lists many entries of fuel and other costs.[19] He notes that the Long Island Railroad ran 48 miles per cord of wood, at a cost of 10 cents per mile. The Boston and Providence ran 36 miles per cord at a cost of 16 cents per mile. The neighboring Boston and Worcester ran 56 miles per ton (anthracite) at a cost of 12.4 cents per mile. The Baltimore and Ohio Railroad in 1843 fired half its power with wood, the remainder with coal, and had a combined average fuel cost of about 6.5 cents per mile. In 1851 Colburn reported fuel cost (wood) at 18 cents per mile.[20] The railroads of New York had an average fuel cost (wood) of 17.9 cents per mile in 1855; the railroads of Massachusetts averaged 20 cents per mile two years later.[21]

By 1860 increased train weights, speeds, and wood prices drove fuel cost as high as 31 cents.[22] A few roads enjoying a good supply of cheap wood, however, could boast (along with the Baltimore and Ohio) of costs as low as 7.8 cents per mile. Even greater savings were achieved by converting to coal where, for this period, costs ranged from 3.6 cents per mile on the Baltimore and Ohio to 9 cents per mile on the Central Railroad of New Jersey.

The Illinois Central Railroad began to experiment with coal-burning locomotives in 1855. The opening of the southern Illinois coal fields promised a local supply of cheap fuel. Comparative tests showed a saving of 32 per cent in favor of coal; wood cost 18 cents per mile while coal cost only 12 cents per mile.[23] It was soon discovered, however, that higher maintenance costs, resulting from shorter firebox and tube life, reduced somewhat the apparent savings on coal, even though its price had declined by more than half by 1859.[24]

In 1874, after many roads had converted to coal, Forney stated that fuel cost per mile was 6 cents; mileage per ton was 38.[25] By 1892, cost per mile was given as 7.4 cents; mileage per ton was 46.1.[26] This compares favorably with Reading's heavy coal burners which in 1857 cost 13.4 cents per mile and averaged only 19.1 miles per ton.[27] It should be remembered, however, that the Reading was moving 700-ton trains (admittedly, mainly downhill) and that anthracite was difficult to burn. More will be said on this subject in the section on fuel.

---

[19] See *Railway and Locomotive Historical Society Bulletin*, No. 13 (1927), for a reprint of Knight's report.
[20] Colburn, *The Locomotive Engine*, p. 105.
[21] Colburn and Holley, *The Permanent Way*, p. 4.
[22] Clark and Colburn, *Recent Practice*, p. 69.
[23] Illinois Central Railroad, *Annual Report*, 1857, p. 13.
[24] *Ibid.*, 1859, p. 6.
[25] M. N. Forney, *Catechism of the Locomotive* (New York, 1879), p. 449.
[26] T. M. Cooley et al., *American Railway* (New York, 1892), pp. 307–8.
[27] Colburn and Holley, *The Permanent Way*, p. 118.

LABOR COSTS

Wages constituted the largest operating cost after fuel. Salaries were relatively stable during the nineteenth century, and for the years of this study there was surprisingly little change. Knight reports (in 1838) that several lines operating out of Boston paid engineers $2.00 a day, firemen $1.00 a day. Roughly figuring, the cost per mile would be 3–4 cents. In 1843 the Baltimore and Ohio noted that wages cost about 4 cents per mile, and Colburn gives 3.8 cents as the cost in 1851. As late as 1892, wages were reported to be only 5.3 cents per mile. The engineer's daily salary was greater than in former years, but the number of miles run also increased, thus keeping the cost per mile low.

REPAIR COSTS

Repairs, generally a third cost contender, were greater than wages or fuel on some lines. This depended largely on the character of the road and the efficiency of the master mechanic. As a rule, repair costs on American roads averaged 10 cents a mile before 1870. This figure is confirmed by numerous annual reports for the period. Of course, costs varied slightly from road to road and from year to year on the same line. The Baltimore and Ohio's annual report for 1854 listed its repair costs for the previous decade. In 1844 cost per mile had been 7.2 cents per mile; in 1854 it was 9 cents per mile. The lowest costs recorded were those for 1852, only 5.1 cents per mile, but the next year repairs jumped to 9.6 cents. The railroad blamed the increase on higher material and labor costs, but a more probable explanation was that the engines were neglected in 1852, thereby making heavy repairs necessary in 1853.

Repair costs showed marked improvement by the 1870's when more careful maintenance and heavier locomotives became the rule. The Erie Railway, for example, cut its repair expenses from 14 cents per mile in 1870 to 3.9 cents in 1880.[28] This economy was largely achieved by replacing many small, worn-out engines with fewer but more efficient machines. Moreover, the new engines were of a standard pattern permitting a reduction in the number of repair parts.

LUBRICATING COSTS

Cylinder oil was one of the smallest running expenses of a railroad, yet operating officials showed a maniacal concern for this trivial cost. Several roads issued elaborate monthly statements on cylinder oil consumption, carefully noting the engine, its crew, and the miles run per pint. Annual reports reveal the same preoccupation; happy was the master mechanic who could report an engine's running 35 miles on a pint of oil. Cylinder oil cost about $1.25 a gallon in the mid-1850's thus making it cost generally less than one cent per mile. The worst engine on the line could average no better than 20 miles per pint and not bankrupt the road, while the best "oil" engine might be the worst "fuel" or "repair" engine and thus drain the road's finances. The absurdity of the oil account was repeatedly commented on in the *Railroad Advocate* and other trade journals. The matter was shelved, at least in part, in the 1880's when petroleum began to replace lard and tallow as a cylinder

[28] Ringwalt, *Transportation Systems*, p. 325.

lubricant. In 1889 animal oil cost $1.00–$2.00 a gallon while petroleum valve oil was only 35 cents a gallon.[29] Even so, as late as 1907 the Baldwin Locomotive Works observed that "the cost of lubricating oil is a subject of great discussion among motive power men, . . . although it is usually not over one percent of the locomotive expenses."[30]

OPERATING COST TRENDS

It can be seen from the preceding remarks that operating costs were high in the nineteenth century. In the 1830's costs were fairly low and probably did not total more than 25 cents a mile, because of modest train weights and low speeds. The highest costs were undoubtedly reached in the 1850's and 1860's when small locomotives, expensive fuel, and poor roadbeds pushed operating expenses as high as 40 cents a mile. Costs dropped after 1870 when more capital was diverted to improving existing lines and larger, more efficient locomotives were acquired. The following tables will illustrate this general trend:

TABLE 3: OPERATING COSTS OF THE BALTIMORE AND OHIO RAILROAD IN 1843[a]

Item	Cost/Year (dollars)	Cost/Mile (cents)
Fuel	33,547	6.5
Wages (Engineer & Fireman)	22,049	4.0
Repairs & Depreciation	35,941	7.0
Oil	3,989	0.8
Cotton Waste	410	.00?
Total	$95,936	18.3±

[a] The Baltimore and Ohio had twenty-seven locomotives with twenty-two in daily use. *Baltimore and Ohio Annual Report*, 1843, Table E. Total mileage for 1843 was 509,765 miles.

[29] *Master Mechanics Report*, 1889, p. 9.
[30] Baldwin Locomotive Works, *Record of Recent Construction*, No. 60, 1907, p. 14.

TABLE 4: OPERATING COSTS OF A PASSENGER LOCOMOTIVE IN 1851[a]

Item	Cost/Year (dollars)	Cost/Mile (cents)	Remarks
Fuel	5,040.00	18.0	1120 cords/$4.50
Wages (Engineer & Fireman)	1,080.00	3.8	
Repairs	1,680.00	6.0	28,000 miles
Water	250.00	0.8	
Interest	480.00	1.6	on 1st cost of engine
Waste	16.80	0.00	
Oil	224.00	0.8	
Total	$8,770.80	31.0	

[a] Zerah Colburn, *The Locomotive Engine* (Boston, 1851), p. 105.

TABLE 5: OPERATING COSTS OF A TYPICAL LOCOMOTIVE, 1874[a]

Item	Cost/Mile (cents)	Remarks
Fuel	6.0	38 miles/ton
Wages	6.0	
Repairs	7.0	
Oil	0.4	
Cleaning	0.6	
Total	20.0	

[a] Forney, *Catechism of the Locomotive* (New York, 1879), p. 448.

TABLE 6: OPERATING COSTS OF A TYPICAL LOCOMOTIVE, 1892[a]

Item	Cost/Mile (cents)	Remarks
Fuel	7.42	46.17 miles/ton
Wages	5.39	
Repairs	2.40	
Cleaning	1.25	
Oil	.32	15.32 miles/pint
Water	.49	
Miscellaneous	.10	
Total	17.37	

[a] *American Railway*, 1892, pp. 307–8.

A final point, and one that has already been alluded to, is the variation of operating costs from road to road. The efficiency of motive-power officials was a major factor in holding down operating costs. A road might have cheap fuel, good shops, and new equipment, but indifferent management could quickly cancel these advantages. G. S. Griggs of the Boston and Providence Railroad was an exemplary master mechanic who by careful management achieved an outstanding record of economy with light, standard engines.[31] Griggs did not resort to complex devices or gimmicks for economy but instead paid attention to the petty day-to-day routine of maintenance. Flues were kept tight in the boiler, valves were checked for proper setting, piston rings and cylinder packings were always steam tight. Thus, without large or fancy locomotives, Griggs by regular attention to details maintained one of the best economy records in the United States.

[31] *American Railroad Journal*, July 16, 1853, pp. 458–60.

FUEL

7.

American railroads have always had access to a plentiful supply of fuel. Vast, seemingly endless tracts of forest offered enormous quantities of firewood during the first six decades of the nineteenth century. Equally large reserves of coal provided cheap fuel when wood became too valuable for locomotive firing. In more recent times, oil was adopted by many western roads because of its abundance and low price.

Wood was the predominant locomotive fuel in the United States for the first forty years of the railroad era. Traditionally it was the accepted native fuel for all purposes because of its abundance, cheapness, and the ease of harvesting. It burned readily and cleanly, and was all but entirely consumed during combustion. In spite of early predictions of rapid forest depletion and a widespread interest in coal-burning locomotives during the 1830's, wood had no serious competitors as the major railway fuel for many years.

The supply of wood seemed inexhaustible; vast forests covered the eastern half of the nation. This fabulous asset was viewed by many pioneers as a liability, as an impediment to land development and farming. Since wholesale land-clearing was accepted, few questioned the use of a "surplus" commodity. Farmers and yeomen lucky enough to receive a wood contract from a nearby railway regarded it as a cash crop and a supplementary income. Cordwood became a large industry composed of many small producers who were naturally insistent that wood remain the standard locomotive fuel. These producers did all in their power to prevent the introduction of competing fuels; they were shippers and possibly even stockholders, and railroad managers were bound to give a partial hearing to their views.

It was common practice for farmers to deliver wood to small fueling stations on the line; however, most large roads maintained one or more large woodsheds where immense quantities of fuel were stored to insure a good supply for the winter months. The annual reports of many railroads listed vast supplies of cordwood as an asset along with equipment, structures, and other property. In 1853 the Rochester, Lockport and Niagara Falls Railroad, not a particularly large line, reported a supply of 12,900 cords (nearly one year's supply) valued at over $28,000 and stored in several narrow sheds ranging from

112 to 208 feet in length. At the larger wood terminals, by the 1850's, steam-powered saws sliced the logs into cordwood lengths. As early as 1838 David Matthew built a special sawmill locomotive, fitted with a large circular saw, that could run up and down the Utica and Schenectady Railroad cutting up cordwood as needed.[1] Despite steam sawing and splitting mills, the cost of preparing and delivering cordwood was high, generally $1.00 per cord or roughly a third of the total cost of that fuel.[2]

Many eastern roads, particularly those operating around urban areas, found local woodlands had long since disappeared. These roads were obliged to import wood from the first years of their operation. In the larger cities the existing industrial and home markets for cordwood had already driven prices to great heights. The Boston and Providence Railroad was forced to pay $7.00 per cord as early as 1838 when most other railroads were paying less than half that price. Finding no cheap timberland nearby, the road eventually bought property in North Carolina and Virginia. Light logging railways were built adjacent to water connections. Cordwood was shipped to Boston by a fleet of scows and schooners.[3] The railroads operating out of Philadelphia also began to import wood by sea. In the 1830's boatloads of "Jersey yellow pine or Carolina pine knots" were brought to Philadelphia for locomotive fuel.[4]

The practice spread as northeastern forests disappeared. As more roads purchased southern woodlands, the prices rose. In 1855 it was noted that wooded property that had "formerly" sold for $3.00 an acre was bringing from $10.00 to $25.00 an acre.[5] Nevertheless, great savings could yet be realized. New York wood dealers received $6.40 per cord, whereas a railroad owning a southern wood lot could cut and ship its own wood for $3.50 a cord.[6]

Western railroads, particularly those of the treeless prairies, faced a more difficult problem than did the eastern lines. Having no access to cheap sea transport and being far distant from any native forest, these roads undoubtedly paid top prices and suffered occasional shortages. At least one midwestern line, the Illinois Central, is known to have planted locust trees along its lines in an effort to create a local wood supply.[7]

Locomotives consumed staggering quantities of wood during the nineteenth century. During the period of intensive woodburning, 1830–65, it is probable that timber from well over one million acres of land was devoured for this purpose. In 1836 the Boston and Worcester, a short railroad with only nine locomotives, consumed 2,687 cords.[8] The Baltimore and Ohio's annual consumption jumped from 3,419 cords to 7,732 cords between 1841 and 1844, even though half of its locomotives were coal-burners.[9] As late as 1851 the Philadelphia and Reading used more than 61,000 cords, although it was operating some coal-burning engines.

Unfortunately, no statistics for total railway fuel consumption are available for the years before 1880, but several estimates can be calculated from the following rule-of-thumb formula: 140 cords per mile per year.[10] The following table is of course only a rough estimation, except for the year 1880, for which census figures are available.[11] It should also be noted that after 1850 the number of roads operating coal-burners

---

[1] *Railway Age,* May 24, 1883, p. 296.
[2] *Railroad Advocate,* June 16, 1855.
[3] A. F. Harlow, *Steelways of New England* (New York, 1946), p. 354; the *Boston & Providence Annual Report* for 1856 mentions 1,753 acres of woodland in Virginia but also notes the remoteness of this property for proper management and the desire to sell the road's holdings and quit the wood business.
[4] "Recollections of G. E. Sellers," *American Machinist,* December 12, 1885, p. 1.
[5] *Railroad Advocate,* June 16 and June 30, 1855.
[6] *Ibid.*
[7] *Railroad Advocate,* November 15, 1856.
[8] J. Knight and B. H. Latrobe, *Report upon the Locomotive Engines . . . Several of the Principal Rail Roads in the Northern and Middle States* (Baltimore, 1838), pp. 20, 21.
[9] *Baltimore and Ohio Annual Reports,* 1841–44.
[10] *Railroad Advocate,* November 15, 1856, suggested this formula; the *Mining Magazine,* November, 1856, stated the same rule but calculated on the basis of number of engines rather than mileage, viz., each locomotive uses 800 cords per year; 5,000 locomotives use 4 to 5 million cords per year. It was further stated that at this rate about 100,000 acres of forests were used per year.
[11] The 1880 census did not total the fuel figures but listed the amounts used by each road.

increased, and while no exact percentages are available a certain allowance has been made for declining wood consumption.

TABLE 7: TOTAL WOOD CONSUMPTION BY U.S. RAILROADS

Year	Miles of Track	Cords
1840	2,800	392,000
1850	9,000	1,260,000
1860 (90% wood-burning)	30,600	3,780,000
1870 (50% wood-burning)	53,400	3,640,000
1880	93,300	1,388,000

Both hard and soft woods were burned. While it might be supposed that any native wood was acceptable, many railroads carefully selected the variety of wood to be used, realizing the different heating values. Hardwoods and pitch pine were favored because of their high heating values, but varying prices and sources of supply caused most roads to periodically switch from one wood to another. The Central Railroad of New Jersey used yellow pine from 1838 to 1846, southern pine from 1846 to 1853, and oak from 1853 to 1857.[12] The condition of the wood was another factor considered by economy-minded railroad managers. Green or rotten wood was rejected whenever possible. However, some roads, such as the Philadelphia and Columbia, were forced to burn "wood of almost any kind, and most procurable, and in every stage of seasoning."[13] This road's wood was so poor that a mixture of coal and wood was burned to keep up steam.

Wood prices varied widely from place to place and from year to year. It is therefore a gross oversimplification to credit the conversion to coal *entirely* to a steady rise in wood prices. This is particularly true for the period of conversion to coal because the known examples of wood prices do not reveal any marked increase. The wood prices noted in the Baltimore and Ohio annual reports show considerable fluctuation in the price of wood per cord: 1840, $2.50; 1854, $1.79; 1857, $3.80; 1862, $3.00. The Boston and Providence experienced an erratic price schedule, paying $7.00 per cord in 1838, about $3.50 in 1855, and $5.41 in 1860. The 1880 census showed continuing wide variations in railroad wood prices, ranging from $1.25 to $6.00 per cord.

George L. Vose commented on the effect of location on fuel price in 1857: "It does not follow that because coke in England, anthracite in Pennsylvania, or wood in New England, is the most economical fuel, that either of the above will be so in Ohio, Indiana, or Illinois, or because wood is the cheapest in some parts of a State, that it is so throughout, or even that one fuel should be applied to the whole length of a single road."[14] For these reasons it made good sense for railroads operating through coal fields to convert to that fuel sooner than a southern road that enjoyed a good supply of cheap wood. Some roads operated coal engines on one section of their line and wood burners on another. The Baltimore and Ohio, for example, was a coal-burning road by 1859 (except for a few odd engines), yet in later years it acquired and continued to operate many wood-burners as it expanded westward by means of the consolidation of several midwestern railroads. There are also occasional instances of roads reconverting to wood. In 1863 the New York Central returned to wood because of inflated coal prices generated by the Civil War.[15] Conversely, a few years earlier, the Philadelphia, Wilmington and Baltimore Railroad forced wood dealers into lowering their prices by experimenting with coal-burning locomotives.[16]

Most southern railroads paid low prices for wood throughout the nineteenth century and were accordingly reluctant to convert to coal. Yet, as early as 1872, the master mechanic of the Western and Atlantic Railroad predicted: "In some portions of

---

[12] *American Railway Times,* March 9, 1861.
[13] Knight and Latrobe, *Locomotive Engines,* p. 28.
[14] George L. Vose, *Handbook of Railroad Construction* (Boston, 1857), p. 325.
[15] *Western Railroad Gazette,* September 5, 1863.
[16] *Railway and Locomotive Historical Society Bulletin,* No. 21 (1930), p. 20, gives data from the Philadelphia, Wilmington and Baltimore Railroad's annual report for 1857.

the South, before long, we shall be compelled to use coal."[17] While this forecast was essentially borne out, wood-burning remained popular in that area for many years, particularly on secondary and logging roads. One small line, the Mississippi and Alabama Railroad, operating in 1949, was surely the last United States common carrier to burn wood.

Another late stronghold of the wood-burner was found in the northern New England states. The extensive tracts of forests and the conservative ways of the area encouraged the use of several wood-burners on such major roads as the Vermont Central as late as 1894.[18] This was viewed as an anachronistic practice at that time because most self-respecting lines had given up wood-burning a decade earlier.

COAL-BURNING

Compared to coal, wood is a bulky, primitive fuel with a low calorific value. In the nineteenth century one ton of soft coal was considered equal to 1¾ cords of wood, or, roughly figuring wood at 3,000 pounds per cord, 2,000 pounds of coal equaled 5,250 pounds of wood.[19] While these figures are not exact, the greater heating value of coal over wood was well understood by engineers at the beginning of the railroad era. Contrary to present erroneous beliefs that wood was the only fuel considered at the time, a surprising number of our first railways initially experimented with coal-burning locomotives, turning to wood only as a last resort. In 1828 the Delaware and Hudson planned to use coal-burning engines for two reasons: one, it was a coal-carrier; two, the several locomotives imported for that service—the *Stourbridge Lion* among them—were copied from British colliery locomotives, which had always burned coal. The failure of this pioneering steam railroad venture was attributable to weak tracks rather than to the use of coal as fuel. The Baltimore and Ohio's first experimental locomotive, the *Tom Thumb*, burned anthracite successfully; the road then specified that all engines entering its 1831 locomotive contest must use the same fuel. In later years this road was a leader in the development and use of coal-burning locomotives. S. H. Long, the renowned civil engineer and bridge designer, built a number of anthracite-burning locomotives in the early 1830's. His locomotives were tested on the Philadelphia and Columbia, the Newcastle and Frenchtown, and the Boston and Providence Railroads. All were unique and highly individual mechanisms, but none was considered a success.[20] Another engineer usually remembered for his civil-engineering contributions, John B. Jervis, a contemporary of S. H. Long, conducted some experiments with coal-burning locomotives on the Mohawk and Hudson Railroad. Jervis carried over this idea from his previous employer, the Delaware and Hudson Canal Company. The *De Witt Clinton*, before entering regular service, was fired with "Lackawanna Coal" in July, 1831.[21] The test, though unsuccessful, did not discourage Jervis from having another hard-coal burner constructed a year later. This engine, the *Experiment*, worked no better than the *De Witt Clinton* and was soon fitted with a wood-burning firebox. The Camden and Amboy Railroad is also known to have been an early investigator of coal-burning engines. A good deal of this interest is attributable to members of the Stevens family of

---

[17] *Master Mechanics Report*, 1872, p. 49.
[18] *Railway and Locomotive Historical Society, Special Bulletin*, "Vermont Central," 1942.
[19] Colburn and Holley, *The Permanent Way*, p. 8. Marks's *Mechanical Engineers Handbook*, pp. 711 and 799, shows 1 pound coal = 13,000 B.T.U.; 1 pound wood = 5,800 B.T.U. R. H. Thurston's *A Manual of Steam Boilers* (New York, 1896), p. 160, offered a different ratio, stating that 1 cord of well-seasoned yellow pine equaled only ½ ton of good coal.
[20] Colonel Long's attempts at locomotive building are covered in *Railway and Locomotive Historical Society Bulletin*, Nos. 79 and 101. Between 1826 and 1833 Long secured several patents for locomotive boilers and running gears.
[21] W. H. Brown, *The History of the First Locomotive in America* (rev. ed.; New York, 1874), p. 178, reproduced a notice from the *Albany Argus*, July 25, 1831.

Hoboken, New Jersey, who were not only chief promoters of the railroad but also early advocates of coal-burning steamships. The road may have experimented with coal-burning as early as 1833, as indicated by the wide firebox boiler engine (Fig. 42), but it is definitely known that the eight-wheel freight locomotive the *Monster*, built between 1836 and 1838 at the company shops, was intended for coal-burning.[22] The road did not make a wholesale conversion to coal as a result of this experiment, but other hard-coal engines were placed on the road in the late 1840's.

Of the several early attempts at coal-burning, only two small anthracite roads in Pennsylvania, the Beaver Meadow and the Hazleton, completely rejected wood. Both lines ran slow coal trains with continuous runs of only 14 miles; thus there was an opportunity to rekindle and otherwise nurse the engines along. Of course, this restricted manner of operation was not practical for an ordinary railroad where passenger and merchandise trains could not be expected to put up with delays occasioned by a dull fire.[23]

Obviously the early attempts to introduce coal-burning locomotives were a failure. A small number of "coalers" continued to work, but, in general, early American roads were powered almost exclusively by wood-burners. The chief difficulty was an inability to *burn* coal. The blame for this falls directly on the type of coal available. Only anthracite, or as it was first known, "stone coal," was mined in this country before about 1840. It was a difficult fuel to burn, particularly in the small locomotive fireboxes of that time. In addition, it was a slow-burning fuel and was therefore particularly unsuited to the needs of the locomotive, where rapid combustion was essential for a rapid production of steam. Had soft coal been more commonly available in the 1830's, it is likely that successful coal-burners would have been developed many years earlier.

The high price and limited supply of coal in this period were other factors that discouraged an early introduction of coal-burning locomotives. Coal cost from $7.00 to $10.00 per ton in the 1830's.[24] The big mines were located in eastern and central Pennsylvania. Transportation costs considerably boosted the price per ton for roads outside this area. As other coal fields opened, particularly the Maryland, West Virginia, and southern Illinois deposits, railroads in these areas were encouraged to adopt coal. But few of these fields were in production before 1850; some were not in full operation until many years later. Anthracite fields, which had been commercially worked before any locomotives were employed in this country, did not achieve large production until after 1840. Only after that time did coal become an important American fuel.[25] Industry, in general, was slow to adopt coal; thus, railroads were not alone in their slow acceptance of this fuel.[26] As production in the old fields grew, new fields opened, and railroads could reach the mines and offer cheap transportation; coal prices accordingly showed a steady decline as the nineteenth century passed. By the mid-1850's coal was down to about $3.00 per ton, and in 1862 the Baltimore and Ohio was able to get coal at 75 cents a ton because of the many mines along its route. It was this *decline* in coal prices, rather than the dramatic increase in wood prices, that brought about the great conversion in locomotive fuel.

To return to examples of railroads making early use of coal for locomotive fuel, no such account would be complete without reference to the Baltimore and Ohio Railroad. This company showed a strong interest in coal burning from its earliest years and was the only major American line to operate coal-burning locomotives continuously in the last century. Following the first experimental locomotives in 1831, a class of coal-

[22] *Master Mechanics Report*, 1885.
[23] Reports on the Beaver Meadow and Hazleton coal-burning locomotives appear in the *Journal of the Franklin Institute*, June, 1847, and in G. W. Whistler, Jr.'s *Report upon the Use of Anthracite Coal in Locomotive Engines on the Reading Rail Road* (Baltimore, 1849), pp. 27–28.

[24] Knight and Latrobe, *Locomotive Engines*, note coal prices at this rate on several eastern roads.
[25] H. N. Eavenson, *First Century and a Quarter of the American Coal Industry* (Pittsburgh, 1942).
[26] The iron industry showed little interest in replacing charcoal with coke until after 1850; the slow conversion of western riverboats to coal is discussed in Louis C. Hunter's *Steamboats on the Western Rivers*. One of the few important industries to make an early conversion to coal was Hudson River and ocean steamers.

burning engines with vertical boilers was perfected for freight service. In later years the road continued to build or purchase new designs of coal-burning freight engines but turned to wood-burners for passenger service in the mid-1830's. Before 1840 anthracite was imported from Pennsylvania at $8.00 a ton.[27] After the road reached the soft-coal fields of western Maryland in 1840, bituminous coal was adopted. In the next year, half of the road's power burned coal exclusively while a few engines burned a mixture of wood and coal. A large number of eight-wheel, coal-burning freight locomotives were purchased during the next several years. Passenger trains were handled by wood-burners until the mid-1850's when new tests were made with coke and coal. In 1858, 14 of 34 passenger engines were converted to coal or coke. By November, 1859, all 235 locomotives were coal-burning except for one freight engine and ten old light machines.[28]

The Philadelphia and Reading was another road obviously interested in coal-burning locomotives because of the immense coal traffic passing over its line. The original minutes of the Board of Managers revealed the earnest desire of the Reading's management to adopt coal as expressed in the following resolution dated April 13, 1835: "Resolved that this board deem it of the utmost importance that the locomotive engines to be constructed for this Company be built with a view to the exclusive use of anthracite as fuel." The road's third locomotive, the *Delaware*, delivered by Winans in 1837, was a coal-burner but it was not a success and was abandoned in 1845. The *Gowan and Marx*, also intended for hard coal, was converted to a wood-burner not long after entering service in 1839. Discouraged by these and undoubtedly other unrecorded failures, the Reading abandoned further attempts at coal-burning until 1847. In that year a mechanical monstrosity named the *Novelty* was built in the company shops after a patented design of G. A. Nicolls. The boiler was carried on a separate eight-wheel car; the flexible pipe necessary for carrying steam to the locomotive leaked. This and other complex auxiliary apparatus were devised by Nicolls. Predictably the engine was a failure. In the same year Winans delivered several eight-wheel connected engines of a more conventional design, but they were not notable successes. This slow beginning was carried forward by James Millholland, who after an unpromising series of experiments perfected a practical anthracite firebox by 1856. By 1859 the road was virtually all coal-burning.

In New England, as already indicated, slower progress was made. The great distance from the coal fields kept prices high. Coupled with this was the resistance of many old-line managers to innovations. One exception to this trend was George S. Griggs, master mechanic of the Boston and Providence Railroad. Griggs was associated with the road when S. H. Long's anthracite engines were purchased in the 1830's. While these engines failed as coal-burners, Griggs was not prejudiced by their poor performance. In 1856 and 1857 he initiated new tests and in the process developed the brick arch and diamond stack, both of which were important contributions to coal-burning. By 1860 the road was said to be powered almost entirely by coal-burners.[29] This statement was somewhat premature, for, according to the road's annual reports, wood-burners were in service as late as 1874 or 1875.

In the Middle Atlantic states the Philadelphia, Wilmington and Baltimore Railroad was an early convert to coal-burning. In 1841 soft coal was used but it was abandoned after only six months.[30] When interest in coal-burning was revived, the road made the mistake of turning to freak firebox designs. In 1856 a Taunton-built engine, the *Essex*, with a Dimpfel boiler, was placed in service. This machine and the several that followed were only moderately successful. Conventional coal-burning engines were acquired and by 1862 the road's annual report stated: "Coal burning in locomotives is no longer an experiment, but a well established fact and a decided economy." Three years later all but seven main-line engines were fueled with coal.[31]

[27] Whistler, *Anthracite Coal*, p. 21.
[28] Letter of Henry Tyson to J. W. Garrett (president, Baltimore and Ohio Railroad), November 9, 1859.
[29] *American Railway Times*, January 28, 1860.
[30] *American Railway Times*, March 9, 1861.
[31] For more data on the Philadelphia, Wilmington and Baltimore's conversion to coal-burning, see *Railway and Locomotive Historical Society Bulletin*, No. 21.

The fuel question followed a similar pattern in the Midwest. The Chicago, Burlington and Quincy was one of the first midwestern railroads to convert. It acquired its first coal-burner in 1855 after reaching the southern Illinois coal fields. The next year, eleven coal engines were in service; in 1859 twenty-five were on the road. The conversion was accelerated by the purchase of coal mines so that by 1868 all of its engines were burning coal. The Illinois Central began experiments with coal-burning locomotives in 1855. At first, poor local coal dampened prospects for an early conversion, but, despite this difficulty, over half of the road's engines were coal-burners by 1861. Five years later only 5 of 151 engines were wood-burners. Not all railroads in this area found native coal satisfactory, and the Galena and Chicago Union's report of 1863, while admitting that wood prices were prohibitive, stated that coal was no ready solution.

Illinois coal was inferior because it contained a high percentage of sulphur. Until more precise methods of processing were developed, eastern coal was imported. In later years Illinois coal was successfully employed, despite early complaints regarding its quality, and western roads kept pace with the other major lines in abandoning wood.

The conversion of locomotive fuel from wood to coal may be summarized as follows: The early interest in coal-burning resulted in no substantial use; only a few coal-field lines regularly employed this fuel. By the 1850's a renewed and substantial interest in coal-burning was thwarted by the mistaken belief that revolutionary changes in firebox design were necessary. It was quickly established that fireboxes of ordinary construction were capable of successful coal-burning, and by the late 1850's several important railroads had adopted coal. During the 1860's and 1870's coal was accepted as the best fuel for locomotives, and all major railroads began abandoning wood. By 1880 more than 90 per cent of railway fuel was coal.[32]

During the next two decades all American railroads, except for a few obscure lines, converted to coal.

Compared to coal and wood, the other fuels considered for locomotive use are of only passing importance. Coke was the most important alternate tested, but its use was extremely limited. As late as 1850 an English technical writer observing American practice stated: "The use of coke is nowhere resorted to. Its expense would make it inadmissible; and in a country so thinly inhabited, the smoke proceeding from coal or wood is not objectioned to."[33] From this comment it can be understood that coke was the required fuel in England because of the smoke nuisance of coal and wood. Coal was adopted in that country during the period that American railroads changed from wood to coal. To turn to the use of coke in the United States, one of the few lines to use the fuel was the Baltimore and Ohio Railroad. In 1854 two passenger locomotives were tested with coke; the results were encouraging for, although it was more expensive than "raw coal," no sooty smoke was given off.[34] The road's 1857 report noted the construction of six new coke-burning passenger engines. The greater cost of coke ($2.00 per ton compared to 75 cents for a ton of coal) discouraged continuance of coke-burning locomotives after about 1862. While coke was for a short time considered ideal for passenger service because of its clean burning, it was soon discovered that careful firing of coal could eliminate a good portion of the smoke. To encourage this practice the Illinois Central placed the following notice in the cabs of its passenger locomotives.[35]

Machinery Department, Illinois Central Railroad.

Weldon, May 7th, 1868.

SPECIAL INSTRUCTIONS TO PASSENGER ENGINEERS.

To prevent your engine throwing out large quantities of smoke, you will see that your fireman is very particular in the manner of firing, and that he observes closely the following rules:

---

[32] The 1880 census lists fuel consumed by individual lines but gives totals for major geographic sections only. The author's totals are 1,388,723 cords of wood; 9,531,080 tons of coal; ninety per cent is an approximate percentage based on 1 ton's equaling 1½ cords.

[33] Dionysius Lardner, *Railway Economy* (New York and London, 1850), p. 336.

[34] Mendes Cohen, *Report on Coke and Coal Used with Passenger Trains, on the Baltimore and Ohio Railroad* (Baltimore, 1854).

[35] *Locomotive Engineering*, May, 1899, p. 238.

Do not throw more than two shovels of coal at one time, and scatter it well over the grates.

Keep the fire as nearly uniform as possible.

Keep the coal in your tender dampened, so that the dust from it will not be blown back upon the train. Whenever the steam is shut off, the blower should be used lightly.

The air openings around the furnace and in the door should be kept open as much as possible.

Much of the annoyance from smoke and coal dust will be prevented and a large saving in fuel effected by attention to the above rules.

S. J. HAYES,
Superintendent of Machinery.

Petroleum was considered as a fuel surprisingly early, but its short supply and high price prevented any extensive use in this country during the nineteenth century. As early as 1864 the United States Navy experimented with oil-fired boilers, and suggestions were made at the time for oil as a locomotive fuel.[36] Ten years later an old engine of the Boston and Providence was altered to burn coal oil, but the experiment ended in failure after a twenty-mile run when oil leaks set the engine on fire.[37] The first regular use of petroleum for locomotive fuel developed in Russia. Thomas Urquhart, superintendent of the Grazi Tsaritzin Railway, began the use of fuel oil in 1882; by 1885, 143 engines on the line were using this fuel.[38] There was, however, little interest in oil-fired locomotives in the United States until the great western oil fields produced large surpluses during the early 1900's.

[36] *American Railway Times,* January, 1864, p. 14.
[37] *Boston and Providence Annual Report,* 1874.
[38] *Institution of Locomotive Engineers Journal,* 1952, pp. 425–515; see also Eugene McAuliffe, *Railway Fuel* (New York, 1927).

COMPONENTS

II

BOILERS AND BOILER ACCESSORIES
8.

BOILERS

The British perfected the basic design for the locomotive boiler before the United States had any steam railways in operation. This design followed the fire-tube plan. The boiler had a separate firebox and smokebox. Numerous small-diameter tubes or flues passed through the boiler shell to connect the fire- and smokeboxes. This arrangement was well established by 1830 and continued unchallenged until the end of steam locomotive construction.

American builders copied the English boiler from the beginning and, unlike their experience with the British running gear design, found little reason to develop a new plan. Nevertheless, several feeble attempts were made to introduce water-tube boilers. John Stevens' tiny experimental locomotive (1825) had such a boiler. A. B. Latta adopted Stevens' water-tube fire-engine boiler for locomotive construction in 1857. At the same time, several locomotives were constructed with water-tube boilers after the patent of F. P. Dimpfel. All of these ventures were experimental, however, and did not influence locomotive practice.

While American builders slavishly adhered to the British arrangement of locomotive boilers, our style of construction rapidly took on a distinctive character during the 1830's. Because wood was our most common fuel, fireboxes were made deep and narrow, a shape well suited to this combustible material. To save weight American boiler plates were about half the thickness of British plates; one-fourth to five-sixteenths of an inch was a common weight, despite our high boiler pressures. The abandonment of copper fireboxes and tubes by 1860 was another major difference between the two systems of boiler construction.

A common criticism of American locomotives before 1870 was that boilers were too small for cylinder size. The criticism was justified and was recognized by most motive-power authorities of the time. However, small boilers were essential to keep total engine weight within the province of our notoriously weak tracks. Because the restrictions on engine weight were such an overriding consideration, boilers showed remarkably little change in size during the early years. From about 1845 through the early 1870's boiler diameter was virtually frozen at a maximum of 48 inches. The length of boiler waists and, correspondingly, total heating surface underwent steady but modest progress. It was possible, within reasonable limits, to lengthen the boiler without greatly increasing its weight. However, 48 inches was the maximum safe diameter for a five-sixteenths of an inch wrought-iron plate at about 100 pounds per square inch. An increase in diameter required a thicker plate with an attendant and rather large increase in weight. In all fairness to the "small-boiler concept," it should be remembered that speed was secondary to power and that the small boiler and large cylinders were a workable combination at low speed.

Another prevailing concept of boiler design which affected boiler size was the idea of keeping the center of gravity as low as possible. This idea was virtually a mania with engine designers and caused boilers to be set down as low as the wheels and frame would permit. In some cases the boiler's belly was below the frame's top rail. This concern for top-heaviness was manifested in the diameter of boilers; a large-diameter boiler would be set higher. As late as 1872 this old argument was reaffirmed in the *Master Mechanics Report* which stated that a 48-inch boiler is as large as can be used "unless you carry it up high."

To produce the desired performance within severe weight and size limitations, the locomotive boiler was of necessity worked very intensely. In stationary, and to a lesser degree, in marine practice, boilers might be as large as required for optimum performance and efficiency. Such boilers were worked "lazily" and were spread over a large area. But locomotive boilers restricted by weight and clearance specifications were limited in size and were worked hard. A reasonable rate of combustion might be set at 100 pounds of fuel per square foot.[1] For the sake of economy, stationary boilers operated at only 15–25 pounds per square foot. Railway engines, because of the tremendous demand for steam and the limited size of their fireboxes, were forced to burn 200 pounds per square foot, a rate far in excess of what was considered reasonable. This explains the poor efficiency, high maintenance cost, and short life of locomotive boilers.

[1] *Modern Locomotives* (New York, 1897), p. 8.

Before discussing boiler construction and the many accessories associated with it, some mention should be made of the various styles of locomotive boilers used in this country during the early years of locomotive development. The Stephenson boiler was one of the earliest types favored by American railroads. A simple, sound design, it became familiar to mechanics wherever Stephenson engines were used. The slightly raised semicircular wrapper over the firebox was the Stephenson's distinguishing feature. The connecting plate was vertical to the boiler waist. This same boiler made with a sloping rather than a vertical connecting plate became the familiar wagon-top boiler of later years.

The Stephenson boiler was copied by many American builders. Rogers and Baldwin used it in their first productions but later abandoned it for the Bury boiler. New England builders, however, were enthusiastic and consistent users of the Stephenson design. They continued building this style of boiler well into the 1850's. Mason, popularly associated with radical mechanical innovations, used a Stephenson boiler on his first locomotive in 1853. The *Robert Fulton* (Fig. 7) illustrates the original Stephenson boiler. The *Champlain* (Fig. 146) shows the same style of boiler as it was enlarged by the New England builders.

The Bury boiler was contemporary with the Stephenson and was also a British design. Edward Bury of Liverpool introduced his hemispherical boiler in 1830. Within a few years some twenty Bury engines had been exported to this country. A huge dome over the firebox and a large space for steam distinguished the Bury boiler. The ample space for steam allowed for dry steam and thus prevented priming, a characteristic detrimental to American roads using poor water. The "Dome" or "Round-top" boiler as it was known in this country was quickly adopted by several major builders. Norris, Rogers, and Baldwin, the largest locomotive manufacturers, used it almost exclusively between about 1838 and 1855. Amoskeag built its first engines with Bury boilers but is thought to be the only New England builder to do so.

For a design with so many disadvantages and only one advantage, steam room, it is surprising that the Bury boiler enjoyed such wide and extended popularity. Its defects were many. Chief among these was its complex and expensive construction. Many small plates and much fitting and riveting were required to fabricate the dome. Only the most skillful boiler-maker could be employed for this work. It was difficult to jacket the dome. This large area remained uncovered and was free to radiate the boiler's heat into the atmosphere.

A final defect of the Bury boiler was its small grate area. In the first years this was not a serious problem, for small wood-burning engines were adequately serviced by small fireboxes. However, the locomotive grew in size and by the late 1840's the tiny D-shaped Bury firebox proved wholly inadequate. The Baltimore and Ohio attempted to increase the grate area of such fireboxes by making the back sheet slope outward. This design was tried on several 0–8–0's in the late 1840's. Another method was to make the firebox and dome elliptical in plan view (see the *Gowan and Marx* [Fig. 124] for an example of this scheme). The attachment of a "box" extension to the rear of the Bury firebox was yet another method of enlarging the grates (see the sketch, Peoples Railway No. 3, in Fig. 126). In all, these attempts to enlarge the Bury's grate area were at once desperate and futile. A few such boilers were built as late as 1857, but the Bury's complexity, expense, and restricted grate area had relegated it to virtual obsolescence several years earlier.

The vertical boiler is a curious chapter in American locomotive history. Although it had no lasting influence on boiler design, it did enjoy a brief period of popularity. It is mentioned here because it was a distinctively American design. The story of the vertical boiler focuses on the Baltimore and Ohio Railroad. The road's first locomotive, the tiny experimental *Tom Thumb*, incorporated the vertical boiler. The Baltimore and Ohio's first road engines were peculiar four-wheel machines with vertical boilers known as Grasshoppers. The vertical boiler allowed for an engine of very short wheel base which was admirably suited to the sharp curves of the Baltimore and Ohio line. Ross Winans supplied a number of these machines to the road and in the process became an advocate of the vertical boiler; he built no locomotives with horizontal boilers until 1843. Winans built a small number of four-wheel engines with vertical boilers for other roads between 1836 and 1837.[2] The eight-wheel engines discussed briefly in the earlier section on 0–8–0's were built by Winans in 1841 and 1842 for the Western Railroad and were the last vertical-boiler engines built for main-line service in this country.

The defects of the vertical boiler were obvious from the beginning. It was difficult to enlarge because overhead and side clearances restricted any substantial increase in size. Grate area and tube length were limited. In an attempt to increase heating surface a large number of small-diameter tubes (400 at 1–1¼ inches) were used; the result was a greater number of tubes to be kept tight. Because the boiler was very short, about 30 inches in length, the better part of the fire's heat passed out through the stack. Sediment settled on the bottom tube sheet thereby insulating the water and causing the sheet to burn out. Considering these drawbacks, it is surprising that more than thirty vertical-boiler engines were built, some as late as 1842.

To return to the main line of boiler development, we should recall that until 1850 there were only two major styles of locomotive boilers used in this country, the Stephenson boiler, favored because of its simplicity, and the Bury boiler, favored for its generous steam room. However, in 1849 or 1850 the celebrated wagon-top boiler was introduced and soon became the overwhelming favorite. The new design was an obvious adaptation of the Stephenson boiler, but the firebox wrapper had been raised to provide more steam room, and the connecting sheet was sloped. The wagon top thus combined the advantages of the old rivals, the Bury and the Stephenson. Thomas Rogers is credited with this design and is said to have first used the wagon-top boiler on the *Madison* of the Madison and Indianapolis Railroad, which was completed in November, 1850.[3] No information can be uncovered to show prior inventions in this country or in Europe.

In addition to the large amount of steam space, the wagon top offered numerous advantages. It put more water surface

---

[2] Winans supplied 0–4–0's to the Paterson and Hudson River, the Philadelphia and Columbia, and the Reading railroads.
[3] Sinclair, *Locomotive Engine*, p. 241.

over the fire, was not likely to prime, and put weight on the drivers (particularly of the 4–4–0's), where it could be used to best advantage. The wagon top also simplified interior repairs (crown sheet and crown bars) because it was spacious.

At this point in the discussion, it would be well to explain the overwhelming emphasis placed on generous steam room. The reason is simple—bad water. Impure water would foam or "boil up" if the boiler was suddenly relieved of a large portion of its steam. In such cases the pressure dropped, thus causing the water, more particularly, impure water, to boil up or foam. Foaming caused priming, the carryover of water to the cylinders, which not only reduced the engine's power but damaged the pistons and cylinders if enough water was carried over. The most effective insurance against priming was a large steam reserve so that the pressure could not be reduced suddenly by the quick opening of the throttle or safety valves. Most American roads were cursed with impure or muddy water and thus faced the constant danger of priming. The water on some western roads was so poor that from one-eighth to three-sixteenths of an inch of scale was deposited inside the boiler per year.[4] No roads are known to have effectively treated boiler water before 1885, although rapid progress was made after that date.[5] The continued use of untreated water was responsible for the use of the wagon-top boiler into the 1890's.

The wagon-top boiler had one major failing; it was weak. The high outer wrapper and the long connecting sheet were difficult to reinforce with stays. The wagon top was further weakened by the preference for placing the steam dome directly over the firebox. The large hole necessary for the dome weakened the boiler's outer shell at the point of least durability.

The straight boiler was the wagon top's only serious competitor after 1855. It was at best a second contender and did not approach the wagon top's popularity. As its name suggests, the straight boiler had no variation in diameter but was uniform from the smokebox to the backhead. It was the strongest, simplest, and cheapest boiler made. It lacked steam room, however, and accordingly was not well suited to the majority of United States roads. Two attempts made to improve the straight boiler's steam room were to increase its diameter or to use two steam domes. The increase in diameter was objectionable on two counts. First, an increase in the 48-inch diameter called for heavier plate and thus caused an undesirable increase in total weight. Second, the boiler diameter when increased to its full length enlarged the front end thereby putting more weight on the truck. The wagon top, it will be recalled, had its steam room over the drivers and had a light, small-diameter waist which put less weight on the truck.

Increasing the straight boiler's steam room by using two domes was considered to be far more desirable because it eliminated the necessity of a large-diameter boiler. The twin domes would facilitate a more even steam "gathering" because steam was collected at two points rather than at one. A British patent was issued in 1837 covering the idea of two domes for straight boilers but it is not thought to have been used in this country much before the mid-1840's. The idea rose to great favor by mid-1855, but its main use was for wagon-top rather than straight boilers. Double-dome straight boilers experienced a modest revival in the 1870's. As with all arrangements, there were objections to the two-dome boiler, be it straight or wagon top. The two holes required for the domes weakened the boiler; the front dome was particularly difficult to stay since, unlike the rear dome (with the crown bars at hand), there was nothing to which to anchor it. The second dome exposed a large area that radiated heat.

The straight boiler had one champion, Wilson Eddy, who conceived a unique design that silenced the common objections to this design. Eddy did not resort to the usual schemes but built a straight boiler without a dome. The boiler was tapered to a smaller diameter at the smokebox in order to reduce as much objectionable front-end weight as possible. The firebox end was kept large for a good-sized grate and weight on the drivers. The steam room problem was handled by the enlargement of the diameter to 50 inches and the introduction of a perforated dry pipe. The dry pipe was perforated by holes or slots throughout its length and diameter, thus serving as a

---

[4] *Master Mechanics Report*, 1872, p. 143.
[5] Baldwin Locomotive Works, *Record of Recent Construction*, No. 60 (1907).

steam reservoir. It was placed near the top of the boiler so as to collect as much of the dry steam as possible. Eddy's boiler was the strongest possible plan having no dome openings, yet it was not adopted by any other builders.

The perforated dry pipe was not Eddy's invention; it had been patented by Hawthorn in England in 1839, but Eddy was its chief advocate in the United States. Rogers is said to have used the arrangement on some engines built for the Erie; William Mason was also an enthusiastic user of the perforated dry pipe, an example of which can be seen on the *Phantom* (Fig. 193). Even though this style of steam collection was eminently successful and was praised by some of the most intelligent locomotive authorities, it was never widely used in this country.

BOILER CONSTRUCTION

There was no appreciable change between 1830 and 1870 in the material and method of boiler manufacture. Wrought iron was the standard material and single-riveting the accepted method of construction. This general scheme was borrowed from the British builders in the 1830's and was found satisfactory for the small-diameter, relatively low-pressure locomotive boilers that remained standard for the next forty years. American builders soon found single-rivet construction dependable for the five-sixteenths of an inch wrought-iron plate and 100-pound-plus pressures, while the British regarded the three-eighths of an inch or half-inch plate and 50 pounds of pressure as the safe limits for this style of boiler.

Telescope construction was universally used for making the boiler barrel or waist. Iron plate was rolled in a ring. The ring nearest the firebox was the largest while the rings progressing toward the smokebox were made slightly smaller in diameter so that one would fit into the next. Each ring was single-riveted. Rivets five-eighths to three-quarters of an inch in diameter on one and three-quarter inch centers were common. If a more uniform waist was desired the diameter of the rings would alternate from one to the next so that the telescoping was not a steadily decreasing diameter from the first to the last course. Each ring had a longitudinal seam that was made by lapping the plate and single-riveting the resulting joint. The lap seam was considered the weakest part of the waist. Single-rivet longitudinal seams were found adequate for pressures up to about 100 pounds per square inch, even though such a fabrication was only 56 per cent as strong as the plate. Larger-diameter boilers and higher pressures called for double-riveting, which was equivalent to 70 per cent of the plate's strength. As the locomotive boiler grew in size after 1870, multiriveted longitudinal joints were found to be necessary. Welt plates were used for additional reinforcement. The lap joint fell into disfavor and was replaced by butt seams by the 1890's.

A minor controversy developed over punched versus drilled holes for rivets. Punched holes were easy to make, but some engineers contended that the plate was fractured in the process. In 1872 the *Master Mechanics Report* stated that riveted seams with drilled holes were 30 per cent stronger than those made with punched holes. Nevertheless, punched holes were favored because of their cheapness. In practice, the best aspects of each method were adopted in the compromise of punching a small-diameter hole and drilling or reaming it to size; thus, the fatigued metal around the punched hole was removed.

Wrought-iron rivets ranging in size from five-eighths to three-quarters of an inch in diameter were most commonly used during the first forty or fifty years of locomotive construction. During the early years rivets were handmade but by 1855 machine-made rivets had become common.[6] Riveting was the only method of fabrication used, although Bury is known to have assembled fireboxes by means of blacksmith-welding in the 1830's. Gas-welding was proposed by A. L. Holley in 1861 but little or no progress was made with welded locomotive boilers until the twentieth century.

[6] *Railroad Advocate*, June 16, 1855, p. 3.

British boiler plate was preferred, particularly Low Moor, because its *quality* was more dependable than stronger American boiler iron. The best Yorkshire plate was rated at 25 tons per square inch; the best American plate (Sligo charcoal iron) was rated at 31 tons per square inch.[7] Boiler iron was judged on its appearance and the reputation of the manufacturer. The Paterson builders were reported to purchase only "warranted boiler plate" and would guarantee to refund the price of plate plus labor in case of a boiler failure.[8] The foremost of the Paterson builders, Rogers, took extraordinary pains to insure that only perfect, first-class plate was used in boilers of his manufacture. Despite careful inspection and a consistent rejection of all imperfect iron, an occasional error was made. One such error ended in tragedy on the Buffalo and Erie Railroad. In 1854 Rogers delivered a giant, six-wheel freight engine named *Vulcan*. Two years later the machine exploded with terrific force, killing the engineer and fireman. An inspection of the wreck revealed that a flaw *two feet long* had caused one of the boiler plates to fail.[9] Part of the blame was laid on the crew because it was regular practice on this road to tie down the safety valves and to work the boilers at dangerously high pressures, in some cases as high as 200 pounds.

No matter how keen the eye of the shop superintendent or how good the maker's reputation, wrought iron was not a superior material for boilers. It was a reedy material fabricated from small strips or rods that were piled, heated, and (with luck) rolled together into homogeneous plates. It was not a homogeneous material and it was liable to fracture and break down under pressure. Boilers were commonly made of flange iron, a good grade of charcoal iron manufactured from selected scrap and charcoal blooms. Because its reedy or grainy texture was well known, wrought-iron plate was set with the grain running around the diameter of the boiler for greater strength.

At first it was possible to obtain plate only in small sizes. The many joints thus required resulted in a weak and leaky vessel. The *Susquehanna*'s boiler, made by Stephenson in 1831, although only 3 feet in diameter and 11 feet long, was made of twenty-two plates.[10] Rolling mills made rapid progress in the first half of the nineteenth century and it was soon possible to procure larger plates, thus reducing the number of joints.

The introduction of cheap steel in the mid-1850's provided a new and wonderful material for steam boilers. The alloy was strong, ductile, and homogeneous. Experiments with steel boilers began early. The Schenectady Locomotive Works built an engine with a steel boiler for the New York Central Railroad in 1860.[11] The *Scotia*, completed in January, 1861, at the Hamilton, Ontario, shops of the Great Western Railway of Canada, is generally credited as the first engine having an all-steel boiler built in North America. Other builders cautiously followed this lead, but in the main there was a good deal of resistance to steel boilers. Steel fireboxes were readily accepted in the 1860's and 1870's, but only a few roads showed much enthusiasm for all-steel boilers. The Baldwin Locomotive Works built its first steel boiler in 1868 and did so undoubtedly at the request of the Pennsylvania Railroad.

Expense and brittleness caused by too much carbon were the chief complaints raised against the steel boiler. The *Master Mechanics Report* for 1875 could account for only 200 steel boilers. The major reason for such a small number was that "the difference in cost of steel over good iron is greater than its utility over iron." The next year a gain of only 76 steel boilers was reported out of a total of 1,690 locomotive boilers included in that year's survey. If this ratio was accurate for all locomotives in the country, it can be seen that only a small percentage had all-steel boilers. In 1870 steel boilers cost from $250 to $500 more than an iron boiler of the same size.[12] By 1875 the difference in cost was only $100 to $200.[13] As the price of steel fell and new, softer alloys were offered, railroad mechanical officers showed greater interest in its use during the 1880's. It

[7] Clark and Colburn, *Recent Practice*, pp. 1, 2.
[8] Holley, *American and European Railway Practice*, p. 17.
[9] *Railroad Advocate*, August 23, 1856, p. 3.
[10] Warren, *A Century of Locomotive Building*, p. 272.
[11] *American Railway Review*, August 9, 1860, p. 71.
[12] *Railroad Gazette*, November 19, 1870, p. 174.
[13] *Master Mechanics Report*, 1875, p. 53.

was recognized that, because steel was stronger, it would be possible to build large-diameter boilers, without thicker plate or much increase in weight. A final consideration was the more uniform and homogeneous character of steel as compared to wrought iron. In 1890, Forney observed that the use of steel had overtaken wrought iron in the construction of boilers "to a very great extent."

The waist or cylindrical portion of the boiler was self-supporting. The flat parts, the tube sheets and the firebox, required additional support or bracing commonly referred to as staying. Staying was also required for steam dome openings. Iron bars, generally 1 inch in diameter, were commonly used in this country for staying. Gusset plates were preferred in Europe at the time but were little used here. Stephenson used stay rods running the full length of the boiler to strengthen the front tube sheet in the early 1830's. An example of this arrangement is shown in an illustration of the *John Bull* (see Fig. 7). It was used as late as the 1850's as indicated by the boiler drawings of the *Columbia* (Fig. 158) and the *Phantom* (Fig. 193). A cheaper and more common method of staying tube sheets involved the use of a short, diagonal rod, one end of which was riveted to the boiler waist, the other to the tube sheet. The boiler tubes supported the middle and bottom of the tube sheets. See the Tyson Ten Wheeler boiler drawing (Fig. 181) for this style of construction.

Wagon-top boilers were notoriously weak vessels and required elaborate staying. Long corner stays riveted to the back-head and vertical stays attached to crown bars were deemed necessary for this fragile boiler design. The wagon top was further weakened by a large steam dome opening which was placed directly over the firebox. In an attempt to support the dome, stay rods were riveted to its side, the lower ends of the rods being attached to the crown bars. A good example of this style of construction is shown by the *Southport*'s boiler (Fig. 198).

The best method for fastening the firebox and the dome to the waist was disputed throughout the nineteenth century. Stephenson preferred angle-iron connections and used this form of construction in the early 1830's. It was copied by many American builders and remained in favor until the 1850's. Some builders before 1850 found that the quality of iron used in angle stock was not trustworthy.[14] These builders flanged dome and firebox plates for the waist connection. It was felt that the joint was more reliable when made with first-class boiler iron. Flanged joints superseded angle-iron connections and remained the standard construction until the 1890's. At that time steam domes were again fastened to the boiler by angle irons.[15] The argument was that in flanging the boiler plate a right-angle bend was necessary and the fatigued metal thus made a dangerous joint.

BOILER TUBES

The boiler waist contains a large number of small-diameter fire tubes intended to increase heating surface and promote steam-making capacity. The tubes are parallel to one another and connect the fire- and smokeboxes.

The earliest American locomotive boilers invariably contained 100–150 copper tubes varying from one and a half to one and three-quarters inches in diameter. Copper was easy to fabricate. Thin copper sheets, generally less than one-eighth of an inch thick, were cut into strips; the strips were rolled and lap-welded into tubes. In about 1860 seamless copper tubes were introduced, but lap-welded tubes continued to be manufactured for many years. The soft metal was easily flanged and a good steamtight joint could be made at the tube sheets with a simple calking tool. When the joint worked itself loose during expansion and contraction, it was readily reflanged and made good for many more miles of use. Copper tubes gave remarkably good service. The Pittsburgh, Fort Wayne and Columbus Railroad reported twenty years of service, while the

[14] Colburn, *The Locomotive Engine*, p. 40.
[15] *Modern Locomotives*, p. 18.

Little Miami Railroad realized 150,000–200,000 miles when copper tubes were used in wood-burning engines.[16]

Brass tubes were first used in this country in 1851.[17] Their use spread fairly rapidly so that within the next four years 800 locomotives were fitted with brass tubes.[18] Greater cost and difficulty in flanging (brass being less ductile than copper) prevented them from superseding copper tubes in America; they were immensely popular in Britain, however.

Iron tubes were used as early as 1831 by the Baltimore and Ohio and were taken up at an early period by the other roads operating coal-burning locomotives. Iron tubes were difficult to flange, but they were considerably cheaper and more durable than copper tubes. The *American Railway Review* stated that copper tubes cost $1,000 per locomotive compared to $400 for iron tubes.[19] In some cases, copper ends were welded on for easy flanging. Iron tubes were fabricated from sheet stock and the joint was brazed. The ability of iron to withstand the erosive action of fly ash led to its adoption for coal-burning engines. After 1860 iron tubes were on the ascent and the use of copper and brass became increasingly rare as the century closed.

Experiments with steel tubes began in the early 1860's. By 1863 a British supplier, Russell and Howells, could report the use of steel tubes by several important railroads. These included the Camden and Amboy, Erie, Hudson River, and other eastern lines.[20] Despite this initial interest, however, few roads adopted steel for locomotive tubes. In 1876 it was reported that lap-welded iron tubes were the general rule; steel tubes were not considered to be worth the extra cost.[21] As late as 1892 Meyer concurred with this observation, stating that steel was only "sometimes" used for tubes.[22]

Steel did not rival iron for tubes during the nineteenth century, even though it surpassed its competitor years earlier in boiler and firebox manufacture. It was difficult and expensive to weld. Because it was stronger than iron, very thin steel tubes could be constructed. These proved to be better heat exchangers because there was less wall thickness to act as insulation. Yet the difficulties of welding and higher costs prevented the general adoption of steel tubes until about 1900 when cheap, seamless steel tubes were introduced.

Tube diameter remained remarkably unchanged throughout the last century. One and three-quarter inch tubes were standard up until 1860 when a movement began for two-inch diameters. With minor exceptions, this size was popular through the 1890's.

BOILER LAGGING AND JACKETS

The heat loss by radiation from an exposed boiler shell was considerable. Estimates of radiation heat loss varied from 12 to 25 per cent for a *stationary* boiler.[23] However, a locomotive when at work was not a stationary boiler and the faster it traveled the worse the heat loss became; no figures can be cited, but the loss was greatly increased by the passage of air over the boiler's surface. This effect was heightened, of course, by high winds or extremely cold weather. A well-fitted lagging, made of a good insulating material, could reduce radiation heat losses by nearly one half.

Engineers recognized the economy resulting from this simple, cheap measure from the beginning of steam engineering,

---

[16] *Master Mechanics Report*, 1870, p. 105; 1872, p. 148.
[17] Colburn, *Locomotive Engineering*, p. 83.
[18] *Railroad Advocate*, May 5, 1855, p. 3.
[19] *American Railway Review*, July 4, 1861, p. 405.
[20] A circular issued by Russell and Howells, in the M. W. Baldwin Letters (Historical Society of Pennsylvania, Philadelphia, Pa.); hereafter cited as Baldwin Letters.
[21] Institution of Civil Engineers, *Proceedings*, Vol. 53 (1878), p. 51.
[22] Meyer, *Modern Locomotive Construction*, p. 437.
[23] J. C. Hoadley's estimate of 12 per cent heat radiation is given in *Locomotive Engineering*, July, 1895, p. 440. The 1920 Carey Manufacturing Company catalog estimated a heat loss of from 15 per cent to 25 per cent. Other sources usually note amount of radiation per square foot.

but the first use of lagging for locomotive boilers is uncertain. It was regularly used for British locomotives by 1830, and within a few years no intelligent builder would deny its value as an absolute necessity for economical operation. Winans was a notable exception to this prevailing opinion and built locomotives without lagging until the end of his production in the 1860's.

Wood was the first material used for lagging. It was, in fact, about the only insulating material available to early locomotive builders. Narrow strips, about two or three inches wide and seven-eighths of an inch thick, were laid horizontally around the boiler's waist. The sides of the firebox and the backhead were not lagged; these parts were so hot that wood lagging would be quickly destroyed. The boards were tongue and groove and were often made with an ornamental beading on one edge. They were held in place by brass hoops. The hoops or bands were brightly polished and often were rolled with handsome moldings. English builders occasionally used expensive woods, such as mahogany, with rich varnish finishes, and while there were probably instances of this practice here, painted pine was the standard in the United States.[24]

Wood lagging rapidly deteriorated when exposed to rain, snow, sun, and sparks from the engine itself. Sheet-metal jackets fitted over the wood lagging not only shielded it from destruction but also reduced radiation by reflecting the heat back into the boiler's interior. The smooth, polished surface of the metal jackets was in effect a mirror. Russian sheet iron was a favorite material for locomotive jackets because of its remarkably rich and lustrous finish. The finish was often a silvery gray but was also available in shades of green or brown.[25] Cheaper, planished iron of American manufacture replaced the authentic Russian product in the 1870's. Russian iron and its imitations were of course ornamental as well as utilitarian and did much to enliven the engine's appearance until about 1900 when it was replaced by ordinary sheet steel painted a dreary black.

Metal jackets apparently were not much used before 1845, but the evidence is sparse and only a few specific details are available. Early mention of a "sheet iron" jacket is found in a letter of W. H. Clement to M. W. Baldwin dated April 7, 1846, regarding the Little Miami Railroad's locomotive the *Milford*.[26] British builders began to use metal jackets about a year later.[27] The good sense of this plan was quickly recognized and judging by several surviving drawings and lithographs it was in general use in the United States by the late 1840's.[28] No contemporary illustrations known to the author show American locomotives without sheet-iron jackets after 1850.

As already mentioned, wood remained the favored insulating material for locomotive boilers throughout the nineteenth century. Other materials were used, however; the best of these was felt, which was the only serious rival of wood until the 1890's. It was a better insulator than wood, having more air space between its loosely packed fibers. Felt was applied in layers of from 1 to 1½ inches thick and was treated with a lime, alum, and sal soda mixture to make it less combustible.[29] In 1855 the cost of felt was about the same as for wood, $10 per engine. Wilson Eddy, Thomas Rogers, and other builders used it on all their best engines. The specifications of the famous Civil War locomotive the *General* show that felt was used for the boiler lagging (see Appendix E).

R. A. Wilder of the Mine Hill Railroad used paper for boiler lagging on two engines built in 1860 by Baldwin. One of these is shown in Fig. 55 of this volume. Pasteboard was cemented together in layers with a glue formulated by Wilder from shellac, soapstone, and plaster.[30] The practicality and economy

---

[24] E. L. Ahrons, *The British Steam Railway Locomotive, 1825–1925* (London, 1927), p. 36.

[25] The manufacture and character of Russian iron is described by Oliver Byrne in *Practical Metal Worker's Assistant* (Philadelphia, 1884), pp. 633–47.

[26] Clement's letter is in the Baldwin Letters.

[27] Ahrons, *The British Steam Railway Locomotive*, p. 36.

[28] Early illustrations of locomotives with metal jackets are the *Ramapo* (Rogers, 1848), the *Champlain* (Taunton, 1849), and the *Massachusetts* (Hinkley, 1849).

[29] *Railroad Advocate*, August 11, 1855, p. 1.

[30] Wilder's paper lagging is mentioned in the Baldwin specifications (September, 1860) at the De Golyer Foundation Library and in *Engineer* (Phila.), October 18, 1860.

of Wilder's plan is questionable, but the idea indicated a dissatisfaction with wood and an attempt to devise new boiler laggings.

Increased steam pressures and the resulting rise in boiler temperatures brought about a growing dissatisfaction with wood lagging as the nineteenth century closed. During one year's service wood shrank and charred to the point that its insulating qualities were seriously impaired, and after four years it was burned out.[31] A fast-running passenger locomotive could generate enough heat to set the lagging afire, thus causing no end of trouble for the hapless crew which was obliged to somehow extinguish a fire under the metal jacket. The situation became intolerable by the 1890's and fireproof lagging was eagerly sought.[32]

Asbestos lagging, tried as early as 1873 on the Fitchburg Railroad, gave good service for sixteen years.[33] Even so, it was not widely adopted until about 1900 when most leading railroads began using asbestos or magnesium laggings. A few years later wood lagging was regarded as entirely obsolete.[34]

FIREBOX CONSTRUCTION

The major patterns of fireboxes have already been discussed in the opening section on boilers. However, the construction and material of this structure require further explanation. The firebox, known as the heart of the boiler, is a box within a box. The space between the inner and outer box is filled with water. This water space partially insulates the inner firebox plates from the destructive action of the fire. In wood-burning engines the water spaces were 2 inches wide or less because of the relatively low heat. Coal-burners, developing more intense heat, were built with 3- or 3½-inch water spaces. The inner and outer firebox plates were held parallel by stay bolts usually set on 5- to 6-inch centers. These bolts were threaded and riveted at both ends for security and steamtightness. Robert Stephenson used this style of construction on the *Rocket* in 1829 and it remained in use well into the twentieth century. Hollow stay bolts were used by F. P. Dimpfel for stationary boilers as early as 1839.[35] Their introduction to locomotive practice is uncertain, but hollow stay bolts were recommended by the master mechanics in their report of 1872 and were apparently in regular use before that time.

The crown sheet or top-flat plate of the firebox required support to prevent its collapse. Flat, iron bars called crown bars were used for this purpose. Five to twenty such bars, depending on the size of the crown sheet, were used. They were set on edge and riveted or bolted to the crown sheet, thus forming a truss. Only the ends of these bars rested on the top of the firebox. Washers were inserted between the bar and sheet at each rivet so that the contact between these two parts was held to a minimum. This was done so that as much of the crown sheet as possible would be covered with water to prevent its burning out. Before 1860 crown bars were fastened transversely or longitudinally, depending on the preferences of the designer. After the introduction of coal-burning, longer fireboxes prevailed and crown bars were invariably placed in a transverse position.

An ordinary crown bar was a thick (2 inches wide by 5 inches deep) piece of iron with a hole drilled through for riveting. However, double crown bars were used in the 1840's (and possibly earlier) as shown by the Winans 4–4–0 in Fig. 129. This form of construction called for two thin bars set close together. The rivets or bolts passed between the bars, thus eliminating the labor of boring holes through the bar.

Wrought iron was used universally for crown bars. The only variant from this rule was Norris' disastrous experiment with

---

[31] *Master Mechanics Report*, 1889, p. 64.
[32] *Locomotive Engineering*, July, 1895, p. 440, in an undoubted exaggeration, estimated that 90 per cent of U.S. locomotives still used wood lagging.
[33] *Master Mechanics Report*, 1889, p. 64.
[34] *Locomotive Dictionary*, 1906, p. 45.
[35] Holley, *American and European Railway Practice*, p. 91.

cast-iron crown bars in the early 1840's. The failure of cast iron for this use is illustrated by the explosion of the locomotive *Richmond* (see p. 302).

Stay bolts were a second method of supporting the crown sheet. These bolts, much like those already described for the firebox side sheets, were longer but were similarly attached to the inner and outer sheets of the firebox. Stephenson used this style of construction in 1829 on the *Rocket* but abandoned it almost immediately for simpler and cheaper crown-bar construction. Isaac Dripps built a number of locomotive boilers, beginning in the 1830's with X-braces or "crow's feet" in place of crown bars. This method was similar to stay-bolt construction but received little or no attention apart from the Camden and Amboy Railroad's limited use.[36]

In 1847 or 1848 Dripps designed a slope-backed firebox with a combination of crown bars and stay bolts to support the crown sheet. Several Crampton engines were built for the Camden and Amboy Railroad in the next few years with this style of boiler. Soon thereafter Ross Winans and James Millholland adopted Dripps's design but eliminated the crown bars. This style of boiler, introduced in about 1852, was undoubtedly the first built in this country to depend entirely on stay bolts for support of the crown sheet. The earliest evidence of a conventional boiler so-built, not a slope-backed affair like Winans', is Henry Tyson's ten-wheeler design of 1856 (see Fig. 181). While stay-bolt crown sheet boilers made an early appearance, they did not succeed the crown-bar boiler for many years, despite the several failings of the latter. Crown bars restricted the free circulation of water over the crown sheet and were a notorious collection place for scale. This not only hampered the boiler's efficiency but hastened the burning out of the crown sheet. Nevertheless, crown bars were considered the best and cheapest construction plan until 1890. After that date their decline was swift. The American Society of Civil Engineers in 1893 reported that crown bars were losing favor. Four years later they were reported to be obsolete and rarely used except on small locomotives.[37]

Several methods were used to seal the bottom of the firebox water space. Stephenson used two angle irons (see the *John Bull*, Fig. 96), but this was a weak and complicated construction. Another method, devised contemporaneously, was to flange the inner firebox sheet and rivet it to the outer sheet. This simple, cheap form of construction was popular for years. It is illustrated by the *Lancaster*, Fig. 106. The above methods were superseded by the foundation ring, which by the 1850's was the approved plan for sealing the water space. According to this plan, the inner and outer firebox plates remained parallel and a thick, square iron bar was riveted between them. See the *Columbia*, Fig. 158, for this style of construction.

The bottom of the firebox water space was the lowest part of the boiler and thus served as the collecting point for loose scale and mud. One or more blowoff cocks, usually fitted to the rear of the firebox, were used to eject all impurities that accumulated at that point. Washout plugs or hand holes (also for cleaning) were located here.

Before 1860, copper and wrought iron were the only materials used for firebox construction. At first iron was favored, but its reedy texture and propensity to blister when under the direct action of fire caused Stephenson to adopt copper for the firebox's inside sheets in about 1832. Early United States builders followed the British example, and copper remained the favored material for this purpose until the 1860's. Because of copper's low tensile strength and increased weakness when heated, it was necessary to make firebox sheets very heavy. Rear tube sheets (the front sheet of the firebox) were generally from two-thirds to three-quarters of an inch thick if made of copper. The crown side and rear sheets were thinner, probably from five-sixteenths to three-eighths of an inch thick. A copper firebox (1,850 pounds) weighed nearly twice as much as an equivalent iron firebox (1,000 pounds) and cost nearly eight times as much ($540 versus $70).[38] The higher cost was justified in part by the longer life of a copper firebox when compared to that of the iron. This was undoubtedly true when wood was the fuel, but soft copper sheets were rapidly worn

[36] *Master Mechanics Report*, 1885, p. 48.
[37] *Modern Locomotives*, p. 7.

[38] Holley, *American and European Railway Practice*, p. 20.

out by fly ash from coal. The Reading complained that copper fireboxes wore out after only fourteen months service when fired with anthracite coal.[39] The Baltimore and Ohio found that the bituminous fly ash did not cut away copper sheets so quickly and that about three and a half years' service could be obtained.[40] The Pennsylvania Railroad also found soft coal to be easy on copper plates and used copper fireboxes for six years.[41] This must be considered a record, although it was not so-reported. Most roads found copper expensive and short-lived for coal-burning.

Thick copper sheets not only increased costs but reduced efficiency. While copper might be supposed a better heat conductor than iron, in practice, iron plates were found equally efficient in transmitting heat because of their thinness. In effect, thick copper plates insulated the water. This defect, added to the others already outlined, led to the general abandonment of copper fireboxes in the United States during the 1860's. In 1870 the Baldwin Locomotive Works produced 280 locomotives. Only 6 of these had copper fireboxes; the rest had fireboxes of steel.

Iron fireboxes were used as early as 1836 by the Beaver Meadow Railroad for coal-burning locomotives. The Reading found this material best suited for coal-burning engines when it began to convert to anthracite in the late 1840's. Iron fireboxes should not be associated exclusively with coal-burning, however. The Baltimore and Ohio preferred copper fireboxes for their soft-coal engines, while many builders produced wood-burners with iron fireboxes, largely because iron was cheaper than copper. Thus, copper or iron was used for either fuel although copper was preferred for wood engines before 1860. In 1860 Colburn stated that "the firebox is always of iron" in American locomotives. This was another indication of iron's triumph over copper.[42]

English plate was preferred by most American locomotive builders. Millholland, however, was not satisfied with commercially produced firebox iron, imported or domestic, and took a direct hand in its manufacture. He procured the largest charcoal wrought-iron blooms available and worked them over with a steam hammer in the Reading shops. After determining that the iron was of the best quality, the blooms were sent out to a rolling mill for manufacture. Exceptionally large plates, from 10 to 12 feet long and 6 feet 10 inches wide, were thus obtained.[43]

Iron had no sooner succeeded copper as the favored firebox metal when it was challenged by steel. In 1860 Millholland was reported to be using steel fireboxes.[44] The Taunton Locomotive Company built a steel firebox for the Erie Railway in the same year, which gave good service for ten to thirteen years.[45] Other builders began to offer steel fireboxes but these were not entirely satisfactory. As with the earliest steel boilers, the plate used was too hard and cracking was common. Softer alloys soon became available and by the mid-1860's steel fireboxes were common. The Pennsylvania Railroad had 400 locomotives with steel fireboxes in 1869; some of these had been in service for six or more years.[46] The success of the steel firebox is further shown by the Baldwin records for 1870, which note that the vast majority of new engines were so-built.

Not all master mechanics found steel satisfactory for fireboxes. Samuel Hayes respected steel's good qualities but found iron more durable on the Illinois Central, where bad water was the rule.[47] Wilson Eddy was much sharper in his dissent and claimed that steel firebox sheets became brittle like "glass." He did not believe that steel was a miracle metal and stated that he would welcome more criticism of the "new-fangled" material. Eddy's viewpoint was undoubtedly shared by other old-time master mechanics and was in line with his conservative approach (viz., his attacks on Consolidation and Mogul locomotives) to the entire idea of locomotive reform. Eddy's

---

[39] Whistler, *Anthracite Coal*, p. 18.
[40] *Ibid.*
[41] Holley, *American and European Railway Practice*, p. 15.
[42] Clark and Colburn, *Recent Practice*, pp. 55, 58.

[43] *Engineer*, February 8, 1861, p. 6.
[44] *American Railway Review*, August 9, 1860.
[45] *Master Mechanics Report*, 1875, p. 23.
[46] Institution of Civil Engineers, *Proceedings*, Vol. 28 (1869), p. 454.
[47] *Master Mechanics Report*, 1872, p. 28.

opinion carried little weight and most roads went ahead with the conversion to steel fireboxes during the 1870's.

The rapid rise of steel for fireboxes must be credited to its long life in such wearing service. With good water a steel firebox would last 300,000 miles or about fifteen years.[48] An iron firebox gave only about three years' service. Thus, under the best conditions a steel firebox would last nearly the entire life expectancy of the boiler.[49]

COAL-BURNING FIREBOXES

Although a few roads operated coal-burning locomotives as early as the 1830's, this fuel was not widely used until many years later. Coal was scarce and expensive, while wood was plentiful and cheap until this time. Moreover, when serious experiments with coal-burning began in the late 1840's, the idea was soon established by ill-advised inventors that only specialized fireboxes could burn coal successfully. The designs offered were highly contrived affairs where novelty and complexity rather than performance appeared to be the goal. Any design, as long as it did not resemble the ordinary firebox, was offered as a solution to the "coal-burning problem." Dimpfel, Boardman, and Phleger patented their plans and saw several locomotives built, but with no practical results. Rogers, Baldwin, and Norris produced their own, somewhat less complex designs but, again, offered little of value. By the late 1850's most responsible builders agreed that the standard firebox with minor modifications was well suited for coal-burning. It was further agreed that the problem was not so much firebox design as good coal and skillful firing. This is not to say that some notable changes were not made, but rather that firebox design was *modified* rather than *revolutionized* for coal-burning.

This view was well stated in the *American Railway Review* of July 12, 1860: "We are at last adopting the belief that the proper combustion of coal can be effected without any structural modifications of the ordinary boiler, and, beyond a few air holes, a hodfull of fire-bricks, or a different form of grate, we are insisting upon the retention of the locomotive boiler as it is, and upon its proper behavior under the discipline of coal-burning."

[48] *Master Mechanics Report,* 1876, p. 76.

The most fundamental modification for coal-burning engines was the increase in firebox size. Ross Winans was unquestionably a pioneer in this field and was the first to build locomotives with large grate areas for coal-burning. In 1847 he built several eight-wheel coal-burners for the Reading, each with a grate area of 17.6 square feet. A common wood-burner in this period had a grate area of only about 10 or 12 square feet. During the next few years Winans devised a firebox of increased size so that by 1850 or 1852 his engines offered grate areas of 24.5 square feet. Winans' firebox and boiler designs are traced in the drawings shown in Fig. 37. In 1844 he abandoned the vertical boiler and adopted the Bury boiler. This boiler did not provide adequate space for the firebox and was modified in 1847 to include a boxlike structure at the back of the boiler, which increased the grate area. In 1848–49 the arrangement was improved by moving the large, Bury style dome forward and making a "step" over the firebox. This helped to balance the boiler by reducing weight at the firebox end but it did not provide a much-enlarged grate. In 1850 or 1852 the familiar slope-backed firebox was adopted. This firebox was not a Winans design as is commonly believed. It can be traced back to Isaac Dripps's 1848 plan for several Crampton type engines built for the Camden and Amboy Railroad (1849–53).[50] Dripps may have borrowed the idea from an earlier design.

James Millholland was another designer who recognized the

[49] *Ibid.* A boiler was well past its prime after twelve years of use. After twenty years wrought-iron boilers became brittle "like very poor cast iron." This is very likely true, but many wrought-iron boilers are known to have given up to thirty years' service.

[50] A drawing of Dripps's 1847 slope-backed boiler is included in the 1884 *Master Mechanics Report.*

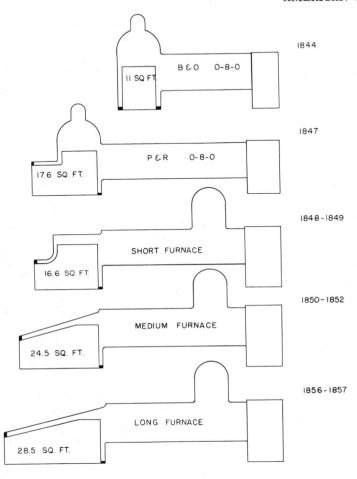

Fig. 37. *Horizontal coal-burning boilers developed by Ross Winans from 1844 to 1857.*

importance of large fireboxes for coal-burning.[51] Benefiting from Winans' experience, Millholland began to perfect a coal-burning firebox for the Philadelphia and Reading Railroad. His early work involved the rebuilding of several engines with enlarged fireboxes (1849–51). Unfortunately, Millholland was sidetracked from any fruitful results for the next four years by adhering to an impractical scheme for central combustion chambers which he patented in 1852. The central combustion chamber was abandoned in about 1855 and a slope-backed design was adopted. Millholland used the slope-backed firebox as early as 1852 but unfortunately combined it with the central combustion chamber. He used two small steam domes, thus making a stronger boiler than Winans' large, Bury style dome. In about 1858 Millholland introduced the water grate as a means of increasing grate life. This form of construction required a water space at the rear of the firebox. Previously Millholland had copied Winans' questionable practice of using no rear water space.

Both the Winans and the Millholland firebox had fallen from favor by 1870 (for reasons unknown to the author) and neither had a lasting effect on boiler design. They did demonstrate, however, that a simple, straightforward firebox of sufficient size was practical for coal-burning.

Winans and Millholland increased the firebox size by lengthening it. Both inventors achieved modest grate enlargements by making the firebox as wide as the boiler frame (42 inches) but this represented a rather small increase in size. Dripps built an engine, the *Monster*, in 1836–38 with a firebox 43 inches wide. Wilson Eddy increased the width of the firebox slightly by introducing slab-rail frames in 1851. Colburn's giant *Lehigh,* built in 1856 for the Delaware, Lackawanna and Western, had a firebox 90 inches in width, but this design lay dormant until 1877 when it was revived by J. E. Wootten. Essentially, firebox enlargement was confined to longer rather than wider units until the 1890's when the above frame designs became more common.[52]

[51] "James Millholland and Early Railroad Engineering," *U.S. National Museum Bulletin*, No. 252 (1967), Paper 69.

[52] Millholland is credited with building the above-the-frame firebox engine, the *Vera Cruz,* in 1857.

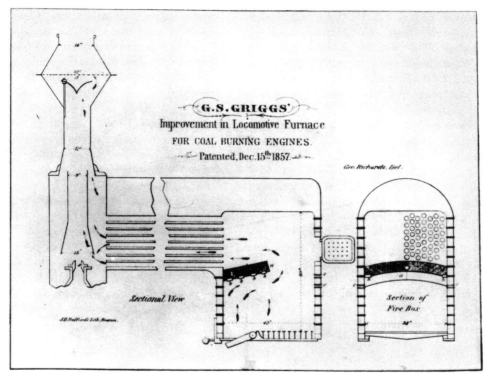

Fig. 38. Griggs's firebrick arch and diamond smokestack for coal-burning locomotives. Copied from an original broadside of 1857.

Other than increased size, the firebrick arch and the combustion chamber were the most important and long-lived alterations in firebox design made before 1860. Both devices were intended to improve combustion and the efficiency of coal-burning boilers. The combustion chamber was an extension of the firebox into the boiler's waist. The purpose was to provide more room for combustible gases and air to mix for burning. It was thought that this could be accomplished more readily in an open chamber than within the boiler tubes. This was a good arrangement for large, modern boilers but on the whole it was self-defeating in the small boilers used before 1890. A combustion chamber of any great length shortened the tubes and thus materially reduced the heating surface. Aside from this consideration the chamber was a common source of leaks, a defect not easily corrected until the advent of modern welding in the twentieth century.

The combustion chamber was in evidence as early as 1832 on the Camden and Amboy Railroad.[53] Dripps also used a combustion chamber on the *Monster* (1838) and on several Crampton locomotives (1849), all on the Camden and Amboy

[53] Sinclair, *Development of the Locomotive Engine*, p. 384.

line. Despite this long-established use, the idea was patented in England by Stubbs and Gryll in 1846. As coal-burning engines became more common, interest in combustion chambers became widespread. Millholland, Winans, and other advocates of coal-burning adopted this scheme at an early date. A. F. Smith, while master mechanic of the Hudson River Railroad, carried the idea to its extreme by equipping eleven passenger locomotives with combustion chambers 5½ feet long. It was claimed that this alteration saved some $60,000 per year in fuel.[54] Few combustion chambers in this period were more than 18 inches long and many were only 6 inches deep.

The fire arch, like the combustion chamber, was designed to enhance combustion by improving the mixture of unburned fuel gases and air in the firebox. This was accomplished by increasing "flame length," but the fire arch did not take space away from any other element of the boiler as did the combustion chamber. Water legs and cast-iron fire arches were used experimentally before 1850, but all such contrivances had a common failing in that they burned out under the direct action of the fire.

This problem was solved by the firebrick arch, which was not readily consumed by fire. George S. Griggs is commonly credited with this invention but Matthew Baird is said to have used the device as early as 1854 on a number of engines built by the Baldwin Works.[55] Baird did not patent the firebrick arch and thus, if the Baldwin history is correct on this point, lost credit for one of the most notable single contributions to locomotive design. Griggs first used the brick arch in 1856 and secured a patent on December 15, 1857 (No. 18883). The patent specification reveals that Griggs apparently was not aware of the chief merit of the fire arch, the fact that the flame is lengthened by its passage *around* the arch; at least there is no mention of this in the patent. The inventor claims instead that the arch is heated by the fire to the point of igniting the air and gases as they pass over it. This matter aside, the firebrick arch was rapidly accepted as part of standard boiler construction. It could be added to existing boilers; in fact, Griggs developed it to convert wood engines to coal. Not all authorities agreed that it effected any measurable fuel economies, but it was recognized by all as an effective smoke preventer.[56]

GRATES

Grates were a simple, trouble-free mechanism in the days of wood-burning. Cast-iron bars, often T- or V-shaped, answered construction needs very well. Each bar was about five-eighths of an inch thick, four inches deep, and as long as the firebox required. The bars were set about one inch apart. Because the wood was all but entirely consumed in burning, rocking grates were not required. The small amount of ash that was not thrown out through the stack filtered through the grate bars to the ashpan.

Coal presented many more problems and required a more elaborate grate. It produced a hotter fire, and cast-iron bars burned out quickly. In 1849 it was reported that ordinary iron grates burned out in one month.[57] Some years earlier Eastwick and Harrison had designed a wrought-iron grate specifically for coal-burning. In this plan a U-shaped slot in the top of each bar was filled with clay.[58] The effectiveness of this arrangement is not known, but apparently it was not successful, because no other reports exist on its subsequent use. A more successful and long-lived arrangement was Millholland's water grate. The grate was formed of iron tubes that connected the front and rear water spaces of the firebox. The water grate is illustrated in Fig. 39. Millholland first used the water grate in about 1858 and saw it used by many other roads burning anthracite.

[54] Colburn and Holley, *The Permanent Way*, p. 160.
[55] *History of the Baldwin Locomotive Works*, p. 57.
[56] *Master Mechanics Report*, 1877, p. 104.
[57] Whistler, *Anthracite Coal*, p. 18.
[58] Harrison, *The Locomotive*, p. 69.

COMPONENTS

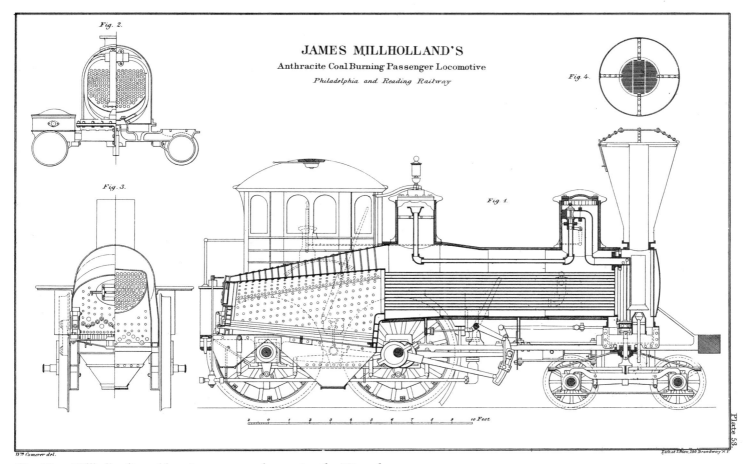

Fig. 39. *Millholland's coal-burning passenger locomotive the* Hiawatha, *built in 1859 at the Reading shops. Note the large firebox, water-grate bars, metal cab, and smokebox superheater.*

Most coal-burning roads found rocking grates cheaper and less complex than Millholland's water grate. The rocking grate was in fact considered indispensable to soft-coal burning. Bituminous coal formed massive clinkers which in turn cut off the air supply. The rocking grate assisted in breaking up these clinkers and reactivating the fire. The exact date of the introduction of this apparatus is unknown, but Ross Winans offered a simple plan for the rocking grate as early as 1847. Winans' arrangement consisted of individual, loose bars that could be tipped from side to side with a jacking bar. This style of grate was used by Winans on his Camel locomotives until the end of their production in the late 1850's; it is shown in the *Susquehanna* drawing, Fig. 169. More complex rocking grates, connected so that all the bars might be actuated by a simple lever, were introduced in the mid-1850's.

GRATE AREA AND HEATING SURFACE

The importance of adequate grate area and heating surface was recognized early in the history of locomotive design. Although there was a tendency to "overcylinder" engines, heating surface and cylinder cubic area were held at a remarkably constant ratio (about 200 to 1) between 1835 and approximately 1880. There was, however, an intelligent movement to enlarge the heating surface proportionately more than cylinders as the locomotive grew in size.

The typical 4–2–0 of the 1830's rarely had a heating surface of more than 400 square feet, which was quite adequate for slow speeds and small cylinders. In the next decade the advent of the heavier 4–4–0 made 500 square feet of heating surface common, with cylinder diameters ranging from 13 to 15 inches. In the late 1840's and early 1850's the development of an enlarged heating surface moved forward rapidly. Colburn observed in 1851: "The heating surface of locomotive boilers has of late years been considerably increased, not only having been extended with the enlargement of the cylinders but in a much higher ratio."[59] Thus, we find engines of the early 1850's with 15-inch cylinders but a heating surface of more than 700 square feet. The enlargement of heating surface showed much slower progress during the next few decades. Yet several designers produced machines with exceptional heating surfaces. In 1851 Wilson Eddy built the *Addison Gilmore* with the unprecedented heating surface of 1,175 square feet. Winans' Camel engines built between 1848 and 1860 had heating surfaces of about 1,000 square feet.

Grate area closely followed the expansion of heating surface. The early British and American locomotives of the 1830's generally had grate areas of about 6 square feet. By the 1840's grate areas of 10 square feet were common. The abandonment of the Bury boiler in the 1850's made grates of 12 and even 14 square feet possible. Again, as with heating surface, little real progress was made until after 1875. Most standard-gauge 4–4–0's rarely had grate areas of more than 16 square feet until the 1880's. Exceptions are of course to be found, particularly with hard-coal or broad-gauge engines. Winans' and Millholland's coal-burners had grates of 24 square feet and the Delaware, Lackawanna and Western's *Lehigh* had a grate area of 45 square feet. But such engines were decidedly peculiar to American practice of that time.

During the 1880's the locomotive experienced a new growth in size which by the end of the decade had become a revolution. By the mid-1890's firebox area had increased in size by 75 per cent. In these same years heating surface increased by 45 per cent. Boiler waists formerly limited to 50-inch diameters jumped to 60 inches, and a few large freight engines had 72-inch boilers.[60]

---

[59] Colburn, *The Locomotive Engine*, p. 57.

[60] *Modern Locomotives*, pp. 7–8.

SMOKEBOXES

At the front end of the boiler is a receptacle known as the smokebox. It serves primarily as a collecting point for waste materials—smoke from the fire and exhaust steam from the cylinders. It functions as a convenient mounting for the smokestack and cylinders. It also forms the front-end support for the boiler.

Credit for this basic element of locomotive construction must be given to Robert Stephenson. The first engine built with a smokebox was Stephenson's *Phoenix,* which was constructed for the Liverpool and Manchester Railway in 1830. Its merits were immediately recognized, and by the time most American builders were in production the smokebox had been established as an essential part of the locomotive engine.

Two styles developed in this country, the round and the D-shaped smokebox. Baldwin used the round style at least as early as 1834 as evidenced by the drawings of the *Lancaster*. He used the round smokebox regularly and only on a few occasions in the 1860's resorted to D-shaped boxes. Dunham, who closely copied Baldwin's designs in the 1830's, also was an early user of the round smokebox. During the late 1840's Souther and several other New England builders adopted the round smokebox for inside-connected engines by making the top of the cylinder a saddle.

Norris and Rogers championed the D-shaped smokebox and used it on their earliest engines. Norris continued its use into the 1860's. Rogers was an even more steadfast patron and used D-shaped smokeboxes until the late 1860's. Most builders abandoned this form of smokebox for the round style when the cylinder saddle became common, in about 1855.

While designers disagreed on the shape of the smokebox, they nearly all agreed that it should be airtight and as small as possible. The thinking behind the small, tight smokebox was that it would promote a good draft for the fire. The draft through tubes and the entrance of air through the bottom of the firebox were induced by the partial vacuum formed in the smokebox by the exhaust steam. The smaller the smokebox, the easier it was to form a vacuum. This logical scheme was upset by the extended-smokebox craze of the 1880's. The history of the extended smokebox will be treated in the section on spark arrestors, but it is mentioned here for purposes of general discussion and description. The extended smokebox was introduced in 1860 as a new method of spark-arresting. The apparatus was placed inside the smokebox rather than in the stack. To accommodate this equipment a sizable enlargement of the smokebox was necessary. The concept of small smokeboxes was abandoned by many designers in the 1880's. Engines fitted with giant, ungainly smokeboxes (in some cases 7 feet long) protruding beyond the pilot beam were seen on many American railroads. Other authorities contended that the extended smokebox was not only useless as a spark arrestor but functioned poorly as a draft producer. In the end the extended smokebox proved as ineffectual as it was ugly and was abandoned in the early 1890's.

BLAST PIPES

The blast or exhaust pipes were located in the smokebox and directed the waste steam out the smokestack. The use of exhaust steam to create a strong draft, so essential to the successful working of a locomotive boiler with its high rate of combustion, was a British invention. It was introduced by Trevithick on his first locomotive in 1804, despite the conflicting claims of later engineers for the honor. The value of the blast pipe was not fully recognized until the 1830's when relatively high-speed locomotives were required for public railways. It became apparent that for such work a locomotive must steam rapidly, yet stay within reasonable size and weight limitations. A strong draft caused by the powerful exhaust of steam through a *contracted* exhaust opening produced the desired results without an elaborate or extensive apparatus.

Unfortunately, the idea was carried to an extreme. Small-diameter blast pipes soon became the fashion. Exhaust nozzles (the top opening of the blast pipe) were generally held to a 1¾-inch diameter for 15-inch cylinders. This restricted opening did indeed create a powerful (and noisy) exhaust. It was reported that "these tremendously sharp nozzles made no end of a row, each steam stroke going off like a rifle, or the roar of a Mississippi steamboat; but nobody in the States ever used the indicator in those days, and there was no knowing what was going on inside the cylinder."[61] Noise was only a by-product of the actual defects of the small blast pipe. Power-absorbing back pressure was the chief problem. Robert Stephenson estimated that at high speed a locomotive fitted with a small-diameter blast pipe lost half of its power.[62] Not all American engineers were ignorant of the back-pressure problem, but since speed was of no consequence and rapid steaming was, small-diameter blast pipes stayed in favor until the mid-1850's. Colburn called for much-enlarged blast pipes as early as 1851 and praised Taunton for building engines with 2⅜-inch blast-pipe nozzles.[63] Nevertheless, most early American engines, because of their small heating surfaces, could not produce sufficient steam without a powerful exhaust. With increased heating surface and more consideration given to scientific design and better-proportioned parts, the diameter of exhaust nozzles increased noticeably after the late 1850's; most builders began to adopt openings of 3 or more inches.

The height of the blast pipe developed inversely, becoming shorter and shorter. Blast pipes made by early British builders protruded into the base of the smokestack. American manufacturers copied this practice, as can be seen from the *Robert Fulton, Lancaster,* and *Dunham* drawings in Figs. 7, 106, and 115. Tall blast pipes persisted until the 1850's when they generally extended no higher than the top row of tubes. The pipe became lower in the next few years. "It has lately become quite customary, however, to place the mouths [blast pipe] even with the lower row, or near the lower row of tubes, and to suspend a short pipe, say three feet long and 8 or 10 inches in diameter, directly over the exhaust pipes."[64] The device just described was the petticoat pipe, Ross Winans' invention of about 1848. Its purpose was to create a more even "pull" or draft on all of the tubes. Without it, the strongest draft was on the upper tubes, the draft on the bottom being so weak that the lower rows often filled up with ash and cinders. This uneven pull reduced the heating surface and the steaming quality of the boiler. The petticoat pipe became quite popular and was much used after 1855. The drawings of Rogers' 4–4–0 (Fig. 222) and Erie's No. 254 (Fig. 218) illustrate this device. The *Southport* drawings (Fig. 198) show a variation on Winans' single petticoat pipe which is made up of three short telescoping pipes.

While the blast pipe was the universal method of creating a draft in locomotive boilers, it might be mentioned that the Baltimore and Ohio Grasshoppers used a fan. These fans were powered by a small turbine operated on exhaust steam. In this system the draft could be varied according to the needs of the engine. It was, however, a complex arrangement and was not widely used.

The steam jet was another device used for draft, but it was only an auxiliary to the regular blast pipe. The steam jet's main function was to keep steam up when the engine was standing at a station. The steam jet was simply a steam line running to the base of the smokestack with a valve in the cab to be opened or closed by the enginemen as required. A. F. Smith, of the Cumberland Valley Railroad, is credited with its invention in 1852.[65]

---

[61] *Engineering,* July 26, 1867, p. 66.
[62] Colburn, *The Locomotive Engine,* p. 67.
[63] *Ibid.,* p. 55.
[64] *Railroad Advocate,* July 7, 1855, p. 3.
[65] Clark and Colburn, *Recent Practice,* p. 71; see also "The Pioneer," U.S. National Museum Bulletin, No. 240 (1964), Paper 24.

COMPONENTS

VARIABLE EXHAUSTS. The contracted blast pipe was a valuable invention that did much to promote the proper working of the steam locomotive, but its size was fixed and was therefore only generally suited to the widely varying workings of the locomotive. At times a stronger blast might be needed to enliven a sluggish fire. Conversely, it might be desirable to increase the engine's power by enlarging the exhaust nozzle, thereby reducing back pressure. A contracted blast pipe naturally created a certain amount of back pressure which under normal circumstances did not materially reduce the engine's power. But in emergencies, such as getting a heavy train over a steep grade, relieving the back pressure by opening the nozzle might provide the extra degree of power required to ease the train over the hill. Just as the steam was governed variably by the throttle, some engineers thought that the exhaust should be subject to the adjustments a working engine required. It was also believed that a worthwhile fuel economy could be effected.

The earliest record of a variable exhaust is a drawing of the locomotive *Pioneer* built in 1832 by Rothwell for the Petersburg Railroad.[66] The drawing shows a conical plug mounted on top of the blast pipe. The plug was raised or lowered by a simple lever arrangement, thus opening or partially closing the exhaust nozzle. Three years later the Paterson and Hudson River Railroad fitted its engine the *McNeill* with a conical-plug variable exhaust.[67] This contrivance was not a success, because the cone, pointing downward, deflected the exhaust steam and thus prevented its free passage out of the stack. In 1836 the famous French engineer De Pambour experimented with a shutter-valve variable exhaust on an engine of the Liverpool and Manchester Railway. His design was not a startling success, but the description of this experiment in his widely circulated treatise on locomotives attracted attention to the variable exhaust as a promising auxiliary for locomotives.[68]

As indicated above, the idea of the variable exhaust was familiar to American builders by 1840 (though it was not widely used); yet Ross Winans was issued a patent on the device in November of the same year. The patent drawing (No. 1868) shows two arrangements of the conical-plug exhaust, one of these being identical to the 1832 variable exhaust of the *Pioneer*. Winans may have been unaware of the *Pioneer* or the *McNeill* arrangement, although he sold engines to the Paterson and Hudson River Railroad. But whether the invention was his by accident or by cunning, Winans made full use of the patent in later years. Probably all of his Camel engines were equipped with it, and the variable exhaust was one of several patents that charged a $750 fee per engine. The Philadelphia and Reading and the Baltimore and Ohio Railroads used a variable exhaust based on Winans' plan, not only on their Camels, but on other engines as well. The *Philadelphia* (Fig. 134) and the Tyson Ten Wheeler (Fig. 183) are so-equipped. Winans' patent was extended in 1854. During the next few years other builders began offering coal-burning locomotives, and variable exhausts were at first considered necessary for coal-burning. Winans, however, was careful to see that his competitors did not infringe upon his invention. Many patent suits developed. As in the case of the eight-wheel car patent, Winans devoted his full energies to the prosecution of any trespassers. In 1861 a United States circuit court ordered Charles Danforth to pay Winans $3,000 for unauthorized use of the variable exhaust on two locomotives.[69]

To avoid Winans' patent, innumerable variable exhausts were devised. Between 1840 and 1915 the United States Patent Office issued 132 patents for such devices. In the end the variable exhaust was not widely accepted in this country except during the early years of coal-burning. Once the great "scare" was over and it was realized that outlandish boiler designs and complex auxiliaries were not required for coal-burning, the variable exhaust was virtually abandoned in the United States. It was another complex device to maintain, rarely was it properly regulated by the enginemen, it clogged

---

[66] A drawing of the *Pioneer* reproduced from an original appeared in the *Railroad Gazette*, April 12, 1901, p. 251.
[67] *American Railway Review*, April 4, 1861, p. 199.
[68] Variable exhausts, including a history of Pambour's and Winans' work, are discussed in a paper by J. S. Bell in the *Master Mechanics Report* for 1915.

[69] *American Railway Review*, April 4, 1861, p. 199.

up with cinders, and deterioration, caused by heat and cinders passing at great speed through the smokebox, was rapid. Finally, the variable exhaust in its closed position could materially increase back pressure in the cylinders.

Variable exhausts had become all but extinct in the United States by 1870, although a few roads, such as the Reading, used them until about 1900. The British were disinterested. Only the French employed them enthusiastically.

SMOKESTACKS AND SPARK ARRESTORS

Abundant evidence exists that the enormous production of embers and live sparks by wood-burning locomotives was an ever-present annoyance and expense to the traveling public. Travel journals and the technical press regularly commented on "the spark problem." One angry passenger protested: "Is there a single person, who has traveled on any other road in the United States, on which locomotives are used, with wood for fuel, that has not been annoyed, and either had his flesh or clothing burnt . . . ? . . . Baggage cars have been burnt, passenger cars have been on fire, and ladies almost denuded."[70]

Less vocal, but of great concern to railway managers, was the immense property loss accountable to locomotive sparks. Railroad companies not only lost their own cars, bridges, and way structures but also paid out large claims to shippers for damaged or destroyed merchandise. Track-side property holders sued for barns, woodlands, fences, and other valuables destroyed by errant locomotive sparks. Few over-all figures are available for such losses, but a few examples can be offered. The burning of $60,000 in paper money when a Newcastle and Frenchtown Railroad car was set on fire by a locomotive spark in 1832 was an early instance of a large and spectacular fire loss.[71] The Philadelphia and Columbia Railroad was reported to have paid out nearly $80,000 in fire claims between 1833 and 1857.[72] In 1881 the Cincinnati, New Orleans and Texas Pacific Railroad paid $33,000 for such claims.[73] Two years later the *Master Mechanics Report* estimated that hundreds of thousands of dollars were paid out annually by railroads in fire damages.

The locomotive's disgorging of sparks was an expensive nuisance but it also provided a magnificent pyrotechnic display at night. A bright storm of sparks could be seen for miles, appearing and disappearing as the train progressed on its journey. Dickens described the sight, as seen through the car windows of a Boston and Lowell train, as "a whirlwind of bright sparks, which showered about us like a storm of fiery snow."[74]

The spark problem was so obvious that hundreds of mechanics, inventors, ordinary citizens, and no small number of cranks were attracted to it. Well over 1,000 patents were issued for smokestacks and spark arrestors during the nineteenth century.[75] Contests with attractive cash rewards were held in the hope that some creative mind would come forth with an effective design.[76] Surely no other single element of the locomotive attracted such a wide, concentrated, and popular interest as did smokestacks. Despite considerable thought and effort, no truly effective spark arrestor was devised. Two conflicting conditions prevented the development of a practical design. An

---

[70] *American Railroad Journal*, July 11, 1835.
[71] *American Railroad Journal*, January 9, 1833, p. 33.
[72] A. L. Bishop, *State Works of Pennsylvania* (New Haven, Conn., 1907), p. 278.
[73] *Master Mechanics Report*, 1887, p. 28.

[74] Charles Dickens, *American Notes* (orig. pub. 1842; New York, 1961), p. 89.
[75] Three articles covering spark arrestor patents are recommended for more data on this subject: *Journal of the Franklin Institute*, January, 1880, pp. 1–16; *Master Mechanics Report*, 1883, pp. 89–185; J. S. Bell, "Locomotive Front Ends," *Western Railway Club*, September 19, 1919.
[76] *American Railroad Journal*, August 17, 1833, p. 513, reported the Franklin Institute's offer of a $250 prize; the same journal, on October 8, 1836, p. 631, commented upon a $500 prize offered by the Ponchartrain Railroad.

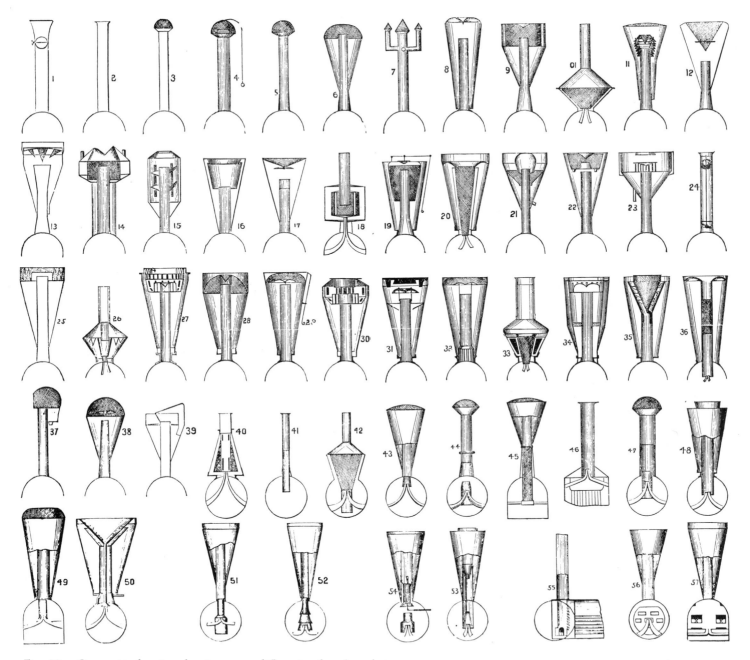

Fig. 40. Composite drawing showing many different styles of spark arrestors devised before 1860. Prepared by the Baldwin Locomotive Works in 1860 (see p. 116 for table).

unobstructed draft was necessary for good steaming; an effective spark arrestor necessarily obstructed the draft. Obviously a compromise was necessary. The successful and popular spark arrestors were admittedly only partially effective in the work of consuming live sparks.

The great bulk of smokestack designs was brought forth by novice or impractical mechanics. These can be dismissed as worthless novelties that either lay fallow in the patent office or received at best a limited test. Nevertheless, the number of designs produced by first-rate engineers is enormous and even a survey history of smokestacks is complex and involved. The present discussion will attempt to touch on only the more important types and examples. Before treating the several major types, however, reference should be made to the composite drawing of 57 representative smokestacks, Fig. 40. This drawing not only presents an idea of the variety of designs developed before 1860 but also illustrates many of the designs subsequently mentioned in the present work.[77] The drawing was prepared by the Baldwin Locomotive Works as part of its defense against a suit of David Matthew for infringement of his patents.[78] The following table identifies the stacks:

DEVELOPMENT OF LOCOMOTIVE SMOKESTACKS, 1831–1857

1. Baldwin, 1831
2. Baldwin, 1832
3. Baldwin, 1833
4. Baldwin, 1833
5. Baldwin, 1834
6. Baldwin, 1834
7. E. A. G. Young, 1833, patented July 22
8. Isaac Dripps, 1834
9. Isaac Dripps, 1835
10. Wm. Shultz, 1836, patented March 31
11. W. Duff, 1837, patent #521
12. Ben. Briscoe, 1838, patent #1037
13. Thomas Reaney, 1839, patent #1447
14. L. Phleger, 1838
15. James Stimpson, 1837, patent #161
16. Wm. T. James, 1838, patent #688
17. H. Waterman, 1839
18. L. Phleger, 1839, patent #1417
19. L. Phleger, 1840, patent #1778
20. E. T. Moore, 1840
21. Wm. Pettit, 1840
22. D. Matthew, 1840, patent #1920
23. Wm. C. Grimes, 1842, patent #2455
24. Germantown R.R., 1841
25. R. French, 1841, patent #2131
26. T. Reaney, 1841
27. French & Baird, 1842
28. A. McCleary, 1848, patent #5541
29. Yankee, 1844
30. Radley & Hunter, 1850, patent #7040
31. S. Sweet, 1853, patent #10172
32. Clark & Baldwin, 1854, patent #10514
33. R. A. Wilder, 1854, patent #11880
34. J. A. Cutting, 1849, patent #6559
35. Wm. C. Grimes, 1845, patent #4046
36. Grand Trunk R.R., 1856
37. Wm. Norris, 1834
38. J. C. Gilpin, 1837
39. C. H. Miller, 1837
40. C. H. Miller, 1839
41. *JOHN BULL* (English), 1832
42. Wm. Shultz, 1836

---

[77] J. S. Bell, in his history of spark arrestors (see the *Journal of the Franklin Institute,* January, 1880), commented on the Baldwin drawing and reproduced a few drawings from it.

[78] *Engineer* (Phila.), December 15, 1860, discusses the Baldwin-Matthew patent case. Matthew's stack, used by the Utica and Schenectady Railroad as late as the 1850's, is illustrated in the *American Railroad Journal,* April 1, 1841, and on p. 46 of this volume.

COMPONENTS

43. E. T. Moore, 1839
44. T. Fagan, 1840
45. Eastwick & Harrison, 1840
46. Ross Winans, 1842
47. John Lotty, 1844
48. Klein (German), 1846
49. Wm. Thomas, 1847
50. M. W. Baldwin, 1848
51. M. W. Baldwin, 1852
52. Experimental, 1853
53. A. F. Smith, 1856
54. Rulter & Parry, 1857
55. Unknown, 1855
56. Winans, 1856
57. Wm. S. Hudson, 1857

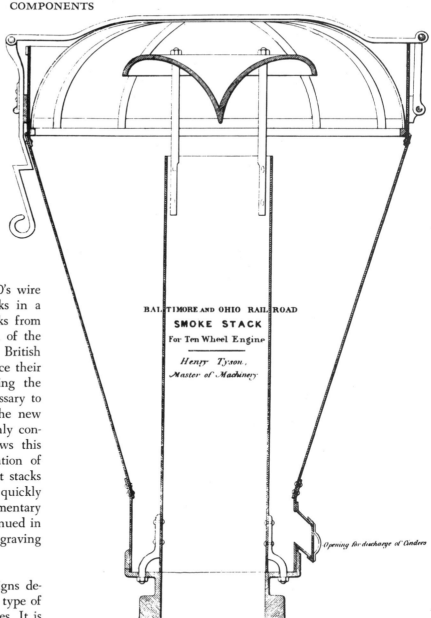

The Baldwin drawing shows that by the early 1830's wire screening had been fitted over the top of smokestacks in a desperate effort to prevent the emission of large sparks from locomotives. This was a simple and obvious alteration of the high, straight stack copied from an early British design. British builders were not concerned with spark prevention since their engines burned coke, but American railways adopting the cheapest native fuel, wood, immediately found it necessary to adapt the straight stack to the volatile character of the new fuel. The *Brother Jonathan* drawing (1832) is the only contemporary drawing the author is aware of that shows this earliest type of spark arrestor (Fig. 6). The application of basket-shaped wire screenings to the prevailing straight stacks of the period was a simple task and was undoubtedly quickly adopted by all wood-burning railroads. While this elementary plan was soon superseded by the bonnet stack, it continued in use into the 1840's as evidenced by the *Virginia* engraving (Fig. 60).

THE BONNET STACK. Of all the numerous designs developed, the bonnet stack was by far the most common type of wood-burning smokestack used on American locomotives. It is

Fig. 41. *Typical bonnet stack for coal- and wood-burning engines; the bonnet screen is not shown.*

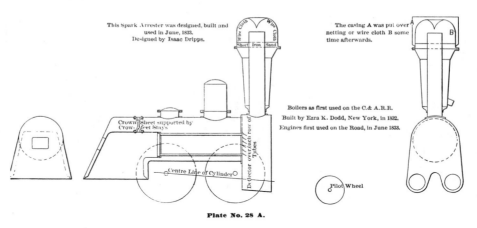

Fig. 42. The earliest known style of bonnet stack is shown in this drawing of an 1833 Camden and Amboy locomotive.

identified by its distinctive funnel shape, but it is named for the bonnet-shaped wire screen or netting arched over its top. Figure 41 shows the construction of a typical bonnet stack. The funnel-shaped outer casing, more than 5 feet in diameter at the top, is a hopper for holding cinders. Note the hand plug at the hopper's bottom; this simple apparatus permits removal of accumulated cinders. The interior straight pipe is the smokestack proper, which directs waste steam and the products of combustion out of the smokebox. The inverted cone fixed atop the stack deflects the smoke downward upon its discharge from the smokebox. The larger sparks and cinders are thus directed into the hopper. The smoke and small sparks swirl around the hopper and eventually discharge out the top through the wire screening of the bonnet. Hopefully, the more dangerous sparks escaping the hopper are trapped or extinguished by the bonnet. A coarse iron-wire screen with openings of less than an eighth of an inch was used. The inverted deflecting cone was the ingenious contribution of the bonnet stack and served not only to trap the larger sparks but also to prevent the wire screen from burning out. Bonnet netting usually burned out within three or four weeks.[79]

Two well-known mechanics, William T. James and Isaac Dripps, are credited with inventing the bonnet stack. James reportedly used such a stack on an engine delivered to the Baltimore and Ohio in 1831,[80] but no drawings or contemporary references to James's stack have been found. It might be observed that this engine was intended for anthracite coal; thus, the necessity of a spark arrestor was unlikely. Dripps is said to have fitted a bonnet stack to the Camden and Amboy's second locomotive in June, 1833. A sketchy outline of this machine, apparently based on an original draft supplied by Dripps, was included in the *Master Mechanics Report* for 1884.[81] The drawing, which shows a stack with all the elements of the bonnet design, is reproduced as Fig. 42 in this volume. Whatever its origin, the bonnet plan was common by 1840 (see Von Gerstner's drawing, Fig. 43).

The bonnet stack that developed in the next few years as a clumsy, top-heavy structure became one of the most distinctive

---

[79] Knight and Latrobe, *Locomotive Engines*, p. 33.

[80] Clark and Colburn, *Recent Practice*, p. 59. The date 1833 is assigned to James's engine in error.

[81] The 1833 Dripps bonnet stack is mentioned in *Locomotive Engineering*, January, 1892. The 1884 drawing, with several views omitted, is reproduced on p. 384 of Sinclair's *Development of the Locomotive Engine*.

COMPONENTS

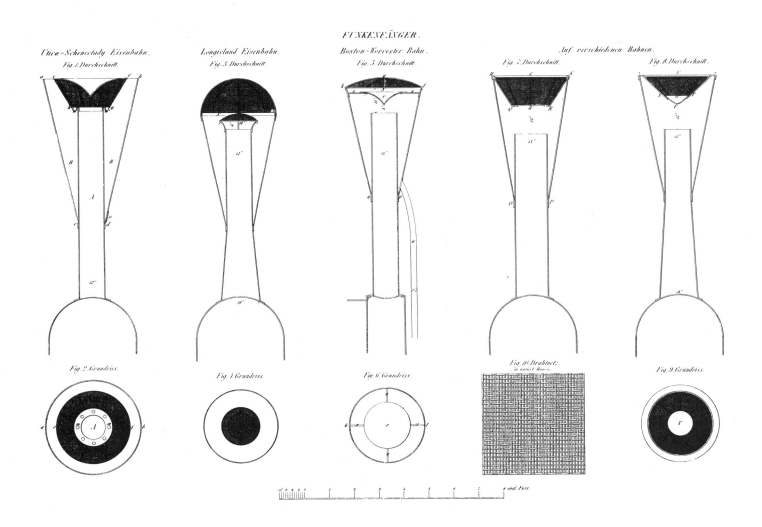

Fig. 43. Wood-burning spark arrestors showing the early development of the bonnet stuck.

features of the nineteenth-century American locomotive. Opinions varied widely on its aesthetic merits, as noted in the following contemporary statement: "Even the ugly locomotive chimney employed in America for wood burning locomotives is admired by some for the 'air of frontal grandeur' and the 'majestic expression' which it imparts to the machine beneath and behind it."[82] Yet designers, such as William Mason, who were concerned with the locomotive's appearance regularly used the ungainly bonnet stack, by necessity if not by choice.

Use of a secondary cinder reservoir, known as the "subtreasury," was developed to reduce the weight of the stack by removing the cinders from the stack's hopper. The subtreasury was usually placed in front of the smokebox. The cinders flowed through a pipe connecting the stack and subtreasury. Leonard Phleger's patent drawing dated September, 10, 1840 (No. 1778), shows a subtreasury in conjunction with the patented feature of his spark arrestors. This is probably the earliest illustration of such a device although it is not possible to credit Phleger with its invention.[83] Through the survival of several contemporary illustrations dating from the late 1840's and early 1850's, subtreasuries are known to have been used by New England locomotive builders, but there is little evidence that they were used later than this period.[84]

The bonnet stacks remained the most popular "sparker" despite the many new inventions offered during the nineteenth century. In 1855 it was stated that "the old bonnet pipe is being used where many of the patented pipes have been tried and abandoned."[85] Fourteen years later the same view was presented in the *Master Mechanics Report*: "The best form of stack for wood burning engines is the 'bonnet' stack. . . . This form of stack gives a better draft . . . than any known to your committee. There are other stacks that more effectively prevent the emission of live sparks, but it is accomplished at the expense of the draft."[86] Shortly after this statement was made, however, coal began to replace wood, and by the mid-1870's the faithful "old bonnet" was rapidly disappearing from the American railroad scene.

In 1856 or 1857 George S. Griggs devised a special form of stack which was essentially the bonnet stack with adaptations for coal-burning locomotives. Griggs's design, more familiarly known as the diamond stack, succeeded the old bonnet and became immensely popular in the United States during the 1870's and 1880's. The deflecting cone, and in some cases the wire netting, was retained, but unlike the bonnet the diamond stack had no cinder hopper. The diamond-shaped outer casing was used only to provide space for the deflecting cone. An early drawing of the diamond stack is shown in the illustration of Griggs's boiler (Fig. 38).

THE CENTRIFUGAL STACK. The centrifugal, baffle-plate stack was the second major class of spark arrestor. Next to the all-popular bonnet, it was the most important type of locomotive smokestack for wood-burning. It was introduced shortly after the bonnet and curiously enough was the last style of wood-burning stack manufactured in this country, some being produced as late as the 1920's.[87] To the unknowing eye the centrifugal stack cannot be distinguished from a bonnet stack externally. Both designs used a funnel-shaped outer casing, but the plan of operation and the internal arrangement were entirely different. In the centrifugal stack, a system of stationary

---

[82] Colburn, *Locomotive Engineering*, p. 2 of the Introduction.

[83] Sinclair, *Development of the Locomotive Engine*, p. 213, claims that William Bullock invented the "sub-treasury"; however, no date or other details are given.

[84] Surviving illustrations of subtreasuries include a line drawing of the Western Railroad's *Massachusetts*, 1839 (based on a daguerreotype made in the 1840's), and lithographs of the *State of Maine*, Portland, 1851; the *Massachusetts*, Hinkley, 1849; the *General Stark*, 1849, and the *Greyhound*, 1850, both by Amoskeag.

[85] *Railroad Advocate*, July 7, 1885, p. 3.

[86] *Master Mechanics Report*, 1869, pp. 34–35.

[87] Bulletin No. 4 of the Lima Locomotive Works, undated but obviously from about 1920, states that the Radley and Hunter stack is the best form available for wood-burning engines. The bulletin's illustrations consist of several "modern" Shay locomotives fitted with Radley and Hunter stacks.

The Rushton stack is mentioned on p. 39 of *History of Baldwin Locomotive Works* as a modern development of the Radley and Hunter stack. The Rushton stack was in production in 1923 and later but was used only by industrial and foreign roads.

baffle plates or volutes converted the upward rush of the exhaust into a whirling, rotary motion. Ideally the heavy sparks would be thrown against the inside of the outer casing, pulverized, and extinguished, and thus would fall harmlessly into the hopper of the outer casing. The smoke and exhaust steam would whirl out of the stack's open top.

The first evidence of centrifugal stacks was possibly the April 17, 1837, patent (No. 161) of James Stimpson, which used a stationary sheet-metal screw to give the exhaust a rotary motion. William T. James followed a year later with a simpler plan for inducing rotary motion. A deflecting cone with curved metal fins (shaped like an impeller) was placed over the top of the stack. A funnel-shaped hopper was used to catch dead sparks. James's patent, though apparently little used, established the basic pattern for all subsequent centrifugal stacks. Other plans for centrifugal spark arrestors were patented but none were widely used until the advent of the popular French and Baird stack in about 1842. This stack, based on two earlier patents, was devised by Matthew Baird of the Baldwin Locomotive Works. Baird's association with one of the nation's most respected locomotive builders afforded him ample opportunity to promote his design in competition with other forms of centrifugal spark arrestors. William C. Grimes's patent of February 12, 1842 (No. 2455), was the basis of the French and Baird design; French's patent of 1841 did not cover any type of centrifugal motion but it did provide a jigsaw-shaped wire netting that permitted a greatly increased surface area of screening. The Grimes, French, and the resulting French and Baird stacks are shown in Fig. 40 as numbers 23, 25, and 27, respectively. The *American Railroad Journal* included a drawing of the French and Baird stack in its June, 1844, issue (p. 178), and commented optimistically: "When in perfect order, . . . in running in the night, there is scarcely ever a spark to be seen." It became a favorite on southern railroads, where inflammable pine forests and cotton fields made spark-arresting an important consideration.[88]

The Radley and Hunter stack followed the French and Baird and eventually replaced the older design as the favored form of centrifugal spark arrestor. James Radley and John W. Hunter secured a patent on January 22, 1850 (No. 7040), for an improved centrifugal stack, the outstanding feature of which was that no wire screening was employed. Wire screening not only burned out quickly but in some cases became clogged, thus blocking the draft. A tangential motion was imparted to the smoke by a series of baffle plates pierced with slots. This arrangement can be seen in the cross-sectional drawing of the *Croton* (Fig. 154). Another cross-sectional drawing of a Radley and Hunter stack (1861) shows modifications and improvements in the original design (see Fig. 44).

The French and Baird and the Radley and Hunter stacks did not long remain competitors, for all the patents involving these sparkers were acquired by Edwin R. Bennett and Company of New York City before 1853.[89] Bennett successfully manufactured and promoted these stacks for many years. By 1861 the maker stated that up to 100 railroads were using the Radley and Hunter stacks.[90] No figures were offered on the actual number in service, but surely the number was comparatively small if figured against that of the bonnet stack. The Radley and Hunter stack, together with the other, better forms of centrifugal stacks, was unquestionably a more efficient spark arrestor than the bonnet stack. The price of this efficiency, however, was increased obstructions to the draft. In addition, more complex construction, greater weight, and increased cost prevented the centrifugal stack from enjoying more widespread popularity.

PERFORATED CONE STACKS. The perforated cone stack was the third major class of spark arrestor. In this style of sparker the deflecting cone and the bonnet screening were combined into one element. William Shultz, a mechanic employed by the Philadelphia and Columbia Railroad, introduced the perforated cone stack in answer to the need for a stack capable of being lowered when passing under the low bridges

---

[88] Clark and Colburn, *Recent Practice*, p. 59.

[89] *American Railroad Journal*, July 2, 1853, Advertisement of Bennett.
[90] *Lowe's Railway Directory*, 1861, pp. 3–4.

Fig. 44. *Radley and Hunter centrifugal spark arrestor patented in 1850.*

prevalent on that line. Shultz, recognizing the impracticability of hinging a common bonnet stack, developed a plan in which the spark-arresting apparatus was placed at the *bottom* rather than at the top of the stack. Thus, the heavy part of the stack remained stationary while the relatively light, straight stack, hinged to the sparker, could be raised or lowered as required. A diamond-shaped casing containing a conical wire screen was fitted directly atop the smokebox.

Schultz patented his idea on March 31, 1836, and within a few years many engines on the Philadelphia and Columbia Railroad were so-equipped. The stack was said to be a more efficient spark arrestor than the bonnet plan, but the wire screening, being closer to the fire, burned out quickly.[91] Nevertheless, several Shultz stacks were in service as late as 1857. Three illustrations of Shultz stacks are presented in this volume, Figs. 11, 40 (drawing No. 10), and 83.

A more conventional design for the perforated cone spark arrestor was offered by Benjamin Briscoe's patent of December 15, 1838 (No. 1037). Briscoe's stack was a simple adaptation of the bonnet stack. Other inventors came forward with varying designs for perforated stacks, several of which are shown in the Baldwin drawing, Fig. 40, but none were widely used according to the best available records. Except for an occasional patent, the idea lay dormant until the 1870's, when the Pennsylvania Railroad adopted a design by James Smith. Smith's design, covered by two patents, August 16, 1870 (No. 106515), and March 7, 1871 (No. 112506), offered no new or particularly imaginative features. A truncated screen extended from the mouth of the blast pipe upward into the smokestack. A straight stack, usually fitted with an ornamental cap, was used. The Smith stack was standard on the Pennsylvania's passenger locomotives for many years. It was also used by several other coal-burning roads. Despite the numerous and ingenious styles offered, the perforated stack was not widely used, for few roads found it to be a more effective spark arrestor than the ordinary bonnet or diamond stacks.

[91] Knight and Latrobe, *Locomotive Engines*, p. 33.

COMPONENTS

Fig. 45. *Modified Radley and Hunter spark arrestor with a section removed to show interior arrangement. Manufactured in about 1920 by Lima Locomotive Works.*

SMOKEBOX SPARK ARRESTORS. The extended smokebox, already mentioned briefly, is the final type of spark arrestor to be discussed. It was the only type not directly associated with the smokestack and was thus a distinctive departure from conventional attempts to solve the spark problem. The merits of this arrangement were widely contested, and a spirited controversy arose where the extended smokebox was "earnestly insisted upon and as savagely opposed."[92]

The Hannibal and St. Joseph Railroad was said to have first used the extended front end in 1858.[93] Rogers claims to have built an engine so-equipped a year later; a drawing of it was reproduced on page 41 of the 1886 Rogers history. The drawing shows a smokebox extension that could not measure much more than 10 inches. However, the smokebox extension was not relied upon for spark-arresting, for the drawing shows that a bonnet stack was used.

The first proponent of the greatly extended smokebox appears to have been John Thompson, master mechanic of the Eastern Railroad. On May 29, 1860, Thompson secured a patent (No. 28520) that covered a plan to arrest sparks simply by increasing the length of the smokebox. The usual wire screens, deflecting cones, and other complex apparatus were dispensed with. Thompson's patent contended that if the smokebox was sufficiently lengthened the sparks would "pass out and beyond the current of smoke, so as to be deposited in the box by the action of gravity and not carried up the chimney." Unfortunately, Thompson's concept was in error. The hurricane-like condition inside a smokebox did not permit many sparks to remain inside without the aid of a screen or other auxiliary equipment.

Jacob Hovey, master mechanic of the Cleveland and Pittsburgh Railroad, was one of many to develop some obvious modifications of Thompson's patent. In a patent dated April 7, 1863 (No. 38111) Hovey suggested the use of a horizontal wire screen running the full length of the smokebox. A horizontal deflecting plate placed at the same level as the screen,

[92] *Master Mechanics Report*, 1883, p. 108.
[93] *Master Mechanics Report*, 1887, p. 17.

but only about one third its length, directed the products of combustion to the front end of the smokebox.[94] According to J. Snowden Bell, Hovey's plan was a dismal failure and was soon abandoned by the Cleveland and Pittsburgh Railroad.[95]

The extended smokebox did not end with the failures of Thompson and Hovey, however. During the 1870's the idea was revived by the New York, New Haven and Hartford Railroad, which apparently reasoned that the earlier smokebox extension of from 16 to 18 inches was not enough; a 7-foot smokebox was called for! New interest was generated, more patents granted, and by the end of the 1870's the extended smokebox *craze* was in full swing. In 1879 the Baltimore and Ohio Railroad made long smokeboxes the standard for their passenger engines; the next year the same measure was taken for all freight engines.[96] An example of one of these long, front-end extensions appears in the Tyson ten wheeler pictured in Fig. 179.

The extended front-end craze reached its height in the 1880's when not only a large portion of new engines were built to this plan but many existing machines were similarly rebuilt. By the early 1890's a reaction had set in, occasioned by the preparation of detailed and scientific tests for front-end designs. Finally, in 1894, extended front ends were condemned by the Master Mechanics Association. It should be noted, however, that smokebox design never returned to the "small" concept; a reasonably long box with deflecting plates and screening remained the accepted plan. The bonnet, diamond, and all other forms of smokestack spark arrestors were abandoned for inside smokebox designs.

Our discussion of smokestack and spark-arrestor development is only a brief and incomplete survey of a complex subject. It is possible to conclude, however, that, despite the concerted efforts of many mechanics, little analytical study of this subject was made during the early part of the nineteenth century and no one design was totally successful in extinguishing the locomotive's sparks without disrupting the draft.

FEED-WATER PUMPS

In 1844 Dr. Lardner stated one of the simple truths of locomotive operation, that "even the worst engineer" would not fail to render the all-important duty of feeding the boiler, even though it involved the use of the meanest, most cantankerous mechanism on the locomotive, the feed-water pump. Complaints about the pump's being an independent, contrary apparatus that always refused to function when most needed were commonly voiced throughout the nineteenth century. Yet the feed pump remained the standard boiler feeding device well into the 1880's, even though injectors had been introduced some twenty years earlier.

A single-action force pump was used for this hard service. Such pumps were well suited for slow speeds but quickly broke down when subjected to the relatively high reciprocating speeds of locomotive service. Other defects included freezing in the winter (despite steam heating lines), valve-sticking caused by hard or dirty water, and the consequential blockage of the boiler's water supply. Aside from these frequent failures, the feed pump was an inconvenient water supplier that operated only when the locomotive was in motion. A heavy train or steep grade could slow the engine's speed to a point where the pumps were not working fast enough to properly furnish the boiler with water. Thus it would be necessary to uncouple the engine and run it up and down the track until the safe water level was restored. Similar inconveniences were experienced at terminals and sidings. Another exasperating situation attributable to pumps was the common abandonment of locomotives

---

[94] Deflecting baffles over the forward end of the tubes were used as early as 1830 by Robert Stephenson (Warren, *A Century of Locomotive Building*, p. 276). Dripps employed a similar arrangement in 1833 as shown by the drawing of the Camden and Amboy's second locomotive (see Fig. 42).

[95] Bell, "Locomotive Front Ends."

[96] *Master Mechanics Report*, 1887, p. 17.

in snowdrifts, not because the engines couldn't eventually work through the snowbanks, but rather because in backing and charging the drifts the pumps weren't worked fast enough to keep the water level up. These defects were common to all styles of feed pumps. Hand pumps were fitted to a few early locomotives to avoid these problems (see the *Brother Jonathan* [Fig. 6] and the *Lancaster* [Fig. 106]), but later engines were not so-equipped.

The crosshead pump was the most common type used on American locomotives. It was the simplest and cheapest arrangement possible. The pump plunger was connected to the crosshead and worked in line with the locomotive's piston rod; the frame, guides, and yoke all provided convenient points of attachment for the pump. A single crosshead pump was used as early as 1828 by Robert Stephenson. By the early 1830's Stephenson had begun using two pumps, one for each crosshead. This became the universal practice on all subsequent locomotives before the injector was adopted. The pump plungers were generally wrought iron, 2 inches in diameter. The stroke was equal to that of a locomotive's cylinders; hence crosshead pumps were often referred to as full-stroke pumps. The pump body was made of cast iron or brass, depending on the purchaser's expense account.

Despite its simplicity, the crosshead pump had several serious defects. The worst fault was the racking of the crosshead; this was attributable to the pump plunger's being attached to one side of the crosshead and thus forcing it out of line. As the racking pushed the guides out of line, the piston rod also worked out of line, thus damaging the packing gland of the cylinders. Since the pump plunger followed the crosshead, the pump's packing gland was also forced out of line. Besides leaking, the reciprocating parts worked with more friction. Norris attempted to avoid this complaint by placing the pumps directly in line with the piston rod. The pump was placed under the cylinder and an arm connected the crosshead to the pump plunger. This arrangement, shown on the *Philadelphia* (Fig. 133), was not widely used, despite its merits. Winans achieved similar results by attaching the plunger rod to the top of the crosshead, in line with the piston rod. The pump was at the engine's rear, however, and an extremely long plunger rod was required.

The high speed of the crosshead pumps, approximately 10 feet per second at 30 miles per hour, forced the water violently through pump valves, connecting pipes, and check valve. All these parts were subjected to hard, pulsating vibrations. The pump did not function well when the valves were worked so forcefully. In addition the pipe joints were loosened.

Centuries earlier, air domes were used on fire engines to cushion the erratic pulsations of piston pumps. Yet it was many years before it occurred to any locomotive mechanics to apply this well-known contrivance to locomotive pumps. The circumstances of the first such application are interesting. In 1845 Walter McQueen applied small air domes to the feed pumps of the old locomotive the *John Bull*, which at the time was being rebuilt as the *Rochester* by the Mohawk and Hudson Railroad.[97] The domes themselves were made from the 4-inch copper tubes of the *De Witt Clinton*, the road's first locomotive, which had been dismantled some years earlier. The pump's performance was noticeably improved by the air domes, and within a very few years they had been adopted by nearly every railroad in the land.

The short-stroke pump was developed to answer the several failings of the crosshead pump. It commonly worked at half the cylinder's stroke but occasionally a one-third-stroke pump was used. Such pumps were driven by return cranks and eccentrics fitted to crank pins or by bell cranks attached to the crosshead. Although short-stroke pumps of various patterns were used in Europe for many years, there is little evidence that they were employed on American locomotives before 1845. In that year Rogers began to use a vertical short-stroke pump actuated by a bell crank. A link connected one arm of the bell crank to the crosshead. While this design produced a slow-moving pump, it was no better than the crosshead pump insofar as racking was concerned, for the power takeoff was from one side of the crosshead. The few surviving drawings of Rogers' engines for this period show pumps of this design; one

---

[97] Clark and Colburn, *Recent Practice*, p. 49.

of these appears in the present volume as Fig. 21.[98] The bell-crank pump was abandoned by Rogers before 1850.

Eccentric-driven pumps were introduced in America during the time of Rogers' bell-crank pump. The Baltimore and Ohio fitted a few of its eight-wheel connected engines with large eccentrics on the connecting rod pin of the rear drivers. This peculiar arrangement was first used in about 1848 but was quickly changed by the use of a return crank in place of the eccentric. The Cumberland Valley Railroad's *Pioneer* (1851) had short-stroke feed pumps driven by eccentrics mounted on the driving-wheel axle.[99] Rogers built a small number of six-wheel freight engines in 1854 and 1855 with axle-mounted, eccentric-driven pumps. Wilson Eddy also used eccentric-driven pumps on the locomotives of the Boston and Albany. The pumps were located between the frames, under the foot plate, and were driven by eccentrics on the rear axle. This arrangement was popular in England at the time but was only rarely used by American mechanics.

Winans used a variable-stroke, eccentric-driven pump in 1855 on his experimental passenger locomotive the *Centipede*. The length of the pump stroke was regulated by a link-motion arrangement similar to that of a Stephenson valve gear. Meritorious as this plan was, it was not repeated.

The return-crank pump was by far the most popular style of short-stroke pump in the United States. Hinkley began to use pumps of this type in about 1847. As previously noted, the crank was attached to the front driver's crank pin with a long pump rod reaching back to the pump, the pump being attached to the rear of the engine's main frame.[100] Other New England builders followed Hinkley's design until the early 1850's when most returned to the exclusive use of crosshead pumps. Smith and Perkins used the Hinkley style return-crank pump as late as 1855, as evidenced by a lithograph of the *Virginia* issued in that year. They were not only the last builders known to retain this style of pump but were also two of the few manufacturers outside of New England to use it. A few years earlier Smith and Perkins used return cranks fitted to the side-rod pin of the *rear* drivers. This arrangement called for a short connecting rod to the pump plunger. It was used by several other makers in the 1850's but enjoyed a much larger employment in the 1860's and 1870's. The *Consolidation* (Fig. 227) is a good example of the later return-crank pump.

All forms of the short-stroke pumps discussed above were superior to the crosshead pump in their smoother working, avoidance of crosshead wear, and placement at the rear of the locomotive where its weight was beneficially added to the drivers. They were also more complex and costly, in certain cases costing as much as $150 more than a crosshead pump of equal capacity.[101] Finally, the air dome successfully cushioned the worst of the crosshead pump's fitful workings, permitting it to remain the leading style of boiler feed pump until all such apparatus was belatedly succeeded by injectors in the 1880's.

The crosshead and half-stroke pumps' speeds were regulated by the motion of the locomotive and not by the requirements of the feed. Thus, if the engine were moving fast, the pump would work rapidly whether a large or small amount of feed water was needed by the boiler. To regulate the quantity of water delivered by the pumps, a valve was fitted to the suction pipe between the tender and the pump. These valves were often placed under the foot plate with a control rod projecting up into the cab. Occasionally the water supply was regulated by a valve on the tender.

Another problem associated with feed pumps was the difficulty in determining if they were feeding water to the boiler. The plunger might be working away but a blocked suction pipe or pump valve would prevent its delivering any water. Therefore, a small try cock was fitted to the discharge side of the pump with a control rod running to the cab. When opened, this simple testing device readily revealed whether or not the pump was delivering water.

---

[98] Other examples of Rogers' engines of the mid-1840's with bell-crank pumps are: a drawing of the Morris and Essex Railroad's *Sussex, Railroad Gazette,* June 13, 1902; and the Erie Railway's *Sullivan,* p. 263 of Sinclair's *Development of the Locomotive Engine.*

[99] See "The Pioneer, a Light Passenger Locomotive," *U.S. National Museum Bulletin,* No. 240, Paper 42, pp. 241–68, for drawings of eccentric-driven feed pumps.

[100] Hinkley's return-crank pump is illustrated by the contemporary engraving of the *Massachusetts* (see Fig. 65).

[101] *Railroad Advocate,* September 1, 1855, p. 1.

Before the advent of the injector, the alternatives to the regular force pumps were incidental to the history of locomotive boiler feeding. Henry R. Worthington, after several years of intense experimentation, developed a thoroughly workable direct-action steam pump in 1849. It was compact, dependable, and relatively cheap. Worthington's pump was rapidly adopted for stationary and steamboat boilers but was viewed skeptically in railroad circles. The *American Engineer*, August 15, 1857, enthusiastically recommended Worthington's pump, pointing out that it worked independent of the locomotive's motion and would supply the boiler when the engine was in motion or stationary. Its cost was not prohibitive, $200, while a pair of ordinary crosshead pumps cost from $75 to $100. Three alternate mounting positions were suggested, as shown by Fig. 46. A second alternative to the ordinary force pump was offered by the Silsby Manufacturing Company, makers of the well-known rotary steam fire engines. This concern regularly employed Holley rotary pumps for feed-water service on their fire engines and offered similar devices for locomotive service in their 1860 catalog. As in the case of the Worthington pump, there was little or no interest in the Holley pump for locomotive feed-water supply.

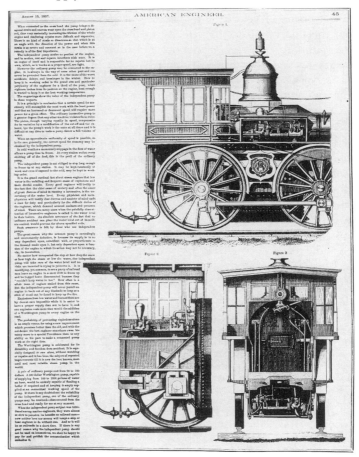

Fig. 46. Worthington feed-water pump, so successful in stationary practice, was frequently suggested but never adopted for locomotives.

INJECTORS

Most locomotive improvements of the nineteenth century were developed by practical mechanics who resolved mechanical questions by empirical rather than scientific methods. These men, if not suspicious of "fancy theories," were in the main unschooled in the abstract principles of engineering and the laws of physics. The invention of the injector, however, was a notable exception to this situation. Henri J. Giffard, a French engineer and a graduate of the Ecole Centrale, based his invention on Venturi's law and on other fundamental theories of pressure, velocity, and acceleration. The injector is based on the conversion of velocity into pressure. A jet of steam drawn from the boiler imparts its velocity to the feed water and by passing through a system of nozzles and tubes achieves a pressure greater than that inside the boiler.[102]

The workings of the injector would appear impossible to an ordinary engineman, and many engineers undoubtedly viewed it as a mechanical paradox or a fanciful scientific novelty when Giffard first projected the idea in 1850. The injector was devised as a lightweight feed-water pump for an airship Giffard proposed to build at that time, but apparently a working injector was not actually produced before May, 1858, when a French patent was granted. Even after a practical demonstration Giffard's invention received little attention. To promote the device, sample injectors were sent to several engineering firms; one of these, Sharp, Stewart and Company of Manchester, England, quickly recognized its value as a boiler feeding device.[103]

The injector possessed several obvious advantages over feed-water pumps; it was compact, it had no moving parts, it worked when the locomotive was stationary, it did not absorb locomotive power, nor did it rack the crosshead or strain any other reciprocating parts. Furthermore, the feed water was preheated and, while no fuel economy could be claimed (for the steam was taken from the boiler), the boiler was saved from the strain of the cold feed water normally supplied by the pumps.

After the sample injector was tested on a stationary boiler at the works of Sharp, Stewart and Company, a locomotive of the St. Helens Railway was fitted with the device in mid-1859.[104] The injector worked so well that all engines on the line were equipped with it, and other English roads began to consider its adoption. By July of 1860 about 25 to 30 injectors were used on English locomotives, but their high cost, $115 each, was a major obstacle to their more general adoption.[105]

The injector's introduction to the United States was occasioned by a visit of William Sellers, the prominent Philadelphia manufacturer, to the works of Sharp, Stewart and Company in 1860.[106] Sellers, immediately recognizing the injector's potential, secured the exclusive manufacturing rights for the American market. A special department for the manufacture and promotion of the injector was established at the Philadelphia plant of William Sellers and Company; in a few years it became a profitable adjunct. The good reputation of Sellers' firm as a manufacturer of machine tools, turn tables, and other products required by railroads provided many established contacts for promoting the new invention. The need for

---

[102] The velocity theory of the injector prevailed in the nineteenth century and is explained in some detail on pp. 221-23 of Forney's *Catechism* (1890 ed.). This theory has been questioned in recent years, however; see *Institution of Locomotive Engineers*, Vol. 40 (1950), p. 664, for the "pressure" theory of injectors.

[103] Sinclair, *Development of the Locomotive Engine*, pp. 578-79, claims that a partner of Sharp, Stewart and Company met Giffard on a sea voyage and urged him to perfect a working injector after hearing of the idea. Presumably this was before the 1858 patent.

[104] T. H. Shields, "A Survey of Locomotive Injector Development," in *Institution of Locomotive Engineers*, Vol. 40, pp. 597-649. This excellent study was freely consulted during the preparation of the present discussion. See also S. L. Kneass, *Practice and Theory of the Injector* (New York, 1894).

[105] *American Railway Review*, December 5, 1861, p. 170.

[106] *Journal of the Franklin Institute*, May, 1905, pp. 372-73.

COMPONENTS

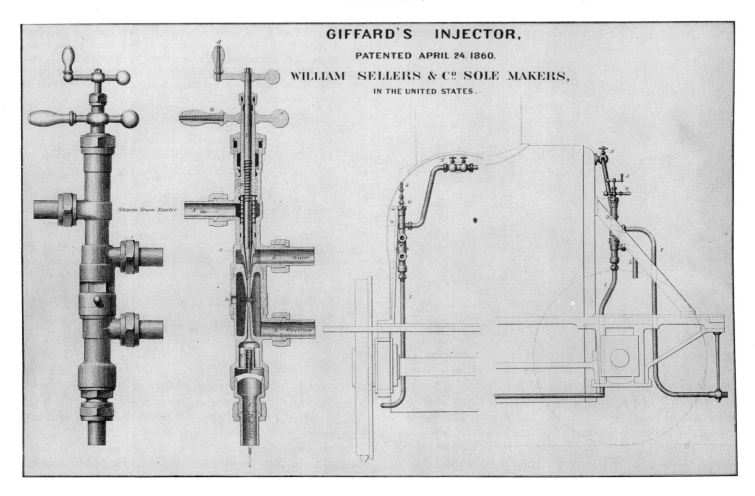

Fig. 47. Injector introduced in the U.S. by William Sellers & Co. in 1860. This drawing shows the earliest vertical type of injector, which was not used after about 1865.

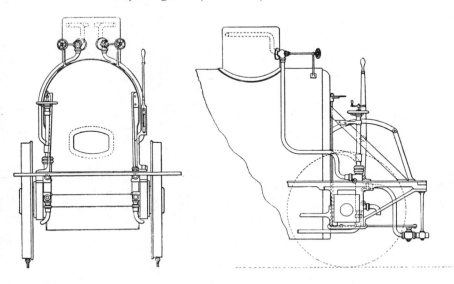

Fig. 48. Injectors, at first enthusiastically accepted for locomotive boiler feeding, soon fell into disfavor.

a boiler-feeder superior to the force pump was evident. The reputation of William Sellers and Company was unquestioned. Therefore, it is not surprising that the injector was enthusiastically received. During the first year of production 2,800 injectors were sold, and while the exact number used for locomotives cannot be determined, a good proportion of the total number was undoubtedly for this use.[107] A circular issued by the manufacturer in 1862 reported the favorable reaction of the country's most respected master mechanics, several of which are given below:[108]

PENNSYLVANIA R. R. Co.

Office of the General Superintendent,

Altoona, Pa., October 14th, 1862.

Messrs. WM. SELLERS & Co., Philada.

Gentlemen:—We have a large number of GIFFARD'S INJECTORS made by you upon our Stationary and Locomotive Engines, and they continue to give us entire satisfaction; upon all the new Engines built for us during the past year, and upon those being built for us at this time we are using them to the entire exclusion of pumps. Wherever pumps require renewal, we use Injectors in place of them; we find them less liable to derangement than pumps and at least equally efficient.

Yours, Respectfully, ENOCH LEWIS, Gen'l Sup't.

PHILADELPHIA AND READING RAIL ROAD CO.

Reading, Pa. Feb. 1st, 1861.

Messrs. WM. SELLERS & Co.

Gentlemen:—I have your letter of the 30th of January, asking for information in relation to the operation of the GIFFARD'S INJECTORS you have furnished this company. In answer to which I am pleased to say they give entire satisfaction, both for Locomotive and stationary purposes. The last new Locomotive, the "Fawn" placed on our Road has no pumps, the water being supplied to the boiler with the Injector only, and is entirely reliable. I intend using the Injector on all Engines we shall rebuild hereafter.

Yours truly,

JAMES MILLHOLLAND.

Illustrations of the earliest Sellers injectors are shown in Figs. 47 and 48. Another early injector is shown in the backhead view of the Pennsylvania Railroad's No. 214 (1861) at the right-hand side of the cab (Fig. 49).

Despite its careful promotion, initial success, and the desire to abandon the feed pump, the injector quickly fell from favor. It proved to be a delicate, sensitive device that required skillful handling and was likely to develop serious disorders when placed in rough locomotive service. The engine crews were chief opponents to the injector and because of their persistent complaints (and in some cases their refusal to operate engines with injectors) many master mechanics abandoned the injector

[107] *American Railway Review,* December 5, 1861, p. 170.
[108] The Sellers circular dated December 1, 1862, is in the data files of the U.S. National Museum, Mechanical and Civil Engineering Division.

COMPONENTS

Fig. 49. *An early injector is visible on the right side of this backhead view of the Pennsylvania Railroad's No. 214, built by Baldwin in 1861.*

as a splendid failure.[109] Recognizing the danger of a boiler explosion, enginemen were likely to panic when the injector failed to deliver water. Its failures increased the longer it was in service. The constant vibration of the locomotive loosened the connecting pipes, thus breaking the vacuum necessary to the injector's working. Leaky valves, caused by prolonged use, let in steam which also adversely affected the vacuum. Bad water, fouling the interior of the injector, was yet another cause of road failures. A final defect was the design of the earliest Sellers injectors; they required constant adjustment as the steam pressure varied, and if this was not quickly or skillfully handled, no water was delivered to the boiler.

The rejection of the injector did not mean absolute abandonment, but few, if any, American locomotives were fitted exclusively with injectors after about 1863. On many roads injectors were abandoned entirely, but the more usual course was to use one injector as an auxiliary boiler-feeder; this proved particularly useful when the engine was standing in a station or at a passing siding. It was common to have two pumps and one injector, but in certain cases the ratio was one to one.

During the next decades improved forms of injectors were developed, not only by Sellers, but by competing manufacturers as well. First, more compact and reliable designs were perfected; the self-acting injector was introduced in 1865, and exhaust steam injectors received some attention during this time. Second, locomotive crews gradually overcame their suspicions of the "French squirt" and came to accept it as a familiar boiler auxiliary. By 1875 the *Master Mechanics Report* stated that "injectors for feeding boilers are again being used." The same group remarked at their next meeting that of 1,361 locomotives surveyed only 22 used injectors exclusively, 508 used pumps and injectors, and the remaining 769 used pumps only. Assuming this to be a representative sampling of American locomotives, it would appear that the injector made slow progress in replacing the pump. Builders' photographs and other available evidence indicate that this trend continued until the mid-1880's. After that time, pumps were rapidly displaced, and by the early 1890's the injector had become the standard water-feeding device.

GAUGE GLASSES AND COCKS

Knowledge of the water level was a fundamental safety requirement in boiler operations because the most common cause of explosions was the lack of enough water in the boiler to cover the crown sheet. Yet American locomotives were equipped with only the most primitive water-gauging devices throughout most of the nineteenth century. Three gauge or try cocks fastened to the back head was the standard arrangement throughout this period. Some roads used as many as seven cocks, to some unknown advantage. The usual arrangement was to place three brass cocks several inches apart at the

[109] *Railroad Gazette*, December 10, 1870, p. 250; the failure of the injector was summarized editorially; chief blame for its rejection was placed on engine crews. The 1887 *Master Mechanics Report*, p. 23, noted that "when the injector first came in . . . they [engineers] did not want to use them."

water line. The top cock would properly show steam only; if it passed water, the boiler was too full and there was danger of priming. The second cock, set exactly at the water line, passed steam and water if the boiler was properly filled. The bottom cock was placed at the lowest point to which the water could fall and still cover the crown sheet. If this cock gave out steam there was imminent danger of burning out the crown sheet or of an explosion. By the early 1850's a drip pan with a drain pipe was secured under the cocks to carry the waste water out of the cab.

A glass-tube water gauge was specified for an American locomotive as early as 1831, as recorded by the contract for the *Experiment* of the Mohawk and Hudson Railroad (see Appendix B). This pioneering application had no influence on American practice, however, even though gauge glasses were commonly used on British locomotives before 1850. American mechanics felt that gauge glasses were an extravagant frill, expensive to install and a nuisance to maintain. In addition, bad boiler water caused foaming, which resulted in a false reading of the water's actual level. Most American locomotives were accordingly fitted with gauge cocks only. Criticism of this dangerous and foolhardy policy was voiced as early as 1847 in connection with the explosion of a Reading locomotive, the *Neversink*. It was stated that "the gauge-cocks, . . . under the most favorable circumstances, are but indifferent indicators of the water level."[110] Glass-tube water gauges were strongly advocated as dependable, convenient, and accurate indicators for measuring the water level. In 1851 Colburn urged the adoption of gauge glasses to counter the rising number of horrible boiler explosions.[111] He reasoned that a properly designed and carefully made glass gauge, such as the type used in England, would silence the criticisms of this device. A thick, well-annealed glass tube, no more than three-sixteenths of an inch in diameter inside, with a proper expansion joint, would make a reliable and practical water glass. Despite this good advice the resistance to gauge glasses continued until the end of the century. In 1860 they were rarely used.[112] Fifteen years later they were only coming into more general use.[113] As late as 1893 glass gauges were considered only a "convenience" and not absolutely necessary for safe boiler operation.[114]

STEAM GAUGES

High boiler pressure was used at an early date on American locomotives in an effort to improve performance without increasing the size or complexity of the machine. The "smart" operation of our early locomotives was credited to high boiler pressures. The Norris engine's remarkable performance on the Birmingham and Gloucester Railway in 1839 was credited to the 100 pounds of steam it carried; only 62 pounds had been reported during its trial run. The first locomotives in this country were built to withstand a working pressure of 50 pounds per square inch. This was considered the safe practical limit for the boilers of that time. United States builders quickly found that higher pressures were possible, however, and by the late 1830's 90- to 100-pound pressures were common. After this initial leap, pressures increased gradually, a few pounds at a time during succeeding decades. Before 1850 few engines carried more than 100 pounds. By 1860 a pressure of 110 pounds was common, with a few roads using 140 pounds.[115] Six years later 120 to 140 pounds was noted as standard while pressures of 180 pounds were only occasionally reported.[116]

Even so, 120 pounds was considered the "safe" pressure as late as 1875 according to the *Master Mechanics Report* of that year. Pressures increased slowly during the years that followed. The *Master Mechanics Report* of 1887 noted that 140–150 pounds was then common, with a few roads employing 160- and 175-pound pressures.

Even 50 pounds per square inch was classed as high pressure in the 1830's, and efforts were made to devise a practical instrument for registering this fearful build-up. The Liverpool and Manchester Railway required that all locomotives entering the Rainhill Trials be equipped with "a mercurial gauge" to show pressures above 45 pounds. The gauge consisted of a long thin column attached to, and running the full height of, the smokestack. The *Rocket* had such a gauge, and, while it was a delicate and expensive device, the Liverpool Line adopted the gauge for its locomotives. According to De Pambour the Liverpool and Manchester was the only railroad to use the pressure gauge. The mercury column gauge was the only

---

[110] *Journal of the Franklin Institute*, July 24, 1847, p. 87.
[111] Colburn, *The Locomotive Engine*, p. 52.
[112] Clark and Colburn, *Recent Practice*, p. 68.
[113] *Master Mechanics Report*, 1875, p. 20.
[114] *Master Mechanics Report*, 1893, p. 161.
[115] Holley, *American and European Railway Practice*, p. 16.
[116] *Engineering*, June 15, 1866, p. 391.

Fig. 50. Back-head view of a Pennsylvania Railroad engine built in about 1850, possibly the Wyoming. *The scene dates from about 1870. Note the steam gauge with coil siphon.*

type available and apparently no other road would stand the expense for so unsatisfactory a device. The Baltimore and Ohio stipulated mercury steam gauges in their 1831 locomotive test. The test and requirements were obviously copied from the Rainhill Trials. Mercury gauges were not used on the Baltimore and Ohio after the trials. Sensing the need for a cheap and more convenient steam gauge, De Pambour devised a manometer for that purpose in about 1836. Although still a mercury gauge, the manometer was more compact than the old column-type gauge. De Pambour's gauge was not widely adopted and undoubtedly received only a few test applications. Interest in mercury steam gauges for locomotive use continued into the 1860's when an American firm, Shaw and Justice of Philadelphia, introduced a small mercury gauge actuated by a rubber diaphragm.[117] As with earlier gauges of this type, there is little evidence that the Shaw and Justice gauge was much used.

The simple spring balance, used to hold the safety valve down, was the most common form of pressure indicator in use before 1850. This elementary device, similar to a butcher's scale, was at best an imperfect gauge and only roughly showed the steam pressure. Spring-balance gauges are shown in the *Lancaster* (Fig. 106), Dunham (Fig. 114), and *Juno* (Fig. 129) drawings.

It was obviously desirable for the engineer to be informed of dangerous pressures in the event the safety valve should stick. But the search for a practical steam gauge was based on more than the desire for safety. With only a crude spring balance to show pressure, a fireman would continue to stoke the engine until the safety valves lifted. The boiler was deprived of a good head of steam, and a sizable amount of fuel was lost. De Pambour estimated that 25 per cent of all steam generated was lost when safety valves blew off. More mischief was caused when the fire door was opened in a effort to cool the boiler and thus to silence the safety valves. Again, more heat was lost, but in addition the cold air made the tubes contract and become loose. These problems prevailed, all for the want of an accurate steam gauge.

In 1849, Eugene Bourdon (1808–1884) of Paris answered this need by perfecting a practical steam gauge. A bent metal tube, usually C- or U-shaped, reacted to the pressure of the steam. Its movement was registered on a dial face. The Bourdon gauge was accurate, compact, and relatively inexpensive. It was introduced in America by E. H. Ashcroft, the American manufacturer who purchased the United States licenses to the invention while attending the 1851 Great Exhibition in London.[118] Ashcroft energetically promoted the gauge during the

---

[117] A circular for the Shaw and Justice gauge is in the Baldwin Letters, January, 1864.
[118] *Railroad Advocate,* August 1, 1857, p. 30.

Fig. 51.   Back-head view of a Grant Locomotive Works eight wheeler of 1871.

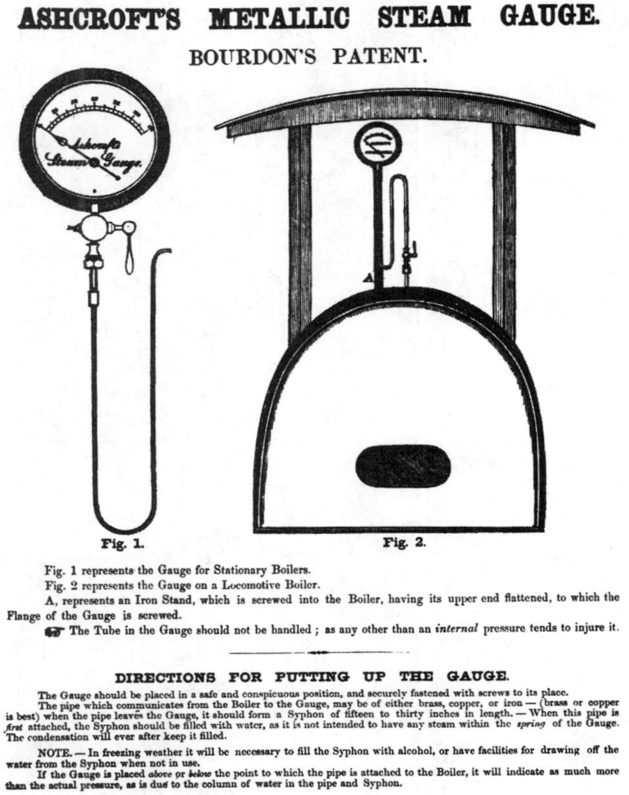

ASHCROFT'S METALLIC STEAM GAUGE.

BOURDON'S PATENT.

Fig. 1. Fig. 2.

Fig. 1 represents the Gauge for Stationary Boilers.

Fig. 2 represents the Gauge on a Locomotive Boiler.

A, represents an Iron Stand, which is screwed into the Boiler, having its upper end flattened, to which the Flange of the Gauge is screwed.

☞ The Tube in the Gauge should not be handled ; as any other than an *internal* pressure tends to injure it.

DIRECTIONS FOR PUTTING UP THE GAUGE.

The Gauge should be placed in a safe and conspicuous position, and securely fastened with screws to its place.

The pipe which communicates from the Boiler to the Gauge, may be of either brass, copper, or iron — (brass or copper is best) when the pipe leaves the Gauge, it should form a Syphon of fifteen to thirty inches in length. — When this pipe is *first* attached, the Syphon should be filled with water, as it is not intended to have any steam within the *spring* of the Gauge. The condensation will ever after keep it filled.

NOTE. — In freezing weather it will be necessary to fill the Syphon with alcohol, or have facilities for drawing off the water from the Syphon when not in use.

If the Gauge is placed *above* or *below* the point to which the pipe is attached to the Boiler, it will indicate as much more than the actual pressure, as is due to the column of water in the pipe and Syphon.

Fig. 52. Bourdon steam gauge advertising circular issued by E. H. Ashcroft in 1852.

early 1850's by sending circulars to various railroads and locomotive builders; one of these is reproduced as Fig. 52. While sound in principle, some difficulty was experienced in the use of the earliest Ashcroft gauges. The *Sam Guthrie,* built in 1853 by Mason, had an Ashcroft gauge that was troubled by the Bourdon tube's leaking or springing out of shape. Reuben Wells, in recalling this incident many years later, stated that many of the early Ashcroft gauges suffered from this difficulty. Experience and more careful manufacture corrected the defect.

Unlike so many improvements, the gauge was readily accepted by the usually wary and suspicious railroad mechanics. Its adoption in standard practice is astonishing. The 1854 annual report of the Baltimore and Ohio stated: "Steam gauges have been placed on the greater number of the locomotives, in order to economise the consumption of fuel." Three years later Bourdon gauges were regarded as a regular part of the locomotive apparatus.[119] Good performance and low cost, about $30 per gauge, were the chief factors in the rapid acceptance of the Bourdon gauge. The 1875 *Master Mechanics Report* stated that "pressure gauges are universally used."

Not long after the Bourdon gauge was introduced in this country, attempts were made to avoid the patent by perfecting alternate designs. Various metal diaphragm gauges were devised. A primitive gauge of this type was tested on Philadelphia and Reading engines from about 1852 to 1855.[120] Reportedly, the *Minnesota,* a Winans' Camel, was fitted with the gauge. A large metal diaphragm was attached between the frames with a steam connection running to the boiler's belly. The movement of the diaphragm was transmitted by a lever to a vertical rod that passed upward into the cab. The upper end of the rod was marked with a scale; a pointer indicated when the pressure on the rod moved up or down. The success of this primitive device is questionable but it may have provided John E. Wootten, a mechanical officer of the Reading, with an idea for a steam gauge he later patented.

Wootten's gauge used two flexible metal diaphragms with a screw mechanism to transfer the motion to a dial indicator. It was compact enough to fit into a case not much larger than a Bourdon gauge. A patent was issued to Wootten on November 17, 1857 (No. 18655). The Reading used Wootten's gauge in preference to other designs and by 1860 had largely equipped its engines with gauges of this type.[121] Examples of early steam-gauge installations can be seen in the several boiler back-head illustrations in this volume (Figs. 49–51).

[119] *Ibid.*

[120] *Railroad Gazette,* July 26, 1907. C. H. Caruthers, in an article on early Philadelphia and Reading locomotives, discusses the diaphragm gauge but gives no source.

[121] *Engineer* (Phila.), October 4, 1860, p. 63.

FEED-WATER HEATERS

It was obvious to the earliest steam engineers that substantial fuel economies could be realized by using waste heat to preheat the feed water. An impressive number of experiments were conducted with such apparatus during the nineteenth century. However, the cost and complexity of the necessary apparatus discouraged large-scale use of feed-water heaters until the 1920's.

A feed-water heater was patented as early as 1804 by Trevithick and Vivian. Other patents followed, but Hackworth's *Royal George,* built in 1827, was probably the first locomotive equipped with a feed-water heater. The *Stourbridge Lion,* the first steam locomotive in North America, had a primitive feed-water heater. A box-shaped heater was mounted under the boiler and received exhaust steam from the cylinders. Whether it was an injection type or a surface type is uncertain, but it appears to have been patterned on Gurney's 1827 surface heater.

Feed-water heaters received no more attention in this country until about 1836 when Winans introduced a small drum-shaped heater on several Grasshopper engines. Winans' heater was a surface heater that worked on exhaust steam. It was patterned after the common fire-tube boiler with steam passing through the tubes. The heater was included in Winans' omnibus patent of July 29, 1837, which covered several features of his "Crab" locomotive. Apparently its use was restricted to locomotives of the patentee's manufacture.

The Baltimore and Ohio was sufficiently impressed with Winans' feed heater to apply such devices to other locomotives and to seek new designs. In 1847 the road received four eight-wheel engines fitted with feed-water heaters from Baldwin.[122] The exact particulars of these heaters are not known but possibly they were based on the plan of Thatcher Perkins, the road's master mechanic. At any rate Perkins received a patent for a feed-water heater on June 26, 1849, which is known to have been used on several Baltimore and Ohio engines. Perkins' heater consisted of two water casings, one in the smokebox, the second running around the outside of the boiler near the firebox. Exhaust steam was directed into these casings; a long pipe under the boiler carried the exhaust to the rear casing. A control valve in the smokebox varied the amount of the exhaust steam to be directed through the blast pipe or the heater. It was altogether a very complex system. After leaving the Baltimore and Ohio in 1851 and forming a locomotive works in Alexandria, Virginia, Perkins applied heaters to several freight engines that he built for the Pennsylvania Railroad.

James Millholland, another early advocate of feed-water heaters, began use of such devices on some of the Reading's locomotives during the 1850's. The heater was a long water pipe, about 6 inches in diameter, with a smaller steampipe running through it. Exhaust steam supplied the heat. Millholland placed the heater on the right-hand side of the boiler just under the running board. An early lithograph of the *Juniata,* built by Millholland in 1855, shows a heater of this type. While not all Reading locomotives were so-equipped, this style of heater was employed by the Reading for more than twenty-five years. The famous 408, the first Wootten firebox locomotive (1877), had a Millholland heater that was more sophisticated than the original design, having thirty-seven $5/8$ of an inch copper steam tubes. The heater was 10 feet 4 inches long, about 8 inches in diameter, and brought the feed water to a temperature of 110°F.[123] The Reading applied this style of heater to a limited number of new locomotives as late as 1881.[124]

In 1859 William S. Hudson, the gifted superintendent of the Rogers Locomotive Works, designed an exhaust steam, surface type feed-water heater that was in essence an enlargement of Millholland's design. The heater was a drum about 14 inches in diameter by 6 feet long, placed under the forward

---

[122] *History of the Baldwin Locomotive Works,* p. 44.

[123] *Engineering,* January 24, 1879, p. 68.

[124] *Master Mechanics Report,* 1918, J. S. Bell's paper on feed-water heaters. This valuable discussion traces the history of feed-water heaters.

end of the boiler. It contained a large number of small-diameter tubes. This low mounting and its proximity to the exhaust pipes, feed pumps, and check valves made it the best-planned heater yet developed, according to the *American Railway Review* for October 4, 1860. Hudson's heater was applied to a number of engines built for the Southern Railroad of Chile and possibly other locomotives built by Rogers. See Fig. 53 for a drawing of Hudson's heater.

A year after Hudson installed his first heater Samuel F. Allen began experiments with a nearly identical unit on locomotives of the Chicago and Rock Island Railroad. Allen's heater differed from Hudson's only in that it was shorter and was placed transversely rather than parallel to the boiler. A circular dated September 15, 1865, describes the heater as a cylinder 16 inches in diameter which contained eighty-four tubes 1¼ inches wide by 56 inches long.[125] The water rather than the steam passed through the tubes. A careful account of the engines with Allen heaters was kept by the Rock Island line between 1861 and 1863. An impossible fuel economy of from 28 per cent to 40 per cent was claimed. Allen received a patent for his heater on February 16, 1864 (No. 45191). Soon thereafter the Chicago Feed Water Heater Company was formed to promote the invention. There is no evidence that the venture was a success or that many roads adopted Allen's heater.

So far, only drum-type exhaust steam heaters have been discussed. The smokestack heater was another style of surface heater that received attention in the nineteenth century. This heater used the heat of exhaust steam as well as the hot gases passing out the smokestack, thus doubling the temperature of such heaters. While it used the engine's waste heat more effectively, the smokestack heater interfered with draft. Its high mounting made the engine top-heavy and added more undesirable weight to the already overloaded front end. Z. H. Mann and L. B. Tyng were apparently the first American engineers to propose a heater of this type. Their idea was set forth in United States Patent No. 628, March 10, 1838. A water jacket surrounded the smokestack; the feed water was delivered to the jacket by the feed pump. The stack was placed centrally on top of the boiler instead of in its usual position on the smokebox. A return flue connected the stack and the smokebox. According to available records, this peculiar arrangement was not tested.

The smokestack feed-water heater was revived in 1852 by David Clark and M. W. Baldwin. A drum heater with a large central flue and surrounded by a number of small tubes formed the base of the stack. The exhaust steam was directed through the large flue by the blast pipe. The waste gases and heat from the fire were drawn through the surrounding tubes by the draft. Patent No. 9312 was obtained by the inventors on October 12, 1852. A second patent (No. 10514, February 14, 1854) described a typical heater of this design as having a central flue 7 inches in diameter with thirty-two small tubes of a 1⅞-inch diameter. The central flue and the tubes were stated to be about 24 inches long. Clark, as chief engineer of the Mine Hill and Schuylkill Haven Railroad, applied heaters of this design to several locomotives on that road. Baldwin interested a few other roads in this heater, but its use was limited to a handful of locomotives.

Two years after Clark and Baldwin's heater was patented, R. A. Wilder, also employed by the Mine Hill Railroad, secured a patent for a similar smokestack heater. Wilder used a water jacket in place of the fire-tube arrangement of the Clark and Baldwin heater. Later he improved the heater by replacing the jacket with a coiled tube. Several engines of the Mine Hill line were fitted with Wilder's heater; the first installation was made in about 1855. The Mine Hill's No. 30, shown in Fig. 55, was equipped with a Wilder heater. Aside from this contrivance, the engine had several other novel devices, including a telescoping smokestack, a center-crown boiler, an iron cab, and Baldwin's flexible-beam truck. The feedwater heater is thoroughly described by a contemporary account that is too detailed to repeat here.[126]

---

[125] The Allen feed-water circular is in the Baldwin Letters, Historical Society of Pennsylvania.

[126] *Engineer* (Phila.), October 18, 1860, p. 77.

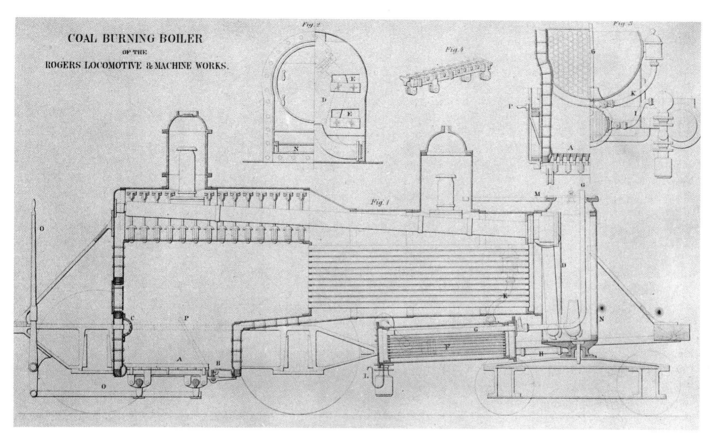

Fig. 53. *Hudson's exhaust-steam feed-water heater of 1859.*

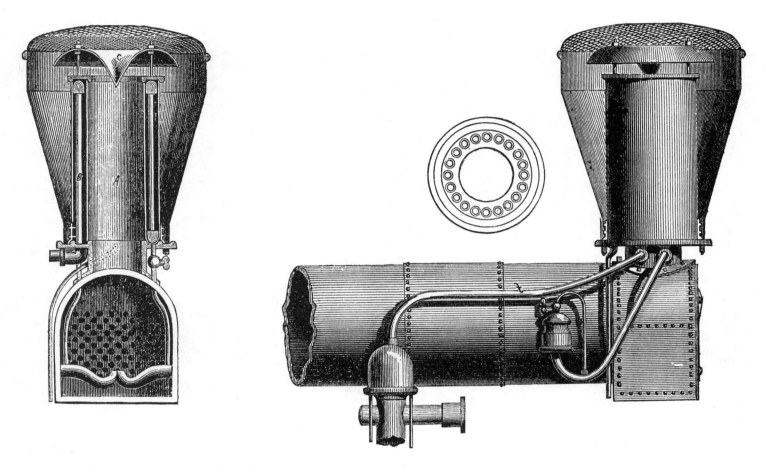

Fig. 54. *Ebbert's exhaust-steam smokestack feed-water heater used by the Chicago and North Western Railway in the mid-1850's.*

Fig. 55. *The Mine Hill Railroad's No. 30 was fitted with a Wilder smokestack feed-water heater. This flexible-beam locomotive, built by Baldwin in 1860, had several other novel features: a metal cab, paper boiler lagging, and brakes on the two rear driving wheels.*

Another smokestack heater was devised in 1855 by Peter S. Ebbert, foreman of the Galena and Chicago Union Railroad repair shops. In Ebbert's heater the feed water was held in a drum with a large central flue that conveyed the exhaust steam and waste gases into the stack. A system of steampipes inside the drum further heated the feed water either by using a portion of the exhaust steam or, if desired, fresh steam from the boiler.[127]

The *Falcon* was the first locomotive to carry Ebbert's heater and was fitted in February, 1855.[128] A year later the Chicago and Galena Union had twenty-six of their locomotives so-fitted and by July, 1857, nearly forty of the road's engines carried Ebbert's heater. The Chicago and Rock Island, the Chicago, Burlington and Quincy, the Illinois Central, and the Hudson River railroads were reported to have tested Ebbert's invention. The heater contained sixteen 2½-inch-diameter iron tubes, held 14 to 16 gallons of water, weighed 500 pounds, and raised the feed water to a temperature of from 130° to 140°F. A fuel saving of 25 per cent was claimed. Ebbert secured two patents on his design: first, No. 14187, February 5, 1856; second, No. 17208, May 5, 1857. The heater worked well for the Chicago road and was viewed with interest by several other prominent railroads. Holley candidly observed, however, "that feedheaters have been most successfully employed on lines managed by their inventors."[129]

Other patents and experiments with feed-water heaters were made during this period. The more important of these devices

[127] Holley, *American and European Railway Practice*. Plate 60 of this work shows a good drawing of Ebbert's heater.
[128] The *Railroad Advocate* supplied several articles on Ebbert's heater, March 15, and May 31, 1856, and July 18, 1857. Sinclair, *Development of the Locomotive Engine*, p. 365, discusses Ebbert's heater but claims that it was a failure and that it was tested only on the *Pioneer*. This claim appears to be entirely in error when compared to the accounts of Holley and the *Railroad Advocate*.

[129] Holley, *American and European Railway Practice*, p. 130.

have been mentioned here, and considering their lack of practical success it is pointless to give other examples. The feed-water heater attracted much attention because it promised a simple, direct answer for fuel economy. A good heater could in theory effect a 15 per cent saving in fuel, but this was rarely realized in actual service.[130] The problem was to design a simple, cheap, and dependable heater, but no one seemed able to produce a design that was acceptable to nineteenth-century railroad mechanics. The general feeling was that fuel economies were incidental and were readily offset by the first cost, patent fees, and maintenance expenses. In fact, the longer the heater was used, the less efficient it became. Impure water soon fouled the interior parts, thereby insulating and dropping its efficiency to a low level. Another complaint common to all surface-type heaters was the severe load placed on the pump. It was required to overcome not only the boiler's pressure but also the resistance of the heater and the additional piping.

No evidence is at hand to show much interest in injection-type feed-water heaters during the last century. The injector was the closest the mechanics of the period came to injection-type feed-heating. In this case, water-heating was only a side effect of the injector's primary function as a boiler feeding device whereby live steam comes into direct contact with the feed water.

SUPERHEATERS

The advantages of superheated steam over saturated steam were understood as early as the eighteenth century.[131] Superheating or "reheating" steam reduced fuel and water consumption and prevented priming by raising the steam's temperature. A second advantage of superheated steam was that it boosted the engine's horsepower because it had more thermal energy than saturated steam.

Several patents were issued and numerous schemes were tested in an effort to develop a practical superheater before locomotives were much more than a novelty seen only in northern England. One of these pioneer efforts was made by Richard Trevithick. In 1828 he successfully tested a fire-tube superheater but he had long since lost interest in locomotives and the invention was used only for stationary work.

Jacob Perkins, the gifted American engineer who did much to advance theoretical steam engineering studies, was probably the first to propose the use of superheated steam for locomotives. Perkins obtained an English patent in 1836 which described a curious single-cylinder locomotive with a boiler that used superheated steam in sealed tubes for heating the water of the boiler. An American patent for the boiler was issued two years later. It should be noted that Perkins proposed only *indirect* use of superheated steam for locomotive use.[132] Two years later R. and W. Hawthorn, locomotive builders of Newcastle, England, obtained a patent for a smokebox superheater. While resulting in no widespread adoption of superheating, the Hawthorn design did introduce the smokebox style of heater which was considered the most promising design in the nineteenth century. Numerous smokebox heaters were tried by British and American engineers, the more important of which will be discussed presently. However, consideration should first be given to the advantages of the smokebox superheater. It could be added to conventional locomotives with only minor alterations to the existing machinery. Of more importance,

---

[130] *Railroad Gazette,* September 11, 1875, p. 376, disputes the exaggerated claims for feed-water heaters and states that 15 percent is the maximum possible saving and that in practice actual savings are much less than this figure.

[131] Much of the information on the history of the superheater presented here is taken from C. D. Young's "Locomotive Superheaters," *Journal of the Franklin Institute,* July, 1914, pp. 1–34.

[132] Some authorities have suggested improperly that Perkins proposed the direct use of superheated steam for locomotives at this early date.

heaters of this type effected only moderate increases in the steam's temperatures and rarely more than a total steam heat of 400°F. This kept steam temperatures within the limits of the animal-fat lubricating oils and cylinder packings available at the time. Proper lubrication was the major obstacle to the general acceptance of superheating. Without high-temperature oils, cylinders quickly scoured and wore out. However, smokebox heaters, while adding only 30°–50°F. to the steam, did offer a compromise between the ideal and practical possibilities of superheating.

To return to a review of early attempts at locomotive superheating, we find the Belgian locomotive firm of John Cockerill applying six superheaters in 1848. Cockerill's heater extended into the smokestack to make better use of the exhaust heat. While European engineers had been busy with superheater developments since the 1830's, the first American attempt was apparently not made for another twenty years. The earlier record of an American locomotive superheater appears to be a smokebox heater designed in 1853 by Uhry and Luttgens, both associated with various Paterson builders and both better known for their patented valve motion which is discussed later in this work.[133] The heater was a shallow cast-iron box, placed high in the smokebox, horizontally, and extending the full width and length of that space, with about seventy fire tubes passing through it. The throttle valve was incorporated within the superheater at its rear.

The purpose of the inventors, as with many other early superheater designers, was to avoid the condensation that might occur during the steam's passage to the cylinders. In outside-connected engines the cylinders were widely separated from the boiler, and relatively low steam pressures and temperatures were employed; thus, condensation and heat loss were a constant problem. The sizable power loss resulting from this situation could be prevented by a moderate reheating of the steam before it left the smokebox.

It is not certain that Uhry and Luttgens' invention was tested. There is no record of a patent being issued, although one was sought. However, other American engineers began to experiment with superheaters at this time. James Millholland applied a number of simple smokebox heaters to various locomotives of the Philadelphia and Reading Railroad. Eight U-shaped iron tubes about the same size as boiler tubes were placed in the smokebox. The heater had only about 30 or 40 square feet of heating surface. One of Millholland's heaters is shown in the *Hiawatha* drawing, (Fig. 39). No definite date can be established for Millholland's heater, nor was a patent granted, but in 1861 A. L. Holley stated that the heater had been used by Millholland "for some time."[134] In the early years of the twentieth century the Baldwin Locomotive Works promoted a smokebox heater on Millholland's plan, but much enlarged. Known as the Vauclain superheater, it was credited with fuel economies of 10–15 per cent. The Vauclain heater could not compete with the highly efficient Schmidt heater and thus represented one of the final attempts to popularize the smokebox superheater.

A. F. Smith, while master mechanic of the Hudson River Railroad, perfected a small but effective smokebox heater in the late 1850's. The heater was a rectangular cast-iron box, 3 inches thick by 27 inches wide, with flue openings passing through. These holes were in line with the boiler tubes so as not to interfere with the draft. Despite the meager 6–8 square feet of heating surface, this heater was said to perform well.[135]

More elaborate than Millholland's or Smith's designs was the smokebox superheater of James Martin, master mechanic of the Grand Trunk Railroad of Canada. Two vertical drums (one for each cylinder) with fire tubes were placed in the smokebox at a slight angle but in a nearly vertical position. The bottom of the drums faced the lower tubes of the boiler; the upper end was connected to the smokestack by a cast-iron hood. Martin's heater offered a much larger heating surface than earlier plans

---

[133] Uhry and Luttgens' superheater is described and illustrated in the *Scientific American*, March 26, 1853, p. 220; see also the *American Railroad Journal*, July 2, 1853, p. 424.

[134] Holley, *American and European Railway Practice*, p. 140.

[135] *Engineer* (Phila.), August 30, 1860, p. 23. Smith's superheater is shown in Plate 58 of Holley's *American and European Railway Practice*.

for smokebox heaters; however, claimed fuel economies of 20–30 per cent were undoubtedly an exaggeration.[136] The inventor claimed that the apparatus cost only $75 per locomotive. A number of Martin heaters were applied to engines on the Grand Trunk line in 1860 and 1861, but no other installations are recorded. One obvious objection to Martin's design was the obstruction of the draft caused by this bulky apparatus.

Except for a test on the Chicago, Burlington and Quincy in 1870, scant attention was given to smokebox superheaters until the issuing of William S. Hudson's patent on March 11, 1873 (No. 136729). Hudson's heater appears to have been an enlargement of the Smith heater. Apparently it was never given a practical test.

Like feed-water heaters, smokebox superheaters were never applied generally to American locomotives during the nineteenth century. The heater's exterior quickly collected a heavy layer of soot which insulated the heater from the relatively low temperature of the smokebox. Smokebox superheaters were also objectionable because they obstructed the draft. The small improvement in operating efficiency was not sufficient to warrant the cost of installation and maintenance.

The fire-tube superheater also was tested during the nineteenth century. According to this plan a number of small tubes or elements placed inside the regular boiler tubes carried the steam into the boiler for reheating. The saturated steam was collected from the dry pipe in a header located in the smokebox. The superheater elements were connected to the header. This arrangement is universally associated with the modern Schmidt superheater introduced in 1897, but its history began nearly fifty years earlier.

A fire-tube superheater, striking in its similarity to the Schmidt plan, was patented in 1849 and 1850 by Jean de Montcheuil, manager of the Montereau-Troyes Railway.[137] Montcheuil's patents were actually amendments to a French patent issued to M. Quillacq for a stationary boiler superheater. The amendments covered modifications for locomotive use. An express engine of the Montereau-Troyes line was fitted with a heater of this design in 1850. It was remarkably efficient, too efficient in fact, for the steam was heated to nearly 700°F., which was far too hot for the low-temperature lubricating oils of the period. Other European engineers perfected other forms of fire-tube superheaters, but none were adopted until better cylinder oils were developed.

In America the earliest evidence of interest in fire-tube superheaters is a patent issued to W. M. Strom in 1857. According to Strom's design the heating elements passed entirely through the boiler and firebox but were insulated from the fire by a long water drum. Only a small gain in temperature could be realized, however, and it is doubtful if Strom's invention was successfully employed. Henry Tyson, master mechanic of the Baltimore and Ohio, is said to have experimented with locomotive superheating at the time of Strom's patent. According to a recollection of M. N. Forney, Tyson added no new machinery to make his superheater.[138] Instead, the existing upper row of boiler tubes were bent upward until they protruded above the boiler's water level. No longer insulated by water, the tubes raised the steam temperature to the point that the oil broke down and the cylinders were badly scoured. Thus, Tyson's experiment ended; however, other experiments continued throughout the century. Many schemes were offered and all were rejected until a gifted German engineer, Dr. William Schmidt, devised a practical unit. Beginning in the late 1890's Schmidt tried several plans; in the end (1902) he adopted the fire-tube arrangement patterned very closely on Montcheuil's 1850 design. Schmidt, however, coordinated high-temperature steam with other elements of the locomotive, namely, packing, lubrication, and valve design. With these issues settled the superheater was enthusiastically adopted; within ten years it was standard equipment for all American road engines.

---

[136] *American Railway Times*, September 21, 1861, p. 364. Martin obtained U.S. Patent No. 30857 on December 4, 1860.

[137] Data on the Montcheuil superheater was drawn from Young's paper "Locomotive Superheaters," in the International Railway Congress's *Bulletin*, April, 1908, pp. 383–88, and from the *Railroad Gazette*, November 1, 1907, pp. 520–24.

[138] *Journal of the Franklin Institute*, September, 1905, p. 237.

THROTTLES

The throttle valve, designed to regulate the supply of steam fed to the cylinder's valves, presented no peculiar problems as long as steam pressure remained low. This was the case during the first decades of the nineteenth century, and a simple, uncomplicated valve was found satisfactory for this function.

The earliest Stephenson engines exported to this country were fitted with a primitive cock valve. The valve consisted of a hollow plug with a narrow slot opening that opened or closed when turned. The valve was mounted inside the boiler at the rear wrapper sheet of the back head. This position, so convenient to the footplate, required no complex linkage; only a stuffing box for the valve stem and a throttle lever were needed. The *Robert Fulton* drawing illustrates this form of valve (see Fig. 7). The *John Bull* (Fig. 97), for all its rebuildings, had a valve of the same type, although it no longer retained its original mounting inside the boiler. For all its simplicity the plug throttle was a crude, imprecise regulator and it was abandoned by Stephenson in about 1833 for the butterfly valve. American manufacturers made scant use of the plug valve; the first few engines of the West Point Foundry are about the only known examples.

The slide-throttle valve, much like a common D-shaped valve with two or three rectangular steam openings, was the standard American throttle valve before 1865. The multiple ports permitted a large and quick opening with only a short valve travel. Its origins are uncertain but it was widely used in the United States by the mid-1830's. The Baldwin 4–2–0 drawing (Fig. 106) and one of the Dunham drawings (*ca.* 1837, Fig. 117) show this style of valve. A study of other drawings reproduced in this volume, particularly those showing boiler cross-sections, illustrate the almost universal use of the slide-valve throttle. While the type of valve was settled early, the proper placement of the throttle valve changed frequently during the first thirty years. At first the valve was centered inside the boiler above the crown sheet. A collecting pipe, extending upward, gathered drier steam from the top of the boiler. From the few drawings available this arrangement appears to have been abandoned by the mid-1840's when throttle valves were mounted atop the collecting pipe. In this arrangement a long lever was necessary to connect the valve with the throttle's lever stem. The Winans 4–4–0 drawing (Fig. 129) clearly illustrates this plan. Beginning in the early 1850's, throttle valves were mounted in the smokebox. By 1860 it was reported that "nearly all" American locomotives were equipped with front-end throttles.[139] The chief advantage claimed was that because the throttle was not inside the boiler it was more accessible for inspection and repair. The chief disadvantage was the likelihood of the valve's overheating and sticking because of high smokebox temperatures. The acceptance of the poppet throttle valve in the 1870's brought obsolescence to smokebox throttles until the idea was revived in the 1920's.[140]

As already stated, the slide-valve throttle remained standard until the 1870's when the poppet valve was adopted by the major locomotive builders. The idea apparently originated with E. A. G. Young of the Newcastle and Frenchtown Railroad in 1833, when the plug throttle valve of that road's *Delaware* was replaced by Young's invention.[141] Young patented his idea, but the patent office fire of 1836 destroyed all record of it. Apparently it was one of a number of patents not reconstructed, for a recent search of the patent office papers at the United States National Archives failed to produce either the specification or a drawing. It is therefore impossible to deter-

---

[139] Clark and Colburn, *Recent Practice*, p. 57.
[140] Bruce, *The Steam Locomotive*, p. 139.
[141] From a transcript copy of the original "Minutes of Newcastle and Frenchtown R.R.," copied in about 1930 by C. B. Chaney from the original manuscript owned by C. L. Winey. The entry for September 13, 1833, noted the engine superintendent's (E. A. G. Young) "late" invention of the poppet throttle. Mention also was made of its use on the locomotive the *Delaware* and its more recent application to the *Pennsylvania* and the *Virginia*.

mine if Young used a double (balanced) or merely a single poppet valve. No other railroads are known to have used Young's valve at this early date, but it is likely that it was used on locomotives built by the Newcastle Manufacturing Company during Young's superintendence of that company (*ca.* 1832–38).

The next evidence of poppet throttles appears in the drawings of the Philadelphia and Reading's *Philadelphia*, published in 1849 and reproduced in the present volume as Figs. 133–136. These drawings show the engine as reconstructed by James Millholland, and, while the exact nature of the reconstruction is uncertain, the poppet throttle is specifically credited to Millholland.[142] The valve is well proportioned and appears to be little different from those used in recent years. Because of its two pistons, the valve was balanced and could be worked easily even under relatively high steam pressures. Millholland, who referred to it as a "double beat valve," used it on all subsequent engines of his design. Whether he used it before the *Philadelphia*'s rebuilding in 1848–49 is uncertain, but it is not unlikely, for he built and extensively rebuilt a number of engines while master mechanic of the Baltimore and Susquehanna Railroad between 1838 and 1848. Millholland was also the originator of ratchet throttle levers. A lithograph of his *Juniatta* [sic], dated 1855, is the earliest evidence of this arrangement. By means of the ratchet and latch, the throttle could be set in any position desired by the engineer for running. It was also a safety feature, since the throttle could be locked shut. Years earlier, thumb screws were used for this purpose but they were far less convenient or positive. The thumb-screw throttle lever was not, of course, immediately superseded by the ratchet, and it continued in use for many years.

By the 1860's other mechanics had begun to use the double poppet throttle; at least one example of this is the Erie Railway's No. 254 illustrated in Fig. 218. In 1871 Weissenborn noted that the "double-beat poppet" was the most widely used form of throttle valve.[143] Three years later Forney referred to the recent universal adoption of the poppet and noted the obsolescence of the old slide-valve throttle.[144] The poppet remained standard for another fifty years.

SAFETY VALVES

An overabundance of steam is occasionally produced despite careful firing and water management by the engine crew. To relieve the boiler of such dangerous surpluses an automatic venting device, commonly known as the safety valve, was introduced to steam engineering practice in the late seventeenth century by Denis Papin. The safety valve was held closed by a weighted lever, the weight being adjusted so that the steam could lift the valve open when the pressure rose above a safe limit. Papin's simple, effective valve was used universally throughout succeeding years and even to this day is occasionally employed. It performed perfectly on the earliest slow-speed steam locomotives. However, the gravity valve was not satisfactory on public railways, where the vibration of the engines at speed caused the valve weight to bounce open accidentally and thus release precious steam.

To correct this defect the counterweighted lever valve was fitted with a spring balance, similar to a butcher's scale, in place of the weight. Mention has already been made of the spring balance regarding its use as a steam gauge. The bottom of the spring balance was attached to the boiler; the top was attached to one end of the safety-valve lever with an adjustment nut. This nut could be adjusted to vary the steam-pressure setting. Tightening the nut increased the tension of the spring and thus required greater steam pressure to open the valve. Loosening the nut reduced the spring's tension and the amount of steam pressure needed to open the valve. Since

---

[142] Emil Reuter, *American Locomotives* (Philadelphia, 1849), p. 17.

[143] Gustavus Weissenborn, *American Locomotive Engineering*, p. 163.
[144] Forney, *Catechism*, p. 155.

it was desirable to use a small spring balance, a long lever, often 36 inches in length, provided the necessary leverage to hold the valve closed with a moderate-sized spring. The valve itself had a beveled edge and was commonly 3 inches in diameter. A good illustration of this arrangement is Winans' 4-4-0 drawing, Fig. 129.

The spring balance–lever safety valve was apparently first used by Stephenson in 1829. It was soon adopted in the United States and remained in favor for the next fifty years. Evidence of its popularity and wide use is exhibited by nearly every locomotive represented in this work.

The Salter spring balance, one of the most popular of several styles used, was housed in a handsome, tubular brass casing. During the 1830's large numbers of Salters were imported to this country by locomotive suppliers. Advertisements of George Salter and Company, West Bromwich, England, appearing in 1931 illustrate a spring balance identical to the type manufactured a century earlier.

Spring balances were made in this country but were generally considered inferior to those of British manufacture. The Bowles Railroad Supply Company announced in its 1860 catalog that the genuine Salter balance, priced at $9, was "very much superior to those made in this country." American balances usually had a flat face engraved with a scale; a pointer noted the pressure. In 1860 the Graham spring balance was introduced as America's answer to not only the Salter balance but all coil-spring balances in general. Graham's balance consisted of a heavy leaf-spring bearing that rested on two short lever arms housed in a compact brass case. It took the place of two Salter balances (two safety valves being standard at the time), and at $24 was advertised as a saving compared to the usual arrangement requiring two balances per locomotive.[145] While Graham's balance did not displace the ordinary coil-spring type, it did receive limited use. Two locomotives discussed in this work, the Rogers 4-4-0 (Fig. 222) and the Erie Mogul (Fig. 218), are equipped with Graham balances.

A second style of safety valve used in the nineteenth century was the spring type in which a spring bore directly on the valve and no lever was used. Blenkinsop used helical-spring safety valves, two per engine, on his pioneering machines of 1812.[146] Hackworth is usually credited for first using spring safety valves on locomotives, but he is not known to have used them until 1827. Although used earlier than the spring balance–lever type, it was not widely used in the United States until after 1875 and then only after marked improvements had been made by Richardson (a development that will be discussed shortly). The spring valve was used mainly as a secondary safety valve fitted so that it was impossible or at least highly inconvenient for the engine crew to adjust it or tamper with it. Commonly referred to as a "secret" or "lockup" safety valve, it was placed in a housing on the forward part of the boiler. A long thin pipe, usually of copper or brass, carried the exhaust steam from the lockup housing to a point above the engineer's line of vision. The early British builders believed that the lockup valve was necessary, and, at first, American mechanics followed the practice without question. However, by the mid-1830's many American locomotives were being built without lockups, in the belief that they were a needless expense. The practice varied from road to road, but by the late 1840's the lockup safety valve was found only rarely on a new American engine. One late example is the *Philadelphia* (Fig. 134). It was considered good practice, however, to use two safety valves, but these were normally on the rear steam dome and under the direct control of the enginemen, who were free to adjust them at will. This surely encouraged the risky practice of "screwing down the safeties" in order to move a heavy train over a grade. But it also permitted the crew members, who were familiar with the engine's workings, to loosen a stuck valve when it became apparent by a sharp exhaust or an unusually "smart" performance that too much steam was on hand.

All forms of spring safety valves, including the spring balance style, labored under a basic mechanical conflict. When

---

[145] Advertisement of Graham's spring balance is on p. 10 (back section) of *Ashcroft's Railway Guide*, 1864. The balance was patented on December 18, 1860.

[146] Marshall, *A History of the Railway Locomotive Engine*, p. 36.

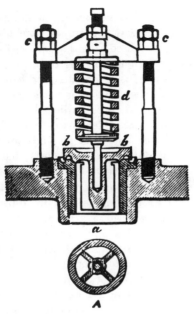

Fig. 56. *Richardson's automatic safety valve, introduced in 1867, soon became standard on American locomotives.*

steam pressure reached an unsafe limit and began to push the safety valve open, the spring holding the valve was compressed. As it was compressed the spring was placed under greater tension and thus, with *increasing* force, resisted the valve's opening. This meant that the valve did not open fully or quickly to provide an instant release of too much pressure. Rather it "wheezed" open, taking too long to vent the surplus steam in certain cases and thus making the danger of an explosion more real. In addition, the valve would be closed by the compressed spring before the boiler pressure was fully relieved. On the first count, danger, the spring balance was the best form of safety valve since the crew could quickly open it if they thought it was not opening fully enough. One direct solution might appear to be simply to make the valve larger, but an increased valve surface called for a stronger spring, the advantage of which was negligible.

During the 1860's a locomotive engineer of the Troy and Boston Railroad, George W. Richardson, turned his attention to solving the safety-valve problem. His invention consisted of a simple alteration to the ordinary spring-loaded safety valve. A cup-shaped groove was cut in the periphery of the valve just at the outer edge of the valve seat. A corresponding groove, but one slightly out of line with the valve groove, was cut in the valve body. The two grooves faced each other making an S-shaped cavity (see Fig. 56). The escaping steam was directed downward by the upper groove into the lower groove, thus "lifting" the valve open. The area that the escaping steam worked against was thus increased without increasing the size of the valve itself. The valve opened quickly and fully, relieving the pressure promptly and closing sharply when the pressure fell 3–5 pounds below that for which the valve was set. The Richardson valve was said to open more than twice as far as an ordinary safety valve.[147] Because of its quick opening it became popularly known as the "pop" valve.

Richardson carried on his experiments in a tiny shop behind his home in Troy, New York. Sometime in the 1860's, after determining a properly proportioned valve to suit his idea for

[147] Meyer, *Modern Locomotive Construction*, p. 384.

the valve seat, he applied to his employer, the Troy and Boston Railroad, for permission to test his invention. The road's locomotive the *Walloonsac* was then fitted with Richardson's valve. The valve proved to be a success and the inventor secured a patent on September 25, 1866. The next year the Pennsylvania Railroad tested the device, but, being somewhat skeptical, insisted upon installing a hand lever so that the valve could be adjusted manually.[148] The hand lever was removed after the valve proved itself fully automatic and entirely dependable. The Richardson valve was reportedly applied immediately to other engines on the line.

The *Master Mechanics Report* for 1872 noted that twenty of thirty-five members of the organization preferred the Richardson valve over all other designs. Two years later the same group observed that the "Richardson valve *alone* of those tried fills the requirements of a good safety valve." However, the excessive and startling noise made by the valve's quick and wide opening was objected to by one member who stated: "I have met with ladies who have expressed themselves as being half frightened to death by that Richardson valve." In later years a perforated casing was placed over the valve to serve as a muffler and it partly silenced the valve's throaty roar.

The merits of the Richardson valve were recognized quickly. Others copied and modified the original design, yet it was many years before the pop valve displaced the spring balance–lever valves. Builders' photographs show many new engines of the 1880's equipped with the old lever-type safety valves. It was not until the next decade that this inefficient safety valve was superseded by the pop valve.

---

[148] *Locomotive Engineering*, January, 1890, p. 3.

RUNNING GEARS

9.

INTRODUCTION

Writing in 1871, Gustavus Weissenborn clearly recognized the main characteristic of the American locomotive in the following statement: "The first and most prominent quality of the American locomotive is its flexibility; in rounding curves, in moving over a rough and uneven track, yielding in all directions, it maintains both its position on the rails and its adhesion to them in a surprising manner." This remarkable agility was the result of a carefully designed running gear, the chief elements of which were a leading truck, a light bar frame, and equalizing levers. While these features distinguished the nineteenth-century locomotive from the standard European model, it has been pointed out in recent years by British locomotive historians that each of these features can be traced to an earlier British origin: the truck to Chapman, 1812; the bar frame to Bury, 1830; and the equalizer to Hackworth, 1827. These claims are valid and there is no reason to question them; however, in the author's opinion the major point is that these devices were perfected and *used* in this country many years before they were accepted in British practice. In this light it appears reasonable to hold that the American locomotive running gear was a distinct departure from early British designs and a leading contribution by American mechanics to this branch of the technical arts.

SUSPENSION

The need for flexible locomotives was quickly met by American mechanics, who devised limber, yet stable, locomotive suspensions. The rapid acceptance of the leading truck after 1832 temporarily eased the demand for improved suspension. The 4–2–0, with its center bearing truck, offered a stable three-point suspension that could adapt itself to the roughest track. However, the need for more powerful, coupled engines called for a new and more complex plan of suspension.

151

Fig. 57. Andrew Eastwick's patent of 1837 was devised to improve the suspension of the new 4–4–0 locomotive. The design was not successful, however, and was superseded by Harrison's plan (see Fig. 58).

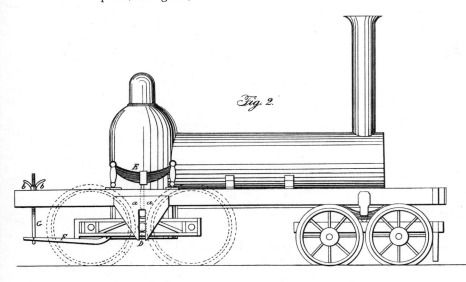

The first plan for improving the suspension of coupled locomotives was that of Andrew M. Eastwick, a partner in the Philadelphia firm of Garrett and Eastwick. Eastwick's idea was obviously prompted by the success of the leading truck; in his plan the driving axles were carried in a separate truck attached to the locomotive's main frame. The first engine built with this arrangement, the *Hercules,* was completed early in 1837 for the Beaver Meadow Railroad;[1] Eastwick's patent drawing is reproduced as Fig. 57. Joseph Harrison, Jr., a partner of Eastwick's, materially improved his associate's plan by discarding the separate locomotive frame. Instead, Harrison devised several methods to connect the driving axles by means of ordinary leaf springs and connecting levers, all of which were attached to the main frame. The object of this arrangement was to distribute the road shocks received by any one driving axle to the other axles, thereby reducing the likelihood of derailments or damage to the running gear. In addition, traction was improved by keeping all driving wheels in contact with the rails. Harrison's patent of April 24, 1838 (No. 706), shows four methods of suspension (see Fig. 58). The first scheme was not used. The second was used at first by Eastwick and Harrison. The Baltimore and Ohio Railroad and Ross Winans favored this simple plan whereby a long leaf spring served as both spring and equalizer. The idea can be traced back eleven years prior to Harrison's patent to Timothy Hackworth, who is known to have used an identical arrangement on the locomotive the *Royal George*.[2] It is probable that Harrison was unaware of Hackworth's earlier use of the spring-equalizer, for it was not adopted in England at the time of its introduction and only came into use in that country after 1851 when it was reintroduced as a "new reform."[3]

To return to Harrison's patent of 1838, we find that the third scheme, a large leaf spring under the frame and attached to a lever connecting the two driving axles, was used by Eastwick and Harrison on some of their later engines, notably on the fast passenger engine the *Mercury.* The Franklin Institute cited an eight-wheel engine, similar to the *Mercury,* equipped with Harrison's equalizing lever in its report for 1839.[4] The suspension described was on the third plan offered in Harrison's patent:

The improvement invented by Messrs. Eastwick & Harrison is designed to obviate this difficulty, by giving to the eight-wheel engine only two bearing points, one on the guide truck, and the other on a frame supported by the driving wheels. The axles of the drivers are placed one in front, and the other behind the firebox, and are confined between pedestals of the usual form, fixed to the main frame of the engine, which allow vertical play, but prevent any horizontal motion.

[1] *Railroad Gazette,* April 22, 1892, p. 294.

[2] Marshall, *A History of the Railway Locomotive Engine,* p. 173 and Fig. 69, shows a contemporary drawing of Hackworth's spring-equalizer.

[3] Colburn, *Locomotive Engineering,* p. 79. R. & W. Hawthorn, a British firm, obtained a patent for compensation levers in 1851.

[4] Harrison, *The Locomotive,* pp. 67–71, reprinted from the Franklin Institute's 1839 report.

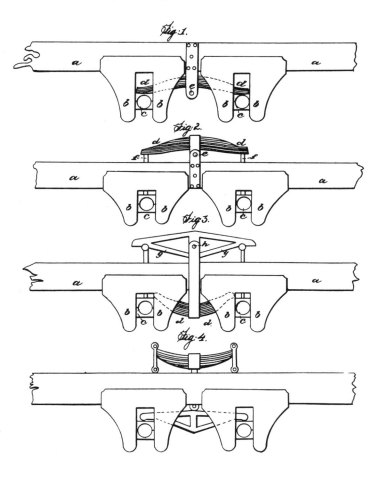

Fig. 58. *Harrison's 1838 patent for locomotive equalizers was a basic and widely used improvement.*

The bearing pins instead of abutting against springs fixed to the frame in the ordinary manner, are jointed to the extremities of horizontal beams of cast iron, one of which is placed on each side of the engine.

To the centre of these beams or levers, are jointed wrought iron rods, which pass down through the engine frame, and carry the springs which support the weight of the engine.

Unfortunately the name of the engine was not given, but it was credited with pulling a heavy train with ease. It was also noted that "the road [was] in such bad condition as to keep the sustaining beam in continual vibration." The final scheme shown in this patent was apparently not used.

Curiously enough the one arrangement of equalizers and springs which became most common was not covered by Harrison. This standard plan called for a leaf spring over each driving axle with a lever between to coordinate the entire suspension into one complementary mechanism. To prevent any possibility of bypassing this patent, Harrison greatly enlarged the original 1838 specification by securing a "reissue" of the old patent on November 21, 1842. The reissue was, in fact, a new patent in which several new equalizing schemes were claimed. Among these was the standard spring-equalizer-spring arrangement, several elaborate bell-crank lever combinations, and one curious chain and spring arrangement. The 1842 reissue firmly established Harrison's claim to this highly important invention. He reportedly received enormous royalties, for few, if any, American locomotives were built without equalizers after 1840.

It is not surprising that other mechanics developed new suspension plans in the hope of bypassing Harrison's patent. One of the first such plans was Henry Waterman's patent of February 10, 1841 (No. 1969). Waterman proposed a separate frame for the driving axles, but unlike Eastwick's, his plan provided for both lateral and vertical movement by means of either hinge or ball-and-socket joints. A long radius rod, attached to the front driving axle pedestal, connected the subframe to the locomotive. This connection was also hinged.

While there is no record that Waterman's invention was tested, a remarkably similar scheme, which did undergo a num-

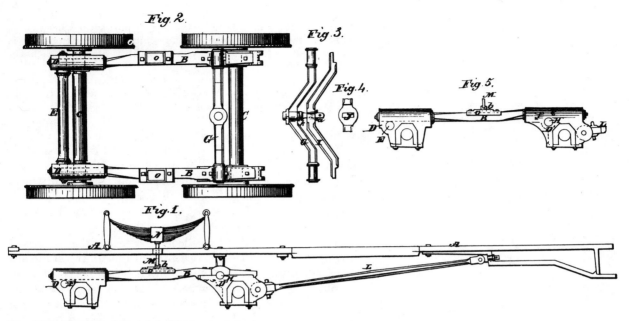

Fig. 59. Norris and Knight's 1843 patent for locomotive suspensions was an attempt to avoid using Harrison's invention. It was not successful and was used for only a few years by Norris.

ber of tests, was patented on February 10, 1843 (No. 2951), by Septimus Norris and William Knight. As in Waterman's patent, a long radius rod connected the subframe to the locomotive; however, Norris and Knight made a second connection with a pivot near the front axle (see Fig. 59). The patentees were forced to recognize the broad claims of the Eastwick and Harrison patents and restricted their claims to the very narrow limits of their peculiar arrangement. For some reason, Norris and Knight were not required to mention Waterman's patent in their patent specification, despite the similarity of the two plans.

Several Norris locomotives are known to have used the Norris and Knight suspension; one of these machines, the *Virginia* (Fig. 60), was delivered to the Winchester and Potomac Railroad some months before the patent was issued.[5] Two large lithographs issued by Norris between 1843 and 1845 show eight-wheel engines so-equipped. A number of European locomotives are also known to have used this type of suspension.[6] Norris advertisements appearing in the *American Railroad Journal* throughout 1845 contain a line cut of a similarly equipped 4–4–0 (see Fig. 132). After the intense promotion effort, however, Norris abandoned the plan for the simpler and more effective Harrison equalizer.

[5] Karl von Ghega, *Die Baltimore-Ohio eisenbahn* . . . (Vienna, 1844), p. 162, reports seeing the new locomotive the *Virginia* on the Winchester and Potomac in May of 1842.
[6] *Railway and Locomotive Historical Society*, Bulletin, No. 79 (1948), p. 62.

COMPONENTS

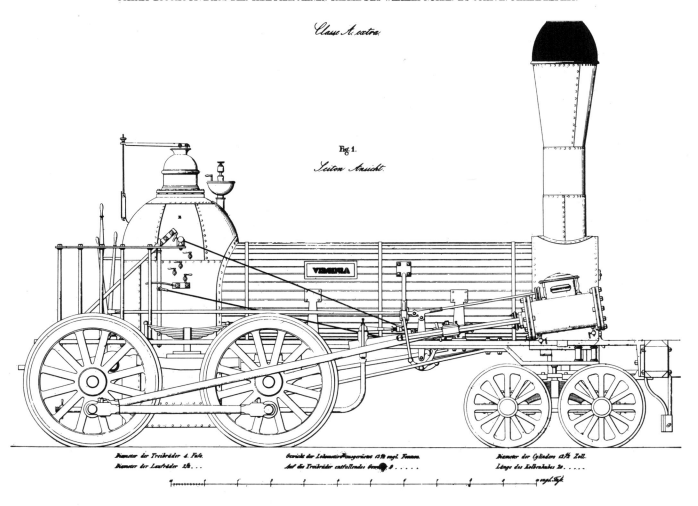

Fig. 60. The Virginia incorporated the Norris and Knight suspension shown in the preceding illustration. This engine was built by Norris for the Winchester and Potomac Railroad in 1842.

In the years that followed, the equalizer lever, also referred to occasionally as the "balance," "vibrating," or "sustaining" beam, underwent certain improvements. Cast iron was abandoned by or before 1845 and wrought iron was adopted. Before about 1855, equalizers were commonly placed under the top rail of the frame; after that date they were placed above the top rail.

SPRINGS. Leaf springs were by far the most common type used for locomotives. The *Lancashire Witch*, built in 1828 by Robert Stephenson, is commonly believed to have been the first locomotive fitted with metal springs; the pattern established by this pioneering machine was followed to the end of steam-locomotive construction. Tempered steel was the only suitable material available, but before the Bessemer process it was an expensive and scarce material. While springs were manufactured at an early date in this country, imported springs were preferred by many companies because of their superior quality and economy. Krupp "cast steel" springs were imported at least as early as 1853 and within a few years were widely accepted as superior to other makes.[7] American makers, such as the Albany Iron and Nail Company of Troy, New York, offered "spring steel" for locomotives of "any thickness" in an advertisement appearing in the *American Railroad Journal* on January 23, 1845. In 1857, McDaniel and Horner of Wilmington, Delaware, made steel springs from the "best Swede Iron."[8] Locomotive springs at this time were made from about fifteen leaves ¼–⅜ of an inch thick, 3½ inches wide, and from 25 to 36 inches long.

Steel coil springs were rarely used, although the *Dr. Ordway* (Fig. 2) was one known example of this type. India rubber "springs" (cylinders of rubber) were occasionally used for very light locomotives.[9] Compared to the steel spring, the rubber spring had limited resilience, a short life span, and absorbed little of the vibrations of the road. After a few miles these springs were often crushed into an unyielding block; freezing weather reduced them to a similar state. India rubber springs were found to be so unsatisfactory in this service that they were entirely abandoned for even the lightest locomotive before 1860 and, in fact, by this time were no longer favored for car or tender-truck suspension.

Steam springs were also considered for locomotive service in the nineteenth century. George Stephenson patented and built several locomotives with steam springs as early as 1816.[10] Ross Winans used a steam spring on the trailing wheels of some eight-wheel engines built for the Reading in 1847. Two years later, he used a steam spring on the driving axle of the experimental passenger locomotive *Carroll of Carrollton*, but in this case the main purpose was to promote traction.[11]

Air springs were briefly considered for locomotive use during the early 1840's. Although several earlier British patents for this idea were granted, Levi Bissell of Newark, New Jersey, obtained a United States patent for an air spring on October 11, 1841 (No. 2307). A set of Bissell air springs was tested satisfactorily on a passenger car for more than a year. These springs consisted of a cast-iron cylinder 10 inches long with a bore of 6 inches; a heavy piston was fitted with a cup-shaped leather seal.[12] A 2-inch level of oil mixed with white lead further insured that the light charge of air pressure—necessary for the spring's action—would not escape. The invention appeared so promising that M. W. Baldwin considered its use for locomotives during the early 1840's. The only known illustration of an engine so-equipped is a contemporary Baldwin advertising cut shown here as Fig. 61; notice that the cylinder is considerably larger than that used for the passenger car previously mentioned. It should also be noted that a rear driving wheel, a second air spring, and a connecting pipe between the two spring cylinders were sketched on the print at the time that it was circulated in 1844. In effect, the pipe was an "equalizer" between the two driving axles. The entire scheme apparently proved impractical, for there is no evidence that Bissell's invention was considered more than an experiment.

---

[7] *American Railroad Journal*, November 19, 1853; Clark and Colburn, *Recent Practice*, p. 56.
[8] *Railroad Advocate*, January 17, 1857, p. 4.
[9] Colburn, *The Locomotive Engine*, pp. 73–74.
[10] Marshall, *A History of the Railway Locomotive Engine*, p. 110.
[11] *Railway and Locomotive Historical Society*, Bulletin, No. 114 (1966); a history of the *Carroll* is given in the article "American Single Express Locomotives."
[12] The *Journal of the Franklin Institute*, 1843, Vol. 6, pp. 168–70, describes Bissell's air spring; a cross-sectional drawing is included.

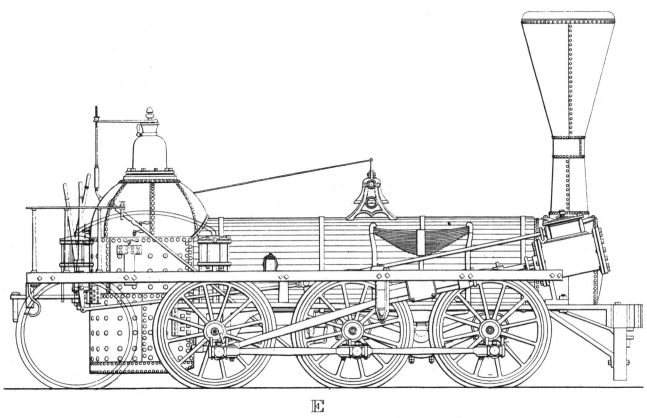

Fig. 61. *A Baldwin flexible-beam-truck locomotive of 1844 shown with a Bissell air spring over the rear driving wheels. The fourth set of wheels and the air spring were added by a contemporary artist to show the method of equalizing a pair of air springs by means of a connecting pipe.*

FRAMES

The frame's primary functions are to tie together the various component parts of the locomotive and to handle the train's weight. Equally important to these functions is the requirement that it also withstand, and largely absorb, the considerable horizontal disturbances set up by the reciprocating parts. Thus, it might be supposed that locomotive frames should be rugged and stiff structures capable of withstanding rough service. Yet, surprisingly weak frames were used on American locomotives of the nineteenth century. The boiler was depended upon for lateral strength, the frame being a light structure, supported by many cross and side braces attached to the boiler. Thus, the boiler was the locomotive's foundation; the frame served as a secondary support to hold the wheels, valve gear, rods, and suspension mechanism in place. This apparent defect in construction was not as serious as might be supposed, for light trains and slow speeds did not necessarily require heavy framing.

It is common to think only of the bar frame when considering nineteenth-century locomotives, but this style of frame, though used by some makers in the 1830's, was not universally adopted for another twenty years. Before this time a fairly wide variety of designs and materials was employed. Certain manufacturers developed peculiar plans of their own and held to these long after the design could be considered obsolete. In the following pages the more important frame styles will be discussed; it should be remembered, however, that many composite designs were occasionally used and that it is not possible to treat all of these variations within the limits of the present history.

Wooden frames, because they were cheap and easy to construct, were used on nearly all early American locomotives. Many of the British imports had wooden frames. Stephenson made both the inside and outside frames of wood; an example of this construction is shown in the *Robert Fulton* drawing (Fig. 7). This British practice obviously impressed American builders and, not surprisingly, Baldwin, West Point, Rogers, and other pioneer domestic makers adopted wooden frames.

While it is assumed that all wooden frames were reinforced with iron plates cleated to either side of the wooden beams, there is evidence that some engines were not so-built. The *Brother Jonathan* drawing, Fig. 6, shows only corner braces of iron. The *American Railroad Journal* for December 15, 1839, indicates a similar failing of some builders (Norris or Baldwin in this case) to plate their engine frames. In reporting on a new Rogers engine, this article states that "The large frame is very strongly plated in the manner of Stephenson's engines, the neglect of which till very lately has been, we are informed, a constant objection to the Philadelphia engines on the Long Island and Troy railroads."

Wooden frames were commonly outside of the driving wheels, thus providing a wide, stable foundation. An auxiliary inside frame was usually provided for inside-connected engines as an added support for the crank axle. Wrought-iron pedestals were used at first, following British practice, but, as usual, American builders substituted cheaper cast iron by the middle 1830's. Within a few years wooden frames were largely abandoned. A wooden frame, despite heavy iron plating, was far too bulky and clumsy. Even on light 7-ton engines 4-by-7-inch beams were required. Baldwin had abandoned wooden frames by 1839 and Norris apparently had done so several years earlier. While most other progressive builders followed this example, the wooden frame enjoyed a limited popularity until the 1850's. The Newcastle Manufacturing Company built engines with wooden, iron-clad frames inside the driving wheels as late as about 1855. Blandy, a small builder of Zanesville, Ohio, is also known to have employed such frames during the same period. The Camden and Amboy Railroad used a massive wooden beam combined with a light bar frame on its several Crampton locomotives (1849–53). While these should be regarded as extreme examples of an obsolete framing practice, it should be noted that several New England builders regularly used outside wooden frames as late as the 1850's.[13] However, it should also be noted that iron frames, inside the wheels, supported the driving axles.

Iron frames had been used before 1820 by George Stephenson. While his son Robert used iron frames on several early locomotives (including the *Rocket*), he turned to wood in about 1830. The apparent reason for this curious reversion to an earlier practice was the expense and difficulty of manufacturing iron frames for the heavier, faster engines required by public railways. Forge or blacksmith welding was extremely difficult for parts over 2½ inches in diameter. Blacksmiths capable of this work were among the most skilled and highly paid workmen.[14] This high labor cost, together with the higher cost of iron, caused many nineteenth-century engineers to avoid the use of large or complex forged iron assemblies.

Despite the desire to avoid such assemblies, the wooden frame when kept to a reasonable size proved to be incapable of supporting the locomotive as it grew in size. It soon became apparent that only iron would constitute a first-class frame.

BAR FRAMES. The first locomotives in this country had iron frames, but these engines, the *Stourbridge Lion* and her sisters, were modeled on the early, slow-moving colliery engines, whose low speeds required only very light framing. The first engine with an all-iron frame used for regular service in America was probably the British-built locomotive the *Liverpool*. This machine, built in 1830 by Edward Bury of Liverpool, was finally sold to the Petersburg Railroad in 1833 after passing through several rebuildings and owners in England.[15] The *Liverpool* was fitted with an inside, iron bar frame, the introduction of which is commonly credited to Bury and his chief assistant, James Kennedy. The other Bury locomotives sent to the United States in the 1830's were undoubtedly fitted with bar frames that not only were carefully developed by the maker but that also became a distinguishing feature of Bury's standard 0–4–0 locomotive design. The bar frame was to become a standard in American railway practice, although it received little attention in England.

The earliest mention of an American-made iron locomotive frame concerns the *Comet*, a four-wheel engine completed in January, 1835, by the West Point Foundry for the Tuscumbia, Courtland and Decatur Railroad.[16] The following description was given: "The outside frame is entirely iron, securely bolted; the inside rails are of wrought-iron—(hitherto in all engines made in this country, they have been made of wood)." This bit of evidence challenges George E. Sellers' claim to the building of the first iron-frame engine in this country.[17] Sellers' first engine was not completed until September of 1835.

Mention of Bury's bar frame has already been made; it should be added that the South Carolina Railroad specified this style of frame for several 4–2–0's built by Stephenson and Tayleur in 1835. One of these machines is shown in Fig. 3 of the present volume; a slightly modified design of the same class of engine built by Stephenson, but one that shows more clearly the frame assembly, is included in Warren's history of Robert Stephenson, *A Century of Locomotive Building*, on p. 308.

It is difficult to determine when American builders first began to use bar frames; even though iron frames are known to have been produced as early as 1835, their exact form is uncertain. The earliest representation of such a frame is the Ghega drawing of the 1839 *Gowan and Marx* (see Fig. 122) which shows a well-formed bar frame, but two other contemporary drawings do not agree in this detail, showing instead an iron frame with plate pedestals.[18] It is known, however, that Eastwick and Harrison, builders of the *Gowan and Marx*, were

---

[13] Photographs and lithographs of many New England builders show wooden, iron-clad outside frames. See the *Massachusetts* (1849, Hinkley), Fig. 65, the *Roanoke* (1854, Souther and Anderson), and the *Globe* (1850, Souther).
[14] Bureau of the Census, *Tenth Census*, II, 50.
[15] *Railway and Locomotive Historical Society*, Bulletin No. 101 (1959), pp. 51–52, gives the history of the *Liverpool*.

[16] *American Railroad Journal*, January 17, 1835, p. 17.
[17] *American Machinist*, October 31, 1885, p. 4.
[18] The other contemporary drawings of the *Gowan and Marx* are discussed in Part III, in the chapter dealing with that machine.

Fig. 62. A Baldwin lithograph of about 1840 showing an outside iron frame. Baldwin is said to have first used a frame of this type on the Philadelphia, Germantown and Norristown Railroad's Fort Erie, *completed in October, 1839.*

using bar frames in 1842, as shown by a lithograph of their engine the *Mercury*.[19]

Norris was another early user of the bar frame, but unfortunately no authentic illustrations of Norris' engines before 1837 have survived, although illustrations of this date have frequently been used to represent earlier machines. The oldest Norris illustration known to the author is of an engine named *Lafayette*. This line cut appeared on an 1838 circular issued by Norris but unfortunately it does not show the frame clearly. At about the same time Norris did issue a finely executed lithograph of a 4–2–0, probably representing the first Norris engine sent to Europe, the *Philadelphia*. This machine, completed in November of 1837, is clearly equipped with a bar frame. Similar machines of a slightly later date are shown in this volume as Figs. 10 and 14.

An early Baldwin bar frame appears in a lithograph issued by the maker in about 1839 and reproduced here as Fig. 62. The striking similarity of this frame to that surviving on the *Pioneer*, presently preserved by the Chicago and North Western Railway, indicates the use of such an arrangement as early as 1836. Baldwin varied from the true Bury bar frame by using cast-iron pedestals in place of integral forged iron bars. Cast iron was used for ease of manufacture and economy. Baldwin

---

[19] The *Mercury's* lithograph is folded in the *Journal of the Franklin Institute*, Vol. 5, No. 3 (1843), Plate I, opp. p. 16.

continued to use cast-iron pedestals as late as 1848 (see the *Tioga* drawing, Fig. 24). Other builders used cast-iron pedestals, particularly those employing riveted frames, as late as 1855. But in 1851 Colburn had remarked that cast pedestals were too weak for locomotive service and were "continually breaking";[20] thus, in 1857, cast-iron pedestals were belatedly reckoned obsolete.[21] Most progressive manufacturers had adopted integral, or bolt-on, wrought-iron pedestals years earlier.

Wrought-iron plate pedestals were rarely used in the United States after 1840. Mention has been made of their possible use on the *Gowan and Marx* in 1839. The *Watson* of 1847 (Fig. 22) and the *Mohawk* of about 1848 (Fig. 23) also illustrate such pedestals.

Another variety of pedestal was the bolt-on bar style. It produced a cheaper frame than the one-piece Bury design in which the pedestals and the top and bottom rails were made as one unit. The bolt-on frame was naturally weaker and undoubtedly required constant attention. Baldwin favored this style of construction for many years. Two late examples of this design are shown by the *Consolidation* (Fig. 228) and the 1870 ten-wheeler drawings (Fig. 233).

Front rails, like pedestals, were offered in several patterns. The most common was the single-rail style made of wrought-iron bars measuring roughly 2 inches in thickness and 4 inches in width. This was the earliest and simplest type devised; it was used as early as 1839 by Eastwick and Harrison. The spliced or divided frame was introduced in about 1851 by Wilson Eddy to facilitate the repair of engine frames in case of a wreck. In Eddy's plan the front rail was bolted to the front pedestal rather than forged solid with the top rail. Hence, should a front-end wreck occur, it was possible to remove the front portion of the frame for repair without disassembling the entire locomotive. This arrangement is shown in the *Phantom, Southport,* and Rogers 4–4–0 drawings (see Figs. 193, 198, and 222).

[20] Colburn, *The Locomotive Engine*, p. 69.
[21] *Railroad Advocate*, May 30, 1857, p. 3.

Walter McQueen used double front rails in about 1848 as shown by the *Mohawk* drawing (see Fig. 23). Two light bars, with cast-iron thimbles bolted between them, formed the front rail. McQueen favored this plan until about 1854 when he adopted slab front rails. Niles and Company and Taunton followed McQueen's double-rail design in the 1850's. This plan enjoyed a substantial revival after 1870 when heavy-pusher and large-cylinder freight engines called for heavier front-end framing.

RIVETED FRAMES. The riveted frame was another early all-iron frame but was distinct from the bar style. It was apparently inspired by the old iron-clad wooden frame but a bar of iron 2 inches square was substituted for the wooden beam. Deep plates, usually 5/8 inch thick, were riveted to either side of the square bar. Cast-iron pedestals were riveted between the outside plates as well. This frame was simpler to make than the one-piece bar frame and undoubtedly owed its popularity to that single advantage.

The riveted frame was extremely popular with the New England builders before 1855; apparently it originated in that area in the early 1840's. Winans was one of the few builders outside of New England to use this style of framing. The Mud Diggers of the Baltimore and Ohio (1841–45) and the eight-wheel connected engines delivered to the Reading in 1847 were fitted with riveted frames. The Reading used this scheme of framing on many of its engines; all of its famous "Gunboats" had riveted frames. One of the last of this class, the 408, built by the Reading shops in 1877, had a frame made from two bars 1¼ inches thick by 4¾ inches deep; cast-iron pedestals 3¼ inches thick were bolted between the bars. A drawing of a sister engine, the 405, is reproduced in this volume as Fig. 30. The Reading continued to use this outdated design at least as late as 1880; drawings of an express eight wheeler showing a similar frame appeared in the *Railroad Gazette* on January 14, 1881. This singularly late use of the riveted frame was not typical, for that scheme of framing had virtually been abandoned by all American builders twenty-five years earlier.

AMERICAN LOCOMOTIVES

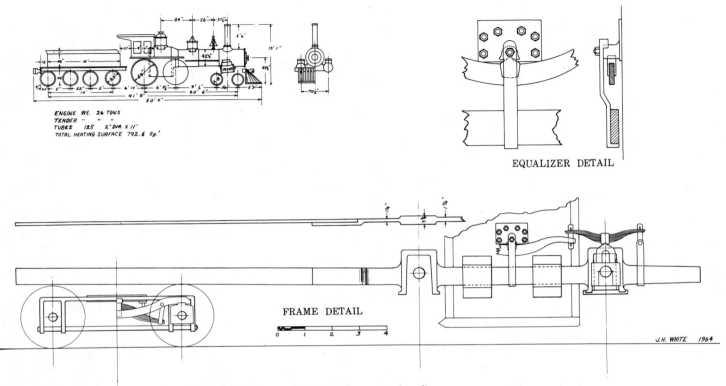

Fig. 63. *Detail of the slab-rail frame on the El Paso and South Western Railroad's No. 1 now exhibited in El Paso, Texas. The engine was built by the New York Locomotive Works in 1857. Slab-rail frames were rarely used on American locomotives.*

SLAB-RAIL FRAMES. Slab-rail frames, a variety of bar frame, enjoyed moderate popularity in the United States after their introduction in the 1850's. The original slab-rail design is credited to Charles Beyer of the Atlas Works, Glasgow, Scotland, who introduced the idea in about 1846.[22] In frames of this design a single thin, but deep, rail replaced the squarish top and bottom rails of the common bar frame.

The New York Locomotive Works of Jersey City introduced slab-rail frames to American practice in 1854. The preference for this rather obscure design might be credited to the firm's head blacksmith, a Mr. Harris, who learned his trade in the British locomotive building shops of Hawthorn and Stephenson before coming to America.[23] The first engines built by the New York Locomotive Works were fitted with slab-rail

[22] Ahrons, *The British Steam Locomotive*, p. 81.

[23] *Railroad Advocate*, January 17, 1857, p. 1.

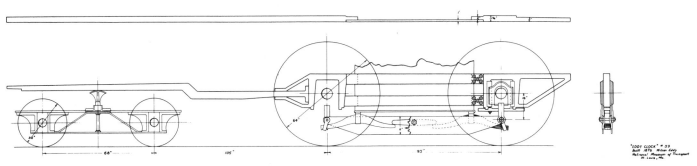

Fig. 64. Detail of the frame on the Boston and Albany Railroad's No. 39, built in 1876 by Wilson Eddy. The use of slab rails and underhung suspension permitted a wider between-the-frames firebox. This locomotive is exhibited by the National Museum of Transport in St. Louis, Mo.

frames; an excellent set of drawings of one of these machines, the *Superior*, is reproduced later in this work (see Fig. 162). Another drawing (Fig. 63) shows a nearly identical frame which still exists on the El Paso and South Western's No. 1 (presently preserved in El Paso, Texas). This machine, built in 1857 for the Milwaukee and Mississippi Railroad, has retained much of its original construction, including the frame. The successive owners of the New York Locomotive Works continued to use slab-rail frames as late as 1860.[24] Other makers who used frames of this design were Amoskeag, Detroit, and Niles and Company.

A variety of slab-rail frame, used before Beyer's design was introduced by the New York Works, was employed on the Baltimore and Ohio Railroad; according to this design, the rear extension of the top rail was made as a thin, deep slab, thus allowing for a wider, between-the-frames firebox. The drawing of a Baltimore and Ohio 0–8–0 of about 1848 shows a frame of this style (see Fig. 34). The Baltimore and Ohio's Hayes (1853) and Tyson (1856) ten wheelers followed the

[24] A lithograph issued by the Jersey City Locomotive Company in 1860 shows a slab-rail frame.

same idea. Another combination bar–slab rail frame was developed in about 1854 by Walter McQueen in which the front rail was made in the form of a slab. The advantage of this plan over McQueen's early two-rail design is not apparent to the author, but it was used on many engines built by Schenectady after 1854.

Wilson Eddy's distinctive combination bar–slab rail frame, already mentioned in the boiler section, is shown here as Fig. 64. Eddy, wishing to develop a wider between-the-frames firebox, began to use this bolt-on slab-rail design in 1851. Its use was continued until 1881 when the last "Eddy Clocks" were built in the Springfield shops of the Boston and Albany Railroad.

The plate frame, so familiar to British locomotives, was rarely used in the United States. Aside from Winans, who used the plate frame on his Camels, it would be difficult to find any other American maker who even considered this form of framing. Plate frames should not be confused with the slab-rail frames just discussed; the former were made of two heavy and deep wrought-iron plates set 6 inches apart with pedestals fastened between. An example of this construction is given in the section on Winans' Camel the *Susquehanna* (Fig. 169).

FRAME CONSTRUCTION. We have discussed the major varieties of frames without commenting on inside versus outside frames or the method of cross-bracing. In general, outside frames were rarely used in the United States after the passing of the wooden frame in 1840. A few examples after this date can be found, of course, but these were in the main restricted to broad-gauge locomotives or a few isolated oddities such as the *Watson* (Fig. 22) and Matthew's 4–4–0 (Fig. 17). Auxiliary outside frames were extensively employed between about 1845 and 1855. However, these assemblies were *not* meant to support the driving axles and accordingly had no bearings. The function of the auxiliary frame was to support the walkways, pumps, yoke, or other machinery that could not be conveniently attached to the inside, or main, frame. Before 1850 such frames were commonly made of wood and were sometimes ironclad; see the *Massachusetts* (Fig. 65). Some builders, such as Baldwin, began to use flat plate or angle iron for auxiliary outside frames in the 1840's. Several such assemblies are pictured in the present volume. Two of the best of these illustrations are the *Amenia,* Fig. 66, and the *Corry,* Fig. 29; both show the elaborate side- and cross-bracing used, together with the walkways and other attached assemblies. Secondary outside frames rapidly fell from favor after 1855, and by 1860 such loyal advocates as Baldwin and Rogers were reported to have abandoned them as well.[25]

Mention was made at the beginning of this section of the dependence on the boiler shell for support of the locomotive frame. This could only be accomplished, of course, by making frequent connections between the boiler and the frame. Light round or flat iron bars riveted to the boiler shell at about three-foot intervals was the most common form of boiler-frame attachment. These bars were attached near the middle of the boiler and ran down at a steep angle to the frame; other such bars were attached to the underside of the boiler and thus ran at a plane nearly level with the frame. The strain imposed on the boiler by this frequent and rigid attachment to the frame was recognized at an early date, but the practice continued until the 1860's when at last it was abandoned and heavy frames, independent of the boiler, were adopted to provide a solid foundation for the locomotive. The major strain imposed on the boiler by a rigid connection to the frame was the expansion (usually a fourth of an inch) of the shell when heated. For this reason it was best to have as few boiler-to-frame connections as possible. The best practice was to attach the boiler to the frame rigidly at the smokebox while providing a sliding connection at the firebox.

The main purpose of the side braces was to add longitudinal stiffness to the frame; their secondary function was to cross-brace the frame. Cross-bracing was effected mainly by the smokebox, front- and rear-end beams, and the foot plate. The front beam was universally a heavy wooden beam; it was cheap, could be replaced easily, and afforded a simple attachment for the cowcatcher.

In closing, it should be emphasized that, despite the wide variety of styles and composite designs, the bar frame had become the standard form of construction in this country by 1855. Although bar frames were introduced more than twenty years earlier, their slow adoption might be explained by the difficulties encountered in manufacturing complex one-piece forgings. However, as steam, drop, and trip forging hammers became more commonplace, the cost and difficulty of making large forgings was markedly reduced. The bar frames offered the following advantages: they were solid yet reasonably flexible; their large openings facilitated access to the valve gear, equalizers, and springs for inspection and repair. Finally, the several bars and their "open" character permitted the convenient attachment of auxiliary parts.

---

[25] *American Railway Review,* March 8, 1860, p. 136, and November 15, 1860, p. 293.

COMPONENTS

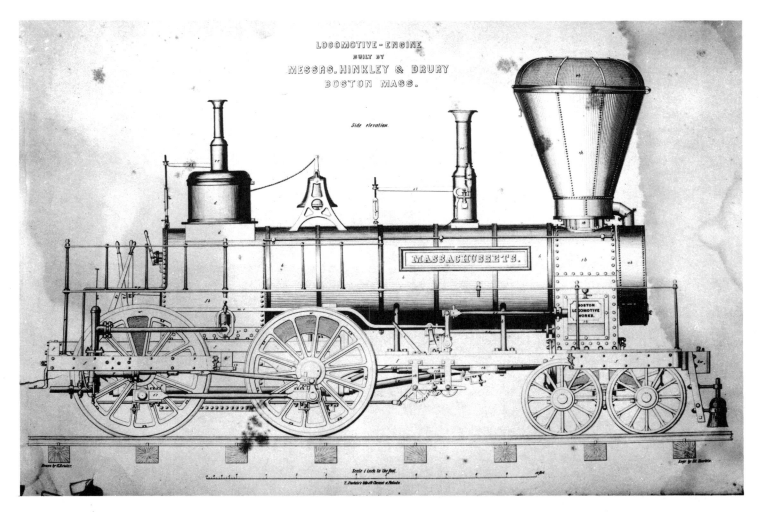

Fig. 65. The Massachusetts *was constructed by Hinkley for the Philadelphia and Reading Railroad in June, 1849. Notice that the outside iron-clad wooden frame is cut away to more clearly show the valve gear. The inside or main frame is of the riveted pattern.*

Fig. 66. *The polished wrought-iron supplementary outside frame is clearly visible in this early view of the New York and Harlem Railroad's Amenia. The engine was built by Rogers in December, 1850.*

TRUCKS

The primary function of the locomotive leading truck is to guide the machine into and around curves. The leading truck was the earliest and most fundamental revision to basic locomotive design effected by American mechanics and must be credited with making steam locomotion practical for our primitive, serpentine railroads.

Locomotive trucks were proposed as early as 1812 in a British patent of William Chapman.[26] A four-wheel swiveling truck, with side-bearing rollers, was placed at the *rear* of the locomotive; a rigid set of wheels led the engine. The patent also proposed a curious method of propulsion. A chain, laid the length of the railway, was wound through a system of drums on the locomotive. A six-wheel machine was built on this plan in 1813 but failed because of the chain's repeated breaking. The next year Chapman built a gear-driven locomotive with two four-wheel trucks. While these engines might rightly be cited as the first truck locomotives, it should be understood that Chapman did fully appreciate the leading-truck concept and that his engines were intended for the slowest possible service on colliery lines.

The truck idea was reportedly revived some years later when the Baltimore and Ohio sent William Knight, G. W. Whistler, and William G. McNeill abroad to make an engineering study of British railroads. The party naturally called upon Robert Stephenson, and during the ensuing conversation the famous engineer reportedly suggested the use of trucks as an aid in working roads with sharp curves. This conversation of 1828 was reported by Stephenson some years later to Zerah Colburn.[27] However, it seems curious that Stephenson did not follow his own good advice but instead built an 0–6–0 for delivery to the Baltimore and Ohio following the Knight-Whistler-McNeill visit. Furthermore, he built no truck locomotives until 1833 when John B. Jervis specifically ordered an engine so-built. Stephenson must have been aware of the failure of rigid British locomotives on American roads for a long time; yet he did not adopt this reform—although it was supposedly his own suggestion—until required to do so by American customers.

The actual evolution of the leading truck was described in some detail by its inventor, John B. Jervis, in a lengthy letter published in 1871.[28] Jervis was careful not to claim the truck principle but, acknowledging its earlier British origins, he properly took credit for the leading truck. In his account he notes that the subject was discussed at length during the summers of 1830 and 1831 between himself and Horatio Allen, who spent both summers at Jervis' Albany, New York, home to escape that season in Charleston while he was working for the South Carolina Railroad. Both men understood the need for light and flexible locomotives. Both were also aware of Chapman's double-truck, eight-wheel locomotive as illustrated in an early edition of Nicholas Wood's *A Practical Treatise on Railroads*. However, they differed on the best way to exploit the truck principle. Jervis proposed a separate leading truck with the driving axle fixed to the main engine frame; Allen contended that this was a "half-way measure" and wanted a two-truck engine with the drivers carried on a separate frame. Allen finished drawings for a locomotive based on his idea during the summer of 1831. A machine, the *South Carolina*, was completed for the South Carolina Railroad in January, 1832, by the West Point Foundry. Several other engines on the same plan were built for the South Carolina Railroad during the next year.[29] All were out of service before 1838, but their failure was not attributed to the running gear.

Jervis did not pursue his idea as quickly as did Allen, and it

---

[26] A well-documented account of Chapman's locomotives, together with several drawings, is given in Marshall's *A History of the Railway Locomotive Engine*, pp. 61–76.
[27] Colburn, *Locomotive Engineering*, p. 96.
[28] *Railroad Gazette*, December 23, 1871, p. 396.
[29] For more details on the Allen articulated locomotive, see *American Railroad Journal*, March, 1890; see also the Annual Reports of the South Carolina Railroad, 1831–38.

was not until August, 1832, that an engine was built on his leading-truck plan. The history of this machine, the *Experiment*, has already been given in the section on 4–2–0's. Although Jervis' plan was introduced about eight months after Allen's, it was the leading-truck concept that triumphed. Far simpler than Allen's articulated locomotive, Jervis' elementary but practical arrangement was at once recognized by other railroad managers who sought flexible locomotives. Jervis explained his invention to the railroad industry not long after his first truck engine was completed:

> One frame embraces four wheels in the same manner as a common wagon: these wheels are small (32 inches) in diameter and of uniform size; one end of the second frame is mounted on the third pair of wheels, which are the working wheels, and the other end is rested on friction rollers, in the centre of the first frame, to which it is secured by a strong centre pin. The small wheels, with their frame, work on the road the same as an independent wagon; and being geared short, they go round a curve with as much ease as a common wagon, and being leaders, they bring round the working wheels, and the large frame on which the whole machinery of the engine rests, with as much ease as practicable. By this method it will be seen the engine may pass a curve with the same ease as a common railroad carriage, having the same weight on the wheels.[30]

There is no question that Jervis deserves full credit for the leading truck, but statements occasionally appear claiming that the Baltimore and Susquehanna Railroad's *Herald* was the first machine to be so-equipped.[31] The *Herald* was completed early in 1832 by Stephenson as a standard 0–4–0 "Sampson" class. After considerable delay owing to difficulties involved in shipment, the engine was tested on the Baltimore and Susquehanna Railroad in August, 1832. The road was originally built for horse traffic and its many sharp curves prevented an easy passage by the locomotive. One of the company's directors, John S. Hollins, was assigned the task of remodeling the engine so that it could negotiate curves. Hollins proposed the substitution of four small leading wheels for the front driving wheels. This arrangement, as built in the company shops, was not a conventional four-wheel leading truck. It consisted of two separate frames, each one attached to the main frame by a center pin, thus very much resembling the flexible-beam truck patented by Baldwin ten years later. The scheme, tested in October of 1832, was a success. It should be made clear that the *Herald* was not fitted with a leading truck until October, 1832—about three months after the Jervis engine began service—and that it was not a true swiveling truck but an odd arrangement rarely employed in later years.

Returning to Jervis, we find that after the outstanding performance of the *Experiment* the inventor was encouraged to promote the use of truck locomotives on other railroads. In 1833, while engaged by the Saratoga and Schenectady Railroad as chief engineer, Jervis prepared plans for another truck locomotive. An engine named *Davy Crockett* was completed by Robert Stephenson from Jervis' design in April, 1833.[32] A lithograph of the *Davy Crockett* was issued at this time and represents the earliest, possibly the first, American locomotive so-illustrated. The distribution of this print (Fig. 67), together with several reports of its performance which appeared in various issues of the *American Railroad Journal* for 1832 and 1833, undoubtedly did much to broadcast the exact nature of, and the many advantages offered by, locomotive trucks. There is no question that all leading American locomotive builders quickly adopted the leading truck and that by 1835 it had become the standard arrangement for locomotives.

The construction details and materials of trucks followed the same path of development already described for the locomotive frame. At first, wooden outside frames with wrought- or cast-iron pedestals were favored. Side bearings, often small rollers, were the rule. Trucks of this type were common into the early 1840's and several examples of them are shown throughout this volume (see Figs. 3, 6, and 8). Spoked, cast-iron wheels, 30–36 inches in diameter with chilled treads,

---

[30] *American Railroad Journal*, July 27, 1833, pp. 468–69.
[31] *Railroad Advocate*, August 4, 1855, p. 2.

[32] Several drawings of the *Davy Crockett* are in existence. A contemporary line lithograph appears in this volume as Fig. 67; remarkably similar drawings are reproduced on pp. 305 and 306 of Warren's *A Century of Locomotive Building*.

COMPONENTS

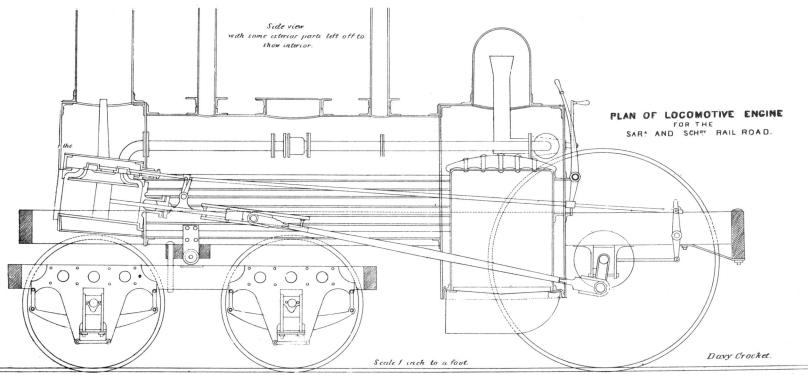

Fig. 67. *Longitudinal cross-section of the* Davy Crockett *built by Robert Stephenson & Co. in 1833 to the designs of John B. Jervis for the Saratoga and Schenectady Railroad. This contemporary lithograph clearly shows the wooden-beam truck and the roller side bearings.*

were used on the earliest Jervis trucks and remained the standard for many years until disc or plate wheels, also cast-iron with chilled treads, came into favor in about 1850. It should be noted that a relatively "wide spread," or long-wheel-base, truck was used at first. Jervis' first truck had a four-foot wheel base, and this remained the accepted spread until the late 1830's when extremely short-wheel-base trucks were adopted. The short wheel base of 34 inches, with the wheel flanges nearly touching, was justified by the argument that it passed around curves with greater ease than did the wider spread trucks. This was undeniable, but short trucks were notably unsteady when running on a straight track, and the violent "chatter" they often developed was the probable cause of many derailments.

IRON-FRAME TRUCKS. Norris was one of the first builders to break away from the old outside, wooden-frame trucks. As early as 1837 Norris is known to have built an inside-the-wheels, all-iron truck as shown by a contemporary lithograph of the locomotive the *Philadelphia*. This truck was center-

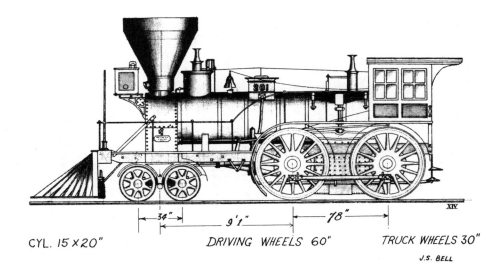

Fig. 68. The inside-connected locomotive No. 201 of the Baltimore and Ohio Railroad showing the "live" spring leading truck. This style of truck was used by a limited number of builders. It proved to be unstable and prone to derail. The 201 was built by Norris in 1853 to a design of the Baltimore and Ohio Railroad.

bearing, that is, no rollers or other side bearings were employed; rather, the weight was borne by the center plate of the truck. The center-bearing arrangement, which had become standard for American locomotives by the 1850's, was claimed by George E. Sellers, who asserted that he first used it on two locomotives built in 1835–36 for the Philadelphia and Columbia Railroad.[33] Baldwin closely followed Norris in abandoning the wooden-frame truck, but the more conservative builder at first retained outside frames and bearings. Baldwin used cast-iron pedestals for trucks as late as the 1850's but some years before he had abandoned outside framing. Other builders followed suit, so that by the early 1840's the standard four-wheel leading truck had an inside iron frame, a short wheel base, and, more often than not, a center bearing. This plan was rarely deviated from until the advent of the spread truck.

A curious variation from the standard pattern was the so-called live truck.[34] It had no frame in the conventional sense; instead, a large leaf spring with journal boxes bolted on either end formed the truck's side frame. A bolster attached to the center of the leaf spring completed the truck. Ross Winans shows a truck of this type in his eight-wheel car patent of 1834 and is known to have used it on the few truck locomotives built at his shop. One of these machines built in the 1840's and fitted with live trucks is illustrated in Fig. 129. Eastwick

[33] *American Machinist,* November 7, 1885, p. 1.

[34] Colburn, *Locomotive Engineering,* p. 97, comments on the spring-side frame truck as the "live truck."

COMPONENTS

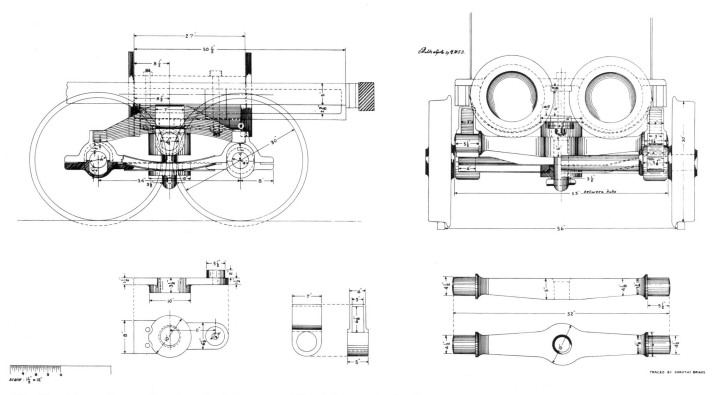

Fig. 69. *Live-spring truck used on the Baltimore and Ohio's No. 201 and other 4–4–0's of the so-called Dutch Wagon class.*

and Harrison built locomotives with live trucks: the *Gowan and Marx* is shown with the springs inside the wheels; the *Mercury* had a similar truck with the springs outside the wheels. Baldwin is also known to have made limited use of live trucks in the 1840's.[35] The Baltimore and Ohio Railroad used this style of truck on many of its passenger locomotives. As late as 1853 a modified form of live truck with a light inside frame

[35] *History of the Baldwin Locomotive Works,* p. 41, shows an 1845 4–4–0 with a live truck. A similar line drawing of a 4–2–0 of the same period issued by Baldwin shows an outside live truck.

to stabilize its working was developed for a class of light, inside-connected passenger locomotives known as the "Dutch Wagons" (see Figs. 68 and 69). The live truck was too common to be dismissed as a mere novelty, but it should not be regarded as a standard design. It was a simple, cheap truck, but its instability and generally "loose" construction resulted in its abandonment by all but a few roads before 1855. The live truck is notable, however, as an early example of an equalized truck; the side-frame spring transferred the movement of one axle to the other.

171

SPREAD TRUCKS A momentous reform in truck design took place in the early 1850's when several builders began to use wide spread trucks. This was the most important single reform in truck design since that device had been introduced some twenty years before. The exaggerated importance of a short wheel base for negotiating curves was at last recognized. In 1855 it was noted that "the adoption of the wide-spread, center bearing truck has been beyond all doubt a means of materially steadying the engine upon the track."[36]

It has been stated that the earliest trucks had a 48-inch wheel base but that, after the late 1830's, 34- to 37-inch wheel bases were adopted. Aside from a minor school that promoted the so-called square truck, where the wheel base was equal to the track gauge, the short wheel base remained in favor until 1850. The spread truck with the 66- to 72-inch wheel base that came in at this time was initiated not so much as a truck reform as apparently an effort to eliminate incline cylinders. Whatever the motivation behind this design change, spread trucks were at once recognized as a meaningful improvement for locomotive running gears.

Unfortunately, it is not clear who first introduced this improvement; several mechanics began using it at the same time, and many others have been proclaimed as its inventor. Thomas Rogers is said to have designed a wide spread truck in 1850. A drawing showing a remarkably modern, equalized spread truck is reproduced in the history of the Rogers Locomotive Works prepared by M. N. Forney.[37] However, it is the author's opinion that this is a representational drawing and not one based on an actual truck design of 1850. Forney admits that no early truck drawings survived and does not present the 1850 illustration as an original document. Furthermore, the few surviving illustrations of Rogers' locomotives of this date show engines with short wheel base trucks. One of these, the *Amenia* (Fig. 66), is shown in the present volume. The Rogers claim should at least be questioned.

Septimus Norris has also been credited as the originator of the spread truck. According to one account, Norris prepared designs for a locomotive with level cylinders and a truck with a 62-inch wheel base in March, 1851.[38] The *Lycoming*, completed in December, 1851, for the Allegheny Portage Railroad, was said to be the first engine so-built.[39] Another twenty-two engines were built on the same plan before Norris, for some unknown reason, returned to the old incline cylinder and short wheel base truck design. Norris' claim is strengthened by the existence of a small lithograph of the locomotive the *North Star*, which—according to the lithograph's printed legend—was built in 1851 for the Syracuse and Utica Railroad. Level cylinders and a spread truck are shown. Presumably this machine was one of the twenty-two spread-truck engines just mentioned.

Wilson Eddy completed an engine with level cylinders and a spread truck early in 1851. Named the *Addison Gilmore*, this machine was something of an oddity, having a single set of high driving wheels. Even so, it entitles Eddy to a place among the claimants of the spread truck. About a year after Eddy's engine was finished, William Swinburn of Paterson completed a spread-truck engine, the *America*, for the Buffalo and Corning Railroad. A lithograph of this machine was issued at the time. William Mason's claim for spread trucks and level cylinders appears unjustified when it is remembered that he completed his first engine in October of 1853—some time after the introduction of these reforms by the mechanics just mentioned.

While the originator of the spread truck is unknown, the universal acceptance of the design by 1855 is certain. A few small builders, such as Denmead, used short-wheel-base trucks as late as 1857.

The introduction of spread trucks brought about no startling changes in truck frame construction. As before, frames were fabricated from numerous small wrought-iron straps and forgings. McQueen had perfected a one-piece side frame as early as

---

[36] *Railroad Advocate*, July 28, 1855, p. 2.
[37] *Locomotives and Locomotive Building*, p. 69

[38] *Engineer* (Phila.), September 13, 1860, p. 33.
[39] A roster of Allegheny Portage engines appearing in *Railway and Locomotive Historical Society*, Bulletin No. 89 (1953), p. 153, indicates that the *Lycoming* was delivered in 1850.

1848. This design was more solid than the fabricated truck frame and was copied by other builders throughout the 1850's. Examples of McQueen's truck are shown by the *Mohawk* (Fig. 23), the *Columbia* (Fig. 158), and the *Superior* (Fig. 162). Norris improved on McQueen's design by producing a one-piece truck frame with top rails, end rails, and pedestals all forged together as a solid unit.[40] Norris was making such truck frames in 1854 and may well have produced them earlier.

Double-bearing trucks received a limited amount of attention in this country during the 1850's. The arrangement of four bearings per axle was achieved by the use of an inside and outside frame. The advantage of this arrangement was that in case of a broken axle the four points of bearing would hold the wheels in line until the engine was stopped. It was considered a necessary safety feature for express locomotives, but the limited number of such machines in the nineteenth century meant that relatively few double-bearing trucks were used. The *Columbia* (Fig. 158) shows a typical double-bearing truck. The difficulty of keeping four axle bearings in line and the added cost and complexity of the double framing were the truck's major disadvantages. The date of its introduction is uncertain, but it may have been inspired by Kite's safety straps, an iron frame inside the wheels that fitted loosely around the axle and was intended to support the axle if it should break. It was used extensively on passenger cars after its introduction in 1838 and may have prompted the development of double-bearing locomotive trucks.

SAFETY TRUCKS. The ordinary center-bearing truck proved satisfactory to most railroad mechanics; however, a few authorities claimed that the "geometry" of the standard trucks was defective. They blamed many derailments on the inability of standard trucks to properly lead the locomotive into curves and urged the development of a safety truck.[41] According to one of these critics, Levi Bissell, the major defect was the central location of the truck's swivel pins. Bissell proposed to move the swivel pin behind the truck by extending the truck's frame. This shortening of the total rigid wheel base of the locomotive permitted an easier transit of curves. According to Bissell, this arrangement also permitted the axles to remain parallel to the radial line of the curve. A second reform that Bissell advocated was the use of an inclined plane centering device. The weight borne by the truck rested on the inclined planes that occupied the position usually taken by the center plate. Bissell patented his idea on August 4, 1857 (No. 17913).

A few months before his patent was issued, Bissell's truck was tested on the Central Railroad of New Jersey. The following is part of a report prepared by two officials of the railroad:[42]

After the said invention of Bissell had been applied the engine was run out onto a curve which she turned apparently with nearly as much facility as she would travel on a straight line, and the forward part of the engine rose on the inclines as the truck entered the curve and remained fixed while running around said curve and then resumed its former position on entering a straight track, and the trial was pronounced by all who saw it as most satisfactory, even by those who before pronounced that it would be a failure.

At one of the trials a bar of iron ¾ x 4 inches was spiked down across one of the *rails diagonally of the track,* . . . and the employees of the company took the precaution to fill in around the track to facilitate getting the engine back again, supposing she must jump off; however on passing over slowly she still kept the track and the speed was increased until she passed over said bar . . . while under a considerable speed.

The Bissell safety truck was favorably received and won a particularly devoted friend in the Rogers Locomotive Works; many of the Rogers engines were equipped with this style of truck. Bissell's success attracted others to the safety problem and within a few years several safety-truck designs had been developed. The most important of these was Alba F. Smith's substitution of swing links for Bissell's inclined planes.[43] Smith admitted in his patent of February 11,

---

[40] *Railroad Advocate,* December 23, 1854, p. 1.
[41] *Introduction of the Locomotive Safety Truck,* United States National Museum Bulletin 228, Paper 24 (1961).

[42] *Ibid.,* p. 123.
[43] Sinclair, *Locomotive Engine,* p. 213, claims that William Bullock

1862 (No. 34377), that swing links for car trucks had been patented in 1841 and that his claim was restricted to their use on locomotive trucks. This plan, sometimes referred to as the swing-bolster truck, quickly superseded Bissell's scheme and was used well into the twentieth century. While swing-bolster leading trucks had been widely accepted by the 1870's, a number of mechanics clung to the stationary bolster believing it to be as effective on curves and cheaper to build than the Smith truck.[44]

Before 1859 any mention of leading trucks referred exclusively to four-wheel trucks. No two-wheel or pony trucks are known to have existed before this time. In 1857 Levi Bissell devised a two-wheel truck at the suggestion of Zerah Colburn. The idea was patented the next year, but no prototype was tested until 1859 when the Eastern Counties Railway fitted its No. 248 with a two-wheel truck. The truck was a great success and other British railways adopted it in later years. The pony truck was tried in the United States late in 1859 but was not enthusiastically received at first. A general satisfaction with the 4–4–0 for all types of service and the tendency of Bissell's two-wheel truck to derail prevented the immediate success of the pony truck. Within a few years, however, the growing need for faster, more powerful freight locomotives revived interest in the two-wheel truck. Freight locomotives without trucks (0–6–0's and 0–8–0's) derailed at all but the most moderate speeds. Because the front drivers were heavily overloaded, considerable damage to the track and the drivers resulted. Four-wheel trucks took too much weight from the drivers, thereby reducing engine traction. A two-wheel truck effectively compromised all of these difficulties.

John P. Laird was one of several skilled mechanics who attempted to improve Bissell's first plan for the two-wheel pony truck. In 1857, while master mechanic of the Marietta and Cincinnati Railroad, Laird rebuilt a 4–4–0 as a chain-driven 2–6–0. Laird equalized the truck with the front driving wheels, thus producing a major improvement over Bissell's truck. Unfortunately his design was complex; two equalizing levers and laterally sliding journal boxes for the front drivers were used. It might be added that in the 1840's Rogers tried equalizing the four-wheel leading truck of ordinary 4–4–0's in a scheme similar to Laird's plan (Fig. 20). During the early 1860's Laird rebuilt a number of engines on the Pennsylvania Railroad with this style of truck. He received a patent for it on March 6, 1866 (No. 53009), but no other railroad or builder is known to have used it.

Another designer who recognized the necessity of equalizing the pony truck was John L. Whetstone of Cincinnati. Whetstone, formerly chief designer of Niles and Company's locomotive department, devised a plan of truck in which the frame and equalizing lever were one unit. The truck wheels and front driving springs were connected by a transverse hanger. Sound as this plan appeared, it was not tested; Niles and Company quit the locomotive business a few months before Whetstone's patent was granted and no other manufacturers are known to have accepted this design.

Four years after Whetstone's patent a practical scheme for equalizing the pony truck to front drivers was developed by William S. Hudson, superintendent of the Rogers Locomotive Works. Hudson's plan consisted of a heavy equalizing lever between the truck and the front driving wheels on the center line of the locomotive. The front end of the lever rested on the truck frame; the rear was attached to a transverse bar connected to the front spring hangers. The equalizing lever pivoted near its center on a fulcrum fitted to the underside of the cylinder saddle. Drawings of this form of truck appear in the *Consolidation* section (see Fig. 229). Hudson's plan was an immediate success. It constituted a major improvement to locomotive trucks and permitted the introduction of Mogul and Consolidation freight locomotives. Hudson obtained a patent on May 10, 1864 (No. 42662); his invention was used on steam locomotives with two-wheel leading trucks until the end of their construction.

A few years after Hudson's patent was granted, the several truck patents of Bissell, Smith, and Hudson came under the

---

used "swinging trucks before Bissell patented that invention." Sinclair erroneously credits Smith's patent to Bissell and fails to mention the year of Bullock's truck.

[44] *Master Mechanics Report,* 1877, p. 196.

control of the Locomotive Engine Safety Truck Company of New York.[45] The patent trust prosecuted both builders and railroads who violated their patents. This activity was carried on with such vigor that many roads were discouraged from employing safety trucks and remained loyal to stationary bolster trucks instead. The trust's power was finally broken in 1880 when the Pennsylvania Railroad successfully challenged the Safety Truck Company in the courts. After this time, swing-motion trucks became more common.

TRAILING TRUCKS. Trailing wheels are normally associated only with wide-firebox locomotives of the twentieth century, yet they were used to a limited extent in the last century. Rogers first used trailing wheels on the *Stockbridge*, a 4–2–2 completed in January of 1842 for the Housatonic Railroad. During the next few years Rogers built several more machines of the same wheel arrangement. In the mid-1840's Hinkley built about six 4–2–2's for various New England railroads. In addition, several old engines are known to have been rebuilt with trailing wheels at this time; two known examples are the *Dover* of the Morris and Essex Railroad, formerly a 4–2–0 (Baldwin, 1841), and a 2–2–0 of the Baltimore and Susquehanna Railroad, possibly the *Baltimore* (Fig. 1), which was rebuilt in 1845 as a 4–2–2.[46] It is unlikely that any of the machines mentioned were fitted with trailing trucks as such; rather, it is probable that trailing wheels were fitted to the locomotive's main frame.

Four-wheel trailing trucks are known to have been used as early as 1849. The Old Colony Railroad acquired two 4–4–4's in that year. Ross Winans' passenger locomotive the *Carroll of Carrollton*, a 4–2–4, also completed in 1849, had a four-wheel trailing truck. In general, however, it must be conceded that trailing trucks were not extensively used on American locomotives until after 1900.

DRIVING WHEELS AND AXLES

A fairly large variety of driving wheel styles—wood, wrought-iron, and cast-iron—were tried by the earliest American locomotive designers. By the mid-1850's, however, the vast majority of American locomotives were being fitted with oval-spoked, cast-iron driving wheels.

Wooden wheels were used on many of our pioneer engines. Such wheels were favored by British makers during the early 1830's, and the majority of their earliest locomotives were so-fitted. The *Stourbridge Lion, John Bull, Robert Fulton,* and several other British imports are known to have had wooden wheels. American builders copied this practice; we find Baldwin, the West Point Foundry, and others equipping their first products with wooden wheels. Such wheels had cast-iron hubs and wooden spokes and fellies. A thin wrought-iron band, fitted tightly around the felly or rim, held the wheel together. A heavier, flanged tire was fitted over the band.

Stephenson used a crank ring to support the wheel and the side pin rod; West Point copied this plan exactly for the *Best Friend's* driving wheels. Other makers used iron for the side pin-rod spoke. With the exception of the cast-iron hub, the manufacture of such wheels was directly derived from long-established wagon wheel techniques.

Economy was the main justification for wooden driving wheels, but this advantage was wholly offset by the miserable performance of such wheels. They were soon found too weak for the heavy demands of railway service, particularly on American lines where sharp curves and rough tracks were the rule. The Camden and Amboy Railroad abandoned wooden driving wheels in 1834 after one year's service on its first locomotive, the *John Bull*. Another Stephenson engine of the same name, but built for the Mohawk and Hudson Railroad, had

---

[45] *Poor's Manual of Railroads*, 1868–69, contains an advertisement of the Locomotive Engine Safety Truck Company on pp. 364–65.

wooden drivers that shrieked and "gave audible complaint of hard service" when passing over curves.[47] Wooden driving wheels were accordingly abandoned very early by American builders, and it is probable that all such wheels disappeared from active service by the mid-1830's in this country.

IRON WHEELS. All-metal driving wheels had been used in England some years before Stephenson advocated the use of wooden wheels in the late 1820's. A return to this form of construction, which in truth had never died out, occurred with renewed interest in the 1830's. At first wrought iron was considered the most promising material for driving wheels, and numerous patterns were developed. In 1832 Stephenson himself began the manufacture of "gas-pipe" wheels; the spokes were made of hollow wrought-iron tubes. A contemporary description of the Newcastle and Frenchtown Railroad's engine the *Phoenix* pictures Stephenson's gas-pipe wheels as follows: "The spokes of the wheels are wrought iron tubes, bell shaped at their extremities; the rim and hub cast on them—the union being effected by means of borax."[48] Wheels of the same design were used by Stephenson's associates at the Tayleur Foundry for the locomotive *Fire Fly* built in 1833 for the Philadelphia and Columbia Railroad. Not long after the *Fire Fly* entered service the railroad reported to the builder the failure of several spokes of the gas-pipe driving wheels.[49] Stephenson was forced to acknowledge the weakness of the hollow-spoke wheel and dropped its manufacture soon thereafter. Along with other British makers, he began to use solid iron spokes.

The first all-iron driving wheels were made in the United States in 1831 by the West Point Foundry for the *De Witt Clinton*. The spokes of round wrought-iron bars 1 inch in diameter were set in a staggered pattern. The hub and felly were of cast iron; a wrought-iron tire was fitted to the wheel. Another locomotive of the Mohawk and Hudson Railroad, the *Experiment*, built a year later by West Point, was fitted with driving wheels of the same construction.[50] The Camden and Amboy was another early user of wrought-iron drivers. The *John Bull* was fitted with fabricated iron wheels in 1834 after the failure of its original wooden wheels. These wheels were made with solid rounded spokes riveted to the hub and felly.[51] The rivets loosened after one year's service, thus making the wheels too unstable for further use. Apparently other roads experienced similar difficulties, and, except for the isolated use of such wheels on the Reading in 1848, fabricated wrought-iron drivers had disappeared from American practice by the end of the 1830's.[52]

While American interest in wrought-iron driving wheels disappeared at an early date, British designers continued to develop new plans for such wheels. By the mid-1840's, elegant, slender-spoked wrought-iron wheels, forge-welded into a solid assembly, had been fully developed by British designers. Wheels of this type soon became standard on British locomotives, but their use in the United States was limited in general to a handful of high-wheel express locomotives built between 1849 and 1854.[53] Exceptions to the prevailing practice can, of course, be found. The Michigan Central, contending that cast-iron wheels broke too easily, ordered a large number of wrought-iron drivers in 1856.[54] The Portland Company and The Nashua Iron Company manufactured forged, wrought-

---

[46] The *Dover* is illustrated in the *Railroad Gazette*, June 13, 1902, p. 440; reference to the reconstructed Baltimore and Susquehanna locomotive is made in the *American Railroad Journal*, November 6, 1845, p. 714.

[47] Stevens, *Beginnings of the New York Central*, p. 48.

[48] Wood, *A Practical Treatise* (1832 ed.), p. 532.

[49] Warren, *A Century of Locomotive Building*, p. 284.

[50] Reference to wrought- and cast-iron wheels is made in the contract for the *Experiment* dated November 16, 1831. See Appendix A.

[51] Memo of Isaac Dripps, September 6, 1885, in the files of the Smithsonian Institution.

[52] *A Century of Reading Company Motive Power* (Philadelphia, 1941), p. 18, shows a drawing of an 1848 4-4-0 with riveted spoked wheels.

[53] Among the high-wheel engines with wrought-iron wheels were the *Lightning*, the *Carroll of Carrollton*, the *Superior*, and the Camden and Amboy's Cramptons.

[54] *Railroad Advocate*, September 27 and October 4, 1856; Michigan Central Annual Report, 1857.

iron drivers during the same period.[55] In addition, small numbers of such wheels were imported from England. However, wrought-iron wheels showed no advantages over ordinary cast-iron wheels. They were difficult and expensive to make—involving twenty to thirty welds[56]—and because of the numerous welds they were considered unsafe. Finally, they were no lighter or stronger than cast-iron wheels.

Cast iron had been used for railway car wheels since the mid-eighteenth century; therefore, it is not surprising to find that it was employed for locomotive wheels at an early date. Cast-iron drivers were satisfactory for the pioneer colliery locomotives, but, as previously mentioned, British designers turned to wrought iron in later years, believing it to be better suited for the faster working engines required by public railways. In America the opposite opinion prevailed; cast iron was considered the best material for driving wheels. Cast-iron wheels had become the favored style of driving wheel in this country by 1835 and remained so until the introduction of cast steel some sixty years later.

The T-shaped spoke was the earliest style used on cast-iron wheels in this country and until about 1850 remained the most popular style. It was simple, light, and easy to cast. The projecting rib of the T might face inside or outside according to the taste of the maker. Baldwin preferred to have the rib facing inside; an example of this design is shown by the *Lancaster* (Fig. 111). An early example of the T spoke with the rib on the outside is shown by the 1835 *Cincinnati* drawing (Fig. 3). H- and U-shaped spokes were occasionally used as well.

Rogers introduced the hollow, oval-shaped spoke in 1837 on his first locomotive, the *Sandusky*. Other builders slowly began to adopt this design. By 1850 it had begun to rival the old T spoke; five years later the oval spoke was the accepted design. Several examples of hollow, oval-spoked wheels are included in the present volume; see the *Phantom* (Fig. 193) and Tyson's Ten Wheeler (Fig. 188). The oval spoke was stronger and lighter than the old T spoke. It was also considered a cleaner and more handsome wheel than the old T-spoke pattern.

The high quality of cast iron used for driving wheels was basic to their long success in this country. Unlike European mechanics, Americans used the finest quality cast iron available. This iron was characterized by an unusually high tensile strength of 30,000 pounds, or about twice that of ordinary gray iron. Aside from superior iron, great care was taken in the manufacture of cast-iron wheels. Particular attention was given to uniform cooling of the casting in order to avoid stresses. Hollow spokes, hubs, and rims not only insured more even cooling but also reduced the weight of the wheels.

Wheel diameter was relatively small during most of the nineteenth century because general service locomotives were the rule—54-inch to 60-inch wheels were common. A moderate number of passenger engines had 72-inch wheels, but the number of machines with larger sizes was exceedingly small and must be classed as variants of the norm.

CUSHION WHEELS. Because the driving wheels and axles were "unsprung weight," consideration was given to cushion wheels in an effort to reduce damage to wheels and track. The basic plan was to insert wood between the wheel and tire, thus providing an elastic layer between the two parts. One of the first plans for this idea was Ross Winans' car-wheel patent of November 19, 1833. A thick continuous segment of wood separated the cast-iron center of the wheel from the tire. Baldwin adapted the idea for locomotive driving wheels but used blocks instead of a one-piece insert. A patent was secured on September 10, 1834.[57] A number of engines were built with this style of wheel—The Utica and Schenec-

---

[55] A lithograph of the *Minnehaha* issued in 1856 by the Portland Company notes the manufacture of wrought-iron wheels. A handbill in the Baldwin Letters, Historical Society of Pennsylvania, dated September 20, 1859, illustrates wrought-iron drivers manufactured by the Nashua Iron Company, Nashua, New Hampshire.

[56] *American Railroad Journal*, July 29, 1853, p. 478; see also *American Blacksmith*, April, 1918, p. 164, for an article on the making of wrought-iron drivers in England in 1855.

[57] Baldwin's cushion driving wheel is illustrated on p. 22 of *History of the Baldwin Locomotive Works*.

tady's No. 1 (Fig. 107) is a late example—but it appears to have been abandoned by Baldwin before 1840. Baldwin obtained a second patent on April 3, 1835, for a cushion wheel, but it is not certain that this design was used.[58] In the same year George E. Sellers devised a wooden cushion wheel for two locomotives being built for the Philadelphia and Columbia Railroad.[59] He inserted wood blocks inside the hollow, cast-iron rim of the wheel so that the tire rested partially on the block and partially on the wheel's rim. Sellers admitted the doubtful effectiveness of the plan and stated that the wood blocks were burned to dust when the tire was shrunk onto the wheel. After the work of Baldwin and Sellers, interest in cushion wheels disappeared except in the case of passenger cars, for it was believed that wooden crossties—which succeeded the earlier rigid stone block ties—effectively cushioned the wheels at low speeds.

The idea was revived in 1857 by George S. Griggs, who obtained a patent (No. 18966) on December 29 of that year. Griggs's main concern, unlike that of earlier mechanics interested in such wheels, was to promote tire life rather than merely to cushion the wheels. Griggs had noted the benefits of wooden cushion wheels on the passenger cars of the Boston and Providence before 1851 and was undoubtedly prompted by this experience to develop a similar wheel for locomotives.[60] The felly of Griggs's wheel was cast with dovetail recesses into which hard-wood wedges were driven. The wheel was turned in a lathe so that the wedges could be trimmed uniformly to receive the tire. The tire was shrunk onto the wheel. To prevent burning of the blocks it was necessary to quickly douse the wheel with water.

Griggs claimed a 50 per cent saving in tire wear for his wheel.[61] A test on the Boston and Providence showed that engines with wooden cushion wheels averaged 35,208 miles while those with ordinary driving wheels averaged only 20,774 miles before the tires required turning.[62] Other railroads, impressed by this excellent performance, adopted Griggs's wheel, and, while it was not universally accepted, it was widely used. The introduction of steel tires, which were better able to withstand wear than wrought-iron tires, brought an end to wooden cushion wheels during the 1870's.[63] A pair of Griggs's cushion wheels have survived to the present day on the locomotive the *Daniel Nason,* built by Griggs in 1858 and exhibited in Danbury, Connecticut.

AXLES. Driving-wheel axles were made from wrought iron and generally measured about 6 inches in diameter. A forging of this size was difficult to make, yet George Sellers recalled making an axle of this diameter in 1835 with only a blacksmith's forge and a 200-pound trip hammer.[64] Colburn described two methods of axle-making as developed in 1860:

> Axles are made with different degrees of care at different forges. Scrap axles are generally drawn out under the hammer at two heats. At some of the forges where new iron is puddled for axles, the blooms are first squeezed by a rotary squeezer, then piled, heated, and rolled, twice in succession, and finally heated and swaged to shape. An excellent but expensive axle is made also from cold blast, charcoal iron, shingled with a hammer, then piled, heated, and hammered three times in succession, and finished with one swaging heat. Considerable difference of opinion prevails as to the relative merits of new iron and of scrap for axles.[65]

Axles, like most other forgings at the time, were made from "piles" of scrap wrought iron. If worked skillfully from selected scrap, a solid, inexpensive axle could be made. However, a defective axle often resulted from the unintentional inclusion of steel or cast-iron scraps in the pile; these foreign bits

---

[58] E. S. Ferguson (ed.), *Early Engineering Reminiscences (1815–1840) of George Escol Sellers,* United States National Museum Bulletin No. 238 (1965), p. 177, illustrates Baldwin's 1835 wheel patent.
[59] *Ibid.,* pp. 165–66.
[60] Colburn, *The Locomotive Engine,* p. 72.
[61] Advertisement in *King's Railroad Directory,* 1867, p. 180.

[62] *American Railway Review,* January 26, 1861, p. 38.
[63] Weissenborn, *American Locomotive Engineering,* p. 172.
[64] Ferguson (ed.), *Early Engineering Reminiscences,* p. 164.
[65] Clark and Colburn, *Recent Practice,* p. 55.

would not weld with the wrought iron. A stronger, more homogeneous axle was made from charcoal wrought-iron blooms in the 1850's by the Freedom Iron Works of Lewistown, Pennsylvania.[66] Before the introduction of steel these were considered the best axles available.

In 1848 or 1849 a set of cast-steel axles, imported from Krupp, were fitted to an engine built for the Pennsylvania Railroad by Baldwin.[67] In 1854, after running some 80,000 miles, the axles were said to show no signs of wear.[68] The good service of these axles did not bring about an immediate acceptance of steel, however; their greater cost, the brittle alloys first offered, and the skepticism surrounding innovation prevented their general use for many years. Moreover, iron was a familiar material which, though not free from faults, had defects that were well known to mechanics of the time and could be worked into a reasonably reliable axle. Steel, on the other hand, was a new, variable material that few mechanical officials cared to trust for so critical a component as axles. The following statement expressed this opinion: "Hammered iron seems to be a favorite material, as its quality is pretty well known and its uniformity can be depended upon."[69] In 1870 the *Master Mechanics Report* stated that only half the roads questioned had any steel axles in service.[70] Wrought-iron axles were still standard with such famous builders as Baldwin seven years later, and even as late as 1891 only a small number of master mechanics were reported to favor steel axles.[71] In the next few years as heavy freight and faster passenger locomotives became more common, steel axles were finally accepted.

Previous to about 1855, driving wheels were driven onto the axles by sledge hammers.[72] The wheel hub was bored with a tapered hole; the axle ends were turned with a corresponding taper. After the keys were in place an ornamental brass cover plate was fitted over the ends of the axles and the hub, which were somewhat battered from the heavy hammering. Such cover plates are clearly shown by the photograph of the *Philip Thomas* (Fig. 18). The hammer mounting method had several defects; the wheels were not securely or accurately mounted, and precise quartering of the crank pins was difficult to achieve.

Parley I. Perrin of the Taunton Manufacturing Company recalled building a machine tool early in 1849 that simultaneously bored the crank-pin holes on a pair of driving wheels, thus assuring perfect quartering.[73] A duplicate machine was supplied to the Philadelphia and Reading Railroad in June, 1849.

By 1855 Norris is known to have used a mechanical press and a quartering machine for precision mounting of driving wheels.[74] An equally elaborate wheel-quartering machine was designed at about the same time by Thomas Rogers.

By 1860 hydraulic wheel presses and quartering machines were commonly used for wheel- and axle-mounting.[75]

TIRES

The tire is the hoop-shaped wearing portion of the driving wheel. It is designed to be removable so that the wearing portions of the wheel, the tread and flange, might be replaced without discarding the entire wheel. Tires were one of the greatest operating dangers, expenses, and nuisances associated with locomotives because of their short life, tendency to break in service, and the work involved in replacing them. A good

---

[66] *Railroad Advocate*, May 5, 1855, p. 3.
[67] *Railroad Advocate*, September 12, 1857, p. 79; *History of the Baldwin Locomotive Works*, p. 47, states that the axles were tested in 1848.
[68] *Ibid.*
[69] Weissenborn, *American Locomotive Engineering*, p. 175.
[70] P. 67.
[71] *Ibid.*, 1891, p. 54.
[72] Clark and Colburn, *Recent Practice*, p. 64.
[73] *Locomotive Engineering*, July, 1892, p. 228.
[74] *United States Magazine*, October, 1855, illustrates both machines.
[75] Clark and Colburn, *Recent Practice*, p. 65.

iron tire rarely gave more than 24,000 miles' service before it required turning, and was considered good for no more than a total service of 60,000 miles.[76] The replacement or turning of worn tires involved a considerable disassemblage of the locomotive in order to remove the wheels for servicing. Thus, the locomotive was out of service for several days. To avoid duplicate time for shopping, it was a rule-of-thumb to give the locomotive a general overhaul when the tires needed turning.[77]

The search for a durable, strong, and easily mountable tire led to the employment of many designs and materials. The major styles of locomotive tires were wrought iron, chilled cast iron, and, after 1850, steel. Wrought iron was the first and most popular material for locomotive tires during the nineteenth century. Tires were first used by British makers in the late 1820's and the merit of the practice was quickly established and became standard before any locomotives were manufactured in the United States. The earliest tires were rarely more than 1½ inches thick, but the short life of such thin tires hastened the manufacture of thicker tires so that by the 1870's 2¾-inch tires were common.[78] Thin tires permitted only a few turnings and were also considered dangerous because they were liable to "flex" and thus work loose from the wheel. Some mechanics disputed the thick-tire scheme, contending that after the first or second turning the best quality iron was removed, exposing the interior flaws and laminations so common to thick wrought-iron forgings, and that it was cheaper to buy thin but sound tires.[79]

The first tires used in this country were imported from England, and for many years American railroads exhibited a strong prejudice in favor of Low Moor and Bowling tires. While this prejudice continued well into the century, one American firm attempted the commercial manufacture of tires as early as 1836. Baldwin, eager to obtain a local supply of tires and crank axles, encouraged Stephen Vail of the Speedwell Iron Works at Morristown, New Jersey, to undertake this difficult task. Vail was about to abandon the project because of defective welds. In a letter of August 3, 1836, Baldwin noted that eight of ten tires supplied broke because of this failing; however, he urged Vail not to give up, saying that if a good tire could be produced large orders would follow.[80] The Speedwell Works persisted and was for a time the only domestic maker of locomotive tires.

Some years later another American firm, Norris Brothers of Philadelphia, admitted that "after numerous trials, the attempt to make tyres for the driving wheels of locomotives [has] been given up" and that English tires were being used on locomotives of their manufacture.[81] By 1855 Norris was manufacturing its own tires (from English iron, however), apparently having solved several years earlier the mysteries involved in the production. The methods used are given in the following contemporary account.

In the yard, is the apparatus for tiring wheels. For which, only the Low Moor iron is used here. There are two furnaces, one straight and the other circular. The straight bars are first heated in the former, and placed on the bending machine, which leaves them in the form of a half circle; two of these are welded together, and thus the tire receives its circular shape. When the tiring commences, the tires are heated in the round furnace, and at the proper time are removed to the forming machine, where, by a contrivance in which the lever, wedge and crew are most successfully combined, they receive their perfect circular shape in the process termed "forming." While hot, they are placed around the iron wheel. There being a quarter of an inch play they are then plunged in a vat of cold water, and when cold the adhesion of the tire to the wheel is so perfect, that the wheel and tire [have] the appearance of one solid piece.[82]

Other American concerns making tires in the 1850's were the Nashua Iron Company, Nashua, New Hampshire; Freedom Iron Works, Lewistown, Pennsylvania; Ames Iron Company, Fall Village, Connecticut; Paterson Iron Company, Paterson,

---

[76] *American Railway Times,* September 21, 1861, p. 365.
[77] American Society of Civil Engineers, *Proceedings,* 1893, p. 387.
[78] *Master Mechanics Report,* 1874, p. 215.
[79] *American Railway Times,* September 21, 1861, p. 364.
[80] Baldwin's letter to Vail is preserved by the Historical Society of Pennsylvania.
[81] *American Railroad Journal,* January 23, 1845, p. 57.
[82] From an illustrated description of the Norris Works appearing in the *United States Magazine,* October, 1855, pp. 151–67.

New Jersey; and Newport Iron Works, Newport, Kentucky. These firms, together with the number of locomotive builders making their own tires, indicate a growing independence of English suppliers.

Because of the likelihood of wrecks, care was taken to secure the best tires obtainable, whether of domestic or foreign making. Even so, tires broke all too often, particularly on northern roads where frozen roadbeds produced a rigid, unyielding track. In the winter of 1855–56 the New York Central reported no less than 200 broken tires on two divisions alone.[83]

CAST-IRON TIRES. The chilled cast-iron tire was the only competitor of wrought iron until the advent of cheap steel tires in the 1860's. Cast iron had proved itself an admirable material for car wheels as far back as the mid-eighteenth century. The life expectancy of such wheels was materially increased in about 1812 when chilled or "case-hardened" wheels were introduced in England.[84] The tread of the wheel was chilled by making that portion of the wheel mold of iron rather than sand. Thus the molten iron was more quickly cooled or chilled when it touched the iron segment of the mold. The metal was crystallized to a depth of about half an inch and offered a very hard, durable running surface. The economy and easy manufacture of such wheels naturally led to an early investigation of chilled cast-iron driving wheels. Seth Boyden is said to have used an all-cast-iron driving wheel with hub, tread, flange, and spokes cast in one piece on his first locomotive, the *Orange*, completed in 1837.[85] A line drawing of the *Orange* made by P. I. Perrin, a former employee of Boyden, shows a wheel of this construction.[86] However, D. M. Harris insists that the *Orange* was originally fitted with wrought-iron tires made by his father.[87] At this late date it is impossible to determine whether or not tireless cast-iron driving wheels were used by Boyden on the *Orange*.

It is known that Winans produced similar wheels in the early 1840's. Production of these wheels was soon abandoned, however, because the lack of separate tires made it necessary to scrap the entire wheel when the tread or flange wore through the chill.

The earliest known use of separate cast-iron tires was on the Camden and Amboy Railroad in 1838.[88] This was only a temporary measure resorted to because wrought-iron tires were not available. However, two years later the Baltimore and Susquehanna Railroad's annual report noted the regular use of such tires. "Cast iron wheels with chilled treads were substituted on locomotives for those with wrought-iron tires which wore so rapidly and required frequent renewals. Chilled driving wheels 4½ feet [in] diameter have been running on the road for six months, answering perfectly. Other railroads have followed and adopted this improvement."

The "other railroads" mentioned above undoubtedly were the Baltimore and Ohio and the Philadelphia, Wilmington and Baltimore. Both roads began the use of cast-iron tires in the early 1840's.[89] By 1847 these three roads were said to use cast-iron tires "entirely."[90] The Philadelphia and Reading was another large user of cast-iron tires; this preference was attributable to the road's chief mechanic, James Millholland, who had been master mechanic of the Baltimore and Susquehanna when cast-iron tires were introduced. A number of western lines, the Little Miami, Central Ohio, Galena and Chicago Union, and others are known to have used cast-iron tires, but the over-all extent of their employment is uncertain.[91] Certainly they did not seriously challenge wrought-iron tires. The use of chilled tires was in fact largely restricted to the Baltimore and Ohio, the Baltimore and Susquehanna, the Phila-

---

[83] *Engineer*, January 3, 1862, p. 6.
[84] Marshall, *A History of the Railway Locomotive Engine*, p. 126. In error, the invention of chilled wheels has been repeatedly credited to Ross Winans.
[85] *Railroad Gazette*, June 6, 1902, p. 408.
[86] Perrin's drawing originally appeared in *Locomotive Engineering*, October, 1893, p. 433.
[87] *Locomotive Engineering*, December, 1893, p. 551.

[88] Memorandum of Isaac Dripps to J. E. Watkins, September 6, 1885, in United States National Museum (Washington, D.C.).
[89] *American Railroad Journal*, August 6, 1853, p. 506.
[90] Ross Winans advertisement, *Ibid.*, June 19, 1847, p. 392.
[91] *American Railroad Journal*, December 3, 1853, p. 773.

delphia, Wilmington and Baltimore, and the Philadelphia and Reading Railroads.

Aside from Winans, L. B. Tyng and Company of Lowell, Massachusetts, and Bush and Lobdell of Wilmington, Delaware, were the major manufacturers of cast-iron tires. Bush and Lobdell differed from their competitors by making a hollow tire, thereby decreasing the dead weight and producing a more even chill.

The first cast-iron tires were made 2 inches thick, but by 1853 they had been increased to 3 or 3½ inches for safety.[92] It was impossible to mount cast-iron tires by shrinkage, because cast iron is inflexible; therefore, bolts or rivets were used to fasten them to the wheel center. The tire was turned to slip snugly on the wheel center—hence the term "slip tire," a common designation for cast-iron tires. One of the best methods for fastening cast-iron tires was patented in 1843 by Thatcher Perkins and William McMahon.[93] Several hook-headed bolts passed through the outer edge of the wheel's rim. The head of the bolt gripped the edge of the tire; a nut at the inner edge of the wheel rim drew the tire tightly against the rim. Thus, no holes were made in the tread of the tire. This method also permitted easy replacement of the tire. Examples of Perkins and McMahon's tire bolt are shown in Figs. 13 and 29 and in the wheel drawing for Tyson's Ten Wheeler, Fig. 188.

Cheapness was the major advantage of cast-iron tires. They were said to cost 50–75 per cent less than wrought-iron tires; this claim is essentially substantiated by the following prices:[94]

```
Set of Four Cast iron tires 60-inches diameter    $176.00
 "   "   "    Wt.   "    "    "    "    "         $372.50
```

Aside from their cheapness, cast-iron tires were easily dismountable and were longer-lived than wrought-iron tires. A set of Bush and Lobdell tires was reported serviceable after 15 years' use; however, mileage was not noted.[95] No specific reports have been found for cast-iron tire mileage, but chilled cast-iron car wheels averaged about 80,000 miles.[96] The chief defect of such tires was their lack of adhesion; the hard, chilled surface, so wonderfully suited to long wear, made a slick, slippery contact with the rail.

The decline of the cast-iron tire was rapid after the acceptance of steel tires in the 1860's. By the end of that decade only 3,900 locomotives (or less than 10 per cent of all locomotives in service) were fitted with cast-iron tires.[97] A few years later such tires were no longer in use for road engines but continued to serve on a limited number of switching locomotives.[98]

STEEL TIRES. The introduction of steel tires had a profound effect on locomotive operations and design. Steel tires outlasted iron tires in a ratio of nearly five to one. They were capable of supporting weights that would crush an ordinary wrought-iron tire and thus facilitated a marked increase in axle loadings and over-all locomotive size.

The steel-clad tires that had developed in England by 1840 may be considered as the direct forebears of the all-steel tire.[99] The body of such tires was wrought iron with a thin, steel plate forming the tread. The steel was welded or clad to the tire itself by a rolling process not unlike the making of Sheffield silver plate. Advertisements for steel-clad tires of English manufacture appeared in the *American Railroad Journal* during 1853, but no evidence of their use in this country has been found.

The first all-steel tires were undoubtedly those produced at the Krupp works of Essen, Germany, in 1851. Krupp exhibited steel tires at the Great Exhibition held in London during the

---

[92] *Ibid.,* August 8, 1853, p. 506.
[93] J. S. Bell, *The Early Motive Power of the Baltimore and Ohio Railroad* (New York, 1912), p. 66.
[94] The price of cast-iron tires is from an 1860 Bush and Lobdell price list in the Baldwin Letters, Historical Society of Pennsylvania; the price of the wrought-iron tires is from the *American Railway Times,* September 21, 1861, p. 365.

[95] J. L. Bishop, *A History of American Manufactures* (Philadelphia, 1866), p. 545.
[96] Figures for car-wheel mileage conflict: Galton (*Report to the Lords,* p. 4) gives 60,000–80,000 miles; Clark and Colburn (*Recent Practice,* p. 54) report 50,000–100,000 miles.
[97] *Master Mechanics Report,* 1869, p. 42.
[98] Weissenborn, *American Locomotive Engineering,* p. 172.
[99] Ahrons, *The British Steam Locomotive,* p. 162.

same year. Shortly thereafter Thomas Prosser and Son of New York City became the American agent for Krupp's cast-steel tires.[100] By 1860 Krupp had made 15,000 steel tires. Nine years later total production had doubled; of this number, 16,273 tires were sent to United States railroads.[101]

At the same time that Krupp began production of steel tires, a similar venture was undertaken by James Millholland at the Reading shops of the Philadelphia and Reading, according to a recollection of E. J. Rauch, an associate of Millholland.[102] In the early 1850's, a small forge near Reading produced solid blooms of steel from scrap; Millholland's chief smith, James Mullen, made a set of tires from these blooms in 1851 or 1852. The tires were a success, but Millholland favored cast-iron tires and would not adopt steel for that purpose.

American railroads were slow to accept steel tires; the price was high prior to the time of cheap steel (ca. 1855), and the early steels had a high carbon content that made them brittle and likely to crack. Nevertheless, experiments with steel began at an early date; the Illinois Central tested some Krupp tires in 1854. Conceivably, other American roads made earlier unrecorded tests. The Erie first tried steel tires in about 1862; ten years later every engine on the line, save two or three, had steel tires.[103] Other roads followed a similar pattern so that by 1869 68 per cent of American locomotives were using steel tires.[104] The acceptance of steel tires is unparalleled for its speed and completeness in an industry noted for its aversion to innovation.

The long life of the steel tire was its chief advantage over wrought-iron tires. A steel tire would last 200,000–300,000 miles compared with 60,000 for wrought-iron.[105] The Eastern Counties Railway of England reported a steel tire so little worn after 80,000 miles that it did not require turning, whereas a wrought-iron tire would have gone through several turnings and long since have worn out after an equivalent mileage.[106] Aside from good mileage, steel tires offered a saving in dead weight and cost; because of their greater strength they could be made thinner than wrought-iron tires.

Declining cost was another factor that led to the adoption of steel tires.[107] In 1865, steel tires cost $140 per ton; within a few years the price was down to $100 per ton. By 1878, steel tires were selling for $55–$75 per ton. During the same years, by comparison, wrought-iron tires underwent a modest decline in price from $160 to $190 in 1865 to $95–$125 in 1878.

A few miscellaneous comments are necessary before ending the discussion of tires. Shrinkage was the principal method for attaching tires to the wheel center. The tire was heated enough (not necessarily to a red heat) to expand its diameter by one-quarter to three-eighths of an inch.[108] It was then dropped on the wheel and quenched. The resulting shrinkage provided a large compressive force—too great a force for the good of the tire or wheel. In later years the Master Mechanics Association advised that a shrinkage of one-sixteenth of an inch was sufficient for a 60-inch wheel.[109] Rivets, bolts, and, more rarely, retaining rings were used for greater security in fastening the tire to the wheel center.

---

[100] *American Railroad Journal*, November 19, 1853, carries a Prosser advertisement for Krupp tires.
[101] *American Railway Review*, April 12, 1860, p. 224; *Locomotive Engineering*, July, 1896, pp. 637–38.
[102] *Locomotive Engineering*, June, 1896, p. 500.
[103] *Master Mechanics Report*, 1872, p. 31.
[104] *Ibid.*, 1869, pp. 41–55.
[105] *American Railway Times*, September 21, 1861, p. 365; *American Railway Review*, April 12, 1860, p. 224.
[106] *American Railway Times*, March 23, 1861, p. 113.
[107] *Railroad Gazette*, December 13, 1878, p. 599; all data in this paragraph are taken from this source.
[108] *American Railway Times*, January 26, 1861, p. 38.
[109] In Forney's *Catechism* (1890 ed.), p. 408, a shrinkage table is given.

LOCOMOTIVE BRAKES

Driving-wheel brakes were virtually non-existent on American engines before 1875. Reversing the locomotive was considered adequate to slow or stop its motion, despite the difficulty of this procedure with the clumsy hook motions so common before 1855. The tender brake was used to hold the engine stationary at terminals and other stopping points on the line. Clearly, driving-wheel brakes were considered unnecessary for the safe operation of the locomotive. In addition, it was widely held that the action of the brakes would force the driving axles out of line and impose a heavy strain on the connecting rods.[110]

Despite a general disinterest in locomotive brakes, such devices were applied to locomotives at an early date. Robert Stephenson included a steam brake in the patent specifications for his 2–2–2 "Patentee" locomotive of 1833.[111] While many engines were built according to this patent for British and Continental customers, none are known to have been equipped with the steam brake.

The earliest record of an American steam brake is that of George S. Griggs, master mechanic of the Boston and Providence Railroad. Two locomotives of the road were fitted with Griggs's brake in January, 1848.[112] An upright brake cylinder with a bore 8 inches in diameter was placed on top of the boiler. Necessary rods and side levers connected the cylinder to brake shoes fitted between the driving wheels. According to George Richards, an associate of Griggs, the brakes worked well but were abandoned after less than a year's service because a boiler explosion, caused by a faulty safety valve, was blamed on the steam brake.[113] After Griggs's steam brake there is no further evidence of such applications until 1860 when Baldwin fitted a hand-powered brake to the rear driving wheels of two 0–8–0's built for the Mine Hill Railroad. One of these machines, No. 30, is pictured in the present volume as Fig. 55.

In 1866 George W. Cushing, master mechanic of the Chicago and North Western Railroad, fitted a steam brake to the *Minnie,* a light inspection locomotive built at the company shops three years earlier. Presumably the brake was a success, but according to available records it was not applied to any road engines.

In the late 1870's driving-wheel brakes became more common as train speeds and weights increased. The old prejudices were forgotten during succeeding years as the necessity for more positive train control became increasingly apparent. In 1889 the Master Mechanics Association noted that 12,000 locomotives were fitted with driving-wheel brakes. This would account for about half of the nation's motive power and would include the bulk of the main-line road engines. Within the next decade locomotive brakes became standard for all classes of American locomotives.

RODS AND CROSSHEADS

RODS. The main and connecting rods of locomotives are alternately subjected to large compression and tension stresses as they transmit the reciprocating action of the piston to the wheels. The rods must at once be strong enough to withstand these stresses yet as light as possible in order to reduce the centrifugal disturbances that are unavoidable in reciprocating steam engines. It is impossible, of course, within practical design limitations, to counterbalance the reciprocating parts

---

[110] *Master Mechanics Report,* 1886, p. 111.
[111] Warren, *A Century of Locomotive Building,* pp. 310–13.
[112] *Railroad Advocate,* March 28, 1857, p. 4.
[113] Sinclair, *Locomotive Engine,* pp. 525–26.

(rods, pistons, crossheads, etc.) for all speeds of rotation within a locomotive's normal range of operation. Hence, it is important to keep these parts light so as to reduce the possibility of severe dynamic action should the normal limit of counterbalancing be exceeded.

The main rod connects the piston, through its connection with the crosshead, to the driving wheels. It must be made very stiff to prevent vertical or horizontal bucking while under the enormous working force of the piston. The side or connecting rods labor under many conflicting forces and are best made if they are vertically stiff and can flex slightly horizontally. The connecting rods are continually being pushed out of alignment by the workings of the running gear as it accommodates itself to uneven track, rail joints, and curves. Further misalignment is caused by loose axle boxes and rod bearings or by wheels of varying diameter (possible when one set of tires is more worn than another). It was found that a horizontally flexible connecting rod could thus absorb many of the stresses that might otherwise break a more rigid rod. It was necessary, of course, that the rod be vertically rigid to maintain the parallelism of the crank pins.

Before 1850 round main and connecting rods were universally favored in this country. There were almost no exceptions to this practice. Round rods were favored by the early British makers, and the design was picked up and championed by pioneer builders in the United States. Ease of manufacture probably accounts for the round rod's popularity. Most small shops were better equipped to perform turning than mill work, lathes being the more common machine tool.

Round rods were not of a uniform cross-section but were larger in diameter at the center for stiffness and smaller at either end for lightness. This was a graceful and pleasing design and continued to be favored for stationary and marine engines long after it was considered obsolete for locomotives.

Flat rods began to appear in the early 1850's and so rapidly surpassed the old, round design that they soon became the standard type of locomotive rod. Few new locomotives were built with round rods after 1855. The flat rod succeeded the older design because it possessed the same *vertical* strength as the round rod while offering a great saving in weight. Also, being of a uniform dimension, round rods could not flex horizontally. Finally, planing and large milling machines were more common by the 1850's; thus, the manufacture of flat rods presented less of a production problem. Undoubtedly, there were other reasons for the rapid emergence of the flat rod. It is certain, however, that the design was almost without competition until the 1890's and that it remained in favor for certain classes of light locomotives until the end of steam locomotive construction.

While round and flat rods were the dominant designs for American locomotives of the nineteenth century, it would be amiss not to mention the variant designs. One of the most curious variants was the double bar side rod used by Norris during the mid-1840's. These rods were fabricated from two small-diameter rods set parallel to each other. They were threaded at the bearing ends and could be adjusted for wear by loosening and resetting the double nuts at each end. This style of rod is shown in the 1845 advertising cut in Fig. 132. While undoubtedly cheap to make, this style of connecting rod was little used for locomotives, although it was favored for marine engines to some extent.

The ordinary round rod underwent a minor alteration when used for extra-long connections. It was fitted with two light truss rods placed one above the other below the main rod to provide stiffness. Examples of this construction, which were rare, are found on the Camden and Amboy's Cramptons (1849–53) and Millholland's two high-wheel express locomotives (1852), the *Illinois* and the *Michigan*. On April 1, 1845 (No. 3981), Holmes Hinkley was granted a patent that employed a side rod of the same design.

Fluted side rods were used by Rogers in 1854 on an engine built for the Buffalo and State Line Railroad.[114] The web of these rods was only half an inch thick and represented a great saving in weight. This pioneering effort did not result in any extensive use of fluted rods until the 1890's.

[114] *Railroad Advocate,* January 6, 1855, p. 3.

Rod bearings were usually of a hard brass alloy and sometimes were made with Babbitt metal inserts.[115] The bearings, constructed in two pieces, were held in place by a U-shaped strap that was fastened to the end of the rod. This simple construction, evident on all but two of the locomotives illustrated in the present volume, was universally favored during the nineteenth century. During the 1830's, gibs, held by keys, were often used to fasten the straps to the rod (see the Dunham, Fig. 116). In succeeding years bolts fastened bearing straps to the rod, and a key was used to adjust the fit of the bearings on the crank pin. Although much criticized, strap-end rods were entirely satisfactory for light locomotives and were not abandoned by road engines until the 1890's.

Solid-end rods, for which a one-piece bearing and no straps were employed, were occasionally used by American builders as early as 1850. A surviving example of this type of rod, with square ends, is illustrated by the Norris-built *Copiapo* (Fig. 142). Ross Winans began to use solid-end rods at about the same time and regularly fitted his Camel engines with this type of rod. Winans made the ends of the rods round, thus giving them a modern appearance for the time. One-piece steel bushings were used in place of the more conventional brass bearings.[116] Compared to the fabricated strap-end rod, the great advantage of the solid-end rod was its superior strength. It was not subject to misadjustment by the engine crew, who were so often tempted to "hammer down" the adjusting wedges on strap-end rods whether they required it or not.

CROSSHEADS. During the nineteenth century three major styles of crossheads were used—the single, the double, and the four-bar guide types. Of these, the four-bar crosshead was by far the most popular. Its origin can be traced to Robert Stephenson's "Planet" class introduced in 1830. In Stephenson's design the two sets of double guide bars were widely spaced; see the *Robert Fulton* (Fig. 7). During the 1830's, American mechanics adopted this design for outside-connected engines; by placing the four guide bars close together, a small, compact crosshead could be used. This arrangement is best shown by the *Talisman* drawings (Fig. 172–175). The front ends of the guide bars were bolted to the front cylinder cover; the rear ends of the same bars were attached to a yoke connected to the engine's frame. The crosshead was invariably of cast iron, with the main rod's wrist pin cast integrally![117] Generally the crosshead was made without gibs and was meant to take the wear; the guide bars were made of steel or case-hardened wrought iron. Shims were fitted under the guide bars at the mounting points for adjustment. Naturally, many four-bar crossheads were made with brass- or Babbitt-lined gibs.

Aside from its obvious simplicity and cheapness of manufacture, the four-bar crosshead owes much of its popularity to the widespread use of 4–4–0 and 4–6–0 locomotives during the nineteenth century. The compact four-bar crosshead arrangement easily cleared the rear truck wheels of such engines. However, it rapidly fell from favor after the 1870's when these wheel arrangements generally became less popular. The four-bar crosshead was found to be too weak for the heavier engines coming into service.

Two-bar crossheads followed the four-bar guide in popularity. Norris used such crossheads during the 1830's with square or round guide bars set horizontally. Examples of this construction can be seen on the 1839 Norris 4–2–0 (Fig. 14) and the *Virginia*, 1843 (Fig. 60). During the 1840's Norris set the two-bar guides, generally round in shape, in a vertical position. Reference should be made to the *Tioga* (Fig. 83) and the *Philadelphia* (Fig. 133) for this arrangement. While Norris had abandoned the design by the early 1850's, the Philadelphia and Reading Railroad continued to favor the double round-bar design until almost 1870. The round bar guide was simple enough, but the difficulty in making adjustments for the uneven wear of the round crosshead bearings prevented widespread use of the guide.

---

[115] Colburn, *The Locomotive Engine*, p. 95, mentions use of Babbitt bearings at this early date. Isaac Babbitt discovered the antifriction tin alloy in 1839.
[116] Clark and Colburn, *Recent Practice*, p. 3.

[117] Cast-iron crossheads are described by Clark and Colburn, *Recent Practice*, p. 62. A crosshead of this type can still be seen on the 1851 *Pioneer* locomotive preserved by the Smithsonian Institution.

The familiar two-bar crosshead with square guides was used by Rogers in 1855 on the *Volcano,* an 0–6–0 built for the Buffalo and Erie Railroad. The crosshead was made of cast iron in the "Alligator" pattern. This type of crosshead was little used until the 1870's, and then it was invariably built up from plate. It was a stable arrangement, but the bottom guide, being near the track, collected an inordinate amount of grit on its oily surface and was subject to rapid wear.

The single-bar, or suspended, crosshead was used regularly, but never widely, by American builders throughout the nineteenth century. The *Cincinnati,* built in 1835 by the Vulcan Foundry, was fitted with a heavy single-bar crosshead placed behind rather than above the piston rod (see Fig. 3). Baldwin used a remarkably similar plan at the same time but made the large guide bar hollow so that it might serve a second function, that of the feed pump barrel. This scheme was used on nearly all Baldwin engines until the mid-1840's. The maker continued to use it occasionally and as late as 1860, but by that time it was employed only on very light four-wheel switching engines.[118]

Some of the first engines to use a true, suspended, single-bar crosshead were the 0–8–0's built in 1847 by Winans for the Philadelphia and Reading Railroad. The large, square crosshead was built up from iron plate; the guide bar was about 3 inches square. Winans used this design in succeeding years for his Camel locomotives.

The Baltimore and Ohio Railroad, probably influenced by Winans, used single-bar crossheads on many of the locomotives designed and built at its Mt. Clare Shops. The guide bar was diamond-shaped, and the crosshead was of cast iron. The 0–8–0 of 1848 (Fig. 34) and the Tyson Ten Wheeler (Fig. 183) had crossheads of this type, as did all of the early Hayes ten wheelers (1853).

The best-known suspended crosshead of this early period was that developed by John P. Laird in the early 1860's when he was serving as master mechanic of the Pennsylvania Railroad. Two slender guide bars supported the top of the crosshead. The *Northumberland* illustration (Fig. 206) is an example of a Laird engine so-rebuilt. The Laird crosshead was widely used in this country after 1870.

VALVE GEARS

The indifference of nineteenth-century American mechanics to the fuel economy achieved by adding auxiliary machinery to the basic locomotive has been mentioned previously in this work. Keeping the locomotive simple and free of complex appendages was a laudable achievement of the pioneer master mechanics. Yet these practical men showed a decided weakness for complex valve motions. The mechanical solution to heavy steam consumption had greater appeal for the plain mechanic than the more abstract heat economizers, the much-rejected feed-water heaters and superheaters. The more links, eccentrics, and bell cranks a valve gear exhibited, the surer the cutoff and the more certain its capacity for saving steam. The majority of these arrangements suffered what George Stephenson was wont to term "the danger of too much ingenuity."[119]

The economy resulting from cutting off the steam supply at an early stage in the piston's stroke was understood in Watt's time. This permitted the steam to expand and thus more fully expend its thermal energy. An elementary, single-eccentric valve gear can effect an early, fixed cutoff if the valve is lapped, that is, if the outside edge of the valve is made longer than necessary to cover the steam ports when at the middle of its travel. Lapped valves, effecting a built-in cutoff, were used on the earliest American locomotives. William T. James used

[118] Photographs of the *Tip Top,* Construction No. 871 (1859), and the *Active,* Philadelphia and Reading Railroad (1860), show crosshead-guide feed pumps of the old Baldwin pattern.

[119] Warren, *A Century of Locomotive Building,* p. 370.

a half-inch lap on the engine the *American,* built for the Baltimore and Ohio Railroad in 1832, and had used lapped valves on other road engines as early as 1828.[120]

Lead was another fundamental of steam engineering well understood at the time locomotives were introduced in the United States. Lead is the amount of valve opening when the engine is on center. By admitting steam, this opening cushions the piston at the end of its stroke and helps in starting the engine when it is on dead center. De Pambour experimented with lap and lead on locomotive valves in 1834. In 1836 the results of the test were published in his widely consulted work on locomotive engines, *A Practical Treatise on Locomotive Engines upon Railways.*

Single-eccentric valve gears were used on most of the locomotives operated on early American railroads. Stephenson favored a single, loose-eccentric valve gear between about 1830 and 1835. Probably all of the machines exported to this country during those years by the Newcastle builder were so-equipped. The gear was reversed by sliding the eccentrics, one for each cylinder, along the axle so as to engage a forward or reverse pin that was fixed to a plate mounted on the axle. The Stephenson gear is described more fully in the section on the *John Bull* (see p. 253).

In the early years of his production Baldwin used a single-eccentric valve motion that was based on the plan of J. and C. Carmichael, engineers of Dundee, Scotland. The eccentric was keyed to the axle. The engine was reversed by raising or lowering a double gab-hook frame to engage pins on opposite ends of a common rocker. Baldwin continued to use this simple valve motion until 1838. It is illustrated and described in the sections on the *Lancaster* and Dunham locomotives (see Figs. 111 and 115).

Eastwick and Harrison was another United States firm that championed the single-eccentric valve gear. This gear, patented by Andrew M. Eastwick in 1835, was reversed by shifting the valve ports, which were constructed as separate sliding pieces under the valve. The Eastwick gear was not used by any other American builder and disappeared from domestic practice when the firm moved to Russia in 1844. The section on the *Gowan and Marx* (p. 288) contains more information on the Eastwick valve gear.

Single-eccentric valve gears enjoyed a brief history; by 1840 nearly all American builders had abandoned their use. Double eccentrics, one each for reverse and forward motions, were necessary for the proper operation of the locomotive. William T. James used double eccentrics on a road engine in 1829.[121] Despite this prior use, Stephen H. Long obtained a patent on January 17, 1833, for valve gears with double cams or eccentrics.[122] The several locomotives built by the inventor in the 1830's were equipped with valve gears of this design. Long's partner, William Norris, presumably continued to use the double-eccentric valve gear. In later years Long defended his patent claim by suing at least one railroad, the Baltimore and Susquehanna, for unauthorized use of his "invention," the double eccentric.[123] The questionable substance of Long's claim aside, the double eccentric was firmly established for locomotives by 1840 and was used by all American builders after that date.

The hook motion was directly associated with the establishment of the double eccentric and remained the standard form of valve motion in this country from about 1840 to 1855. It was a simple, reversible valve motion entirely suitable for slow-speed locomotives where the economies realized from cutting off the steam beyond that effected by the lap were inconsequential. This form of valve gear derived its name from the hook or gab formed on the end of the eccentric rod. The lower arm of the rocker was made double. A pin on one arm was engaged by a hook for forward motion; a pin on the other arm was engaged by the second hook for reverse motion. Engagement of the hooks was accomplished by raising or lowering the eccentric rods.

---

[120] Colburn. *Locomotive Engineering,* p. 44.

[121] *Ibid.*

[122] The original patent was destroyed by the 1836 patent office fire. Long resubmitted the patent in 1843. These papers are in the United States National Archives.

[123] A letter of 1850 noting this suit was published in *Engineer,* May 29, 1914, p. 601.

The simple-drop, or D-shaped, hook was favored at first; however, the difficulty in engaging it when one of the pins was slightly out of line led to the adoption of the more costly V hook in the mid-1840's. The V hook with its long, widespread jaw could easily guide itself onto a misaligned rocker pin and thus eliminate the secondary "starting" levers and rod required by the D hook. See the Norris 4–2–0, 1841 (Figs. 11 and 13), for an example of D hook motion. V hooks are illustrated by several locomotives in this volume, including the *Croton,* the *Copiapo,* and the *Philadelphia.*

Because power was the primary concern of the early locomotive designers, steam was admitted for roughly 70–90 per cent of the stroke. Thus, little expansion was possible and fuel and water consumption was great. As already stressed in the discussion of boilers, the need of keeping locomotive weight within the loading limits of the track made it necessary to "work" a small engine to produce the desired power. This was achieved by means of small, fast-steaming boilers and minimum steam cutoff. Fuel economy literally went out the smokestack, propelled as it was by the enormous amount of exhaust developed by working the steam at nearly full stroke capacity.

INDEPENDENT CUTOFFS. The hook motion satisfied most railroad managers. Yet a few mechanics believed that a more economical valve gear should be developed for locomotives. The essential idea was to have a separate, independent valve to govern the cutoff once the train was under way and the need for maximum power was past. The scheme was particularly advisable for passenger trains where loads were light, speeds high, and, in general, power requirements relatively modest after the train was in motion.

While many elaborate variations were developed, the essentials of all separate cutoff gears were as follows: The main valve was operated by an ordinary hook motion. The separate cutoff gear was effective only in forward motion and was set to admit steam for half the stroke. A third eccentric on the driving axle or a crank on one of the side-rod crank pins drove the cutoff valve. A second rocker was usually employed for the cutoff valve. The cutoff valve was always above the main valve and was either in a separate valve box, with a partition plate between it and the main valve, or rode directly on top of the main valve—hence the term "riding cutoff." The merits of partition plate and riding valves were hotly contested. With a partition plate the steam was partially expanded when it passed from the upper valve box to the lower. The excessive wearing of the cutoff valve riding atop the main valve was a troublesome maintenance problem. Neither system triumphed, but the partition plate gained the favor of the larger builders. The cutoff valve itself was usually an open-top slide valve with two or more ports.

The history of the separate cutoff, like that of so many elements of locomotive engineering, can be traced back to Robert Stephenson. In 1828 Stephenson applied a curious, gear-driven, half-stroke cutoff with plug valves to the *Lancashire Witch.*[124] The scheme had no widespread effect but was reported at the time by two German engineers who thought its merits should be more widely known. William Norris and S. H. Long are credited with a patent issued on December 30, 1833, for a riding cutoff valve gear.[125] Unfortunately, no record of this patent exists; assuming that it was granted, all traces of it perished in the patent office fire of 1836. The only other evidence that such a plan was used is a statement in a Long and Norris printed prospectus dated November, 1833, claiming that their locomotives were worked by "expansive force" at five-eighths of the stroke.[126] The Camden and Amboy Railroad's locomotive No. 6, built in 1833 at the company shops, was equipped with a cutoff valve gear after the design of Robert L. Stevens. Among the several early engines of this line similarly fitted was the No. 9, pictured in the present volume as Fig. 94. Credit for the riding cutoff, however, is usually given to Isaac Adams of Boston, whose May 17, 1838, patent has survived. David Matthew of the Utica and Schenectady Railroad was another early user. Matthew first used such a valve gear in 1837 and continued its use for some years there-

---

[124] Warren, *A Century of Locomotive Building,* p. 147.
[125] Clark and Colburn, *Recent Practice,* p. 49.
[126] *Proposals of the American Steam Carriage Company,* p. 16.

after.[127] The Utica and Schenectady's No. 11, fitted with an independent cutoff, is shown in this volume as Fig. 17.

Ross Winans developed a separate cutoff at the same time that Matthew put his to work. Winans' gear, patented on July 29, 1837, was a distinct departure from the usual plan of American cutoff gear in that only a single valve was used. The patent drawing shows two forward cams, each set at a different point of cutoff; a third cam was used for the reverse gear. The eccentric straps, made like a Scotch yoke, could be shifted by foot treadle or large hand lever from one eccentric to the other, thereby shifting between the two points of cutoff or reversing the engine as desired. In later years Winans modified this plan by using two eccentrics and a single cam with drop hooks, thus eliminating the shifting eccentric strap. This valve gear was used on the Camel locomotive for many years. It is illustrated and described in the *Susquehanna* section of this volume (p. 354).

By the early 1840's the independent cutoff plan was attracting more attention, and the major builders, probably at the prodding of specific purchasers, began to offer various plans of such valve gears. Baldwin adopted a half-stroke cutoff driven by a third eccentric, with a separate rocker placed behind the main valve rocker. A partition plate was used. The gear was so arranged that the cutoff valve could be disengaged when its operation was not desired. Illustrations of the Baldwin cutoff appear here as the *Tioga* (Fig. 24), the 1848 4-4-0 (Fig. 26), and the *Cumberland* (Fig. 31).

The first Baldwin locomotive built with the separate cutoff was the *Atlantic* of the Western Railroad (Massachusetts). The machine was completed in September, 1844, but it was another year before Baldwin built a second cutoff engine. This machine was the *Champlain* of the Philadelphia and Reading Railroad.[128] In 1846 Baldwin built only eight locomotives with separate cutoffs; by 1850, however, slightly more than half of his production had been so-equipped. Within a few years the separate cutoff was replaced by the variable cutoff, which will be discussed shortly.

In 1845 Thomas Rogers offered a separate cutoff gear that was a modification of Horatio Allen's cutoff patented on August 21, 1841 (No. 2227). Rogers received a patent on May 1, 1845 (No. 4028), for a complex motion with the cutoff valve driven from a bell crank attached to the crosshead. The main valve was actuated by an ordinary hook motion. A partition plate was used. At least one locomotive is known to have used Rogers' gear. This machine, the Morris and Essex Railroad's *Sussex*, was completed in May 1846.[129] The following year Rogers adopted a more conventional three-eccentric riding cutoff valve motion. The complexity of the arrangement whereby only two points of cutoff were possible undoubtedly prevented any widespread interest in the new design.[130]

By the mid-1840's the cutoff mania was established, and the Norris Brothers, being the largest American locomotive building firm, could no longer ignore the demands for locomotives with separate cutoff valve gears. According to an advertisement in the 1845 *American Railroad Journal*, locomotives were offered "with their Patent arrangement for variable expansion." Such a patent cannot be found at the patent office, and it is possible the device was never granted a patent, although one may well have been sought. In any event the *Philadelphia* drawing (Fig. 133) indicates that Norris had developed a cutoff gear by 1844. As shown, it was a riding cutoff; the cutoff valve was driven by a separate rod attached to the forward eccentric rod. A separate rocker was used. The most striking feature of the Norris cutoff was the closed, double-V hook visible just behind the crosshead in the *Philadelphia* drawing (Fig. 133). The cutoff valve was driven when the upper V engaged the forward rocker. When the double V was raised so that the bottom V engaged the pin attached to the main valve's stem, the cutoff valve traveled with the main valve and no cutoff was effected.

---

[127] Colburn, *Locomotive Engineering*, p. 53.
[128] Baldwin registers list types of valves used on locomotives. These records are the basis for the remarks offered on Baldwin valve gears.

[129] The *Sussex* is described and illustrated in the *Railroad Gazette*, June 13, 1902, p. 441.
[130] Rogers' 1847 valve gear is illustrated on p. 52 of the Rogers history, *Locomotives and Locomotive Building* (New York, 1886).

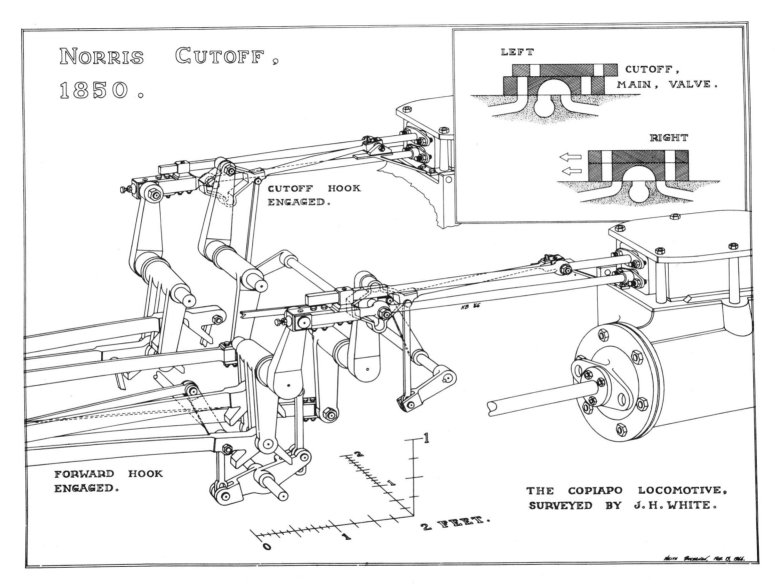

Fig. 70. Norris' independent cutoff, introduced in about 1845. The upper, or cutoff, valve was engaged by the double V hook. This drawing is based on the Copiapo drawings.

In later years a third eccentric was used to drive the cutoff valve (see the *Copiapo,* Fig. 70). Norris continued to manufacture the separate cutoff until the mid-1850's when it was finally abandoned for link motion.

In New England, the Hinkley works was a leader in adopting separate cutoffs. Little can be found on the arrangement of the Hinkley cutoff, but a partial idea of its construction can be gained from the *Massachusetts* drawing (Fig. 65). The first engine so-equipped was the *Boston,* completed in April, 1843, for the Western Railroad (Massachusetts). Within a few years nearly half of Hinkley's engines were furnished with separate cutoff valve gears; in 1850 about 90 per cent were so-equipped.[131] A variable cutoff gear on Horace Gray's plan was offered next, but the link motion rapidly succeeded both the separate and variable cutoff gears, few of which were fitted by Hinkley after 1853.

Many other firms and designers offered cutoff valve gears that will not be included in this discussion. Among these are the Taunton Locomotive Manufacturing Company and Walter McQueen. McQueen's Croton valve gear is illustrated and described in the *Croton* and *Columbia* sections of this work (pp. 328 and 337).

VARIABLE CUTOFFS. The popularity of the independent cutoff led to the development of the variable cutoff. With such a mechanism it was possible to select several points of cutoff and to achieve substantial steam economies, particularly at higher speeds. The earliest such arrangement was perfected by the British mechanic John Gray and was fitted to the Liverpool and Manchester Railway's *Cyclops* in 1839. The complexity of the plan prevented its further use but it contained the basic element of all subsequent variable cutoff valve gears, the curved link with a sliding block.

The variable cutoff was first applied to an American locomotive by the Cuyahoga Steam Furnace Company of Cleveland, Ohio. The firm did not consider locomotives its major line and engaged in the trade only between 1850 and 1857. The introduction of the Cuyahoga cutoff brought national attention to the company's products and prompted several major eastern builders to copy the arrangement. Ethan Rogers, superintendent of Cuyahoga's marine engine department, conceived of the valve gear idea in the fall of 1849.[132] Rogers' invention was used on the *Cleveland,* Cuyahoga's first engine, completed in March, 1850, for the Cleveland, Columbus and Cincinnati Railroad. The valve gear was regularly applied to Cuyahoga locomotives for the next five years; after that the link motion was used.

The inventor did not patent his invention and no complete mechanical description or drawings are known to exist.[133] Therefore, a reconstruction drawing of the Cuyahoga gear was prepared especially for this volume (see Fig. 71). The reconstruction is based on several illustrations of Cuyahoga locomotives. Certain aspects of its construction, notably whether two or three eccentrics were used, are in question. There is no question about the external plan of motion, which is plainly visible in surviving photographs, nor is there doubt that it was a riding cutoff gear. As the illustration indicates, the main valve was driven by an ordinary hook motion. A curved link with a sliding block, attached to the outside end of the main valve rocker shaft, drove the cutoff valve. The position of the link block controlled the travel of the cutoff valve in exactly the same manner as in the link motion. It was possible to achieve a minimum cutoff of 12.5 per cent, or about two-thirds more than that possible with the link motion.[134] While the opportunities to operate a locomotive with so small an admission of steam were extremely limited, the Cuyahoga gear was celebrated as a wonderful steam-saver. The performance of the *Rocket* on the Cleveland and Pittsburgh Railroad is an eloquent testimonial to the efficiency of the Cuyahoga gear. In 1854 this engine consumed only three-quarters of a cord of

---

[131] Hinkley Locomotive Works, Register, 1841–56, in Boston Public Library; hereafter cited as Hinkley Register.

[132] *Railroad Advocate,* January 12, 1856, p. 2.
[133] The diagram drawing in Sinclair's *Locomotive Engine,* p. 453, is incorrect in showing two rockers.
[134] Clark and Colburn, *Recent Practice,* p. 65.

COMPONENTS

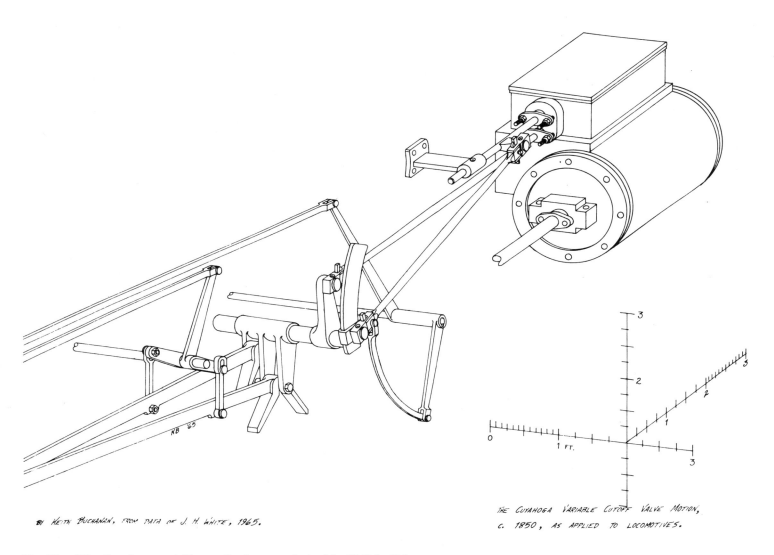

Fig. 71. The Cuyahoga variable cutoff valve gear devised in 1849 by Ethan Rogers of the Cuyahoga Steam Furnace Co. It was the first variable cutoff gear used in the U.S. This is a reconstruction drawing based on photographs of Cuyahoga locomotives.

wood in pulling a four-car express train 104 miles between Wellsville and Cleveland, Ohio, including twenty-one stops en route.[135]

In the spring of 1850, not long after the first Cuyahoga gear was placed in service, Horace Gray of Boston contrived an almost identical valve motion.[136] It was tested on an engine of the Fitchburg Railroad in the summer of the same year.[137] The good performance of the gear prompted its use by other New England roads. The Hinkley, Amoskeag, Taunton, Mason, and Lawrence works are all known to have built engines with valve gears of this type. As shown by contemporary illustrations, Gray's cutoff differed from the Cuyahoga gear by using a separate rocker for the cutoff valve which was driven by a third eccentric.

Taunton became a strong advocate of the variable cutoff after first using it on a group of engines delivered to the New York and Erie Railroad in 1851. It was reported that "the Taunton Company takes a warrantable pride in the superiority of their engines with independent variable cutoff, and we expect it will be long before they will adopt the link motion."[138] In 1856 Taunton at last accepted the link motion. The Taunton variable cutoff is shown in the *Gasconade* photograph, Fig. 145.

Baldwin was another eastern builder who offered a variable cutoff on the Cuyahoga-Gray plan. He began to apply gears of this type in 1852.[139] On September 13, 1853, he received a patent (No. 10007) for this form of valve gear which covered the trifling and doubtful "improvement" of shifting the link block by replacing the simple radius rod with a geared quadrant. A chain was also tried, but by 1857 Baldwin had abandoned all such contrivances for the conventional radius rod.

The awesome complexity of Baldwin's variable cutoff is shown by the detailed drawing of the *Memphis'* valve gear (see Fig. 72). A greater number of parts for so little purpose is difficult to imagine. Surely any saving in fuel would be canceled by maintenance costs and the inconvenience of having the engine out of service for repairs. The general arrangement of this mechanism will be better understood from the following list of parts:

1. Reach rod—V hooks
2. Reach rod—variable cutoff
3. Link block—variable cutoff
4. Link (upper arm of rocker)
5. Lifting arm—link block
6. Rocker—V hooks
7. Valve-stem rod—main valve
8. Valve-stem rod—cutoff valve
9. Lifting arm—V hooks
10. Eccentric blade—V hooks
11. Eccentric blade—variable cutoff
12. Rocker (lower arm of variable cutoff)
13. Guide block (stationary)
14. Rocker (lower arm of V hooks)

The Baldwin history claims that link motion was adopted "exclusively" after 1857, but, while this is in the main correct, there is evidence that a few engines continued to be fitted with the old variable cutoff as late as 1860. One known example, which very likely was attributable to the order of the purchaser, was the Mine Hill Railroad's No. 30 (see Fig. 55).

LINK MOTION. We turn now to the most important valve gear of the nineteenth century and one of the most successful single improvements ever devised for steam locomotives, the celebrated Stephenson link motion. This "exquisite motion," to quote D. K. Clark, combined all the desirable features of a

---

[135] *Boston Evening Transcript*, August 31, 1854.
[136] Clark and Colburn, *Recent Practice*, p. 65.
[137] *Railroad Advocate*, January 12, 1856, states that Gray's cutoff was first applied to either the *Littleton* or the *Leominster* of the Fitchburg Railroad.
[138] *American Railroad Journal*, September 24, 1853, p. 620.
[139] The Baldwin registers list the *New Orleans*, completed in March, 1852, for the New Orleans, Jacksonville and Great Northern Railroad, as having a variable cutoff; *History of the Baldwin Locomotive Works*, p. 54, notes that an engine built for the Western and Atlantic in 1852 was similarly fitted.

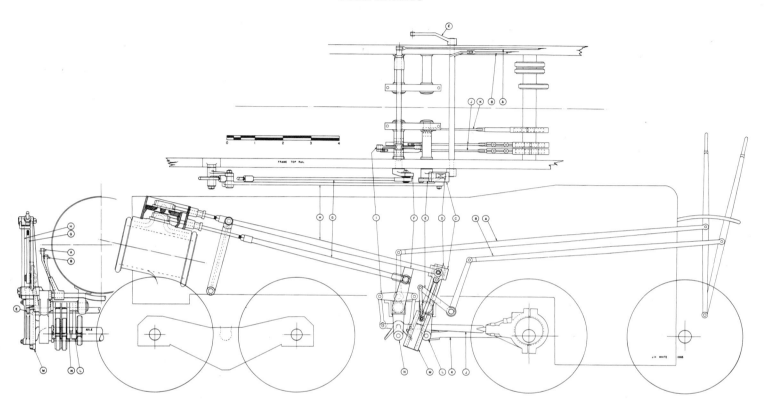

Fig. 72. *The Baldwin-Cuyahoga variable cutoff valve gear was introduced in 1852, and Baldwin continued its production until 1860. The drawing shown here is after an original drawing of the Virginia and Tennessee Railroad's Memphis, completed in August, 1857.*

locomotive valve gear in a remarkably simple and rugged arrangement. It was reversible and offered a variable cutoff in both forward and reverse—all with less than half the number of parts required in most hook-motion variable cutoff gears. Some years after its general adoption Colburn said of the link motion: "No innovation, in respect of locomotive engines, ever divided practical and professional opinion so completely; and none has at last so firmly established itself in general favor."[140]

The controversy centered on the link motion's characteristic variable lead. Too small a valve opening in advanced cutoff

[140] Colburn, *Locomotive Engineering*, p. 64.

positions, too early an exhaust, and "wire-drawing" the steam were defects cited by critics of the link motion. These charges were answered by Colburn in the following statement:

> Experience has shown that most of these objections to the link are really unimportant. When working the valve so as to cut off very early in the stroke, the link may no doubt be charged with a certain loss of expansive effect due to wire-drawing the steam; but when it is considered that the full boiler pressure may be, and, indeed, generally is, admitted upon the piston at the beginning of its stroke, and this, too, even with the least travel of the valve, it is clear that the expansive effect may be very good after all. Nor at high speeds does the early opening of the exhaust port, the link being supposed in nearly mid gear, appear to be attended with any considerable loss, inasmuch as the steam is generally already expanded to a point not greatly above that of the constant back pressure upon the piston.[141]

It might be added that the variable lead in later years came to be regarded as one of the most valuable features of the link motion.[142] A more serious defect of the link motion involved the geometry of the arrangement, particularly the angular motion of the eccentric rods.[143] This condition was accentuated by short eccentric rods and often presented serious design problems with close-coupled, multidriver locomotives, such as the 2–6–0 and the 2–8–0, where only short rods were possible.

The origins of the link motion are clouded by conflicting claims and the lack of solid contemporary evidence. William T. James of New York is credited with devising a valve gear on this plan in 1832 and applying it to a small, four-wheel locomotive, the *American*, sold to the Baltimore and Ohio Railroad. The *American Railroad Journal*, in its October 20, 1832, issue, noted the demonstration of this machine behind the maker's workshop but failed to describe the valve gear. It was stated, however, that the engine ran "a distance of about fifty feet forward and backward, eight times in sixty-three seconds, including stops." As Colburn suggested, such a remarkable performance could only be possible with an extremely efficient reversing gear such as the link motion.[144] The statements of Samuel B. Dougherty, an employee of James at the time the locomotive was under construction, are the only other source on James's valve gear. In a letter of May, 1858, to Zerah Colburn, Dougherty claimed to have fitted up a link motion under James's direction.[145]

In later years Dougherty corresponded with, and was interviewed by, several locomotive historians, including Sinclair and Pangborn. The aging mechanic held firm to his recollection but could not produce any original documents or drawings.[146] It is the author's opinion that Dougherty, a respected master mechanic, was correct in his recollection and that his statements on the James valve gear should not be dismissed too casually.

The James valve gear made no impression on American railroad mechanics. The locomotive exploded in 1834 and the link motion was temporarily lost to the engineering world. The idea was revived eight years later in England by William Williams and William Howe, two young employees at Robert Stephenson and Company's locomotive works. The link was surely devised independent of James and appears to have been inspired by the long, V hook gear used at the time by Stephenson. Williams was a draftsman and Howe a pattern maker. The co-inventors, who in later years sought exclusive recognition, began work on the link motion in August, 1842.[147] Williams' original concept was reportedly a straight

---

[141] *Ibid.*, p. 66.
[142] Jacob H. Yoder and George B. Wharen, *Locomotive Valves and Valve Gears* (New York, 1917), p. 83: "The lead increases as the link block is made to work nearer the center of the link when the position of the link is changed by means of the reverse lever. This 'variable' or changeable lead is an advantage, since a large lead is desired when running on a short cut-off or when running at high speed; while at low speed an early admission is not necessary; besides it is then undesirable, as it tends unduly to hold back the piston while the crank pin is passing dead center."
[143] *Ibid.*, p. 66.
[144] Colburn, *Locomotive Engineering*, p. 44.
[145] *Ibid.* Dougherty's letter is partially quoted by Colburn.
[146] Sinclair, *Locomotive Engine*, p. 119, claims that the illustration of the James locomotive (Fig. 63) is based on an original drawing. The accuracy of this statement is questionable since the illustration agrees with a reconstruction drawing prepared by Dougherty in 1893 which is now preserved by the Smithsonian Institution.
[147] N. P. Burgh, *Link Motions . . .* (London, 1872), p. 1.

link attached almost directly to the eccentrics.[148] The authenticity of this mechanical absurdity was questioned by Colburn, who believed it was prepared by Howe in later years to discredit Williams as an engineer and, in effect, to minimize his claim to the invention.[149] In any event, the curved shifting link was developed soon thereafter as witnessed by a working drawing prepared by Williams and dated September 15, 1842.[150] The new valve gear was first applied to a 2–4–0 freight locomotive, the North Midland Railway's No. 71, completed on October 15, 1842.[151] The merits of the shifting link motion were immediately apparent to manufacturers, who so vigorously advocated the new valve gear that it soon became known as the "Stephenson" valve gear. It might be added that the motion was not patented by Stephenson as is occasionally reported but, like the leading truck and other basic improvements, was given over freely to the industry.

Unlike British railways, American railroads were slow to adopt the link motion. The first link motions manufactured in the United States were applied to locomotives built for export. In May, 1845, Baldwin completed three flexible-beam locomotives with link motion for the Wurtemburg State Railway. The eccentrics were placed on the rear driving-wheel axle behind the firebox. The efficiency of the valve motion was no doubt impaired by the short eccentric rods. One of these machines, the *Necker*, together with a detail of the valve gear, is shown in a contemporary German lithograph reproduced in the present volume as Fig. 82. Norris built three 4–4–0 locomotives for the Württemberg State Railway equipped with link motion in the same year. The eccentrics were mounted on the forward driving axle, and, because of the long eccentric rods, this valve gear was undoubtedly more effective than the Baldwin arrangement.[152] In 1846 Norris delivered four engines of the same wheel arrangement to the Baden State Railway. A detailed drawing of the shifting link motion used on these machines is reproduced in a well-documented history of German locomotives.[153]

In spite of the direct involvement of two major American builders with the link motion, neither is known to have produced such valve gears for a domestic road until many years later. The first recorded application of the link motion to an American locomotive was the Eastern Railroad's *Courier*, completed in 1847 in the company shops by H. W. Farley.[154] Farley used a stationary link rather than the Stephenson type. This pioneering effort attracted little attention and resulted in no further tests until Thomas Rogers took an interest in link motion two years later. Rogers, often improperly credited with the first application of the link motion in the United States, did apply such a motion to the *Pacific*, a high-wheel "single," completed for the Hudson River Railroad in October, 1849.[155] Like Farley, Rogers used the suspended link. The following year he began using the shifting link and soon became a strong advocate of the link motion. Rogers' reputation as a first-rate practical mechanic and the position of his company as one of the largest locomotive works had an unquestioned bearing on the successful introduction of the link motion to American practice.

Other major builders reluctantly followed Rogers' lead and did so, according to available evidence, only at the specific request of the purchaser. Baldwin resisted the application of the link motion longer than most builders but was at last "forced to succumb"; in 1854 he used the link for the Central Railroad of Georgia's *Pennsylvania*.[156] Three years later it was adopted

---

[148] A drawing of the first arrangement of the Williams-Howe link motion is shown on p. 63 of Colburn's *Locomotive Engineering*. A wooden model, made by Howe in 1842 and preserved by the Science Museum, London, shows a curved link and long eccentric rods. See E. A. Forward, *Handbook . . . Land Transport* (London, 1931), Part II, p. 69.

[149] Colburn, *Locomotive Engineering*, p. 63.

[150] Williams' drawing is reproduced on p. 367 of Warren's *A Century of Locomotive Building*.

[151] Ahrons, *The British Steam Locomotive*, p. 63.

[152] E. H. Helmholtz and W. Staby, *Die Entwicklung der Lokomotive* (Munich and Berlin, 1930), picture these Norris engines on p. 220.

[153] *Ibid.*, p. 216; see also *Railway and Locomotive Historical Society Bulletin*, Nos. 79, 81, and 109.

[154] Clark and Colburn, *Recent Practice*, p. 65.

[155] Frederick Moné, *A Treatise on American Engineering* (New York, 1854), p. 7 of the chapter entitled "Description of the Columbia."

[156] *History of the Baldwin Locomotive Works*, p. 55.

by Baldwin as standard practice. Hinkley was more daring and first used the link motion on the Erie Railway's No. 112, completed in June, 1851. By 1853 about 90 per cent of the firm's engines were fitted with the Stephenson valve gear.[157] The sudden and universal acceptance of the link motion was commented upon by Colburn and Holley in the following words: "It encountered the strongest prejudice, and in New England, especially, locomotive builders and railway mechanics would have nothing whatever to do with it, *until* about 1855, when all opposition suddenly gave way, and each vied with the other in its immediate adoption."[158] This acceptance marked the return to simple, straightforward mechanics and permanently ended the widespread use of complex valve gears in this country.

The construction of the link motion remained relatively static during the many years of its popularity (1855–1910). The eccentrics were universally mounted on the axle, inside the frame. An exception to the rule was the outside placement of the link motion by the New Jersey Railroad and Transportation Company in the 1860's. Nearly every engine on the line was so-fitted, but, while this mounting allowed free access for maintenance, the awkward appearance and the instability of two eccentrics on the crank pin did not lead to the employment of outside link motion by many other roads. The eccentric and straps were universally of cast iron, sometimes made with Babbitt liners. The link was of hardened wrought iron to promote long wear; in later years steel was of course substituted. In 1865 the Baltimore and Ohio Railroad fitted its last class of eight-wheel connected road engines with cast-iron links.[159] Because these massive castings weighed 434 pounds each, this plan did not attract the following of any other builders. On better-made engines the rockers were one-piece forgings, and this practice was followed by nearly all makers after 1860. It was necessary to provide a counterbalance so that the link motion might be reversed manually. In the early years a simple cast-iron weight on a lever fitted to the reversing shaft was used. As long as clearance was available this plan was satisfactory, although a bit crude. More elaborate spring balances became the rule after the mid-1850's. Slender leaf springs between the frames with a lever connection to the reversing shaft were popular. Mason used a coil spring (in a neat casing) with a chain attached to the reversing lever. The other constructional details of the Stephenson gear are so obvious that no further comment is needed. The following drawings show the construction of the link motion: the *Superior* (Fig. 164), the *Phantom* (Fig. 193), and the Rogers 4–4–0 (Fig. 223).

A minor controversy developed during the first years of the link motion's use in this country over the merits of stationary links versus shifting links. The stationary link, invented in 1843 by Daniel Gooch of the Great Western Railway, was identical to the Stephenson, or shifting, link in all particulars except the method of moving the valve. According to the Gooch plan, the link remains stationary while the link block is shifted; the reverse is true for the Stephenson scheme. This seemingly inconsequential detail gave the stationary link motion a constant lead over the shifting link. As already noted, this feature was not in the end regarded as particularly desirable, and by about 1860 the advocates of the stationary link were a singular minority.[160]

As might be expected, certain mechanics educated in the old cutoff school would not accept the link motion; instead, they carried on experiments to perpetuate the separate cutoff by combining it with the link. The general scheme was to use the link motion only for a reversing gear and to add a third eccentric and a separate valve for the cutoff. The additional mechanism was justified by the contention that cutoffs of more than 35 per cent, the practical limit of the link motion, were obtainable and desirable.

An early promoter of the combination link–separate cutoff valve gear was C. C. Dennis, superintendent of the Buffalo and State Line Railroad, who in 1854 prevailed upon Rogers to equip two new engines for his road with the combination

---

[157] Hinkley Register, Boston Public Library.
[158] Colburn and Holley, *The Permanent Way*, p. xxii.
[159] Bell, *Early Motive Power*, p. 130.

[160] Clark and Colburn, *Recent Practice*, p. 66.

gear.[161] A second rocker, with the upper arm formed as the link, was obviously based on the Cuyahoga gear.[162] In 1859 Rogers fitted the same gear to the New Jersey Railroad and Transportation Company's *J. J. Chetwood*. This is the only other machine known to have been fitted with a combination valve gear by Rogers. Uhry and Luttgens, employed by the New Jersey Locomotive and Machine Works, also of Paterson, began development of a combination valve gear about the time Rogers was building the Buffalo and State Line engines. The arrangement underwent several modifications; the initial plan, patented on March 22, 1855 (No. 12564), employed two links but is not believed to have been used. A second plan, which used a single rocker and valve, the cutoff being driven by a cam through a combination lever, was applied to a small number of locomotives. Clever though the plan was, its small steam economies did not offset the extra costs involved. The Uhry and Luttgens gear is illustrated and discussed in the *Talisman* section (see Fig. 174).

Another ingenious combination gear was designed in about 1870 by William A. Foster of the Fitchburg Railroad.[163] Foster used a single stationary link with *two* link blocks. Separate rockers were used for the main and the cutoff valves. The plan was tested but not repeated.

The combination link–separate cutoff gear was revived in 1882, very much in its old form, by A. J. Stevens, master mechanic of the Central Pacific Railroad.[164] Three eccentrics, a link-rocker, and riding valves reminiscent of the 1854 Rogers plan were incorporated into this gear. Stevens abandoned this motion the next year for a radial cutoff gear (to be discussed later in this chapter).

The last and possibly the most enthusiastic proponent of the combination link-independent cutoff was David Clark, an old-line master mechanic who was active in locomotive construction from the 1850's until his retirement in 1892. Clark's valve gear was not particularly original; in fact, it was a near-perfect copy of the 1854 Rogers gear. It was first applied to the Lehigh Valley Railroad's *W. C. Alderson*, a 4-4-0 passenger locomotive built by Clark in about 1883 at the road's Hazleton Shops. A patent was somehow obtained for this old idea on May 1, 1883 (No. 276773). The number of Lehigh Valley engines so-fitted is not certain, but Clark was actively applying the gear as late as 1888.

In 1885 Clark furnished the Master Mechanics Association with a detailed drawing of his valve gear showing the arrangement as fitted to the No. 120, a 4-6-0 built under his supervision at the Hazleton Shops. The drawing was reproduced in the Association's report of that year, together with several beautiful indicator diagrams. Impressive as these diagrams were, no saving in fuel could be established and Clark's gear disappeared, the last rebellious flicker of the ancient independent cutoff valve motion.

RADIAL VALVE GEARS. The radial valve gear was not an important class of valve motion in the United States until after 1900. It is difficult to trace the early history of this mechanism since no single definition can be found that suits the various conflicting designs credited as being radial gears.[165] The radial valve gear is conventionally thought to be a gear employing a combination of two right-angle motions such as those in the well-known Walschaert gear. Yet the Hackworth, Joy, and others, gears that use only a single motion, are regularly classified as radial gears.

---

[161] *Engineer* (Phila.), August 23, 1860, p. 11.
[162] The Rogers link-separate cutoff gear is illustrated on p. 54 of the Rogers history, *Locomotives and Locomotive Building*.
[163] *Locomotive Engineering*, October, 1893, p. 437.
[164] Stevens' combination link-cutoff gear is illustrated in *Recent Locomotives* (New York, 1883), Fig. 78.

[165] According to Yoder and Wharen, *Locomotive Valves*, p. 107, any gear where the link block is stationary may be classed as radial; this would presumably include the Gooch stationary link. A conflicting definition is given in C. H. Peabody's *Valve Gears for Steam Engines* (New York, 1900), p. 88: "The name radial valve-gear has been applied to a number of reversing-gears that differ widely in detail and in general appearance, but agree in that they derive the mid-gear motion of the valve from some source that is equivalent to an eccentric with 90° angular advance, and they combine with this motion another that is equivalent to that of an eccentric with no angular advance."

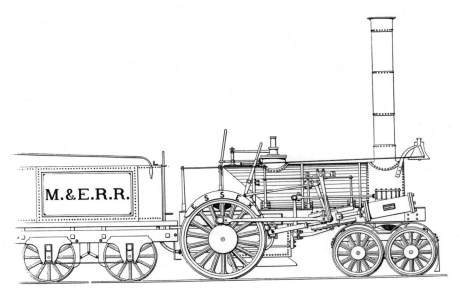

Fig. 73. The Morris and Essex Railroad's Essex *showing the crosshead-drivers valve gear devised by Seth Boyden in 1837–38.*

Presumably the first radial gear used for locomotives was that devised in 1832 by John Melling, locomotive superintendent of the Liverpool and Manchester Railway. The arrangement consisted of a straight link driven by a pin fitted to the connecting rod.[166] Except for the drive, the arrangement was identical to the general plan of the Hackworth (1859) and Joy (1879) gears which attracted the attention of the railway world later in the century.

Seth Boyden was the earliest American proponent of the radial gear, if his arrangement can be properly designated as such. His second locomotive, the *Essex*, completed in 1838 for the Morris and Essex Railroad, was fitted with a curious crosshead-driven valve motion that appears to fall under the broad definition of radial gears. The gear was reversed and the cutoff was varied by shifting the block in the short box-link mounted high on the engine's side as shown in Fig. 73. Boyden built only one more locomotive and his valve gear was not copied by other builders.

The next instance of an American radial valve gear was that designed by John L. Whetstone of Cincinnati for the three Sellers grade-climbing locomotives built in 1850–52 for the Panama Railroad.[167] The complex running gear of these loco-

[166] Sinclair, *Locomotive Engine*, pp. 438, 440.

[167] See J. H. White, Jr., *Cincinnati Locomotive Builders, 1845–1868* (Washington, D.C., 1965), for a complete account of the Sellers grade-climbing locomotives. Errors in Sinclair's description of Whetstone's gear are noted in the above work.

motives precluded the employment of an ordinary link motion inside the frame. Whetstone devised a special valve gear to overcome the cramped clearances. It was mounted outside the frame, except for one eccentric and its connections. A lever, actuated by the crosshead, worked in combination with the eccentric. In short, Whetstone's gear was an exact copy of Walschaerts' original 1844–48 scheme, but it is believed that Whetstone developed his design independently. Aside from its limited use by Niles and Company, Whetstone's radial gear was not recognized by other American mechanics until many years later.

Surely the most famous radial valve gear was that invented by Egide Walschaerts (1820–1901) of Brussels, Belgium.[168] Walschaerts developed the plan for his valve gear while a young man employed at the workshops of the Belgian State Railway. In 1844 the first rudimentary scheme was patented in the name of his mentor, M. Fisher. A much-revised design was developed, and in 1848 the new radial gear was first tested. While employing the same principles of the modern Walschaert gear, the arrangement was different in that an eccentric mounted inside the frame on one of the driving-wheel axles was used in place of the familiar outside crank. The outside crank attached to one of the side-rod pins was not used until about 1860.[169] The gear rapidly gained favor on Continental roads from this time forward and eventually spread to England and North America.

Acceptance of the Walschaert gear is barely within the scope of this study, but its emergence as the successor to the link motion makes it desirable to outline its progress. William Mason first used the Walschaert gear in 1874 on a Mason-Fairlie double-truck tank engine, the *Wm. Mason*, built for the Boston, Clinton and Fitchburg Railroad. For several years Mason had been building machines on this plan with Stephenson link motion and undoubtedly had experienced difficulty in fitting the inside motion around the massive center bed plate necessary for these peculiar locomotives.[170] Because the Walschaert gear was entirely outside the frame, it did not interfere with the bed plate or exhaust pipes. Mason used the Walschaert gear only on his double-truck locomotives and never once applied it to a standard machine. This indicates that he was a less enthusiastic champion of the outside radial gear than is often set forth. Incidentally, it is possible that Mason was inspired to use the radial gear after seeing two Niles locomotives on the Lehigh Valley Railroad fitted with Whetstone's valve motion.[171] Mason fitted about 140 locomotives with the Walschaert gear between 1874 and 1889.

Other American builders showed little interest in the radial gear. Baldwin built an engine so-equipped in 1878 for export. A. J. Stevens began applying a radial gear in 1883 to the locomotives of the Central Pacific Railroad, but no other roads copied his plan. As with so many worthy schemes, the industry did not adopt outside radial valve gears until it was forced to do so. The introduction of heavy locomotives after 1900, resulting in the need for stronger cross-bracing of frames, made the inside valve gear a mechanical impossibility. The large axles of the heavier engines represented an all but impossible mounting for eccentrics of any reasonable size. The radial gear was not adopted because of any superior economy of steam distribution over the link motion but rather because it was a more convenient mechanical arrangement. By 1910 the Walschaert gear, most popular of all the radial gears, was recognized as standard.

CORLISS VALVE GEARS. One valve gear curiosity worthy of consideration before closing this discussion is the Corliss valve gear. In 1851 George H. Corliss of Providence, Rhode Island, built the *Advance* to demonstrate the utility of his valve gear for locomotives. Corliss began the manufacture of locomotives a few years earlier as an adjunct to his large stationary-engine business. The construction of the *Advance*

---

[168] Data on Walschaerts is from a history of the Walschaerts gear, *Railroad Gazette*, October 24, 1902, p. 809, and from W. W. Wood, *The Walschaert and Other Modern Radial Valve Gears for Locomotives* (New York, 1920).

[169] Forward, *Handbooks of the Science Museum*, Vol. 3, Part I, p. 27.

[170] A history of the Mason-Fairlie locomotive is given in *Railway and Locomotive Historical Society Bulletin*, No. 41 (1936), pp. 15–22.

[171] *Railway and Locomotive Engineering*, July, 1909, p. 302.

was prompted by a desire to prove the superiority of his patented rotary valve gear over the link motion. The machine as completed was a light, inside-connected 4-4-0 with 15- by 20-inch cylinders and 66-inch drivers. As in the stationary Corliss engine, two admission valves and two exhaust valves were used for each cylinder. The four valves were driven by an outside crank mounted on the forward driving wheel's crank pin. The reverse was effected by a stationary link. According to a note made in the 1920's by the late Ornam L. Patt, a former employee of the Rhode Island Locomotive Works and long a student of railroad history, the *Advance* was acquired in about 1853 by the Providence and Worcester Railroad. It was renamed the *Orray Taft* and was refitted with ordinary slide valves. This is partially confirmed by the 1853 annual report of the Providence and Worcester, which lists a 25-ton Corliss locomotive of this name.

A line drawing of the *Advance* is reproduced in F. H. Colvin's memoir *60 Years with Men and Machine*. Colvin misdates the machine by 30 years and fails to comment on its history in his text, but clearly the drawing is more than a whimsical representation of a New England locomotive of the 1850's. It has the clear ring of authenticity and is, in my opinion, a sound draft very likely based on original drawings no longer in existence.

Corliss' valve arrangement, so successful in stationary practice, was predictably a failure in railway service because of its great complexity. The *Advance* was not copied. Alexander L. Holley recalls his three years of demonstrating the *Advance* in the following jocular recollection:

This locomotive was possessed of a certain inborn cussedness, which could hardly be the attribute of a mere machine—her spiritual nature was a sort of Mephistophelian cross with a Colorado mule—and as to her physical constitution and membership, a cotton-factory 'mule' was simple in comparison. The Old Jigger had, as nearly as I can remember, 365 valves, one to break down every day in the year. And as a valve *motion,* well, nobody ever counted the number of its pieces. They were as the sands of the sea-shore. Most of them used to jar off, the first few trips of the week, after which all the men in the shop could comparatively keep track of the rest of them. I will say for the Old Jigger that she made the best indicator-card I ever saw from a locomotive; clean cut-off, almost a theoretical expansion curve, and an exhaust as if she had knocked out a cylinder head. Well, once in a while, after she had been jackassing over the road about four hours behind time, and we had pinch-barred her into the round-house, we used to pull out these indicator-cards of hers, and talk them over right before her, and we would look at her and ask one another why in thunder an engine that could make a card like that would act as if the very old—chief-engineer was in her. And next morning she would rouse up and pull the biggest train that had ever been over the road—ahead of time.[172]

The failure of the *Advance* did not discourage other mechanics from experimenting with Corliss valve gears for locomotives. A number of passenger engines were successfully fitted with this type of valve gear on the Paris-Orleans Railway in the 1890's.[173] O. W. Young tested a two-valve, Corliss-type gear on the Chicago and North Western and on several other United States lines at the turn of the century, but no large-scale use resulted.

VALVES AND VALVE PORTS

The plain, straightforward D valve, so-called because of its silhouette in longitudinal section, was the unquestioned favorite for locomotive service during the nineteenth century. Its simplicity and cheapness account for its popularity. Except for a few early Stephenson engines with double D valves, every locomotive in the present study has the plain valve. Such valves were made of cast iron, although, in a few early cutoff gears where wear was excessive, brass was occasionally used.

[172] American Society of Mechanical Engineers, *Transactions,* Vol. 4 (1882–83), p. 47.

[173] *Engineer,* May 15, 1896, p. 490; *Railway and Locomotive Historical Society Bulletin,* No. 88 (1953), p. 48.

The ordinary D valve suffered from one major defect. That was the rapid wear of the valve seat and the friction generated because of the great pressure exerted on the top of the valve body by the steam. An ordinary locomotive might have a "load" of from 8 to 10 tons on each valve and expend 25–30 horsepower to work each valve.[174]

It was calculated that even in a small stationary engine this force absorbed nearly 1–2 per cent of the engine's power.[175] According to a test performed by the Central Railroad of New Jersey in 1886, the rapid wear of ordinary slide valves was given as one thirty-second of an inch for every 6,000 miles.[176] Such an extraordinary amount of wear shows that the savage cutting action of an unbalanced slide valve was not unlike that of a planing machine. The same railroad recorded only one thirty-second of an inch of wear for up to 65,000 miles when balanced valves were used. Other more minor side effects caused by unbalanced valves were: wear and strain on all valve-gear connections; bent valve stems; and difficulty in reversing.

Clearly there was need for a slide-valve design in which the steam would not bear down on the top of the valve. Valves of this type were commonly known as balance valves since they sought to balance or counteract the destructive downward pressure of the ordinary slide valve. A large number of balanced valves were patented and an even larger number were tested. It will be possible to treat only a small portion of these experiments here.

The earliest record of a balanced valve appears to be the patent granted to Hiram Strait of East Nassau, New York, on June 25, 1834. Possibly the first use of a balance valve for railway service occurred in England on the Grand Junction Railway in 1844. Wilson Eddy was a pioneer in the use of such valves in the United States. In about 1851 he began their use on the engines of the Western Railroad (Massachusetts).[177] In 1856 Rogers fitted a balance valve (patented by John Gleason on August 29, 1854, No. 11607) to an engine of the Buffalo and Erie Railroad.[178] The valve was a failure and was soon abandoned. Thomas Winans used a balance valve on the experimental passenger locomotive the *Centipede*, built a year prior to the Rogers engine just mentioned.

George W. Richardson, already mentioned for his invention of the pop safety valve, introduced a balance valve for locomotives in 1872. Richardson's modification of the slide valve consisted of four spring-loaded packing strips, fitted in the top of the valve, which bore against a supplementary plate attached to the underside of the valve-box cover plate, thus sealing the top of the valve from the steam. Richardson received a patent on January 31, 1872 (No. 111382). His invention was received with moderate enthusiasm by the industry.

The Master Mechanics Association in the same year of Richardson's patent reported that balance valves were coming into more widespread favor. This opinion appears a bit optimistic, for many mechanics remained unconvinced that the balance valve was necessary or perfectible. As late as 1886 several master mechanics challenged the reported merits of such mechanisms and commented on their greater first cost, $10 for a plain valve versus $90 for a balance valve.[179] The usual cranky complaints about excessive maintenance and mechanical humbuggery were voiced as well. Another thirteen years did not see a complete adoption of balance valves although it appeared that they were "more generally adopted."[180] As the locomotive grew in size, and large slide valves and ports were required to distribute steam to the huge cylinders, balance valves became an absolute necessity. The rise in steam pressure also had a decided bearing on the conversion to balance valves.

Before the balance slide valve had firmly established itself, it was superseded by the piston valve in about 1910. The origins of piston valves for locomotives can be traced to the Liverpool

---

[174] R. H. Thurston, *Manual of Steam Engines* (New York, 1897), p. 253.
[175] Forney, *Catechism*, p. 350.
[176] *Master Mechanics Report*, 1886, p. 126.
[177] Clark and Colburn, *Recent Practice*, p. 67.

[178] *Master Mechanics Report*, 1874, p. 191; Forney, *Locomotives and Locomotive Building*, p. 58, illustrates Gleason's valve but incorrectly gives the patent date as 1864.
[179] *Master Mechanics Report*, 1886, p. 124.
[180] *American Engineering and Railroad Journal*, June, 1899, p. 109.

and Manchester Railway's *Atlas,* a four-wheel freight locomotive completed by Stephenson in 1832.[181] Difficulty was experienced in maintaining the packing, and this form of valve was not repeated by Stephenson. The earliest known test of a piston valve in the United States occurred in 1856 when the Boston and Worcester Railroad fitted a piston valve designed by Joseph Marks to its locomotive the *Mercury*.[182] The valve was not a success, because of defective packing, but is known to have been tested by one other road.

Marks's valve was followed by a piston valve patented on February 5, 1861 (No. 31303), by Thomas S. Davis, superintendent of the Jersey City Locomotive Works. The valve was used successfully on stationary and marine engines before it was tried for locomotive service.[183] One of the first machines so-fitted was the Atlantic and Great Western Railway's *Telegraph,* completed in July, 1864, at the Jersey City shop. Several years later, when he became master mechanic of the Milwaukee and St. Paul Railroad, Davis rebuilt several engines with piston valves.

In 1869 the Boston, Lowell and Nashua had at least one locomotive, the *Medford,* fitted with piston valves. It is possible that other early examples are to be found, but it can be stated safely that only a handful of American locomotives were so-equipped before 1905. The greater first cost and the difficulty of maintaining packing rings prevented the more general use of piston valves until higher steam pressure and superheating forced their adoption.

The proper size of valve ports was a subject of some debate among locomotive designers. Most authorities agreed on the advantages of the large valve port. Generously proportioned ports and steam passages permitted an unrestricted flow of steam between the valve box and the cylinders. The good sense of this belief was tempered by practical considerations of size and weight. A finite objection to large ports was that a very large slide valve was required to cover them. Unless the builder wished to go to the expense of providing balance valves—and most did not—the friction of working such large valves countered the efficiencies offered by large ports. While the ideal port length would equal the cylinder diameter, a compromise of roughly seven-eighths that size was usually agreed upon.[184] The large-port idea was carried to an extreme by Niles and Company; it manufactured locomotive cylinders 15 inches in diameter with ports measuring 18½ inches in length.[185]

At the opposite extreme of the controversy was Wilson Eddy, the independent and gifted master mechanic of the Western Railroad of Massachusetts. On some of his freight locomotives with 18¾-inch by 28-inch cylinders, ports only 8 inches long were used.[186] The main concern here seemed to center on keeping the valves small so that balanced valves might be avoided, although Eddy employed balance valves on many of his engines. Eddy's engines were notable performers and were well regarded, despite their several peculiarities of design.

VALVE AND CYLINDER LUBRICATION

The cost and consumption of valve oil have already been discussed in the section on operating costs. The type of oils employed and the means of application are also of interest and will be treated briefly here.

The favorite tool of valve and cylinder lubrication during the greater part of the nineteenth century was the simple oil cup attached directly to the valve-box cover. At first, open cups fitted with a stopcock were used. Such cups are clearly

---

[181] Ahrons, *The British Steam Locomotive,* p. 21; a drawing of the piston valve is given on p. 285 of Warren's *A Century of Locomotive Building.*
[182] *Railroad Advocate,* June 14, 1856, p. 3.
[183] Davis' piston valve is discussed in *Scientific American,* November 11, 1865, p. 291, and in the *Railroad Gazette,* May 5, 1876, pp. 191–93.
[184] *Master Mechanics Report,* 1877, p. 31.
[185] Clark and Colburn, *Recent Practice,* p. 66.
[186] *Locomotive Engineering,* March, 1891, p. 41.

shown in the *Lancaster* and Dunham sections of this volume, in Figs. 110 and 114. By the mid-1850's more fastidious makers were placing covers on the cups in an attempt to keep dirt and cinders out of the lubricant. In either case it was possible to open the stopcock and thus lubricate the cylinders only when the engine was not working steam. Otherwise the steam would obviously blow out through the oil cup. It was the fireman's duty to climb out to the front of the engine when it was in motion and open the oil-cup valve as it was required. It was usual to give the valves a shot of oil before starting, and, because stops were frequent, the fireman was called upon to make his perilous journey less often than might be supposed.

Wilson Eddy, a former trainman and undoubtedly sympathetic to the trials and discomforts of engine crews, devised a simple method of placing the cylinder oil cup in the cab. An oil line running at an angle outside the boiler jacket carried the lubricant to the cylinders. Aside from the convenience and safety for the engine crew afforded by this arrangement, the oil was kept warm inside the cab and was less exposed to dirt and water in this protected position. Eddy used this plan on his first locomotive, the *Addison Gilmore,* completed at the Western Railroad's Springfield, Massachusetts, shops in 1851.[187] The idea was slowly adopted by the industry during the next few years and by the 1860's was well established. It was the usual practice to place the oil line under the jacket. Some new locomotives, however, were fitted with cylinder-mounted oil cups as late as 1914.[188]

The need for a steady, automatic supply of oil to the cylinders led to the development of the displacement, or hydrostatic, lubricator. A tiny jet of steam was allowed to enter a closed vessel containing oil. After condensing, the water sank to the bottom of the vessel and thus displaced the oil out into the lubricating feed lines. The rate of feed could be varied by the amount of steam admitted. The invention of the hydrostatic lubricator has been credited to John Ramsbottom, locomotive superintendent of the London and North Western Railway.[189] While it might be questionable that Ramsbottom was the originator of this device, he did obtain a patent relating to the concept as early as 1858. A year earlier he received a patent for a mechanical lubricator, but, finding this approach unsatisfactory, he continued the development of an automatic lubricator on the hydrostatic principle. A patent was granted on October 10, 1860 (British, No. 2460). The patent drawing shows a small, bulb-shaped reservoir connected directly to the valve-box cover of each cylinder; thus, in order to adjust or refill the mechanism, it was necessary to climb out onto the engine when in motion or to wait for station stops. In later years a single lubricator mounted in the cab with oil lines running to the cylinders was used. The hydrostatic lubricator was well received in England. The Midland Railway first employed one in 1862 and soon had all its engines so-equipped.[190]

Its adoption for locomotive service in the United States was much slower; there is little evidence that hydrostatic lubricators were generally employed before the 1880's. One of the earliest American manufacturers of such an apparatus was the Detroit Lubricator Company, which produced its first locomotive hydrostatic lubricator in 1877.[191]

Displacement lubricators were used with considerable success for stationary engines; however, the impractical connection of both the condensing and delivery tubes to the boiler side of the throttle in a locomotive meant that the oil would be siphoned out of the reservoir when the throttle valve was closed.[192] In 1882 this problem was solved by the addition of a simple equalizing tube fitted with a choke valve.[193] During the next ten years hydrostatic lubricators succeeded in all but replacing the antiquated cylinder oil cups. Aside from temperamental operation, the main disadvantage of the hydrostatic

[187] *Ibid.*
[188] Alco builders photograph (1914) of the Boston, Revere Beach and Lynn's No. 24 shows the cylinder-mounted oil cup.

[189] Forward, *Handbooks of the Science Museum,* Part II, p. 73.
[190] Ahrons, *The British Steam Locomotive,* p. 166.
[191] Catalog 118, Detroit Lubricator Company, 1924.
[192] M. M. Kirkman, *The Science of Railways* (Chicago, 1904), Vol. 14 (Supplement), p. 249, mentions hydrostatic lubricators.
[193] *Locomotive Engineering,* March, 1895, p. 177.

lubricator was that the supply of oil was not automatically varied to the needs of the engine. It worked at a constant rate set by the operator. The so-called mechanical lubricator was developed in answer to this defect. Operated by the motion of the engine, it pumped the lubricant at a rate corresponding to the speed of the engine. One of the earliest applications of such a device in this country was made to a locomotive of the Chicago and North Western Railway in about 1900.[194] It was another thirty years before mechanical lubricators were regularly employed on American locomotives.

Tallow, refined from beef fat, was the major type of cylinder oil used before the adoption of petroleum in the 1880's. Sperm and lard oils were also used. The chief objection to animal oils was the acid, destructive to the valves and cylinders, formed as the oil vaporized in the steam.[195] Before the commercial production of petroleum in the 1860's, there was no alternative, yet even after this time petroleum was used mainly for running-gear lubrication.[196] As the price of petroleum cylinder oils declined and its composition improved during the next decade, petroleum was readily adopted in standard practice.

CYLINDERS

Cylinders, unlike most other locomotive components, remained remarkably unchanged in basic design throughout the nineteenth century. Cast iron was the only material employed; valve boxes were always separate; cylinders were made universally with open ends, and cylinder heads were required for both ends; right- and left-hand cylinders were cast separately, thus permitting easier machining than would be involved in one massive casting. In addition, only one pattern was necessary for both the right and left castings. Cylinders were the most complex and expensive casting in locomotive construction. Involved pattern and core work was required for the internal steam and exhaust passages, as well as for the valve ports. The production of a flawless casting was only the first step in making a cylinder; finishing the raw casting was a major machining project. Despite the introduction of "boring engines" by Wilkinson in 1775, most American locomotive manufacturers depended on ordinary lathes for this heavy finishing work until the 1860's. Coleman Sellers recalled that in 1856 two days were required (about 20 hours) to bore a locomotive cylinder 15-inches in diameter with a 36-inch lathe fitted with a horizontal boring bar.[197]

Aside from makeshift equipment, the hard grade of cast iron favored for cylinders to insure long wear was another factor that contributed to the lengthy machining operation. It is likely that other manufacturers developed more efficient boring machines earlier, but the Rogers Locomotive Works acquired a mill in 1871 that could bore a 16- by 24-inch locomotive cylinder in seven to eight hours.[198] The mill was designed by William S. Hudson and was built by William Sellers and Company. Several years later Sellers was producing a machine that would bore, face, and counterbore an 18-by-24-inch cylinder in only three and a half hours.[199] It can be understood from the progress made in cylinder-boring mills how the industry was able to produce heavier and more efficient locomotives, yet hold down its prices.

The proper method of fastening cylinders to the locomotive was a much disputed matter among designers. Before the spread truck (introduced in 1855) cylinders were normally set on a steep incline, high on the smokebox, so as to clear the truck wheels. This was in general a poor and weak arrangement that placed a great strain on the smokebox. Naturally,

---

[194] Kirkman, *The Science of Railways*, p. 279.
[195] *Master Mechanics Report*, 1874, p. 241.
[196] *Ibid.*, 1877, p. 40.
[197] *Journal of the Franklin Institute*, January, 1900, p. 13. Sellers was recalling his experiences at Niles and Company but stated that Baldwin used a lathe for cylinder-boring at the same time; presumably this was common practice.
[198] *American Artisan*, March 15, 1871, p. 163.
[199] Bureau of the Census, *Tenth Census*, II, 55.

ample mounting flanges, cast integral with the cylinder, and reinforcement plates inside the smokebox helped to strengthen this mounting. Some builders (see the *Philadelphia*, Fig. 135) made a direct attachment to the frame by use of elaborate mounting flanges.

The most secure form of cylinder fastening is direct bolting to the frame. This is made most conveniently with a bed plate called a "saddle" which can be firmly attached between the frame under the smokebox. This device was used by Walter McQueen as early as 1848 on the engine the *Mohawk* (Fig. 23). Other early examples of the McQueen style of saddle are shown by the *Columbia* and *Superior* drawings (see Figs. 159 and 163). Several years later William Mason developed a more substantial but heavier saddle that was cast as a hollow box. Examples of Mason's saddle are shown by the *Phantom* and Tyson Ten Wheeler drawings (Figs. 194 and 183). While most builders had accepted the saddle by 1855, a few manufacturers resisted its adoption, claiming with justification that its great weight, 1,000 pounds, represented an additional burden to the already front-heavy eight-wheel locomotive. One such builder, Rogers, continued to use the old D-shaped smokebox into the 1860's but provided a substantial fastening in the following plan: "The smoke box is made of sheets ⅜ inch thick, with a bar of two inch square iron riveted around the forward edge, on the inside; the sheets are of double thickness, or ¾ inch, where the cylinder is bolted on, the flange of the cylinder rests upon the frame, while another wide flange reaches high up on the side of the smoke box, affording all the permanence of attachment that can be obtained by any plan."[200]

The so-called half-saddle, where half of the saddle and one cylinder were combined into a single casting, was a natural development of the three-piece cylinder-saddle-cylinder arrangement. The two half-saddles joined at the locomotive's center and provided a neat and strong assembly. Baldwin was apparently the originator of this important improvement; a sketch of 1857 showing this plan was reported incidentally by J. S. Bell in his history of early Baltimore and Ohio locomotives.[201] The half-saddle was universally used in this country by the 1870's and continued in favor until the 1920's when one-piece cylinder castings became more common.

The insulation of cylinders might appear to be obvious for thermal economy, yet, from existing evidence, apparently it was not employed regularly until the 1850's. The earliest known example of cylinder-jacketing is the 1843 Rogers 4–4–0 drawing (Fig. 20). The good sense of this practice was belatedly recognized, but, by the 1850's, builders' lithographs invariably showed cylinder jackets. Wooden or felt lagging was aided by highly polished brass or planish iron jackets. Usually jackets were fitted over the cylinder heads, but polished cast-iron heads without jackets were known. Mason's first locomotive was so-finished, as was Tyson's Ten Wheeler pictured in this work (Fig. 177).

Piston construction, like that of cylinders, showed little change from the beginning of locomotives in this country in the 1830's to the end of the century. During this period, pistons consisted of a thick, hollow, cast-iron body with a flat rear face and a flat bolt-on front. Three to six single-leaf springs with adjustment nuts were fitted inside the body and exerted pressure on the rings. Two, and occasionally three, split rings were used. Plain or Babbitt-lined cast-iron and brass rings were used before 1860.[202] Brass appears to have been the favorite material for rings during most of this period and it continued to be favored as late as the 1890's.[203] Piston rods were made with tapered ends to fit the piston and were secured by wedges, lock nuts, or both. Early drawings show the taper running from front to back with only a wedge and the front cover plate securing the rod (see the Dunham [Fig. 116] and *Philadelphia* [Fig. 136]). In later years it was more common to run the taper from back to front and to thread the extreme end of the rod for a lock nut; see the Rogers 4–4–0, 1865 (Fig. 223) for this style of construction.

Wrought-iron pistons are known to have been used occa-

---

[200] *American Railroad Journal*, October 29, 1853, p. 700.

[201] Bell, *Early Motive Power*, p. 54.
[202] Clark and Colburn, *Recent Practice*, p. 62.
[203] Forney, *Catechism*, p. 270.

sionally on American locomotives although the practice appears to have been rare. One such example was Winans' heavy 4–8–0 passenger locomotive of 1855, the *Centipede*. Wrought-iron pistons could be made lighter than an equivalent cast-iron piston and thus reduced reciprocating disturbances. There is also some evidence that one-piece "cast-steel" piston and piston-rod assemblies were used in the 1860's. An advertisement in the 1864 *Ashcroft's Railway Directory* of Naylor and Company, Philadelphia, agents for Naylor, Vickers and Company, Sheffield, England, shows a one-piece steel piston and piston-rod assembly. No specific reference to the use of such assemblies is known to the author, however.

Hemp or wool remained the standard packing material for piston-rod stuffing boxes until about 1880, although metal packing was advocated by Baldwin as early as 1840. Higher steam pressures and faster working engines eventually brought an end to hemp packing for locomotive service.

INSIDE CONNECTION. One of the most controversial points in basic locomotive arrangement was the nineteenth-century argument concerning the placement of the cylinders inside or outside of the frames. Both plans were used on the earliest locomotives constructed and both had strong advocates. The more prominent British builders had agreed on inside-cylinder engines by about 1830. This preference naturally influenced American makers, who, without exception, built their first locomotives with inside connections. West Point, Baldwin, Norris and Long, and Rogers all at first copied the inside scheme. This preference had disappeared by the late 1830's, however—although it was not entirely abandoned—when the simpler, more direct outside connection was adopted in this country. The inside-connection engine enjoyed a revival after 1845 when George S. Griggs completed the highly successful *Norfolk* at the Boston and Providence shops. This machine was virtually the prototype of hundreds of similar inside 4–4–0's so strongly favored by New England builders during the next ten years. The good performance of these machines prompted builders in other parts of the country, including the Paterson and midwestern builders who were in general opposed to the inside connection, to copy the New England plan. The migration of New England mechanics, often as superintendents and managers, might in part account for the spread of the inside-connection prejudice to regions outside of its home area.

The advantages of the inside connection were remarkably few. The cylinders, placed between the frames inside the smokebox, were kept warm and insulated. In this position the cylinders and all the reciprocating parts were close to the locomotive's longitudinal center line and thus the engine ran steadily. Both of these advantages were equaled by an outside-connected engine with insulated cylinders and carefully balanced reciprocating parts.

The disadvantages of the inside connection were many. Simply stated, it was a more complex and expensive plan than the outside-connected engine. The crank axle was its most important single defect. It was expensive ($500), heavy (1200 to 1500 pounds), about twice the size of an ordinary driving axle, and dangerous because of frequent failure. The breaking of the crank axle resulted in not only the expense of a new axle but also considerable shop time for replacement and possibly the loss of life and property if the fracture occurred when the engine was moving at speed. Crank-axle life was estimated to be one year in heavy service or two to three years in light service.[204] This estimate is supported by the following examples: Boston and Worcester, 1835–37, four engines, eleven broken cranks;[205] Boston and Providence, 1852, fourteen engines, eleven broken cranks;[206] Michigan Central, 1856–57 (winter only), twenty broken cranks.[207]

After the crank, the greatest objection to the inside connection was inaccessibility of the machinery for inspection and repair. The cylinders, valves, crosshead, guides, and main rods were all crammed inside the frame and under the boiler. It might be noted that in more recent years this same objection

---

[204] *American Railway Times*, March 16, 1861, p. 108.
[205] *Railway and Locomotive Historical Society Bulletin*, No. 23 (1930), pp. 9–21.
[206] *American Railroad Journal*, October 29, 1853, p. 700.
[207] *Engineer* (London), January 3, 1862, p. 6.

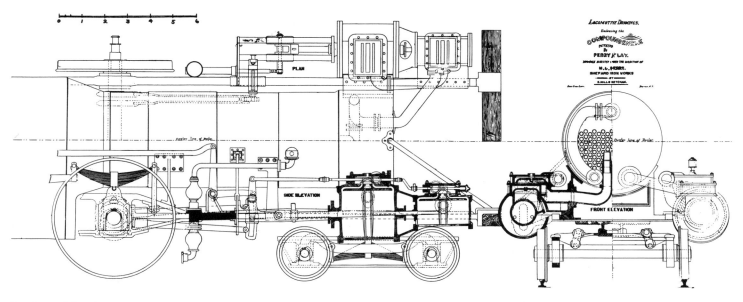

Fig. 74. The Erie Railway's No. 122, rebuilt in 1868 or 1869 by the Shepard Iron Works, is thought to be the first compound locomotive built in America. It was reconstructed on the tandem plan patented by J. L. Lay in 1867.

was basic to the rejection of three-cylinder engines that at first appeared so promising in the 1920's.

Another objection to inside-connected engines was the necessity to raise the boiler high enough to clear the throw of the crank axle. Ten inches of extra clearance was required, and this was not viewed favorably at a time when only low-set boilers were accepted and top-heavy locomotives were widely feared.

Considering the several serious objections to the inside-connected engine, it is not surprising that they were rejected by about 1855. Colburn was their most severe critic; he began his well-reasoned attack on inside-connected engines in a series of articles appearing in the 1853 *American Railroad Journal* and continued this campaign in other journals. By the late 1850's such engines were considered obsolete and their pro-

duction was abandoned by commercial builders. The Boston and Providence Railroad, loyal to the memory of its former master mechanic G. S. Griggs, built an inside-connected engine, the *Viaduct*, in 1873. While this machine was a freakish anachronism in this country, inside-connected engines remained popular in England throughout the nineteenth century. Several inside-connected engines are pictured and discussed elsewhere in this work (see the sections on the *Champlain* [p. 322] and the *Croton* [p. 328]).

COMPOUND LOCOMOTIVES. Compound locomotives might logically be discussed following the commentary on cylinders. Compound locomotives were not widely used in the United States before the 1890's and are in general outside the scope of our study, but, interestingly enough, one such machine was constructed in the late 1860's. This locomotive, shown in Fig. 74, was rebuilt in 1868 or 1869 as a tandem compound by the Shepard Iron Works, Buffalo, New York, from the Erie Railway's No. 122. The original machine had been built by Hinkley in June, 1851. The October 29, 1867, patent (No. 70341) of J. L. Lay was followed in the rebuilding; one of the major points of the design was that existing locomotives could be converted to compounds simply by replacing the cylinders. The high-pressure cylinder was 12 inches by 24 inches, the low-pressure cylinder was 26 inches by 24 inches; two slide valves were actuated by a common valve rod.[208] The tandem system was sound, the parts were well designed, but the 122 was not repeated; it was the only machine built on this plan. No records of its service or of any fuel economies are known to exist. Two reasons have been advanced for its failure: first, it would not steam, because of the weak draft resulting from low-pressure exhaust, the steam being expanded twice;[209] second, the size ratio of the high- and low-pressure cylinder (1 to 4) was too great for the low-pressure steam used.[210]

---

[208] Two accounts of the Shepard compound differ on several points; *Locomotive Engineering,* May, 1891, shows a drawing based on Fig. 74 of this volume and notes that slide valves were used (cylinder size being that given in the text). The *Railroad Gazette,* September 19, 1890, shows incline cylinders and a short-wheel-base truck; cylinder size is given as 11½ inches by 26 inches and 24 inches by 26 inches. Rotary valves were shown in an accompanying sketch which agreed with Lay's patent drawing. Both accounts are based on data supplied by mechanics associated with the engine's reconstruction.

[209] *Railroad Gazette,* September 19, 1890, p. 646.

[210] F. H. Colvin, *American Compound Locomotives* (New York, 1903), p. 7.

MISCELLANEOUS CONSIDERATIONS

10.

COWCATCHERS

Cowcatchers, also known as cattle guards or pilots, were one of the most prominent and distinctive features of early American locomotives. Cattle regularly presented a great hazard to trains by straying onto the tracks, for it was many years before capital-poor American lines could afford to fence their rights of way. Stalled or runaway wagons and carriages at grade crossings were another regular cause of disaster. The purpose of the cowcatcher was to clear such obstructions from the track, to prevent their getting under the wheels of the locomotive. If by some chance the obstacle was in the center of the track and was passed over by the axles, the low-hanging wood-burning firebox as often as not would strike it, thus causing a derailment.

The Camden and Amboy Railroad was the first to use cowcatchers on its locomotives. Isaac Dripps fitted such a device to the *John Bull* in about 1833 and it was soon used on other locomotives on the line. Dripps's cowcatcher differed from the classic plow-shaped structure fastened directly to the front beam of the frame in that it carried an abbreviated, wedge-shaped cattle guard on the end of two stout wooden beams supported by a pair of small wheels. This type of cowcatcher is illustrated in the *John Bull* section of the present volume (p. 265). A similar cowcatcher is shown by the drawing of the Utica and Schenectady Railroad's first locomotive (Fig. 107). In his 1838 account of *Civil Engineering in North America* the British engineer David Stevenson commented on the Dripps cowcatcher in the following statement: "I experienced the good effects of it upon one occasion on the Camden and Amboy Railway. The train in which I travelled, while moving with considerable rapidity, came in contact with a large waggon loaded with firewood, which was literally shivered to atoms by the concussion. The fragments of the broken waggon, and the wood with which it was loaded, were distributed on each side of the railway, but the guard prevented any part of them from falling before the engine-wheels, and thus obviated what might in that case have proved a very serious accident."[1]

Many railroads adopted cowcatchers during the earliest years

---

[1] David Stevenson, *Sketch of the Civil Engineering of North America* (London, 1838), p. 261.

of railroading in this country. The Board of Managers of the Philadelphia and Reading Railroad, for example, in the minutes of its October 14, 1840, meeting ordered that "cow catchers be attached to each engine." At the same time there are an impressive number of contemporary locomotive illustrations dating from before 1855 which fail to show cowcatchers. Such major builders as Baldwin and Norris invariably illustrated their engines before this date as being fitted with simple wheel guards (see Figs. 26 and 132). The consistency of this failure to show cowcatchers would seem to indicate that they were not universally used before 1855 or that such devices were to be fitted up by the purchaser. After this date the reverse was true; hardly a single builder's lithograph fails to show a cowcatcher.

Reports of locomotives derailed by cattle were common. One of the earliest of these was a notice in the *American Railroad Journal* on July 27, 1835, regarding an accident on the Schenectady and Saratoga Railroad: "The locomotive passed over the cow and was thrown off the track with considerable damage." Several passengers were injured in this accident, but the cow was the lone fatality. That a good-sized cow weighing upwards of 1,000 pounds could derail a 10- or 20-ton locomotive is not too surprising. The derailment of a Baltimore and Susquehanna Railroad locomotive "by a large hog getting under the wheels behind the cow-catcher" showed that far smaller beasts were capable of the same destructive work.[2]

A well-made cowcatcher was capable of throwing a full-grown ox (2,000 pounds) 30 feet.[3] A weak cowcatcher would fold under the locomotive on hitting such a heavy object and therefore was more often the cause than the preventer of accidents. Such an incident is related in the following notice: "RUNNING OVER CATTLE NOT ALWAYS SAFE.—An express train on the Pennsylvania Central Railroad was thrown off the track, on Friday last, by running over a bull at a place sixteen miles east of Pittsburgh. Two or three cars were turned completely over, and thrown down an embankment. Twenty-three passengers were injured, and the engineer was so much hurt that he is not expected to recover. We presume the cow-catcher must have broken. No doubt the cost of this accident will be equal to the sum necessary to protect the whole road from such accidents in [the] future."[4]

Although the earliest forms of cowcatchers were largely made of wood, iron was widely favored during succeeding years. Nearly all of the pre-1855 illustrations show cowcatchers made of round or flat iron bars. Several examples of these appear in the present volume as the *Mohawk*, Fig. 23, the *Croton*, Fig. 153, and the *Superior*, Fig. 161. The handsome wooden pilot, shaped much like the moldboard of a plow, rapidly came into favor after about 1850 and had become the accepted design by the middle of that decade. A. S. Sweet of the Buffalo and Erie Railroad was credited with building a particularly strong wooden cowcatcher that "killed over 300 head of cattle . . . and never had a hoof under the engine."[5]

Effective and necessary though such structures might have been, they added much undesirable weight to the front end of the locomotive. The cowcatcher often weighed half a ton or more and was cantilevered far in front of the supporting truck wheels. The *Master Mechanics Report* of 1870 comments on this condition and recommends fenced rights of way to eliminate the need for that formidable appendage, the cowcatcher. It was another twenty years before enough railroad mileage in this country was fenced and shorter and less-pretentious cowcatchers became common.

BELLS AND WHISTLES

For safety in moving around terminals it was found that a bell served as an effective but not too boisterous alarm. Bells for locomotives were suggested as early as 1834 after a crossing accident on the Boston and Worcester Railroad, but it was not

[2] *American Railroad Journal*, November 6, 1845, p. 714.
[3] *Engineering*, July 26, 1867, p. 66.

[4] *Engineer* (Phila.), October 4, 1860, p. 63.
[5] *Railroad Advocate*, July 26, 1856, p. 3.

until the next year that they are known to have been affixed to a locomotive.[6] The need for an effective warning device at crossings prompted the state of Massachusetts to enact a law in 1835 requiring bells for locomotives. Railroad companies of the state were thus forced to adopt warning devices at an early date and to see that their employees observed their use. One of the earliest operating rules of the Boston and Worcester Railroad stated that "the Enginemen . . . are strictly charged and exhorted to ring their bell, at all road crossings, at least 80 rods before they cross." One of the first known instances of such an application was cited in a newspaper account of April 17, 1835, concerning a bell on the *McNeill* of the Paterson and Hudson River Railroad.[7] During the same year three locomotives with bells were delivered to Boston and Lowell Railroad by the Locks and Canals machine shop.[8] Reportedly, the bell from one of these machines, the *Patrick*, is still in existence. Now on loan to the New Jersey Historical Society, Newark, the owner, Thomas T. Taber, states that the bell came into his hands some thirty years ago. The word of the former owner appears to be the only evidence that this bell is from the *Patrick*. Another early bell with only slightly better documentation is that from the *Rahway* of the New Jersey Railroad and Transportation Company, completed by Baldwin in 1837. The bell, now in the United States National Museum, was cast by E. Force of New York City in 1838; it measures 15½ inches at the mouth. In a note on early bells is the stipulation for a bell of "not less than 45 pounds" in the order books of Rogers for the Norwich and Worcester Railroad's *Norwich* completed in June, 1839.[9] Bells apparently had become commonplace on American locomotives by this time. Among the contemporary illustrations showing early applications of bells are the *Gowan and Marx*, 1839 (Fig. 122), the *Baltimore*, 1839 (Fig. 1), and the Baldwin iron-frame engine, 1839 (Fig. 62).

Locomotive bells in 1860 were said to weigh from 60 to 215 pounds and could be heard for a quarter of a mile.[10] The diameter at the bell's mouth varied from about 12 to 18 inches. The early bell stands were very simple wrought-iron frames such as the one on the *Gowan and Marx*. By the late 1830's an ornamental stand with the side members made in the form of an A, was introduced. This style of frame was used by Baldwin, Norris, and Hinkley between about 1840 and 1850. It is shown clearly in the *Copiapo* drawing (Fig. 142).

Bell stands with two turned columns became popular after 1850; several engines pictured in this volume illustrate this simple form of bell stand. Rogers in about 1860 adopted what was possibly the most fanciful bell stand used by any American maker. A baroque composition of rich scrolls, it was used on both freight and passenger engines produced by this builder until about 1870. A stand of this type is shown on the Rogers 4–4–0 of 1864 (Fig. 221) and the Mogul (Fig. 210). The National Museum of Transport (St. Louis) has a Rogers bell stand of the same design in its collection.

The Cuyahoga Steam Furnace Company of Cleveland, Ohio, used an odd combination bell and whistle stand on a number of locomotives it built in the 1850's. The cast-iron stand was oval-shaped with the bell mounted inside the opening. The legs of the oval were hollow, thus providing a steam passage to the whistle, which was fitted to the top of the stand. A somewhat less-elaborate combination stand was used by Mason in the 1860's. A drawing of such a stand, reproduced here as Fig. 75, indicates that the arms were extended outward so as to serve as a hand-rail support.

The modern U-shaped bell stand appears to have originated with Baldwin in the early 1860's. Several Baldwin builders' photographs of this period show bell stands of this type; among them are the United States Military Railroad's *General Dix* and the Northern Central Railroad's No. 47, both completed in 1862. A stand of the same design appears in the *Consolidation* photograph, Fig. 225. The U-shaped stand was soon adopted by all other domestic builders. Automatic bell ringers were not

---

[6] *American Railroad Journal*, September 6, 1834, p. 557.
[7] Lucas, *From the Hills*, p. 286.
[8] *B & O Magazine*, April, 1940, p. 26; data from C. E. Fisher.
[9] *Railway and Locomotive Historical Society Bulletin*, No. 49 (1939), p. 82.

[10] Clark and Colburn, *Recent Practice*, p. 68.

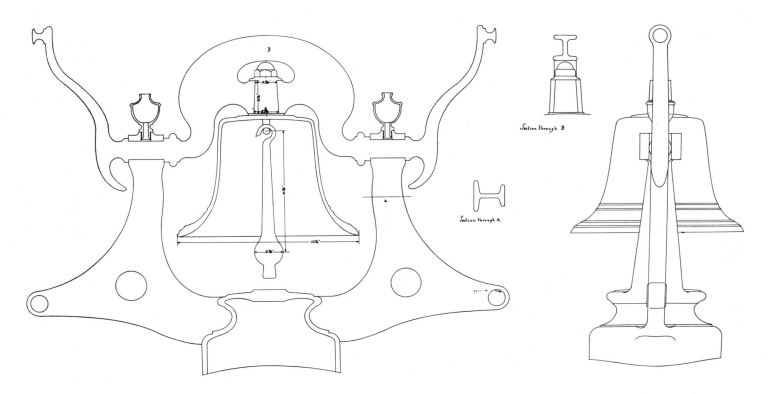

Fig. 75. Combination bell and handrail stand devised by the Mason Machine Works, 1864.

generally used until after 1910, but it may be of interest to note the use of a steam-powered bell ringer on the Erie Railway in 1854.[11]

The steam whistle was a more forceful alarm than the bell and was adopted for signaling on the line. Because of the many hazards associated with single-track railroading, unprotected grade crossings, and unfenced tracks, a resounding whistle was required to warn of an approaching train. An 1836 notice of the Wilmington and Susquehanna Railroad's locomotive the *Susquehanna* commented on the machine's "steam whistle . . . , an instrument whose piercing shrill sound may be heard at a distance of at least a mile."[12] In later years, when larger, more sonorous whistles were adopted, they were described as "a howling reservoir of sound, with an unearthly roar instead of a shriek."[13]

[11] *Railroad Advocate,* January 20, 1855, p. 3.
[12] *American Railroad Journal,* December 10, 1836, p. 779.
[13] *Engineering,* July 26, 1867, p. 66.

The origins of the steam whistle are British and a full account of its invention in 1832–33 is given in a detailed paper published by the Newcomen Society.[14] First used on an English locomotive in 1835, the device was probably first applied in the United States by the Locks and Canals machine shops on several locomotives completed in 1836. One of these machines, the *Susquehanna,* has just been described. Another machine known to have a steam whistle, built in 1836 by the same maker, was the *Hicksville* of the Long Island Railroad.[15]

William Norris has been credited with first applying steam whistles to locomotives, but the earliest account known to the author concerns a steam "trombone" developed by Norris in 1837.[16] Thomas Rogers' first locomotive, the *Sandusky,* completed in 1837, has also been credited as the earliest locomotive with a whistle. As shown by the foregoing account, both statements are in error.

HEADLIGHTS

The headlight was regarded as a necessary safety device for American railroads as nighttime operations of trains became more common after 1850. It lighted the way over twisting single-track roads where broken rails, misaligned switches, and washed-out bridges were ever-present dangers. It served as a warning to opposing trains and was a sure supplement to the inadequate signal systems used on early American railroads.

Horatio Allen's account of the primitive beginnings of the locomotive headlight is a familiar story, but it cannot be avoided here, much as the significance of the achievement is questioned. In about 1832 it was decided to attempt nighttime operations in order to relieve the unexpected heavy traffic that had developed on the newly opened South Carolina Railroad. Two small flat cars were coupled in front of the locomotive. A bright fire of pine knots on a bed of sand on the first car and a simple sheet-iron reflector on the car behind comprised the headlight. The arrangement was obviously a crude expedient. Unfortunately, it is not known if night operations were continued at this time, but if they were, surely a more convenient plan of illumination was adopted. On most roads during the early years, night runs were a rarity. George E. Sellers, for example, recalled that in 1834, because of the necessity of an after-dark run on the Philadelphia and Columbia Railroad, several ordinary hand lanterns were hung on the front bumper of the locomotive.[17]

Not many years later the headlight came into being on the Auburn and Syracuse Railroad. A sheet-metal box with cast parabolic reflector was fitted to one engine on the road by the master mechanics N. W. Mason and Allen Sweet of Auburn, New York.[18] In 1838 Nathan Rogers and William Wakeley, also of Auburn, began the manufacture of locomotive headlights on the plan just described.[19] Two years later the Boston and Worcester Railroad was reported to have adopted "a very bright headlight" for the night operation of freight trains.[20] The idea was to clear the line for daytime passenger trains.

Other isolated notes on the use of headlights undoubtedly could be uncovered for succeeding years, but the best statement on the more general employment of headlights was that given in the *Railroad Advocate* on December 22, 1855: "But night trains have become more frequent than formerly and as it cannot be told when any engine may be called on to a night train . . . it is not considered prudent to leave any first class

---

[14] Charles E. Lee, "Adrian Stephens: Inventor of the Steam Whistle," *Newcomen Society Transactions,* Vol. 27 (1956), pp. 163–73.

[15] *Railway and Locomotive Historical Society Bulletin,* No. 101 (1959), p. 40.
[16] *American Railroad Journal,* December 2, 1837, p. 654.
[17] *American Machinist,* December 12, 1885, p. 2.
[18] *Official Catalog of the National Exposition of Railway Appliances* (Chicago, 1883), p. 21.
[19] *Ibid.*
[20] Ringwalt, *Transportation Systems,* p. 105.

engine unprovided with a good head-light." The same journal went on to report that some four hundred headlights were manufactured in this country each year. By 1860 it is unlikely that any main-line road engines were not so-equipped.

Having briefly traced the history of the headlight, a few comments on its construction are in order. The basic classification for headlights was the diameter of the reflectors. Sizes ranging from 18 inches to 23 inches were offered at a cost of roughly $100 per light. The square or box case was the most familiar style of lamp used during the nineteenth century. The earliest cases appear to have been extremely simple according to early advertising line cuts appearing in the trade journals. By the late 1850's more elaborate cases with paneled sides and ornamental ventilators had appeared. Single ventilators were standard but double chimneys were occasionally offered; an example of the latter appears on the *Essex* (Fig. 151). Round cases were offered in the 1860 Bowles Railroad Supply Catalog. An earlier example of this style of lamp case is shown by contemporary lithographs of Rogers' *Victory* (1849) and Schenectady's *Lightning*. The cylinder-shaped lamps in these drawings are very small, however—not much larger than a good-sized carriage lamp—and it might be asked if they are not of European origin. The round-case lamp did not supersede the box headlight until almost 1900; nevertheless, the Pennsylvania Railroad continued to use box lamps on new engines until the First World War.

The parabolic reflector was used on the earliest locomotive headlights manufactured in this country. By means of a deeply dished reflector of this peculiar shape, a relatively weak light could be directed in a straight and powerful beam more than 1,000 feet long.[21] The reflectors made in 1838 by Rogers and Wakeley were of a soft, cast metal, probably silver plated. Speculum metal, a bright alloy composed of tin and copper, was used by some headlight manufacturers during this early period.[22] In 1850 Henry Ward, a headlight-maker of Rochester, New York, was reported to be making silver plated, cast,

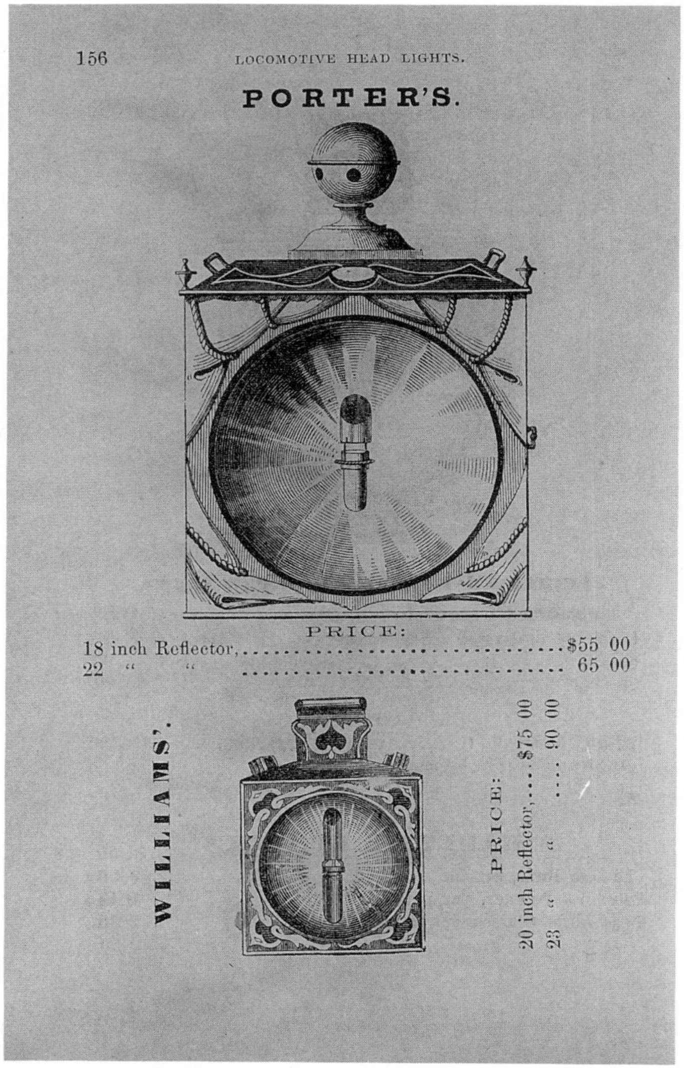

[21] Clark and Colburn, *Recent Practice*, p. 68.
[22] *National Car Builder*, April, 1874, p. 51.

Fig. 76. Square-case headlights, such as the one shown here, were the standard style of lamp used on American locomotives from about 1840 to the end of the nineteenth century.

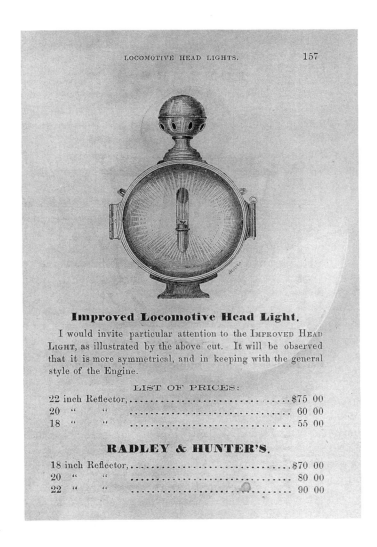

Fig. 77. Round-case headlights were rarely used in the nineteenth century, but they superseded the square-case headlight after 1900.

britannia metal reflectors.[23] While no original headlights from this period are known to exist, a well-detailed patent model, almost half-size, has been preserved by the Baltimore and Ohio Railroad. The model represents T. Snook and S. Hill's December 12, 1852, patent (No. 9490). The most significant feature of the lamp is the heavy cast reflector, apparently a tin alloy, which indicates that this form of construction was still in fashion at that time. The great defect of the cast reflector was its tendency to melt if the glass chimney of the lamp was broken. The tin alloys used had a very low melting point. Silver-plated, spun copper reflectors were used at this time and earlier. They soon superseded the cast reflectors because of their greater strength and ability to survive broken lamp chimneys.

Oil was the universal fuel for headlights in the nineteenth century. However, a variety of other fuels were tested and used. One of the earliest known examples of such a test involved a calcium lamp fitted to a locomotive of the Camden and Amboy Railroad in 1849.[24] The invention was credited to a Professor Grant. The lamp was said to throw a beam half a mile in length and to burn without adjustment for four hours. Various types of gas lamps were tried from the early 1850's onward with some success; the gas headlight manufactured by Radley, McAlister and Company of New York City was possibly the most favored. Lime lamps were tested on the passenger locomotives of the Pennsylvania Railroad in 1866 but were abandoned after a year because of their intensity.[25]

Electricity attracted attention surprisingly early. A Russian locomotive was fitted with a battery-powered lamp in 1874.[26] Nine years later a French railway experimented with a steam turbine generator for electric headlights.[27] The idea was revived in the United States by George C. Pyle in 1897, and its universal adoption was established by federal law in 1915.

[23] *Scientific American*, March 16, 1850, p. 204.
[24] *Scientific American*, October 27, 1849, p. 41; November 17, 1849, p. 65.
[25] *Railroad Gazette*, February 22, 1907, p. 235.
[26] *Annual Record of Science and Industry*, 1875, p. 151.
[27] *Scientific American*, December 1, 1883, p. 342.

DECORATIVE TREATMENT AND FINISH

The popular abandon lavished on highly decorated locomotives is a curious contradiction of the Spartan insistence that the same machine be a simple, practical mechanism devoid of all unnecessary engineering contrivances. The extent of this abandon is neatly summarized in the following statement made by Colburn: "The ornamental character of American engines is indeed remarkable. Besides the profusion of heavily moulded and highly polished brass in the cylinder and steam-chest covers, on the domes, the boiler-bands, whistle-stand, air-vessels, name and number-plates, &c., all the unfinished work is painted in showy colours, and the most ambitious efforts of decorative art are occasionally exhibited upon the panels along the sides of the tenders."[28] The fact that locomotive designers and builders were so ready to expend considerable effort in putting work *upon* rather than *into* their products might be explained simply by the wish to glamourize the new "work horse" of society, the locomotive engine. The railway and, more particularly, the steam locomotive were regarded with pride and veneration in the nineteenth century as a carrier of civilization, a noble product of man's genius, and a symbol of the new industrial age that hopefully promised to end manual toil. To finish this admirable machine in lead-gray or funeral-black was unthinkable. In addition, the whole spirit of the times was expansive and optimistic. This exuberance found expression in rich ornament, gay colors, and luxurious, bright finishes. Locomotives were accordingly finished to the taste of the times.

It will be noted that the degree of locomotive ornamentation varied during the nineteenth century. The earliest machines were rather plainly finished. It is clear from surviving illustrations that the American locomotives of the 1830's and 1840's were remarkably plain compared to those constructed ten years later. The early locomotives were straightforward machines devoid of fanciful bell stands, headlight brackets, and other such trimmings. Aside from ornamental steam domes, these little engines were structurally unadorned. The great era of wild ornamentation flourished between 1850 and about 1865. These were the years of bright brass and ever brighter colors. It is believed that the tragedy of the Civil War, together with a more sober view of industrial America, brought in an era of increasingly somber hues. Bright red became vermilion, then maroon, and finally black. By the 1890's the gay colors and most of the polished brass had been forgotten. More will be said on the decline of locomotive decoration later in this discussion. For the present we will turn to varieties of the decorative treatment itself.

It might be well to add a note here on the erroneous impression created by surviving locomotives that these machines were crude and imperfectly finished. Nearly all preserved locomotives of an early date are exhibited after many years of hard service. Polished bright parts are often deeply corroded or painted over. Paint piled layer upon layer has created rough uneven surfaces. Much of the decorative brass work has been stripped; planish iron jackets have rusted out or have been painted over. The original paint finishes have been lost and are often done over by well-meaning but inexpert shopmen whose clumsy imitations of the original schemes are artistically wanting. Surviving antiques are therefore a poor and unreliable record of early locomotive finishes.

Color would appear to be the most obvious decorative feature of the nineteenth-century American locomotive; it was, in truth, only one of several techniques employed for this effect. To the author's knowledge no detailed color specifications have survived, and it is impossible to reconstruct the color scheme for any given locomotive with much precision. One of the earliest notes on locomotive finishes is the recollection of Walter De Sanno regarding the early locomotives of the Phila-

---

[28] Clark and Colburn, *Recent Practice*, p. 68.

delphia and Columbia Railroad.[29] De Sanno recalled that the small cover domes on Bury firebox boilers were of polished copper or cast iron. He also noted that wooden boiler lagging was painted alternately black and green. This style of finish for wood lagging is shown in the contemporary color lithograph of the *Philadelphia* issued in about 1838 by Norris. A large-scale model built in 1842 by the same maker and now preserved in the Conservatoire des Arts et Metiers (Paris) has a wood-lagged boiler of dark and light strips.

While surviving specification sheets might appear as a promising source, a survey of these records has revealed little of value. Usually no mention of finish was made. The most illuminating instruction for a finish in the Baldwin specifications checked was this gem: "Finish—Plain and smooth, wheels and cow catcher to be painted Indian red."[30] From the available Rogers specifications it was found that the *Seminole*, built in 1867 for the Union Pacific Railway, Eastern Division, had a wine-colored tender tank with gilt lettering.[31] The following painting specification, again for a Rogers locomotive, the *Kittatinny*, completed for the Delaware, Lackawanna and Western Railroad in 1855, presents a vivid picture of mid-century colors:[32]

Wheels and Cowcatcher	— dark brown with vermilion stripes
Frames, belly band, braces, wheel covers and reversing shaft	— Green with black stripes
Pumps	— vermilion with black stripes
Suction pipe	— Green
Smoke-box front	— Dark brown with scrolls
Smoke-box door	— Green with vermilion stripes around it

[29] De Sanno's recollection was published in *Locomotive Engineering*, January, 1893. De Sanno's father, Frederick, was superintendent of William Norris' locomotive works in the early 1830's and was later master mechanic on the Philadelphia and Columbia Railroad.
[30] Baldwin Specification, Cleveland and Pittsburgh Railroad 17-ton "D" class, October 30, 1856, in De Golyer Foundation Library (Dallas, Texas).
[31] Bruce, *The Steam Locomotive*, p. 41.
[32] *Railroad Gazette*, July 18, 1902, p. 566.

Aside from such incidental notes on colors, we are entirely dependent upon the brilliant color lithographs issued by locomotive builders for such information.[33] These handsomely executed views were often prepared by the chief draftsman of the works and surely represent as accurate and finite a record of color, striping, and other details of finish as can be found for early American locomotives. About two hundred prints were issued; nearly all show that primary colors were favored, green and red particularly. Red wheels and flat, black smokestacks and smokeboxes are all but universally shown in these prints. Variants of the general red, green, and black motif can be found, however. Blue is the predominant color on the *Phantom* (Mason, 1857) and *Washington* (Souther, 1853) prints. The *Thomas Rogers* lithograph issued in about 1856 shows white driving and truck wheels with red striping. The counterweights on the driving wheels are done in robin's-egg blue. The cowcatcher is painted to match the wheels. Another rather startling variant is the *Tiger* print issued by Baldwin in 1856 where the tender is shown as a bright pink.

Elaborate striping and lettering reinforced the bright colors used. All painted surfaces were striped or scrolled. Often a cab or tender panel was outlined in several stripes of different colors and thicknesses. Letters were bold in character, shaded, outlined in fine stripes, and thrown into relief by shadow highlighting. Rich, heavily shaded scrolls such as those on the Mason tender (Fig. 196) were favored by some builders. Panels were illustrated with vignettes of factories, eagles, landscapes, pastoral scenes, or even portraits. One example of such portraiture is given in the following description of the Lackawanna and Bloomsburg Railroad's *Wyoming*, completed in 1857 by Norris: "The engineers house . . . [is] built in the Gothic style of oak and mahogany, handsomely painted, and adorned on one side with a well executed likeness of Mr. Phleger, the inventor of the boiler, and on the other with a boy sleeping upon the bank of a stream while a large dog stands by his side watching."[34]

[33] "Locomotives on Stone," *Smithsonian Journal of History*, Spring, 1966, pp. 49–60.
[34] *Mining Magazine*, 1857, p. 279.

Natural wood finishes for cabs were known and in fact appear to have been popular between about 1855 and 1865. The *Wyoming* had a naturally finished cab with painted panels as shown by a surviving lithograph of that machine. One of the first engines turned out by Danforth, Cooke and Company had a cab grained to imitate oak.[35] Light Gothic columns on either side of the windows had gilded capitals and bases. A number of lithographs issued by Norris in the mid-1850's show naturally finished cabs. Colburn noted that rosewood, mahogany, bird's-eye maple, and other rare woods were used in cab construction.[36] Possibly the most elaborate example of the use of wood was the 1867 exhibition locomotive the *America,* whose cab was made of ash, maple, black walnut, mahogany, and cherry.[37]

Thus far in our discussion, some emphasis has been placed on paint, varnish, and color, but it should be noted that the greater portion of the locomotive, discounting the tender, was decorated in bright metals. Planish iron, polished cast iron, steel, brass, and copper constituted most of the locomotive's exterior surface. The boiler jacket and occasionally the cylinder and valve box covers were of planished iron. This material has already been described in the section on boiler jackets. Polished steel was used for side rods, valve gears, boiler braces, throttle levers, and occasionally for other small parts such as bell yokes or whistle valve levers. When no jacketing was employed, cast-iron valve boxes and cylinder heads were often polished. Another instance of polished cast iron was the turned bell-stand column like those on the Cumberland Valley Railroad's *Pioneer* (1851).

Polished brass is by far the most familiar and celebrated bright work associated with American locomotives. Surely there was a plenitude of this material on the early engines. Bells, whistles, valves, cylinder jackets, steam-dome and sandbox covers, handrails, jacket bands, and even feed pumps were of golden brass. The spun dome and cylinder covers were the most elaborate of these fittings and were said to cost from $150 to $350 per locomotive, depending on their size and elegance.[38]

In addition to the lavish use of color and polished surfaces the decorative effect was heightened by the architectural design of many fittings. Classical and antique designs were freely borrowed and incorporated into locomotive design. Whistle and bell stands were made in the shape of fluted columns. Sandboxes and steam domes were given richly sculptured bases adapted from ancient Greek or Roman patterns. Taunton produced an acorn-shaped sandbox during the early 1850's. Locomotive cabs were a favorite point of interest for architectural embellishment. Gothic, Tudor, and Egyptian designs were employed. Heavy moldings and dentil work abounded. Windows were even occasionally fitted with tracery. Headlight brackets and bell stands were frequently fanciful cast scrolls. Smokebox fronts, smokestack bases, and other utilitarian fittings were often made with moldings or decorative contours not necessary to their proper functioning.

Aside from the almost unspoken general agreement that highly decorative engines were desirable, only a few arguments justifying the expense of this practice can be found. One prevailing opinion was that the engineman was inspired to give his charge more care if it was an elegant, highly finished machine.[39] A more elaborate defense was offered by Gustavus Weissenborn, who contended in 1871 that "bright engines will cost less in the long run than the plainer one."[40] Concentrating on the merits of bright brass work, he noted that such fittings lasted the life of the engine whereas painted iron jackets rusted out in a few years. During the engine's lifetime, brass jackets, being soft and ductile, could be repaired easily if dented. When the engine was retired the brass work was not a

---

[35] *American Railroad Journal,* July 9, 1853, p. 442.
[36] Clark and Colburn, *Recent Practice,* p. 57.
[37] "Grant's Silver Locomotive," *Railway and Locomotive Historical Society Bulletin,* No. 104 (1961), p. 55.

[38] *Railroad Advocate,* August 11, 1855, p. 3, reported these prices and noted that Nathaniel Lane and Moses and Sons, both of Paterson, N.J., were the two largest manufacturers of ornamental brass works for locomotives.
[39] *Railroad Advocate,* February 24, 1855, p. 2.
[40] Weissenborn, *American Locomotive Engineering,* p. 125.

dead loss but commanded a high price as scrap. Polished brass was an efficient fuel saver; because of its reflective surface it did not radiate heat. A final, convincing argument was that polished brass was a safety device—it made an approaching locomotive more visible on rainy or dark days!

Reactions to the logic of decorative engines were voiced early but were slow in coming. In 1843 it was said in defense of Winans' freight engine that "we do not look for beauty in a freight engine or a mudscow."[41] Winans had from an early time exhibited a complete indifference to the idea of fanciful finishes or even pleasing designs and appeared to find perverse delight in producing coarse and ugly locomotives. His "crab" and Camel locomotives were indeed as plain as mud scows and because of their unprepossessing appearance and unorthodox design found few champions.

James Millholland was another advocate of plain engines, but his concept of simplicity, as opposed to clumsy, blunt designs, was more successful than Winans'. Millholland built on smooth, restrained lines. His machines were not unlike European machines of the same period, and, while some bright work was employed, they were in the main extremely severe products for the mid-1850's.

While Millholland left no statement on his opposition to gaudy locomotives, his contemporary John B. Jervis wrote these words on the subject: "It would be much better for us, if we . . . left off the brass ornaments, which are not in sound taste on a machine that is made to perform work, and not for show."[42] Other authorities eventually came to hold the same view, and during the next twenty years the brightly finished locomotive steadily lost ground. In 1879 the Chicago and North Western and other lines were reported to be stripping their motive power of its brass decorations.[43] At a cost of $1,500 to $2,500 per engine, such ornamentation was considered an unnecessary luxury. The institution of pooling, whereby locomotives were no longer assigned to individual crewmen, ended the "ownership" of locomotives by cranky engineers who were prone to make pets of their personal engines. Another cause for the decline of bright engines was the cost of cleaning such machines. The Philadelphia and Reading Railroad realized a saving of $285 per day by dismissing the wipers required to polish its 410 locomotives.[44] These several factors combined during the latter part of the century to drive the brightly finished locomotive from the American railroad.

CABS

Because of severe weather and the democratic inclination of the United States, a "house" or cab was provided to shelter the crew. The earliest record of a locomotive cab is a drawing prepared in 1831 by David Matthew showing a simple canopy roof, with open sides, fitted to the *De Witt Clinton*.[45] Matthew was engineer of the *Clinton* at the time. Some years later he prepared a curious series of primitive illustrations showing other early locomotives of the Mohawk and Hudson Railroad fitted with similar canopy cabs.[46] A letter in the Baldwin Papers dated November 11, 1836, confirms Matthew's statements regarding the early use of cabs in upstate New York. The Utica and Schenectady Railroad improved the open canopy cab by enclosing its sides. Baldwin said of this plan that they "have their engines completely boxed up thereby keeping the works and engineer from the weather."[47] Undoubtedly there were many other unrecorded instances of engineers fitting simple cabs to their locomotives for protection against the elements, and it would appear logical that northern roads would provide

[41] *Proceedings of the Western Railroad* (Boston, 1843), p. 36.
[42] Jervis, *Railway Property*, p. 169.
[43] *Railroad Gazette*, March 27, 1879, p. 149.
[44] *Master Mechanics Report*, 1877, p. 100.
[45] Matthew's drawing and a letter of 1876 describing the cab are now in the University of Michigan Transportation Library.

[46] David Matthew, *Pictorial History of Pioneer Locomotives* (San Francisco, 1887).
[47] Letter of M. W. Baldwin to W. Woolsey, November 11, 1836, in Baldwin Papers, Historical Society of Pennsylvania.

Fig. 78. The earliest form of enclosed wooden cab is shown by the New Hampshire Central's Reindeer, *a product of John Souther, 1854.*

their enginemen with enclosures. However, this matter seems to have been handled by the crews and railroads rather than by the builders. Cabs first appeared on the Western Railroad (Massachusetts) in the late 1830's according to the recollection of an old engineer on that line.[48] A drawing of the Western Railroad's *Vulcan* dated November 28, 1843, shows a small wooden cab built on and around the old handrail of the engine.[49] This simple structure was obviously an addition to the original open platform.

Some roads, however, were slow to adopt engine cabs; the Philadelphia and Reading had none until 1846 and the Erie acquired its first locomotive cab a year later. By the early 1850's cabs apparently had become standard, but by no means universal, appendages of American locomotives. At this time some large builders, such as Norris and Hinkley, were publishing lithographs depicting locomotives without cabs. This may well be an indication that, like cowcatchers and headlights, cabs were optional equipment. Both builders were delivering locomotives with cabs in the early 1850's but apparently not as standard equipment. By 1855 cabs had been accepted universally.

The early enclosed cabs were extremely plain, small structures built of tongue and groove boards. A small window was the only side opening. A cab of this type is shown in the *Reindeer* photograph (Fig. 78) and in the *Champlain* and *Copiapo* drawings (Figs. 146 and 142). Within five years, larger, more elaborate cabs were in vogue. The rare woods and decorative designs used for engine cabs of this period have already been discussed in the section on locomotive decoration.

---

[48] *National Car and Locomotive Builder,* July, 1888, p. 112.
[49] The drawing is reproduced in *Railway Age,* July 27, 1946, p. 136.

These showy cabins were further embellished by proud crews with prints, photographs, and mirrors. Cabs were regarded by some engineers as if they were "offices of professional gentlemen rather than the posts of duty of hard-handed enginedrivers."[50]

Wooden cab construction prevailed almost unchallenged until 1900. The Philadelphia and Reading's experimental locomotive the *Novelty* (1847) was probably the first locomotive with a metal cab. Built of sheet iron with small rectangular panels, this cab is shown in a daguerreotype of the *Novelty*. The *Philadelphia* of the same road had an identical cab as shown by the 1849 Reuter drawing (Fig. 135). Unfortunately, only the end elevation is shown, but the metal construction of the side panels is obvious from this view. Some ten years later James Millholland introduced a more elaborate style of metal cab on the Reading. These bizarre structures were oval in design, with a metal, dome-shaped roof. A spherical ventilator topped the roof. A general idea of Millholland's round cab can be gained from the *Hiawatha* drawing (Fig. 39).

After Millholland left the Reading in 1866, the road returned to wooden cabs. About the only other American railroad to show an interest in the use of metal cabs before 1900 was the Manhattan Elevated Railroad. Wooden cabs were found by this road to last only five years. In 1885 a sheet-iron cab was made in the company's repair shop.[51] It might be of interest to note here that the first steel cab was made at the Baldwin Works in 1880 for the Mexican Central Railroad's No. 3.[52] Despite these early attempts to introduce metal cabs, wood remained the favored material for that purpose until 1900.

TENDERS

One of the most neglected topics of locomotive history is the tender. This indispensable auxiliary to locomotive operation was largely ignored by the contemporary technical press as well. Owing to the disproportionate demand of the locomotive for water as compared to fuel, tenders were essentially a water tank. The ratio of water to coal consumption was roughly 7 to 1. Tenders were generally designed so that the fuel and water supply were equivalent. In 1851 Colburn noted that an engine on the Boston and Lowell Railroad consumed 925.6 gallons of water during a 26-mile run.[53] A locomotive during this period could expect 25 miles per cord of wood. A good average capacity for a tender of this period would be 1,000 gallons of water and one cord of wood; it can be seen that both commodities were adequate for about a 25-mile run.

The tender was commonly regarded as the worst tracking railroad vehicle. It often weighed as much as the locomotive; 20 tons was about average for an eight-wheel tender of the 1850's. This great weight was spread over a short wheel base.

The variable weight of the tender, generally a 100 per cent difference when empty versus when full, made truck-suspension design a difficult matter. A good riding truck for a full tank would be too rigid when the tank was empty. The tender was altogether an unstable, hard-riding carriage that could inflict considerable damage to the track.

The earliest surviving drawing of an American locomotive tender is that of the South Carolina Railroad's *West Point*. The tender was completed in 1831 by the West Point Foundry of New York City. A line drawing based on the original West Point drawing, presently preserved by the Associated Engineering Societies Library in New York City, is shown here as Fig. 79. The water was carried in a large cask; the lower portion of it is just visible between the two wooden frame side pieces at the rear wheel. The cask was set transversely on the frame. The paneled wood sides and iron handrailing held the

---

[50] *Engineering*, July 26, 1867, p. 66.
[51] *Railroad Gazette*, July 31, 1885, p. 491.
[52] A Baldwin builder's photograph of the Mexican Central's No. 3 shows a squat, arch-roofed metal cab.

[53] Colburn, *The Locomotive Engine*, pp. 134–35.

Fig. 79. *Tender for the* West Point, *built in April, 1831, by the West Point Foundry for the South Carolina Railroad. Water was carried in a large wooden cask at the rear of the car. This drawing is based on an original wash drawing preserved by the Engineering Societies Library, New York.*

cordwood in place. This simple four-wheel tender had no influence on tender design but it does illustrate the first thoughts for such a vehicle.

Small four-wheel tenders with horseshoe or U-shaped iron tanks were introduced by Robert Stephenson in 1830.[54] The open space between the two legs of the tank's U was used for fuel. This very basic, neat design remained standard for tender tanks until well into the present century. Several early but undated Stephenson drawings preserved at the Science Museum in London show tenders of this construction. One of these drawings shows a tender very similar to that in the *Pioneer* photograph (Fig. 80). This machine was originally believed to be the *Meteor*, built for the Boston and Worcester Railroad by Stephenson in 1832 and sold to the Bangor and Piscataquis Canal and Railroad Company in 1835. The engine was scrapped in 1867; the photograph was presumably taken in the last years of its existence. It should be noted that the initials of the second owner, "B&PC&RR" appear on the axle-box pedestals. This could indicate that the entire tender was fabricated by that road, but considering the difficulties of producing such a car by so small a road it is the author's opinion that only the running gear, possibly only the pedestals themselves, were added in later years by the Bangor and Piscataquis. A comparison of this tender with the one shown by the *Baltimore* (Fig. 1) lithograph of 1840 will show the similar construction of both.

No complete drawings of early four-wheel tenders could be uncovered, although a number of side elevations that agree with the general plan favored have been found.[55] A reconstruction drawing was prepared so that a representative tender of this type might be better understood. The drawing is reproduced here with the *Gowan and Marx* in Fig. 125. Four-wheel

---

[54] Stephenson appears to be the originator of the U tank. He first used it on the Liverpool and Manchester Railway for the *Northumbrian*'s tender.

[55] Among the other contemporary illustrations of four-wheel tenders are the P. R. Hodge drawing (1840) of the *Victoria* and the *Juno*, the Boston and Worcester's *Mars* (1840), and the Baltimore and Ohio's Nos. 7 and 17.

COMPONENTS

Fig. 80. *One of the best photographs of an early* U *tank four-wheel tender. This Planet class locomotive was built by Robert Stephenson & Co. and is shown as it appeared on the Bangor and Piscataquis Railroad and Canal Co. at about the time of its retirement in 1867. It is thought to be the Meteor, built in 1832 for the Boston and Worcester Railroad.*

tenders generally carried from 500 to 700 gallons of water and weighed between 5 and 7 tons when full. Winans supplied what were probably the practical limits in size for four-wheel tenders with a capacity of 1,140 gallons to the Western Railroad of Massachusetts in 1841.[56] Because of their simplicity and cheapness, four-wheel tenders undoubtedly remained in favor until the mid-1840's when heavier engines with increased fuel and water requirements called for larger tenders. At 7 tons the axle loading for four wheels was considered to be at its maximum limit; it was necessary to add more wheels for heavier tenders. Six- and eight-wheel tenders were introduced to meet this need.

The six-wheel tender was surprisingly popular, although its existence is barely recorded in most locomotive histories. The peak period of its favor appears to have been the mid- and late 1840's. Baldwin's specification records for 1847 account for 42 tenders; 37 of these were on six wheels.[57] Two examples of six-wheel tenders are shown in the sketches taken from the same maker's records (Fig. 81). A more detailed illustration of a Baldwin tender of the period is shown by the *Necker* lithograph, Fig. 82. This six wheeler was supplied to the Württemberg State Railway in May, 1845. It held one cord of wood and 900 gallons of water. Baldwin built six-wheel tenders for domestic service through the 1860's and continued their production for export until a much later date.

The diagram of the *Licking*, 1846 (Fig. 16) is evidence that Rogers also produced six-wheel tenders. In this case the four-wheel truck was leading while a larger, single set of wheels fixed rigidly to the frame carried the rear of the tender. The usual plan was for the truck to be at the rear where it could better support the heavier weight of the tank.

New England builders also produced six-wheel tenders. The 1849 annual report of the Fitchburg Railroad listed a number of Hinkley locomotives with 1,400-gallon six-wheel tenders. G. S. Griggs favored rigid six-wheel tenders, similar to the one used on the *Croton* (Fig. 156), and many of the Boston and Providence engines were so-equipped. One of these tenders may be seen today at Danbury, Connecticut, where Griggs's *Daniel Nason* is on exhibit.

Norris built a peculiar style of six-wheel tender where no truck was used and all three axles were non-swiveling. The earliest known illustration of such a tender is the rare view of the *Tioga* shown here as Fig. 83. The photograph was made not long after the machine's completion in 1848. A fine mechanical engraving of an identical tender was given in Frederick Moné's work on American engineering published in 1854 (see Fig. 84). The design is credited to Norris-Schenectady, but it can be seen that precisely the same tender was in production several years earlier at the Norris Philadelphia works. Edward S. Norris opened the Schenectady plant in 1849 but failed after building only a few engines. The general designs used, including the above tender, were unquestionably taken from the Philadelphia works. The long springs resting on all three axles permitted some distribution of road shocks to all axles and represented an early attempt at equalizing tender running gears. This form of tender was offered by Norris as late as 1852 (see the *Beaver*, Fig. 141).

In general it appears that the six-wheel tender had been superseded by eight-wheel tenders by 1855. Some forty years later the idea was revived when a number of high-speed passenger locomotives were supplied with rigid six wheelers very similar to the British design. The supposed saving in dead weight and complexity was offset by the increased axle size and the necessity of cross- as well as side equalization.[58]

The brilliant success of eight-wheel cars on American railroads in the 1830's undoubtedly suggested the merit of this plan for tender running gears. The load was spread over eight wheels. The use of two four-wheel trucks eased the tender's ride and permitted safe, easy passage around curves. It was also found that the highly stable three-point suspension could be achieved in eight-wheel tenders. The front truck was made center bearing, thus providing one point of suspension.

---

[56] *Proceedings of the Western Railroad*, p. 113.
[57] Baldwin Specification, De Golyer Foundation Library. Baldwin built only thirty-nine locomotives in 1847; three of the tenders were built on separate order.
[58] *Locomotive Engineering*, February, 1895, p. 75.

COMPONENTS

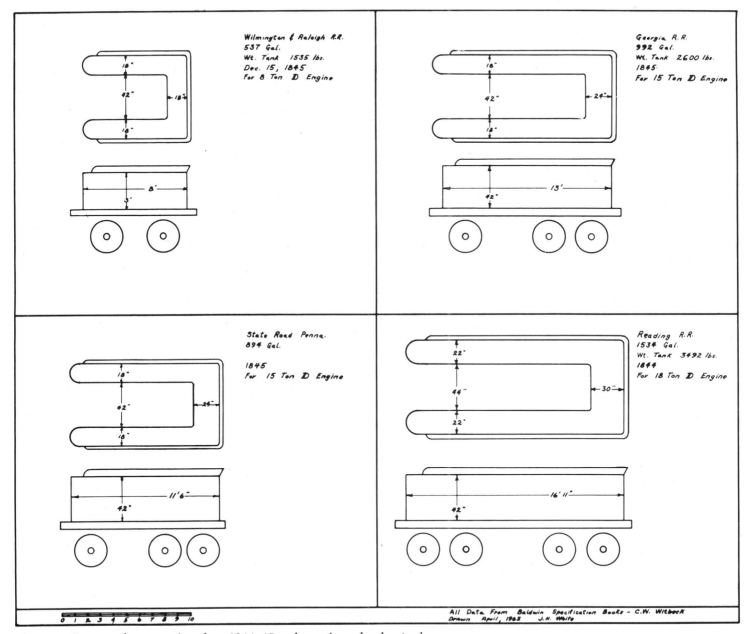

Fig. 81. Diagram drawings of tenders, 1844–45, redrawn from sketches in the Baldwin specification books.

AMERICAN LOCOMOTIVES

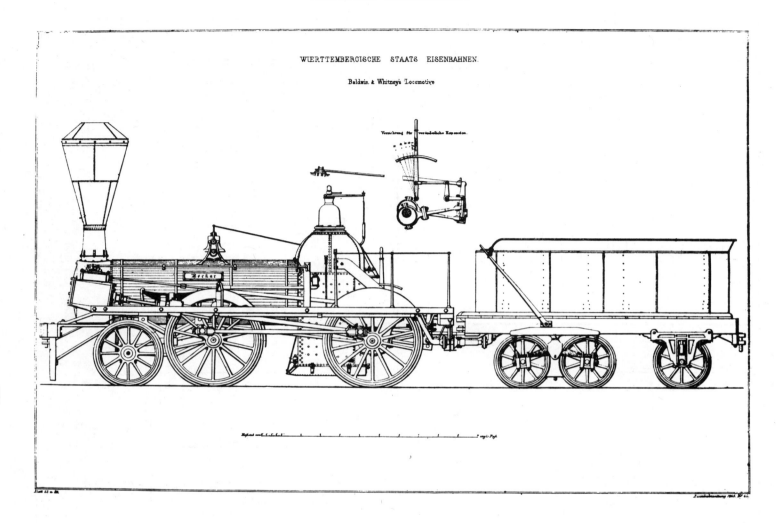

Fig. 82. *The six-wheel tender accompanying the Württemberg State Railway's* Necker *was built by Baldwin in May, 1845. Capacity: one cord of wood and 900 gallons of water.*

Fig. 83. *The Philadelphia and Columbia Railroad's Tioga is shown with a rigid wheel-base six-wheel tender. The locomotive was built by Norris in 1848.*

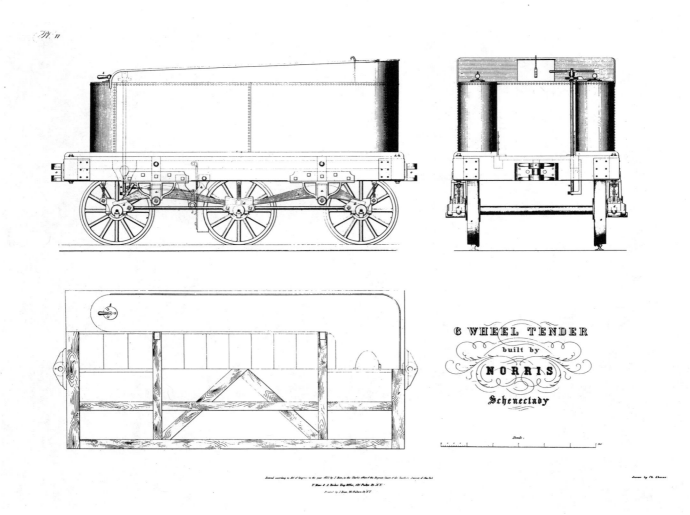

Fig. 84. A six-wheel tender built in Schenectady by E. S. Norris in about 1849–51 after a design developed earlier in Philadelphia.

Fig. 85. A standard eight-wheel tender built by Baldwin in about 1860 for the Cleveland and Pittsburgh Railroad. Note the all-iron trucks and the side bearings on the rear truck.

Side bearings were fitted to the rear truck, thereby providing the other two points of suspension. This plan was much admired and generally followed although some authorities favored side bearings for both front and rear trucks. While earlier applications were undoubtedly made, the first record of an eight-wheel tender is a note in the Rogers 1838 specifications for the New Jersey Railroad and Transportation Company's *Uncle Sam*. Baldwin built an eight-wheel tender at about the same time.[59] No particulars on these pioneers have survived, but Baldwin is known to have supplied a giant eight wheeler to the Western Railroad of Massachusetts in 1844 with a capacity of 1,550 gallons of water and 2½ cords of wood. The Western Railroad was experiencing difficulty in making runs between water stations with four-wheel tenders; even the 1,100-gallon tanks supplied by Winans were inadequate.[60] In 1843 the road reported an eight-wheel 1,600-gallon tender supplied with one of its Lowell-built freight engines; yet the road's engineer, James Barnes, felt that even larger tenders were desirable. It was not many years before 2,000-gallon tenders were common. A tender of this capacity weighed about 20 tons when fully loaded.

As the need for larger tenders (brought about by heavier engines and the desire to avoid frequent water stops) became more obvious, the eight-wheel tender gained in popularity. By the mid-1850's it had completely replaced four- and six-wheel tenders for road engines. The earliest surviving detailed drawings for an eight-wheel tender are those for Winans' 4–4–0 of 1843–49 (Fig. 130). It will be noted that the trucks did not swivel but, like Winans' eight-wheel iron hopper cars, depended on parallel motion for negotiating curves. In the case of the tender trucks, the live spring serving as the side frame could pivot slightly. It should be mentioned that conical axle boxes were employed, although this arrangement is not thought to have been standard practice and was unquestion-

---

[59] *History of the Baldwin Locomotive Works*, p. 26.
[60] *Proceedings of the Western Railroad*, pp. 100, 113. See also Commonwealth of Massachusetts, House of Representatives No. 65, February, 1843.

ably peculiar to Winans. Other, more conventional eight-wheel tenders are illustrated by the drawings of the *Consolidation* (Fig. 230), Tyson's ten wheeler (Fig. 189), and the Baldwin flexible-beam locomotive (Fig. 205).

A notable variant to the standard eight-wheel tender was the peculiar ten wheeler used by the Camden and Amboy Railroad and its sister road, the New Jersey Railroad and Transportation Company. A six-wheel truck carried the greater weight of the tender's rear portion while a four-wheel truck carried the front. Six-wheel trucks had been used on passenger cars before 1850 and were not strangers to the Camden and Amboy. It should be added that not all the tenders on these roads were on ten wheels; the Camden and Amboy also used double-truck eight-wheel tenders as shown by the *John Bull* (Fig. 104) and in its earliest years small, single-truck tenders were employed. The strange, shedlike cabin was a design characteristic of the Camden and Amboy and was used on all the tenders of this road from the 1830's to the time of its merger with the Pennsylvania Railroad in 1871. The New Jersey Railroad and Transportation Company favored a more conventional plan with a horseshoe tank. A detailed drawing for one of these tenders is given with the Rogers Mogul in Fig. 215.

The introduction of the horseshoe tank by Stephenson has already been mentioned; a few comments on its construction now appear to be in order. Heavy-gauge sheet iron was used to form the tank. The side walls were usually one-eighth of an inch thick while the bottom plate was three-sixteenths of an inch thick or more. Despite this heavy construction, tanks rusted out within ten years on the average.[61] Judicious patching could extend tank life far beyond this limit. A strong sheet-iron fender or dicky was riveted around the top of the tank. This was more than an ornament; it provided a barrier so that cordwood could be stacked on top of the tank. This practice is aptly illustrated in the *Tioga* photograph (Fig. 83). The flow of water from the tender tank to the feed pumps was controlled by valves mounted at the bottom of the forward section of the tank. The valves, one on each leg of the tank, were controlled by a rod passing up through the interior of the tank. A handle or wheel was attached to the top of the rod which projected through the top sheet of the tank. A strainer was fitted at the valve to trap large pieces of rust from the tank's interior and other such objects which might foul the pumps, check valves, or boiler.

A flexible connection was needed to carry the water between the engine and tender. Leather or canvas hoses were used at first, but by the early 1850's there was evidence of the regular use of India rubber hoses. The *Croton*, built by Lowell in 1851, is described as having such a hose, and the specification for the Souther engine built in the same year also calls for a rubber hose.[62]

The horseshoe tank was not modified until 1892 when J. B. Barnes, master mechanic of the Wabash Railroad, introduced a sloping deck for the coal pit.[63] This not only made the coal more accessible to the fireman but also increased the water space of the tank. The dicky at the forward end of the tank was raised to increase the amount of coal space.

Wooden frames were used universally before 1870. After that date, iron frames rapidly came into favor. Stout timbers, often measuring 5 inches by 12 inches, were trussed together by iron rods. A continuous draw bar was formed by running rods or a heavy iron plate along the center sill throughout the entire length of the frame.

Simple wood-beam trucks were long favored for tenders. As with so many other long-favored designs, their simplicity and low first cost was the basis of their popularity. For this reason the Baltimore and Ohio Railroad used an extremely simple wood-beam truck on many of its tenders and freight cars through the 1880's. One of these primitive trucks is shown under the tender of the Tyson Ten Wheeler in Fig. 179. By

---

[61] Weissenborn, *American Locomotive Egineering*, p. 184.

[62] See Moné's *An Outline of Mechanical Engineering* (New York [?], 1851), for a description of the *Croton*.

[63] *National Car and Locomotive Builder*, January, 1895, p. 11, refers to an earlier article (1892) regarding Barnes's improved tender-tank design.

the mid-1850's iron-frame trucks had begun to find more regular employment, and within the next few years they became the standard form of construction. Several examples of early all-iron tender trucks are given in connection with the *Talisman* (Fig. 176), the flexible-beam locomotive (Fig. 205), and the *Phantom* (Fig. 195). An interesting variety of iron tender trucks, unfortunately not represented in this work, is the diamond-shaped arch-bar truck reportedly first made by Isaac Dripps in about 1857.[64] Originally intended for freight car service, the arch-bar truck was widely used for tenders after about 1870.

Inside-bearing tender trucks were occasionally used during the nineteenth century. Their general construction was similar to that of a locomotive leading truck, but their advantages for tender service are questionable. Examples of inside-bearing tender trucks are shown here by the *Amenia* (Fig. 66), the *Copiapo* (Fig. 138), the *Beaver* (Fig. 141), and the drawings for the Tyson Ten Wheeler, Fig. 190-A. While few illustrations of inside-bearing tender trucks dating after 1860 are known to exist, the *Master Mechanics Report* for 1877 noted that the following roads used this style of truck: the Terre Haute and Indianapolis, the Central Railroad of New Jersey, and the Cumberland and Pennsylvania Division of the Pennsylvania Railroad.

The tank locomotive is a class of engine which did not employ a tender. This type of engine has not been discussed previously because it was little used in the United States before the mid-1870's. The small numbers used on American roads were either special machines for very light passenger service or contractors' engines. Machines of this type contributed little to the development of the representative road engine. However, it might be well to comment briefly on the history of the tank locomotive as a reasonable digression from the more general subject of locomotive tenders. One of the earliest mentions of a tank engine in this country was the following description of Norris and Long's *Pennsylvania,* built in 1833 for the Philadelphia, Germantown and Norristown Railroad: "The engine weighs four tons and three quarter with her fuel and water, carrying no tender, as her water tanks are on the top of the machine."[65] Long's design apparently won few converts, for little interest was shown in tank engines until the early 1850's. Many roads found that their light passenger trains would be more economically pulled by small, single-axle tank engines. Griggs built two such machines for the Boston and Providence in 1851. Wilmarth built four engines of this type for the Cumberland Valley Railroad between 1851 and 1854.[66] Danforth, Cooke and Company built a fair number of 4–2–2 tank engines for various railroads between about 1855 and 1870. The well-known *C. P. Huntington* of the Southern Pacific Railroad is a surviving example of the Cooke tank engine. Unfortunately, the engine has undergone some reconstruction, including a new boiler.

Matthias N. Forney became a staunch advocate of tank locomotives after 1866 when he secured a patent for locomotives on this plan. Forney argued that the dead weight of the tender could be usefully employed to increase traction if his design for the tank engine was adopted. The high axle loadings and limited water supply possible with tank locomotives countered Forney's argument. The inventor was a respected and influential authority in railroad circles, but his promotion of tank locomotives for road service was not adopted. Forney locomotives were used with great success on the New York elevated railways for nearly thirty years before the electrification of these lines began in 1902. After this time, tank engines were built mainly for contracting and industrial service; occasionally a unit was used for special service on a main-line road.

---

[64] Sinclair, *Locomotive Engine,* p. 254.
[65] *Proposals of the American Steam Carriage Company,* November, 1833, p. 23.
[66] One of the Cumberland Valley locomotives, the *Pioneer,* is exhibited in the Smithsonian Institution. A description and mechanical drawings of this machine are available in the United States National Museum Bulletin 240, Paper 42 (1964).

THE SANDBOX AND OTHER TRACTION-INCREASERS

It was found that a small quantity of dry sand would stop the slipping of driving wheels on wet or oily tracks and that it could assist in starting a train under such conditions. The sandbox was unquestionably the simplest, most effective traction-increaser developed for railway service.

The origins of the sandbox have been confused by a story that the device was not introduced until 1841 when Jordan L. Mott, a New York stove manufacturer, obtained a patent for such a device. We find that, in fact, sandboxes were in use some five years earlier and were probably common at the time of Mott's patent. The earliest-known mention of such apparatus was the note in the August 1, 1836, report of the Tuscumbia, Courtland and Decatur Railroad stating that "it is contemplated to apply sand boxes to the locomotives."[67] The following year the annual report of the Philadelphia and Columbia Railroad noted that "sandboxes have been placed in front of the driving wheels which can be made to discharge their contents at pleasure by means of a tube extending close to the rail."[68] The addition of sandboxes was noted in the same report as a recent application necessitated by defective adhesion caused by dust, oil, and snow on the tracks.

Because of the previous employment of sanders, Mott was forced to admit in his patent specification of August 28, 1841 (No. 2228), that his "invention" was merely a modification of an existing device. Mott's plan was to combine water with the sand and drop the mixture directly on the driving-wheel tire. The patentee claimed that this mixture would hold to the rail and tire better and thus would promote traction more effectively than would dry sand, which could be swept aside by the wheels. Mott's patent gained some notoriety by being reprinted in the *American Railroad Journal*[69] and the *Journal of the Franklin Institute*.[70] No widespread adoption of Mott's patent resulted from this publicity, but others were encouraged to patent sanders. One such patent was issued to Elisha Tolles on October 9, 1841 (No. 2283). Tolles placed an urn-shaped sandbox on top of the boiler. The sand was dried by an internal steam dome; a separate steam line running inside the sand pipe was intended to assist in blowing the sand under the wheels as well as melting the ice on the rails. Tolles' idea attracted little attention and probably did not progress beyond a few tests.

The term "sandbox" has its origin in the box-shaped sand reservoirs first popular for this purpose. Nearly every known locomotive illustration dating before 1850 shows a square sandbox. Several examples appearing in the present volume are the *Ontalaunee* (Fig. 120), the *Amenia* (Fig. 66), and the *Copiapo* (Fig. 139). It might be well to note that in this early period the use of sandboxes was by no means universal; many illustrations show locomotives devoid of such mechanisms.

By the mid-1850's the use of sandboxes had become more widespread. The old, square box was replaced by a more ornamental, cylinder-shaped reservoir, except in the case of Wilson Eddy, who retained the old-fashioned sandbox until about 1875. Some were mounted on pedestals, but the favored plan was to mount them directly on top of the boiler jacket. The top and base of the box were of cast iron, the body of sheet iron. Occasionally the top cover was of polished brass. The sand valves were worked from the cab by a reach rod.

Aside from the sandbox, various mechanical solutions were offered to improve locomotive traction. The most successful of these, the equalizing lever and the use of coupled driving wheels, have already been discussed, but it would not be amiss to describe briefly the less-successful methods tried.

Aside from various forms of rack railway, never intended for main-line service after 1830, the favorite methods for improved traction were various schemes to *temporarily* place more

---

[67] *American Railroad Journal*, December, 1836, p. 731.
[68] *Report of the Canal Commissioners of Pennsylvania*, 1837, p. 50.
[69] 1841, p. 109.
[70] 1841, pp. 348–51.

weight on the driving wheels. Most of these traction-increasers were meant for single-driver locomotives, which were notoriously deficient in adhesion when starting trains. E. L. Miller obtained a patent in 1834 for a plan whereby part of the weight of the tender could be thrown onto the rear of the locomotive's frame, thus adding weight to the drivers. In 1835 Charles and George E. Sellers patented a plan in which the weight of the train was used to promote adhesion. Both plans were designed to improve the slippery 4-2-0. Miller's patent was employed on a number of early Baldwin engines; the Sellers plan was probably used only on the few engines built in their own shops.

With the advent of the 4-4-0 in the early 1840's, the traction question was temporarily answered, but the interest in light, single-axle passenger locomotives, which developed in the last years of the decade, brought a new series of patents and designs.[71] Baldwin built five engines on a modified Crampton design between 1849 and 1850. The weight on the single pair of drivers could be varied by shifting the fulcrum of the equalizing lever that connected the driving axle and a pair of carrying wheels just ahead of the firebox. The shifting fulcrum was used again by the Baldwin Works in 1880 on a 4-2-2 built for the Philadelphia and Reading Railroad. This machine is better known by the name of its second owner, Lovett Eames. Ross Winans developed a unique scheme for the same purpose on his *Carroll of Carrollton,* completed in 1849. A pair of vertical steam cylinders bore down on the axle boxes of the driving wheels and acted as a combination suspension- and traction-increaser. The load on the main axle was varied by the steam pressure in the spring cylinders. Winans patented this scheme in 1851, some two years after the *Carroll*'s road test.

Master mechanics were not idle on this subject and at least two prominent locomotive superintendents are known to have developed traction-increasers. In about 1848, while master mechanic of the Albany and Schenectady Railroad, Walter McQueen perfected a simple plan for a traction-increaser which was applied to a small, single-axle passenger locomotive built at the company shops.[72] A long-leaf spring, doubling as the equalizer, rested on the axle boxes of the driving and trailing wheels. A heavy screw, bearing on the spring and adjustable from the cab, was used to compress the spring; it thus put the added weight on the drivers required for starting. Once the engine was moving, the screw was reversed, tension on the spring relaxed, and the rear weight of the locomotive was distributed more equally to the driving and trailing wheels as the engine gained speed.

George S. Griggs secured a patent on June 17, 1851 (No. 8166), for increasing the tractive power of light, single-axle locomotives. In Griggs's plan the weight of the train, relayed through the draw bar and a series of levers, was applied to the driving axle. This design was probably used on the *Roxbury* and the *Dedham,* both of which were light, single-axle tank locomotives built at the Boston and Providence Railroad shops in 1851.

By far the most bizarre proposal for improving locomotive traction was that proposed by Edward W. Serrell, a prominent civil engineer and one-time director of the Hoosac Tunnel project.[73] In 1859 Serrell proposed that locomotive traction be improved by magnetizing the driving wheels. The idea had been tried several years earlier in Germany, France, and England without success. A coil of wire, placed in a transverse position, encircled the bottom of the driving wheels. One side was charged negatively, the other positively. Batteries were used, but the inventor hoped to develop a generator. The apparatus was tested on the Fitchburg Railroad's *Anthracite* and the Central Railroad of New Jersey's *Lebanon*. An increase in traction of 75 per cent was claimed. A patent was obtained and other tests were probably conducted, but the cost, an estimated $400 per locomotive, and the bulky, unsatisfactory batteries of the period ended the magnetic-traction scheme.

---

[71] A history of American high-wheel singles is given in *Railway and Locomotive Historical Society Bulletin*, No. 114 (1966), pp. 27-36.

[72] *American Railroad Journal*, July, 1893, p. 354.

[73] Data on the Serrell magnetic traction-increaser are from the *Journal of the Franklin Institute*, November, 1859, p. 289; *Scientific American*, September 3, 1859, p. 153; and *Transactions of the American Institute*, 1859, p. 460.

REPRESENTATIVE AMERICAN LOCOMOTIVES III

STOURBRIDGE LION, 1829

Failure to mention the first locomotive that turned its wheels on the North American continent would constitute a serious oversight in any study of early locomotives. Although the history of the *Stourbridge Lion* has been set down in several other accounts, certain significant details have come to light in recent years.[1] Most of the earlier accounts dwelt on the primacy of this engine's American operations, neither paying attention to the *Stourbridge Lion* as a machine nor emphasizing that it was one of the first locomotives to operate outside of England and thus was worthy of international as well as national attention.

The Delaware and Hudson Canal Company was organized in 1823 to develop the anthracite coal deposits found in the Lackawanna Valley near Carbondale, in northeastern Pennsylvania. The original plan was to transport the coal to New York City by canal. However, a short section of the route between Honesdale and the mines, about 16 miles, was too mountainous for a canal. As early as 1825, consideration was given to a railway to connect the mines with the canal. The directors of the Delaware and Hudson received authorization from the state to build the line in April, 1826. John B. Jervis became chief engineer of the Delaware and Hudson in March, 1827, and, since the canal was nearing completion, he began to press for the construction of the railroad. He recommended construction of a series of inclined planes with connecting railroads between them. The level sections were to be operated by locomotives. Jervis' plan was adopted by the directors, but with certain reservations. Even so, the surveys for the line were completed in the summer of 1827.

Horatio Allen had worked with Jervis as an engineer during the construction of the canal. Allen, only twenty-five years old at the time, left the Delaware and Hudson's employ late in 1827 to make an inspection tour of railways in England. Jervis,

---

[1] New facts on the *Stourbridge Lion* have been found in the papers of J. B. Jervis housed in the Jervis Library, Rome, N.Y. Pertinent notes were published in *Railway and Locomotive Historical Society Bulletin*, No. 52 (1940), pp. 41–47. George C. Maynard (1839–1918), a curator at the United States National Museum, interviewed old residents in Carbondale and Honesdale, Pa., in 1904. His efforts, while not conclusive, did uncover several facts concerning the disposition of the *Lion*. Maynard's notes are housed in the Smithsonian Institution, Washington, D.C.

learning of Allen's trip and appreciating his technical knowledge, asked him to act as agent for procuring the rails and locomotives needed for the railroad. A lengthy letter of instruction, dated January 16, 1828, was prepared by Jervis, in which he stated that the weight of the locomotives should not exceed 5½ tons when on four wheels, or 7 tons if on six wheels.[2] They should be able to work at from 3½ to 5 miles per hour and should have two cylinders, 8 inches in diameter with a 27-inch stroke, to work at forty strokes per minute. The steam pressure was to be 60 pounds per square inch. If at all practical, the engines should be equipped with water tanks so that a tender would not be necessary. It was further noted that only one engine was to be ordered in England and that the other three engines might be purchased in this country if the English builders charged more than $1,800. Jervis "supposed" that locomotives could be built in the United States for that figure and estimated that interest and transportation, plus other charges, would add 45 per cent to the cost if the British built the locomotives.[3] While the specifications were closely followed, Jervis' estimates of the cost and his supposition about American-built engines were unfounded. All four locomotives were built in England at a cost well above Jervis' estimate.

A week after receiving Jervis' letter, Allen sailed for England and arrived at Liverpool on February 15, 1828. The exact events of his visit are unknown, but Allen did become acquainted with such leading railway men as George and Robert Stephenson, Nicholas Wood, and John U. Rastrick. Allen wrote to the Delaware and Hudson on July 19, 1828, reporting that four locomotives had been ordered—three from Foster, Rastrick and Company and one from Robert Stephenson and Company.[4] The Stephenson engine cost $3,663.30, and the three Foster-Rastrick engines were nearly $3,000 apiece.

The *Stourbridge Lion* was one of three engines built by Foster, Rastrick and Company of Stourbridge, England, located in west-central England about 10 miles west of Birmingham. John U. Rastrick (1780–1856) was the better known of the two principal partners in the Foster-Rastrick firm. At the time of Allen's visit he enjoyed an enviable reputation as an authority on railways, even though in 1825 he had testified in parliament against the use of locomotives.[5] He had been vindicated of the testimony and had re-established his eminence as an authority by being appointed a judge at the Rainhill Trials.

Rastrick's only experience in locomotive building prior to Allen's order was the Trevithick 1808 London exposition engine. Foster, Rastrick and Company were essentially founders and stationary engine builders. After constructing the three Delaware and Hudson engines, they built only one other locomotive, the *Agenoria*, for the Shutt End Railway in 1829. This machine, believed to be a duplicate of the *Stourbridge Lion*, is preserved, in a somewhat modified state, at the York Railway Museum (see Fig. 87).

As already mentioned, one locomotive was to be built by Robert Stephenson. This machine, often referred to as the *America*, but in truth named the *Pride of Newcastle*, was completed before any of the Rastrick engines and reached the United States on January 15, 1829.[6] The *Stourbridge Lion* was completed a month after the Stephenson engine and arrived in New York on May 13, 1829. It was set up on blocks in the yard of the West Point Foundry and tested under steam. After this trial it was shipped by canal to Honesdale for a road test. The excitement, hope, and failure of this first operation of a commercial locomotive in North America has been told too often and well to bear repeating here. The trip of August 8, 1829, and the second trip of September 9 proved beyond a

---

[2] M. N. Forney, *Memoir of Horatio Allen* (New York, 1890), pp. 8–13, obituary notice reprinted as a pamphlet from the *American Engineer and Railroad Journal*.
[3] Ibid.
[4] *A Century of Progress—History of the Delaware & Hudson Co.* (Albany, N.Y., 1923), p. 50.

[5] See Newcomen Society, *Transactions*, VI, 48, for more details about J. U. Rastrick.
[6] See *Railway and Locomotive Historical Society Bulletin*, No. 52, and Marshall, *A History of the Railway Locomotive Engine*, pp. 139–41, for more details on the *Pride of Newcastle*.

doubt the inability of the lightly built hemlock tracks to support a 7-ton locomotive.

After the trials the *Stourbridge Lion* was stored in a rough shed near Honesdale. Before this was done James Renwick made a scale drawing of the engine which was published in his *Treatise on the Steam Engine* (1830). This is the only existing contemporary illustration of the engine and is reproduced in the present volume as Fig. 86. It might be mentioned that Renwick reviewed Jervis' plans for the Delaware and Hudson railway in 1827 and so had a connection with the project from its inception.

The only evidence that any attempt was made to sell the *Stourbridge Lion* and its sister engines is a letter of Charles S. Wurts to the Pennsylvania Canal Commission, dated February 7, 1834, suggesting that these machines might be purchased cheaply.[7] Apparently they were considered ill-suited for use on the Pennsylvania Public Works Line. They would have been too slow, and their single-flue boilers and walking beams had become antique arrangements by 1834. With little chance of resale or use, the abandoned engines proved a ready source of "bar stock" for the nearby repair shops of the Delaware and Hudson Railroad; good quality English wrought iron was always in demand. In about 1845, after the engine was stripped of its other parts, the boiler of the *Stourbridge Lion* was moved to the Carbondale, Pennsylvania, foundry of John Simpson.[8] Simpson used the boiler for about five years in his shop to operate a stationary engine. He abandoned the enterprise in the early 1850's and went to California in search of gold. The Simpson foundry was reopened in 1856 by John Stuart and William Linsay. In 1869 Patrick Early bought out Stuart, thus making Linsay and Early proprietors of the foundry. By 1871 the boiler of the *Stourbridge Lion*, already over forty years old, had become obsolete and a fuel waster. Accordingly it was placed outside the shop. The owners were aware of its value as a relic and were reported to have asked $1,000 for it in 1874,[9] but no buyers were forthcoming. However, in 1883 Samuel H. Dotterer of the Delaware and Hudson borrowed the boiler for an exhibit at the Exposition of Railway Appliances in Chicago. While in transit, or at the fair itself, souvenir seekers broke off any loose parts possible, in some cases resorting to chisels and sledge hammers. Early hoped to make a profit with the relic by displaying it in Scranton, Pennsylvania, at 10 cents a head. This venture failed, however, and the boiler at last found its way to the Smithsonian Institution in October, 1890, where it is presently preserved.

Aside from the boiler, several other relics have survived, namely, one cylinder, both walking beams, and four tires. There is little question as to the authenticity of the walking beams and the cylinders, but the tires might be from the *Pride of Newcastle* as they measure 48 inches in diameter (they defy absolute identification since both machines have 48-inch wheels). The tires were in fact exhibited with the crank rings of the *Pride of Newcastle* at the Centennial Exhibition (1876) and unwittingly as parts of the *Lion* in the Smithsonian until recent years. The cylinder was reportedly used to operate a pumping engine at the Delaware and Hudson coal mine No. 6. When the cylinder was incorporated in the pumping engine, the upper mounting flange, which originally was bolted to the locomotive's boiler, was removed. This portion of the cylinder casting is now missing. All of these parts had been collected by the Smithsonian by 1913, together with one cylinder and three crank rings of the *Pride of Newcastle*.[10] An attempt was made to re-erect the engine, but with so few pieces extant and the confusion of the Stephenson parts, the reconstruction was unsuccessful. The present display shows the relics as such and includes a scale model to indicate the general arrangement of the original machine.

---

[7] I am indebted to Thomas Norrell for the source of this letter, the Pennsylvania State Archives.

[8] Nearly all of the post-1830 history of the *Stourbridge Lion* is based on the notes of G. C. Maynard (see note 1).

[9] Brown, *First Locomotive in America*, p. 22.

[10] Linsay and Early incorrectly claimed that this cylinder was from the *Stourbridge Lion*; it was so-pictured in the *Locomotive Engineer*, September, 1888, p. 3.

DESIGN AND MECHANICAL ARRANGEMENT

"I am fully of the opinion too that the present locomotive engine is an imperfect machine compared with what it will be 10 or 12 years hence." Horatio Allen wrote these prophetic words in February, 1828.[11] At the time of its construction the *Stourbridge Lion* incorporated the most approved design features available for locomotives. It was representative of the final development of the slow and ponderous colliery engines. The idea of public railways was only being introduced, and the main scheme for locomotives was restricted to the economical transport of minerals, usually at speeds of under 5 miles per hour. The essential elements of the modern steam locomotive, such as direct connection, multitubular boilers, separate firebox, and smokebox, had not yet been developed, although it should be noted that these revolutionary design changes were all adopted by 1830. The *Stourbridge Lion*, however, was built before these reforms were introduced and was essentially a copy of the engines built by William Hedley in 1813.

The practicality of the *Stourbridge Lion* is best proved by the long service of its sister engine, the *Agenoria*, which remained in use for thirty-five years. During this time it hauled 130-ton trains at 3½ miles per hour and ran light at 11 miles per hour. There would seem to be little question that if the Delaware and Hudson had built a line suitable for locomotive use the *Stourbridge Lion* could have performed equally well.

The boiler is built on the Lancashire plan and thus is characterized by a large, single flue (firebox) that branches into two smaller tubes. The tubes unite in the forward end of the boiler forming a base for the smokestack. The boiler shell is 48 inches in diameter, about 10½ feet long, and is fabricated from ½-inch wrought-iron plate. The firebox tubes and the back head are constructed as a unit and are bolted to a heavy flange on the rear end of the boiler shell; this permits the unit's removal for inspection and repair. The manhole in the center of the boiler allowed access for unbolting the stack when the firebox-tube unit was to be removed.

Two safety valves were employed on the *Stourbridge Lion*. One (undoubtedly spring-loaded) was housed in the small dome cover just behind the stack. The other, accessible to the engineer for adjustment, was a simple, counterweighted valve placed at the boiler's center. A single feed pump, mounted on the left side, was driven by the left walking beam. It could be worked by hand when the engine was standing.

The box between the axles and below the boiler is a feed-water heater. Exhaust steam was used to heat the water. Note that after passing through the heater the exhaust moves on to the stack to serve as a blast for the fire.

The cylinders measure 8½ inches in diameter with a 36-inch stroke. The stroke, by way of the walking beams, is reduced to a 27-inch throw at the wheel crank. The parallel motion is that patented by William Freemantle in England in 1803 and interestingly enough was used by Oliver Evans on his famous *Columbia* stationary engine.[12] The use of Freemantle's parallel motion eliminated the need for crossheads. The Grasshopper engines of the Baltimore and Ohio have repeatedly been compared to the *Stourbridge Lion* when, in fact, one end of their walking beams was fixed and crossheads were employed.

The walking beams are hollow and are fabricated from two iron plates. Thus, the piston-rod and main-rod connections are made neatly between the plates. The valves are driven by single eccentrics on the rear axle; they were worked by hand when the engine operated in reverse. The wheels were 48 inches in diameter and were set to a gauge of 4 feet 3 inches.

---

[11] *Railway and Locomotive Historical Society Bulletin,* No. 61 (1943), p. 47.

[12] G. Bathe and D. Bathe, *Oliver Evans* (Philadelphia, 1935), p. 113.

REPRESENTATIVE LOCOMOTIVES

The fellies and spokes were wooden; the hub was cast iron. The spoke used as the crank was undoubtedly forged iron. Possibly the most interesting feature of the wheels, if not of the entire engine, is the counterweighting of the rear drivers. This is the earliest known use of counterweights for locomotives and is substantiated by the Renwick drawing, Fig. 86.

The reconstruction drawings in this section (Figs. 88–90) were prepared by the Delaware and Hudson Railroad in 1932 for its carefully executed replica of the *Stourbridge Lion*. Unless the original working drawings are found, it seems unlikely that a more precise reconstruction can be made. The replica is presently exhibited at Honesdale, Pennsylvania.

Fig. 86. Engraving of the Stourbridge Lion *prepared by James Renwick from the original locomotive. This is the only contemporary illustration of the* Lion.

AMERICAN LOCOMOTIVES

Fig. 87.   The Agenoria *was built by Foster, Rastrick & Co. on the design of the Stourbridge Lion. Photograph shows the engine as modified in later years. It is presently on exhibit at the York Railway Museum.*

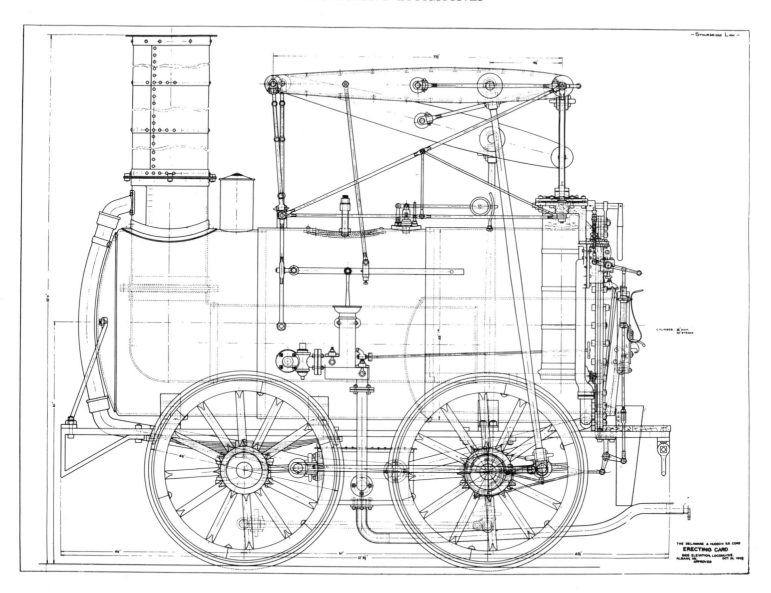

Fig. 88. *Reconstruction drawing of the Stourbridge Lion prepared by the Delaware and Hudson Railroad in 1932.*

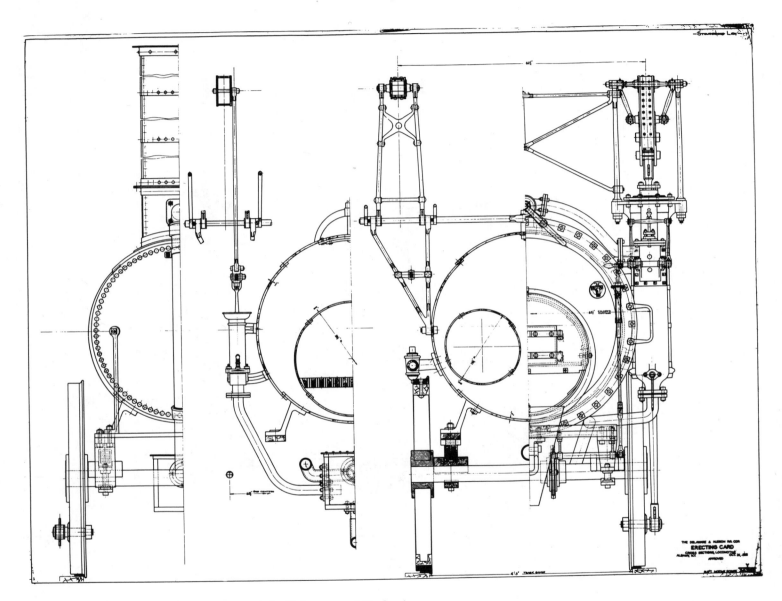

Fig. 89. End elevations and cross-section of the Delaware and Hudson's reconstruction drawing for the Stourbridge Lion.

REPRESENTATIVE LOCOMOTIVES

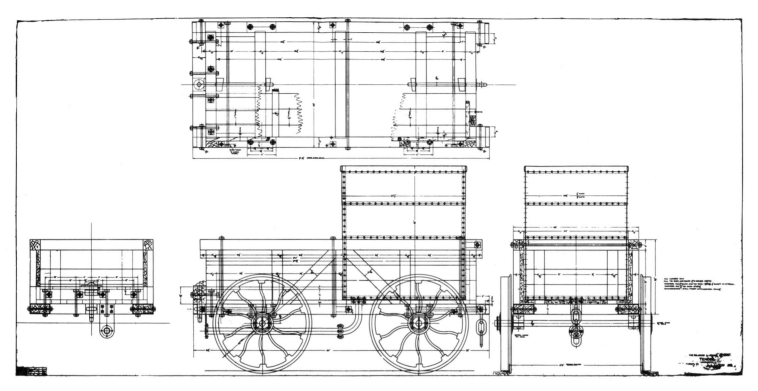

Fig. 90. Drawing of the Stourbridge Lion's *tender as reconstructed in 1932.*

JOHN BULL, 1831

Late in August of 1831, the ship *Allegheny* put ashore the first locomotive for the Camden and Amboy Railroad at Philadelphia. Today thousands of tourists annually view the remains of this venerable machine, the *John Bull,* at the Smithsonian Institution.

The Camden and Amboy Railroad and Transportation Company was chartered on February 4, 1830, to build a 4-foot 10-inch gauge line across New Jersey to connect New York City and Philadelphia by rail and water. Robert L. Stevens (1787–1856), chief engineer of the road, sailed for England in October of that year to procure rails and a locomotive. Early in December of 1830 Stevens visited Robert Stephenson at Newcastle upon Tyne. The result of this meeting was an order for a 10-ton four-wheel locomotive and the following memorandum:[1]

[1] A. D. Turnbull, *John Stevens* (New York, 1928), p. 504.

Mr. Stevens' Engine
Liverpool Railway Office   Dec. 6th, 1830

Jones's Wheels if not found objectionable and 4 ft 6 in. in diameter and wholly wrought iron.

The Boiler 2 ft 6 in. in diameter and 6 ft long. The fire box a vertical cylinder of as great a diameter as convenient. The Plate in the smoke box to be as light as practicable. The cylinders 9 inches diameter and 20 inch stroke, not more than ⅜ thick. The passages to 6 × 1⅛. The exhausting port to be larger and to have a chamber close to the cylinder with a large pipe to free the piston from any pressure—say, 4 in. in diameter. The piston rods to be partly steeled and 1⅜ diameter. The back covers to the cylinders to be of boiler plate.

The joints in the working gear to have straps and keys, and the pin of the eye to be forged in. The guides to be of steel instead of pullies. The cranked axles to be 3¼ inches. The brasses to have more bearing sideways. The Boiler to be kept as low as possible.

The depth of the fire box to be 3 feet. The top of fire box to be braced up to the dome. The boiler to be scant ¼ inch thick but the vertical part to be a full ¼ inch. The chimmey to be a good size.

Four wheels to be alike and coupled on the outside. The side rods to be open work if it be found adviseable.

The cylinders placed to work underneath the angle next the cylinder, the same as the large Engine for Inclined Plane. The tubes to be made of iron as thin as possible. This engine to be completed and delivered in time for the March Packet, 1831.

R. Stephenson
Robt. L. Stevens.

N.B. The crease of the wheels to be 1½ in. deep. The Rails are 5 feet from centre to centre and 2¼ in. broad. Inside breadth 4 ft 9¾ in. To work expansively at ½ stroke.

This agreement was in fact an outline for the specifications given later in this discussion. The mention of the "large Engine for Inclined Plane" refers undoubtedly to the *Sampson*, then under construction for the Liverpool and Manchester Railway. This refutes in part the often-repeated statement that the *John Bull* was copied from the *Planet*, a 2-2-0 completed by Stephenson in October, 1830, for the Liverpool and Manchester Railway. It was the prototype of numerous other engines built by Stephenson, a number of which were exported to this country. Stevens did not witness "the first trial of the *Planet*" as noted by J. E. Watkins in his history of the Camden and Amboy Railroad[2] for the engine had long since entered regular service on the Liverpool and Manchester. Stevens did see it at work in December, 1830. The *John Bull* was in fact a Sampson, or coupled 0-4-0. This class, like the Planet, a standard design with Stephenson, was also known as the "four coupled Planet."

The *John Bull* was completed on June 18, 1831, at a cost of just under $4,000. It was probably the fifth locomotive exported to the United States by Stephenson. The construction number assigned was 25, but it must be remembered that Stephenson's early records are uncertain and that a new number series was started some time in 1831. It is known that by the end of 1831 Stephenson had built about fifty locomotives since the start of production in 1825.[3]

The specifications reproduced below were furnished through the courtesy of Mr. H. Windale, chief draftsman, Stephenson Works of the English Electric Company, Darlington, England.

The *John Bull* was shipped from Liverpool on July 14, 1831, and, after landing in Philadelphia, was transshipped to Bordentown, New Jersey, arriving there on September 4, 1831. Only the boiler and cylinders had been shipped as a unit; all other parts were crated. The task of assembling the locomotive fell to a young mechanic named Isaac Dripps (1810–1892) who later became a respected locomotive mechanical officer for several roads as well as a partner in the Trenton Locomotive Works.

Matthias W. Baldwin inspected the *John Bull* during its assembly at Bordentown and made a few notes on the details of its parts. He did not copy the engine's general arrangement; instead, he followed, on a smaller scale, the Planet design of Stephenson. Several months after his inspection of the *John Bull*, Baldwin supervised the assembly of the *Delaware* for the Newcastle and Frenchtown Railroad; this was unquestionably the engine that Baldwin copied when he built his first locomotive, *Old Ironsides,* in 1832.

According to one report, Dripps had the *John Bull* ready for a trial by September 15, 1831, only eleven days after starting the assembly.[4] On November 12 the *John Bull* pulled several cars containing members of the New Jersey state legislature up and down a short length of track. The purpose of this trip was to convince the distinguished passengers of the utility of steam railway transport.

The *John Bull* was the only complete locomotive supplied to the Camden and Amboy by Stephenson; however, between July 7, 1832, and February, 1833, Stephenson supplied parts for other locomotives.[5] The boilers, frames, and wheels were manufactured by the Camden and Amboy at their Hoboken

---

[2] J. E. Watkins, *Ceremonies Upon the Completion of the Monument Erected by the Pennsylvania Railroad Co. at Bordentown, New Jersey* (Washington, D.C., 1891).

[3] Marshall, *A History of the Railway Locomotive Engine,* pp. 128ff.
[4] *Railroad Gazette,* March 9, 1877, p. 105.
[5] Robert Stephenson and Company, Order Book, 1832–42; transcript supplied by Mr. H. Windale of English Electric.

AMERICAN LOCOMOTIVES

TAKEN FROM A DESCRIPTION BOOK DATED 1831, WHICH STARTS WITH THE "LANCASHIRE WITCH" AND ENDS WITH LOCOMOTIVE NO. 202

DESCRIPTION OF MR. STEPHEN'S [SIC] LOCOMOTIVE CONSTRUCTED JUNE 1831

		Ft.	in.
Boiler			
	Diameter	2	6
	Length	6	9
Smoke Box			
	Length	2	2
	Breadth	4	1
	Depth below Boiler	2	—
Fire Box			
	Outside Diameter	4	1
	Depth below Boiler	1	4
	Total Height	5	3¼
	Inside Diameter	3	7
Fire Grate			
	Area of Fire Grate	10.07 sq. ft.	
Chimney			
	Diameter	13½	
	Area	143 sq. in.	
No. of Tubes	82		
	Diameter	1⅝	
	Area	145 sq. in.	
Heating Surface			
	Heating Surface of Tubes	261.7 sq. ft.	
	Heating Surface of Fire Box	34.8 sq. ft.	
	Total	296.5 sq. ft.	
Cylinders			
	Diameter	9	
	Stroke	20	

REPRESENTATIVE LOCOMOTIVES

		Ft.	in.
Double Slide Valves			
	Wheels (coupled)		
	Diameter	4	6
	Centres	4	11
Frame			
	Length	14	9
	Width	6	3
Wood Inside Frames			

shops, where, with the Stephenson parts, complete locomotives were assembled. One of these machines, nearly identical to the *John Bull,* appears in Fig. 94.

In the summer of 1833 the *John Bull* and three of its sister engines were transported to the South Amboy end of the railroad to work construction trains. It was at about this time that the *John Bull* broke its crank axle (3¼ inches in diameter) and was put out of service until a heavier crank, having a 3¾-inch diameter, could be procured in England.

Finally, in September, 1833, the road between South Amboy and Bordentown opened and regular passenger service with locomotives began. Through service from Camden to Amboy did not begin until almost a year later. When the full line did open it was beautifully built with T rails and a level and nearly straight track (no curves of less than an 1,800-foot radius). Trains were operated as fast as 25 miles per hour, even with the light engines employed. The easy grades and good track permitted light locomotives such as the *John Bull* and its sister engines to run satisfactorily for thirty years without being replaced by heavier power.

A few years after the road opened, the *John Bull* was a much altered machine. Aside from the many mechanical changes, to be discussed in detail later, a cab, bell, whistle, and headlight had been added. By the late 1840's the *John Bull* was used only for light, passenger work. In 1849 it was assigned the humble task of testing the boilers of new engines at the Bordentown shops.[6] The *John Bull* was jacked up so that its wheels cleared the rails. A pipe was run from one of its feed pumps to the boiler of the engine undergoing the hydrostatic test.

The *John Bull* was very likely the first engine to be set aside as a historic relic in the United States. A diploma from the New Jersey State Agricultural Society dated 1858 was awarded to "The Camden & Amboy Railroad and Transportation Company as a Premium for the Original Locomotive Engine *John Bull.*" This indicates that by 1858 the engine was already considered an antique. As early as April 25, 1861, *The American Railway Review* stated: "The Camden and Amboy Railroad Company have preserved with scrupulous care and at some cost considering the value of old iron—such iron as the early machinery was made of—a number of their original engines and much of their original plant. It is to be hoped that they will lay up in ordinary the old "John Bull," the first locomotive, and that they will not break up any historical thing until it has been carefully copied and described."

The exact date and circumstances of the engine's retirement are not known, but it was out of service by 1866. In 1876 the *John Bull* was renovated for operation and exhibited at the centennial celebration in Philadelphia. An attempt at restoration was made at this time which involved removal of the cab,

[6] *Perth Amboy Evening News,* September 21, 1906.

bonnet stack, and Russia-iron boiler jacket, rebuilding the eight-wheel tender into a four-wheel tender, and the addition of the present straight stack—in short, the creation of a hybrid that is not a true representation of the locomotive as it was at any one time in service. In 1883 the *John Bull* was displayed at the National Railway Appliances Exhibition in Chicago.

Two years later it was placed on exhibit in the Smithsonian Institution by the Pennsylvania Railroad, which had acquired the Camden and Amboy in 1871. In April, 1893, the *John Bull* made its well-known trip from New York to Chicago. It has not been exhibited outside of the Smithsonian since 1940.

MECHANICAL DESCRIPTION

The *John Bull* was built as a four-wheel inside-connection locomotive and, with the exception of the dome firebox, was of the standard Sampson class. Figure 91 is a reconstruction of the engine as it was originally built. This drawing is based on Fig. 96, a drawing of the original boiler, and the unidentified elevation of a dome-boiler Sampson shown in Warren's *A Century of Locomotive Building*.[7] It should be noted that this reconstruction differs in the following details from that made by J. E. Watkins in 1891 for the Bordentown Monument: style of the steam dome, placement of the manhole and safety valves, lack of boiler lagging, and shape of the smokebox. The Watkins reconstruction was used for a model now in the Smithsonian collection. It differs in several important particulars from Stephenson's general practice, which is so clearly indicated by drawings reproduced in Warren's history of Stephenson.

Figure 97 shows the *John Bull* as it appears today. This drawing and the details following it were prepared by the Pennsylvania Railroad in 1939 for the construction of a full-size replica that was built the next year at the Altoona Shops.

The dome boiler of the *John Bull* should not be confused with the renowned Bury boiler which was characterized by a high dome and a D-shaped firebox. The *John Bull*'s dome is low and the firebox is circular. In fact, the boiler of the *John Bull* lacks the one merit of the Bury design, ample steam room. It retains the defects of the dome boiler, namely, a difficult and expensive construction with a limited grate area.

While Stephenson apparently disliked the design specified by Stevens, it was repeated for at least one other buyer, the Newcastle and Frenchtown Railroad (see Fig. 95). The boiler was fabricated from ¼-inch wrought-iron plates and is 2 feet 6 inches in diameter.[8] Eleven, long stay bolts running from an iron saddle on the back head to the forward tube sheet offer longitudinal strength to the shell. This curious arrangement survives on the present boiler. The firebox and tubes were of iron; Stephenson did not begin to use copper for these parts until 1832. Eighty-two tubes of 1⅝-inch outside diameter were specified for the original boiler, while seventy-four of 1¾-inch outside diameter were found in the present boiler during a recent inspection.[9]

The original steam dome, an ornamental brass casting, and the internal throttle and dry pipe were removed and replaced by a high dome on the forward end of the boiler. A second round plate was riveted over the hole once covered by the original steam dome. Similarly, a small hexagonal plate now covers the hole for the original throttle. The present throttle and dry pipe are mounted at the rear of the steam dome outside of the boiler shell, without any insulation. This remarkably poor arrangement not only reduces thermal efficiency but also invites priming by cooling the steam on its passage to the cylinders.

[7] P. 302; details of the drawing were taken from those of various drawings in Chapter 21.

[8] Watkins, *Ceremonies*, p. 34, incorrectly gives the boiler diameter as 3 feet 6 inches, an error that has been repeated in subsequent accounts.

[9] *Ibid.*, p. 35, states that the boiler contains sixty-two tubes 2 inches in diameter.

The original stack was straight and, as shown by Fig. 91, closely followed the style used on other Stephenson products of the period. Bonnet stacks were fitted to Camden and Amboy engines as early as 1833. The *John Bull* was then equipped with such a stack (see Fig. 92). The present straight stack with its serrated top is very similar to that of the *Locomotion* and was added, as stated earlier, in 1876 in an attempt to "antique" the engine's appearance.

The feed pumps are of the common style driven by crossheads. The feed-water pipes, and most of the other pipes, were hand-rolled from sheet copper, then lap-welded. The check valves are of a very old pattern and may well be original. They correspond closely to other contemporary drawings reproduced in Warren's Stephenson history.

The cylinders are between the frames at the bottom of the smokebox. The smokebox was hot and thus warmed the cylinders and increased efficiency. The walls of the smokebox also helped to insulate the cylinders. The original cylinders were 9 inches by 20 inches. This size is verified by the memorandum of Stevens and Stephenson (December, 1830) and the engine's specifications. Furthermore, the ten cylinders ordered by Stevens in 1832 were all 9 inches by 20 inches, thus indicating that this was the standard size adopted by the Camden and Amboy. The annual report of the road for 1853 noted the cylinder size of the *John Bull* as 11½ inches by 20 inches. The present cylinders measure 11 inches by 20 inches. It is the author's opinion that the original cylinders were replaced for one of the following reasons: the originals, only three-eighths of an inch thick, wore out or cracked in service, were broken by a head-end accident, or were replaced in an attempt to increase the engine's power with cylinders of a larger bore. The present smokebox is 3 inches deeper than that shown in the original boiler drawing (Fig. 96); this indicates an enlargement for bigger cylinders. One last bit of evidence is that the original cylinders had "double slide valves" (see specifications) with short steam passages; the present valve is a single slide valve with long passages to the cylinders.

The original valve gear was actuated by two eccentrics mounted on the rear axle between the cranks. The eccentrics could be shifted to the right or left by a foot treadle through a simple rod and lever mechanism. The eccentrics engaged a right or left lug fastened to the axle at a 90° angle. One lug would drive the engine forward, the other would reverse its movement. The large levers at the back head permitted hand operation of the valves so that the locomotive could be "jacked" when engaging the lugs. This simple, fixed cutoff valve gear was used regularly by Stephenson until about 1835. A better understanding of this mechanism can be gained from the drawings of the *Robert Fulton* (see especially Figs. 7 and 105).

The present valve gear is an adaptation of the original arrangement. Only two eccentrics are used, but they no longer slide and are keyed to the rear axle. Two eccentric rods are attached to each eccentric. One passes over, the other under, the forward axle through the smokebox to the rocker, which is bolted to the front of the smokebox. To reverse the engine the gabs are disengaged from the rocker by the rod connected to the bell crank at the top of the smokebox. The valve is then shifted to the opposite port by the reversing lever attached at the rear of the firebox. Each cylinder or engine must be reversed separately. Needless to say, there is no way to adjust the cutoff. See Fig. 98 for more details.

An inside frame and an outside frame are used. The inside frame is built on both sides of each crank and is so-placed as a safety feature in case a crank should break. In that event the axle would be supported until the engine could be stopped. The crank axle is supported by six bearings, including those on the outside frame. Originally, the inside frame, like the present outside frame, was wooden. The wooden inside frame is clearly illustrated in Fig. 105.

The driving-wheel springs were originally mounted under the axles as shown in Fig. 91. Later they were moved above the axles atop the frame. When this was done the pedestals were cut off and a truss was added to the outside frame above the wheels.

Isaac Dripps recalled the problems experienced with driving wheels on the early locomotives of the Camden and Amboy in a manuscript dated September, 1885, now on file at the Smithsonian. In this memorandum Dripps stated that wooden wheel centers were too weak and had been abandoned after one year's service. Wheels were fabricated from wrought iron with round spokes. The rivets worked loose, however, and the road decided to try cast-iron wheel centers in 1835. These were successful and were adopted as standard. The present cast-iron wheels, the spokes of which are cast in an H shape, probably date from this period.

A nonswiveling two-wheel leading truck was added to the *John Bull* in about 1832. The truck was devised by R. L. Stevens and is described in the annual report of the Camden and Amboy for 1833:

> The ends of the axle extend beyond the carriage, and run in boxes, inserted in strong frame work which is projected in front of the carriage, and to which two smaller wheels are attached, adapted to the track. As the carriage is propelled forward, these guide wheels, which are of sufficient weight only to keep them on the rail, (so that the friction is comparatively nothing,) follow the track, and necessarily give the proper direction to the wheels of the carriage, and always preserve their parallelism with the rails. By this simple device the difficulty above alluded to is completely obviated, as the flanges of the carriage wheels can never come in contact with the edge of the rail. Both carriages of the American engines are of this construction. Repeated experiments have been made with them, and the success uniform and complete. With one of them, a train of ten cars, containing stone and iron equal in weight to three hundred and forty passengers, has been propelled at a very high speed along the portions of the line, in which not only the greatest curvature, but the highest grade exist, without the slightest impediment or difficulty. In running them from Amboy to Bordentown, the flanges, in no instance, were found to come in contact with the rails, or to produce the least resistance, although the engine was purposely driven with the greatest velocity around the greatest curves in the line, in many instances at a speed exceeding thirty miles per hour. A patent will be obtained by Mr. Stevens, for the purpose of securing the benefits of this invention to himself and the Company.

The truck is attached to the engine on the extended ends of the front driving axle which formerly mounted the cranks for the connecting rods. Dripps claimed that the side rods were never used because they restricted the engine from taking sharp curves. The front driving wheels now have a 1-inch play for curves. The removal of the side rods materially reduced the tractive effort of the engine since only the rear driver is powered. The set screw and spring attached to the front beam of the main frame and bearing on the truck's frame were used to adjust the amount of weight on the rear driving wheel. The truck was made easily removable (only the bearing caps were bolted) so that the engine would fit the short turntables employed on the road. Note the handles on the rear end of the truck's side frame used to assist in uncoupling the truck from the engine (Fig. 100).

The *John Bull*'s first tender was quickly built from a small four-wheel car used by one of the contractors in building the railroad. It was replaced by a double-truck box tender, a standard of the Camden and Amboy. The main purpose of the roof was to keep the fuel dry and prevent sparks from igniting it. The gig top protected the brakeman, who, from this position atop the tender roof, could apply the tender's brakes by pushing the foot lever attached to the rear of the tender frame. The eight-wheel tender was in existence as late as 1876 and was probably rebuilt into a four-wheel car at the time of the United States Centennial. After the *John Bull* returned from the Columbian Exposition (1893) the tender was stored outside of the Smithsonian. By 1910 it was so badly rotted that it was dismantled. In 1930 the Pennsylvania Railroad reassembled the tender and it is now exhibited with the locomotive. The cylindrical iron water tank, 70 inches in diameter, holds 1,150 gallons. The tender carries 2 tons of coal, its total weight with fuel being about 10 tons.

The *John Bull* is an excellent example of what happens to a locomotive during thirty years of service. In this case the boiler shell and possibly a few other details are all that remain of the original. For this reason caution must be used when citing full-size specimens as accurate representations of locomotives of a given date. The present *John Bull*, if a poor example of locomotive construction in 1831, is an excellent illustration of how little of the original machine survives long years of use.

REPRESENTATIVE LOCOMOTIVES

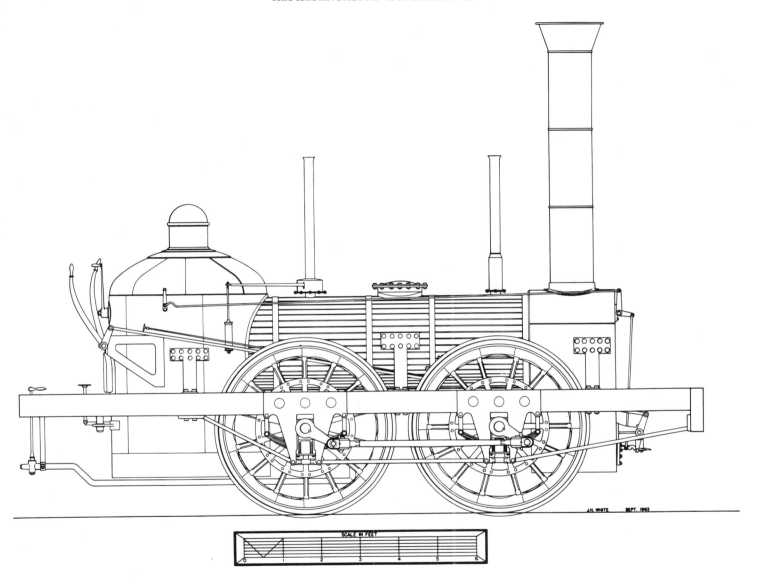

Fig. 91. *This reconstruction shows the* John Bull *as built by Robert Stephenson in June, 1831.*

AMERICAN LOCOMOTIVES

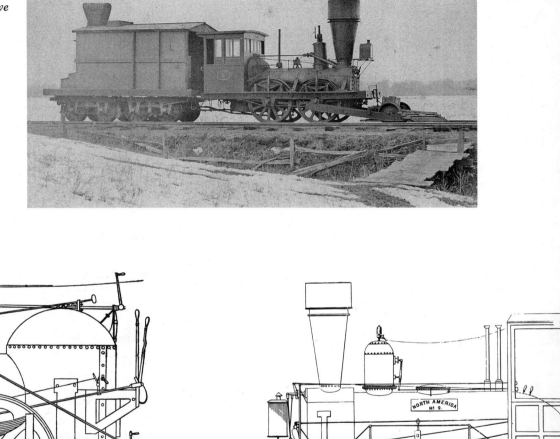

Fig. 92. The John Bull *as it was modified and rebuilt by the Camden and Amboy Railroad. The photograph is believed to have been taken in about 1865.*

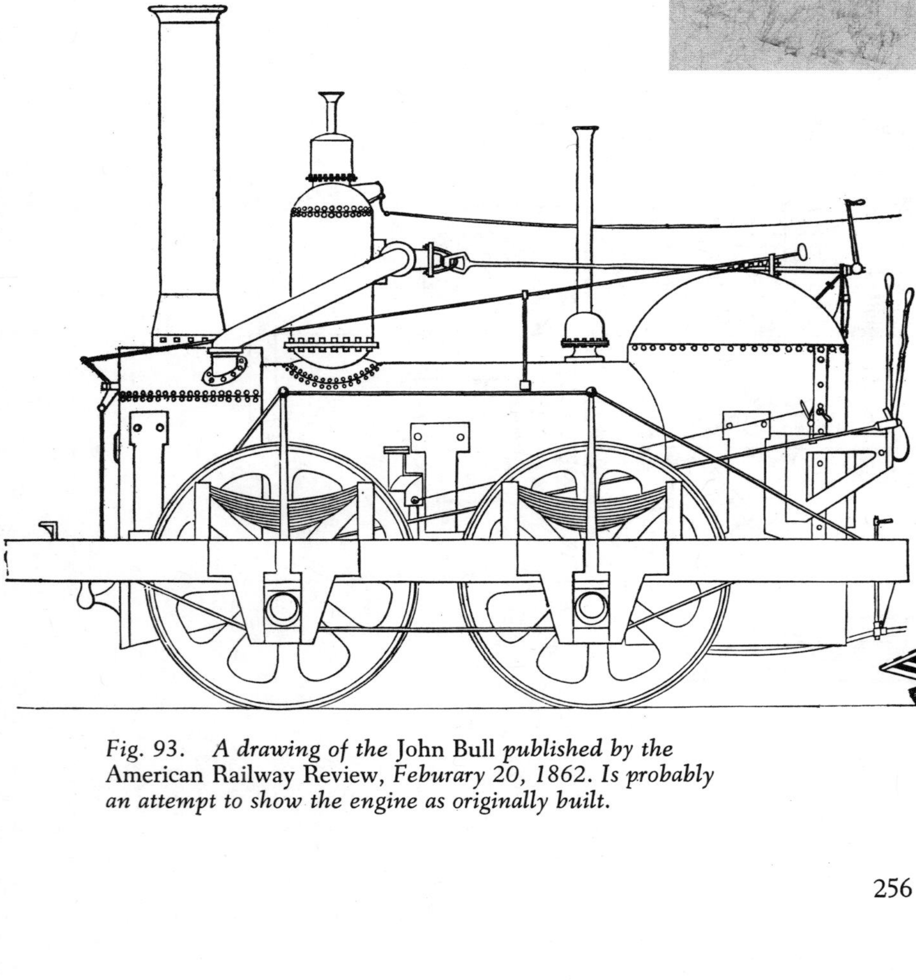

Fig. 93. *A drawing of the* John Bull *published by the* American Railway Review, *Feburary 20, 1862. Is probably an attempt to show the engine as originally built.*

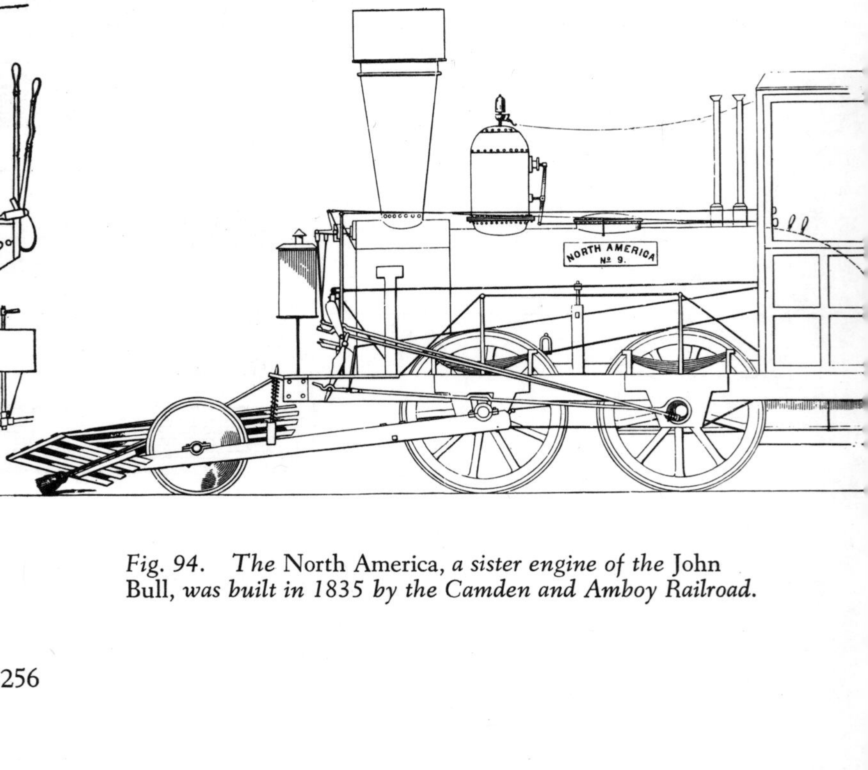

Fig. 94. The North America, *a sister engine of the John Bull, was built in 1835 by the Camden and Amboy Railroad.*

REPRESENTATIVE LOCOMOTIVES

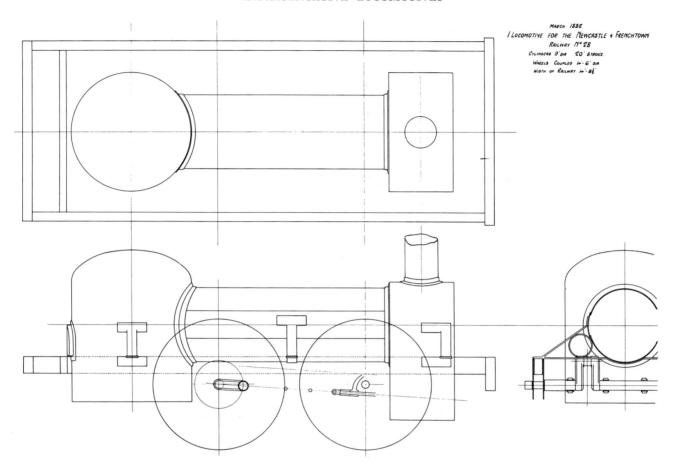

Fig. 95. *A drawing of the Maryland, built in 1832 by Robert Stephenson & Co. for the Newcastle and Frenchtown Railroad. The engine's specifications and general arrangement are identical to those of the John Bull.*

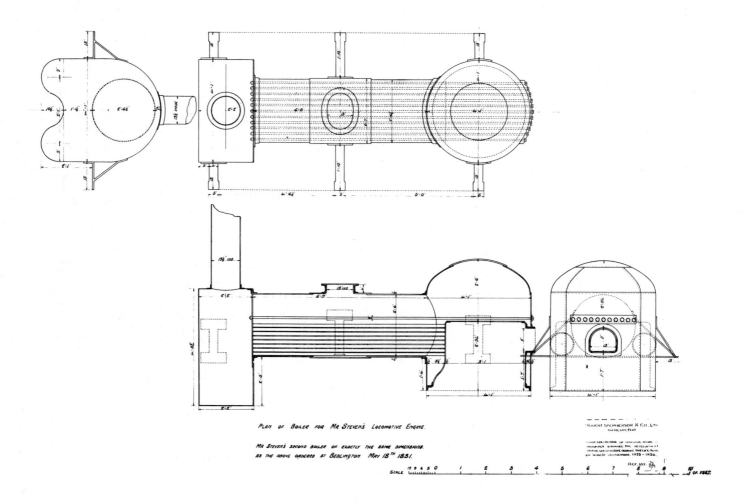

Fig. 96. This boiler drawing is believed to have been used in the construction of the John Bull.

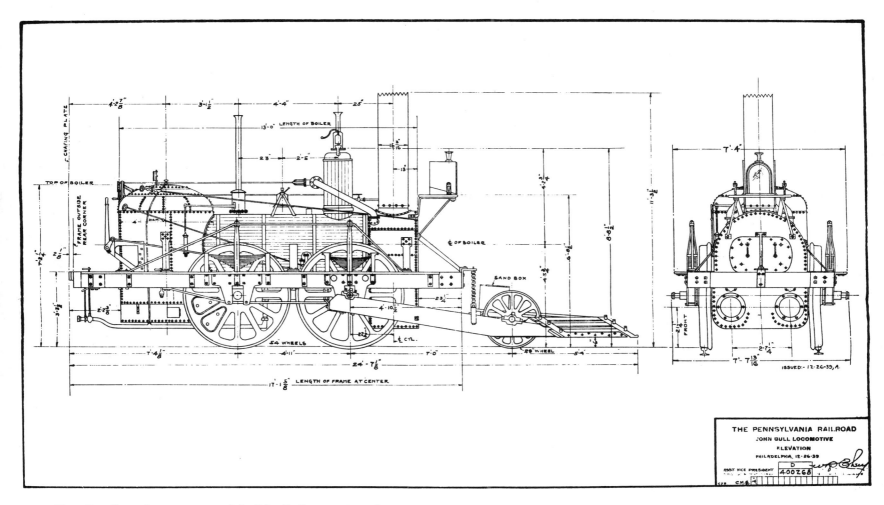

Fig. 97. *General arrangement of the John Bull as it exists today.*

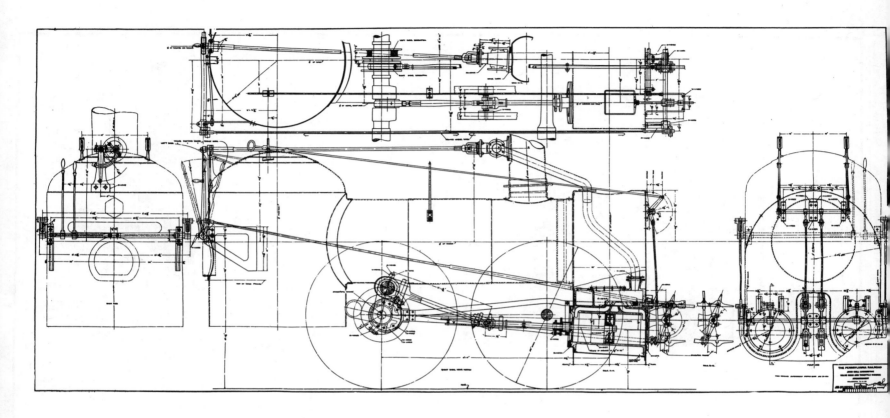

Fig. 98. Arrangement of the valve gear, cylinder, and throttle of the John Bull.

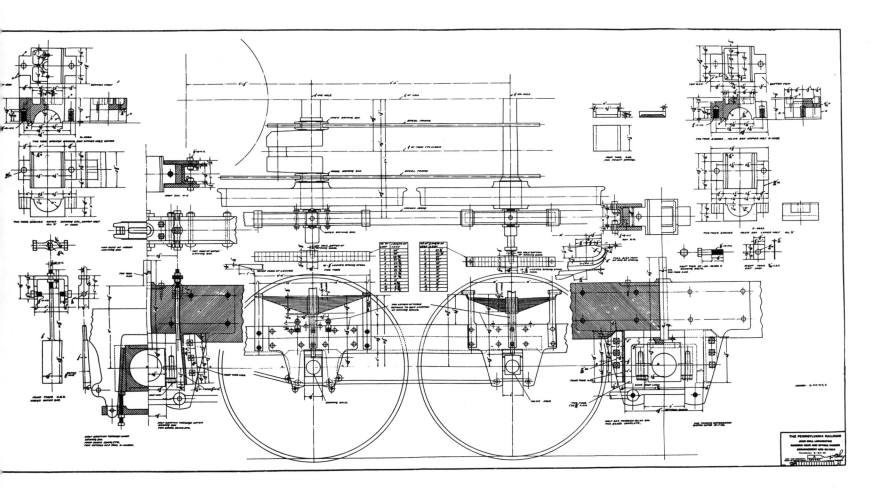

Fig. 99. *Detail of the outside and inside frames of the John Bull.*

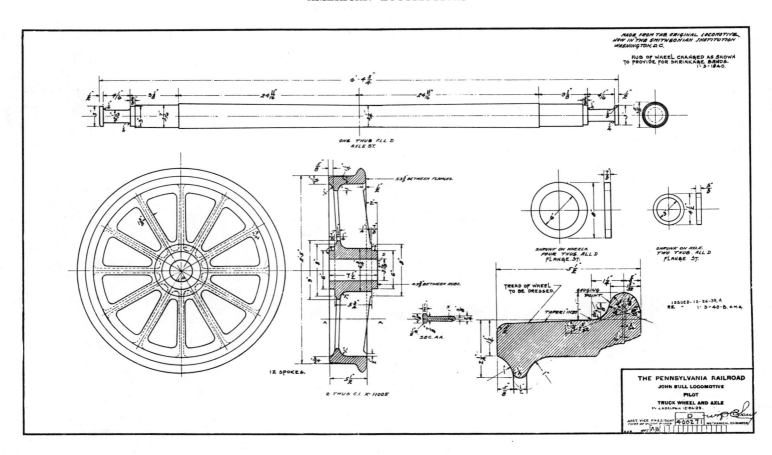

Fig. 100. *Detail of the leading wheels and axle of the John Bull.*

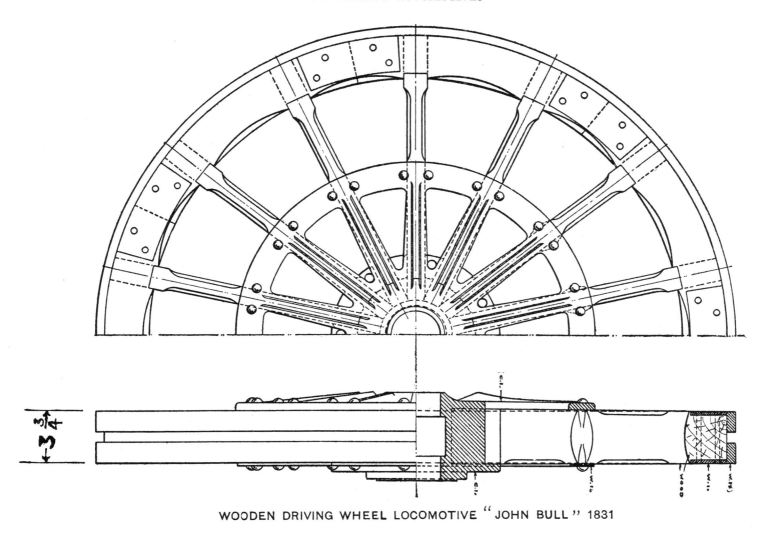

Fig. 101. *Rear view detail of a wooden driving wheel believed to be from the John Bull as originally constructed. Note that the tire is missing.*

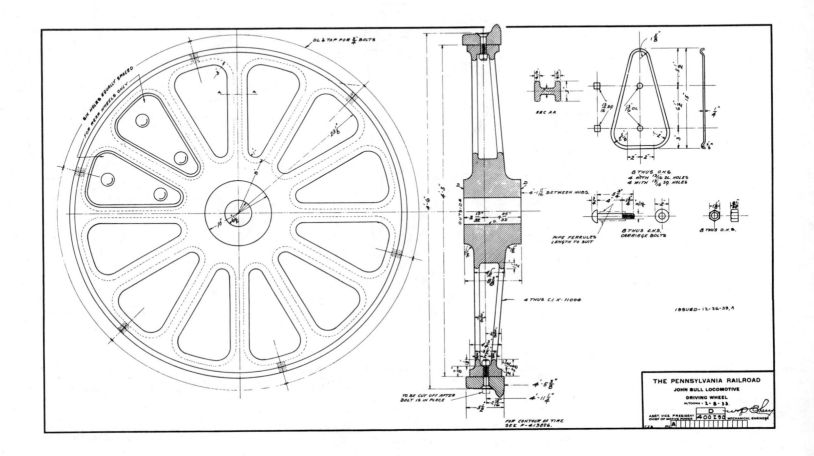

Fig. 102. *Detail of the present cast-iron driving wheel of the John Bull.*

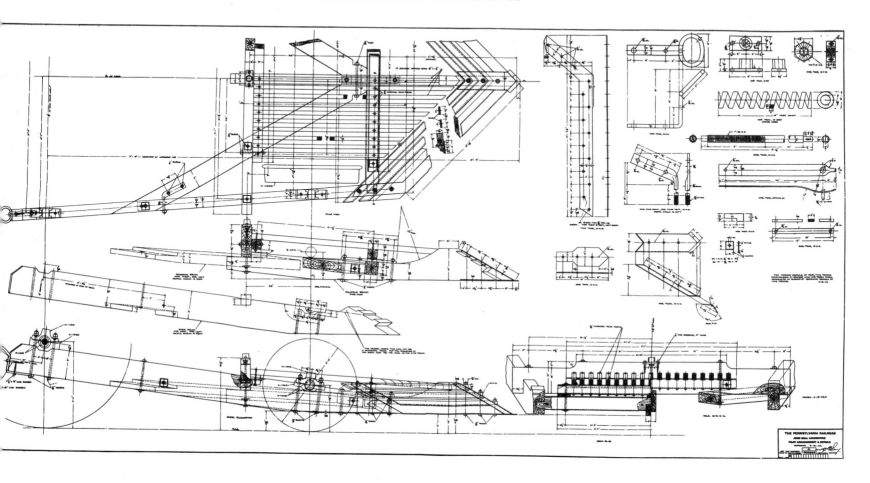

Fig. 103. *Arrangement of the pilot of the* John Bull.

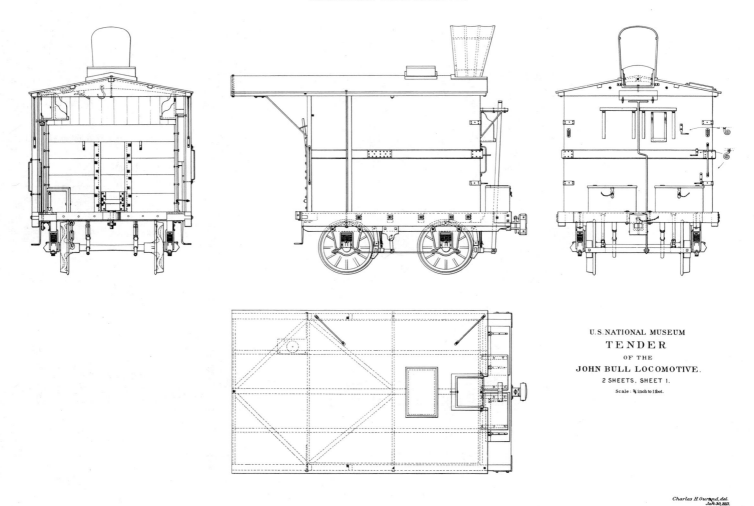

Fig. 104. *General arrangement drawing of the present four-wheel tender of the John Bull. The over-all dimensions are 15 ft. 2½ in. long, 12 ft. 1 in. high, and 8 ft. 6 in. wide. The wheel base measures 61 in. and the wheels are 33 in. in diameter.*

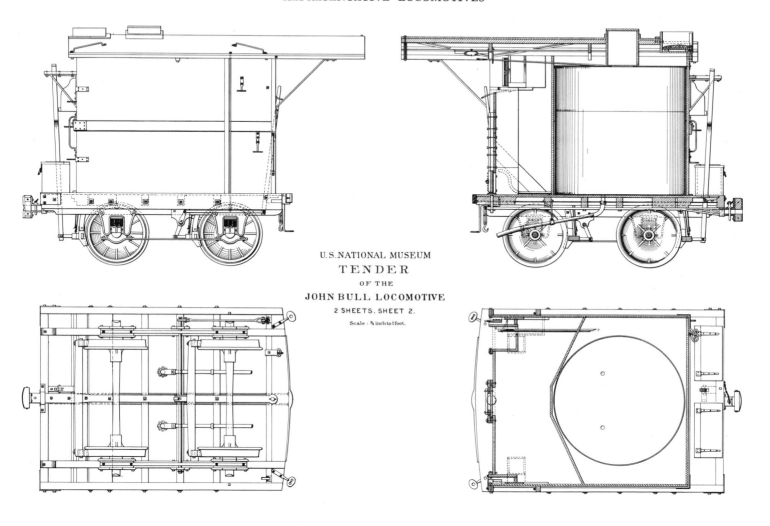

Fig. 104–A. *General arrangement drawing showing left elevation and cross-section of the four-wheel tender of the* John Bull.

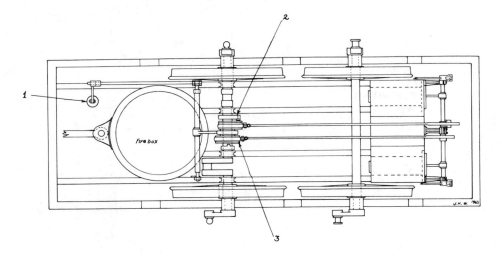

Fig. 105. The original wooden frame and valve gear of the John Bull. Item 1: foot treadle used to shift the eccentrics. Item 2: the gab. Item 3: the right-hand eccentric.

LANCASTER, 1834

The *Lancaster* was Baldwin's third locomotive and was the first engine to operate successfully on the Philadelphia and Columbia Railroad. Early in 1834 the Philadelphia and Trenton Railroad ordered the engine from Baldwin, but since the road was not completed until November of that year it was decided to relinquish the engine to the Philadelphia and Columbia. A memorandum copied from the cash book of the Philadelphia and Trenton explains this agreement:

June 2, 1834

In a conversation with M. W. Baldwin, he stated that the locomotive he was now making, originally intended to be used on the Philda & Trenton Rail Road, and now to be placed on the Pennsylvania Railway, and similar to which he will construct another for the P. & T. R.R. Co. will be as follows

    Weight 6½ tons without water
    add 1 [ton] for water
    7½ tons on the Road

Power equal to 30 horses at 15 miles an hour, drawing 16 cars with 24 passengers each; each car estimated at 4½ tons = 384 passengers or 72 tons. This means common working power—may be increased with good fuel . . . etc.

The Engine to have 6 wheels—will turn with safety at slow speed on 300 feet radius. . . .

    Price, Engine complete $5,000
    Tender   500
    $5,500 ready to use

Terms half the price when the boiler is completed, the work then being considered half done, the remainder on the entire completion.

The *Lancaster* was delivered to the Philadelphia and Columbia at a cost of $5,580. Placed in service on June 28, 1834, it was used to haul material for building the second track of the line. On October 7, 1834, the *Lancaster* pulled one of the special trains conveying the governor and other state officials to celebrate the opening of the two-track line.

The *American Railroad Journal*, October 25, 1834, printed a statement on the operation of the *Lancaster* as reported to E. L. Miller by E. C. Whiting, the machine's engineer. The *Lancaster* was said to perform exceedingly well, considering the character of road, which was said to be ". . . almost made up of curves . . .," some as sharp as 450 feet in radius. Her normal capacity was given as 35 tons, but on one occasion she pulled a 75-ton, sixteen-car train, fourteen of these loaded with rails, at an average speed of from 12 to 14 miles per hour. She had not missed a day's work and required only modest repairs. In all it was a glowing testament, but it reflected the true feelings of the road's management, which loyally patronized Baldwin in later years.

In 1848 the annual report of the road noted that the *Lancaster* was in poor condition. By 1850 the engine was out of service and in 1851 it was broken up for old iron.

The *Lancaster* was the second 4–2–0 built by Baldwin. A year before its construction, Baldwin visited the Mohawk and Hudson Railroad at the suggestion of E. L. Miller to study the truck engines developed by J. B. Jervis. The operation of these 4–2–0's over light track and sharp curves impressed Baldwin favorably on the merits of this design for American roads. Accordingly, Baldwin's second engine was a 4–2–0, the *E. L. Miller,* completed in February, 1834, for the South Carolina Railroad. Baldwin became a militant advocate of this wheel arrangement and, with only a few exceptions, built 4–2–0's exclusively until 1842. In his early correspondence Baldwin continually stated his preference for the truck engine. A typical letter expressing this view reads in part: "I make but one kind of machines and am persuaded from actual experiment that they are the best, they have been tried with English Engines and are pronounced to be more efficient and simple than any they have sent to this country. This being the fact I shall be obliged to decline making any machines according to the plans sent."[1]

Between 1834 and 1842 Baldwin built well over one hundred 4–2–0's. As late as the mid-1850's Baldwin advertised to build 4–2–0's, as evidenced by the lithograph reproduced in Fig. 109.

The chief advantage of the 4–2–0 over the English four-wheel engine was obviously its good tracking qualities. Its chief defect, recognized from the beginning, was a woeful lack of traction. E. L. Miller suggested that a device be added whereby the weight of the tender could be concentrated at the rear of the locomotive when needed for heavy grades or starting the train. Miller patented the idea in June, 1834, and it was used on the *E. L. Miller* and the *Lancaster* as clearly stated in the *American Railroad Journal* of October 25, 1834, quoted earlier in this discussion. The Baldwin history of 1923 claims that the Miller traction-increaser was not used until Baldwin's eleventh locomotive, the *Black Hawk,* in May, 1835.

The reconstruction of the *Lancaster* (see Figs. 110–112) is based on a print in the United States National Museum, reproduced in this section as Fig. 106. This print is from a tracing made in about 1890 of an original drawing at the Baldwin Locomotive Works.[2] Although marked as the *Lancaster,* it may well be a 4–2–0 built in the late 1830's. The style of wheels shown was not used by Baldwin in 1834, and the cylinders have a 10½-inch bore by 16-inch stroke, a common size for third-class Baldwin 4–2–0's of the late 1830's, while the *Lancaster* had 9-inch by 16-inch cylinders.[3] The general arrangement and many details were reconstructed from Figs. 107 and 108.

---

[1] Letter of M. W. Baldwin to a Mr. Hofman, June 17, 1836, in Baldwin Letters, Historical Society of Pennsylvania.

[2] The Pennsylvania Railroad built a model based on this drawing in 1892–93 for the Columbian Exhibition; the disposition of the original is unknown.

[3] See the catalog of *Baldwin Vail & Hufty,* 1840, reprinted in 1946 by Grahame H. Hardy.

MECHANICAL DESCRIPTION

The boiler was a simple Bury type with a D-shaped firebox. The copper safety-valve column over the dome could be unbolted, thus opening a manhole for throttle and crown-bar maintenance. The throttle, located at the base of the dome, was a simple slide valve with three small rectangular openings. No records exist stating the number and size of tubes employed by the *Lancaster,* but a second-class Baldwin 4–2–0 of 1838 was noted as having 118 copper tubes 1½ inches in diameter.[4] The smokebox is round, an interesting departure from the common idea that round smokeboxes did not come into being until the introduction of the cylinder saddle. The boiler of the *Lancaster*—probably also those of all Baldwin locomotives until about 1836—was built by Moses Starr and Sons of Philadelphia. Baldwin's orders were so large that Starr was unable to handle boiler orders of other small Philadelphia locomotive builders such as Coleman Sellers and Sons.

The frame was wooden and was mounted outside of the wheels. No inside "safety" frames were used as on the English crank-axle engines. The frame was not clad with iron as was the practice for later Baldwin engines. This feature is substantiated by the drawing of the Utica and Schenectady's No. 1, Fig. 107, and the comment made by the *American Railroad Journal* on December 15, 1839, regarding Rogers locomotives, which read, "the large frame is very strongly plated in the manner of Stephenson's engines, the neglect of which till very lately has been, we are informed, a constant objection to the Philadelphia engines on the Long Island and Troy railroads."

Baldwin's celebrated half-crank axle was used on the *Lancaster.* One half of the crank axle is omitted and its place is taken by the driving wheel. The *Journal of the Franklin Institute,* May, 1835, noted the advantages as follows: "the power of the engine is applied directly to the wheel, without the intervention of an arm of the crank thus diminishing the strain upon the axle, and consequently, lessening its liability to be broken. By this means, also, Mr. Baldwin has, in some measure, obviated the tendency of the driving wheels to twist upon the axles, and become loose; a very general and troublesome defect of locomotives." Other advantages were the possibility of wider boilers and the placement of the driving wheels behind the firebox since the connecting rod could pass between the boiler and frame. As good as this arrangement might appear to be, it defeated the two major advantages of inside connection—steady running and insulated cylinders. The cranks were at a great distance from the engine's center line and the cylinders were outside of the smokebox.

In 1838 Baldwin claimed that only one half-crank had broken on the 100 engines he had built.[5] Until about 1840 Baldwin pushed the sale of engines equipped with half-crank axles, but few were built after that time. The *Susquehanna* (1850) of the Hudson River Railroad was the last engine built by Baldwin with this style of crank axle. William Swinburne of Paterson, New Jersey, and an undetermined firm of Liverpool, England, built some locomotives with half-cranks in about 1850. It is probable that Baldwin's patent (September 10, 1834) had expired by the time these rival builders began using his invention.

The truck was wood-framed with outside bearings. The forward weight of the engine was transferred to the truck by the bearing pin of the leaf spring attached to the main frame. The truck center pin was merely a pintle and did not carry any weight. It was placed a few inches behind the center of the truck in order to shorten the rigid wheel base.

The wheels were of cast iron in a T spoke pattern, with the ribs facing inside. The driving wheels had wrought-iron tires. The fact that the old Baldwin drawing (Fig. 106) shows this style of wheel, rather than the patented wooden cushion wheel

---

[4] Knight and Latrobe, *Locomotive Engines,* p. 33.

[5] *Ibid.,* p. 32.

used on the early Baldwin engines, provides additional evidence that this drawing is not of the *Lancaster*. If it is not the *Lancaster*, it is at least a contemporary document and the most detailed that this author knows of for a Baldwin engine of 1838 or earlier.

The valve gear is based on the motion devised by J. and C. Carmichael of Dundee, Scotland, in 1818. Joseph Harrison stated that a similar mechanism was used on early Delaware River ferries, and this may be where Baldwin got the idea for his locomotive valve gear.[6] It is about the simplest motion that provides forward and reverse movement. A single fixed eccentric on the driving-wheel axle drives a girder-frame eccentric rod. The rod has two gabs, one facing the other. The rocker has two corresponding pins, one of which the upper or lower gab engages for forward or reverse motion.

If the pins would not line up with the gabs, a jacking lever (shown by the dotted line in Fig. 110) could be slipped over the top of the rocker to adjust its position as necessary. The same lever could be used to work the valve manually. Baldwin continued to use this style of valve gear for a number of years and did not adopt double eccentrics until about 1838.

The crosshead and guide are set back a considerable distance from the cylinder. This allowed convenient attachment to the frame and boiler and also helped to shorten the connecting rod. It should be noted that, since the driving wheels are behind the firebox, an unusually long connecting rod would be required if the crosshead guide were conventionally placed.

The diamond-shaped crosshead guide also served as the feed-water pump body. The check-valve body at the end of the guide feed-water pump is held together by a stirrup. This permitted quick disassembly for cleaning out any obstructions clogging the check valves.

The crosshead and guide were massively made, not only to accommodate the pump, but also to withstand the racking strain imposed on the mechanism because it was not in line with the piston rod. This curious arrangement was used regularly by Baldwin through the 1840's. It was used as late as 1860 on the *Active,* a small four-wheel locomotive built for the Philadelphia and Reading Railroad.

[6] Harrison, *The Locomotive*, p. 33.

Dimensions of the *Lancaster*

Cylinders	9" × 16"
Wheels	54"
Weight	7½ tons
Steam pressure	120 lbs. per sq. in.

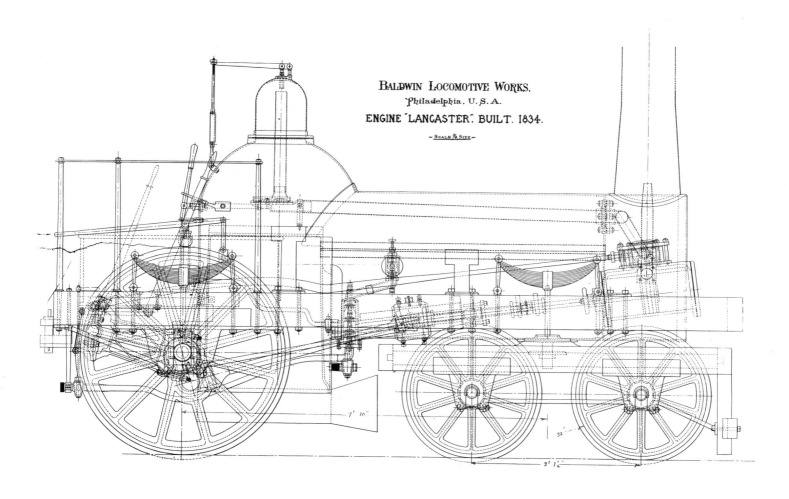

Fig. 106. *Thought to be of the Lancaster, this drawing was traced from an original at the Baldwin Locomotive Works in about 1890. Though the copy was labeled the Lancaster, it may be a drawing for an engine of about 1836–38.*

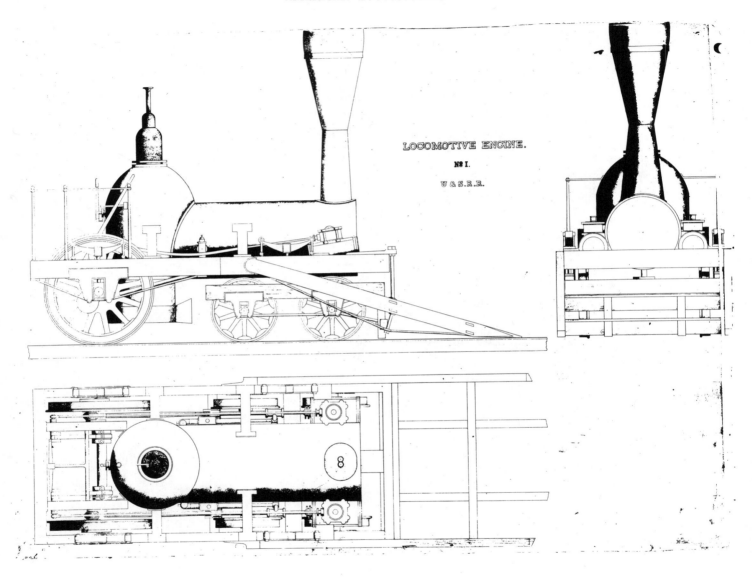

Fig. 107. The Utica and Schenectady Railroad's No. 1, built by Baldwin in 1836.

REPRESENTATIVE LOCOMOTIVES

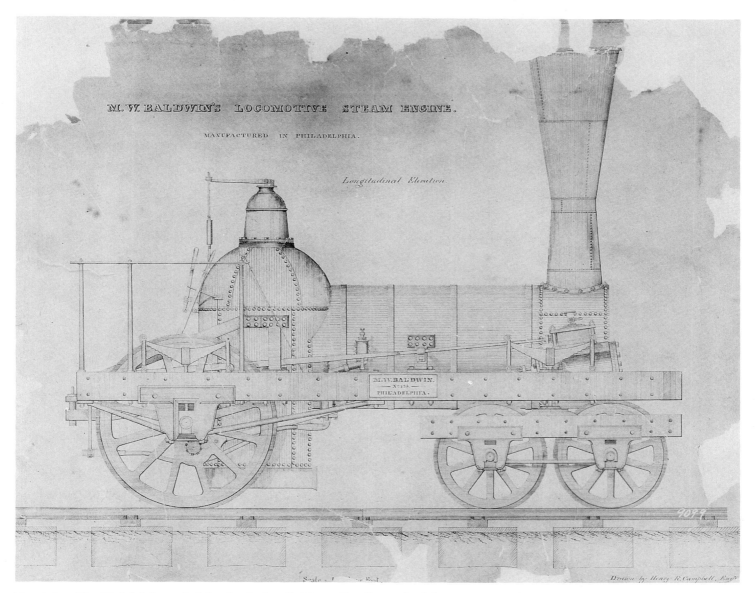

Fig. 108. The Philadelphia and Columbia Railroad's Martin Van Buren, *built by* Baldwin in 1839. This lithograph is commonly used to represent earlier Baldwin locomotives.

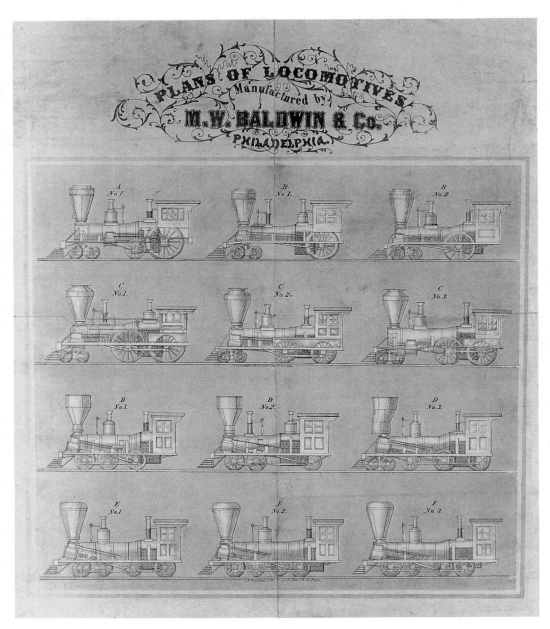

Fig. 109. An advertising lithograph issued by Baldwin in about 1855. Note that 4–2–0's were still being offered.

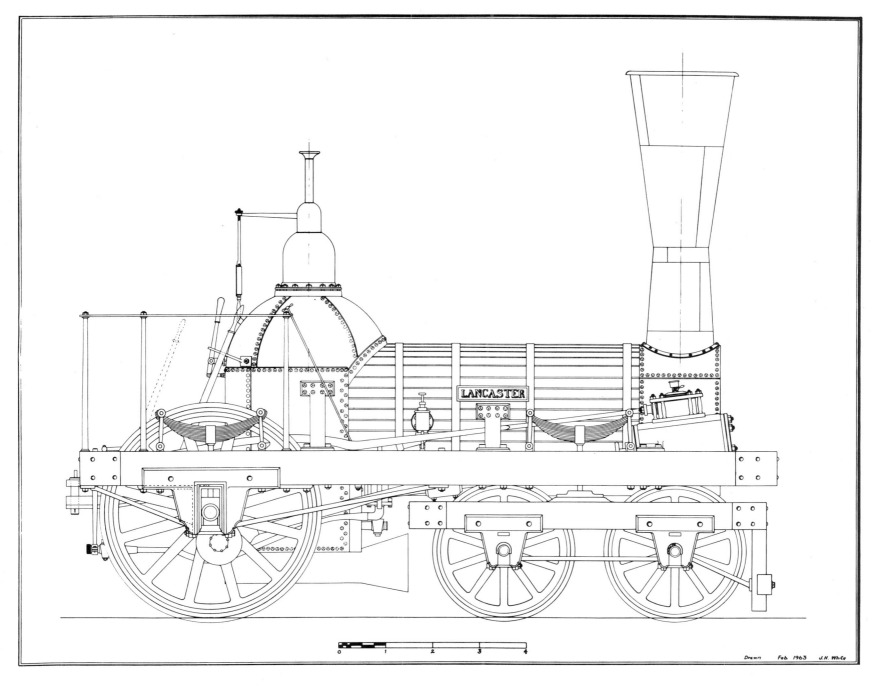

Fig. 110. A reconstruction of the Lancaster, 1834.

AMERICAN LOCOMOTIVES

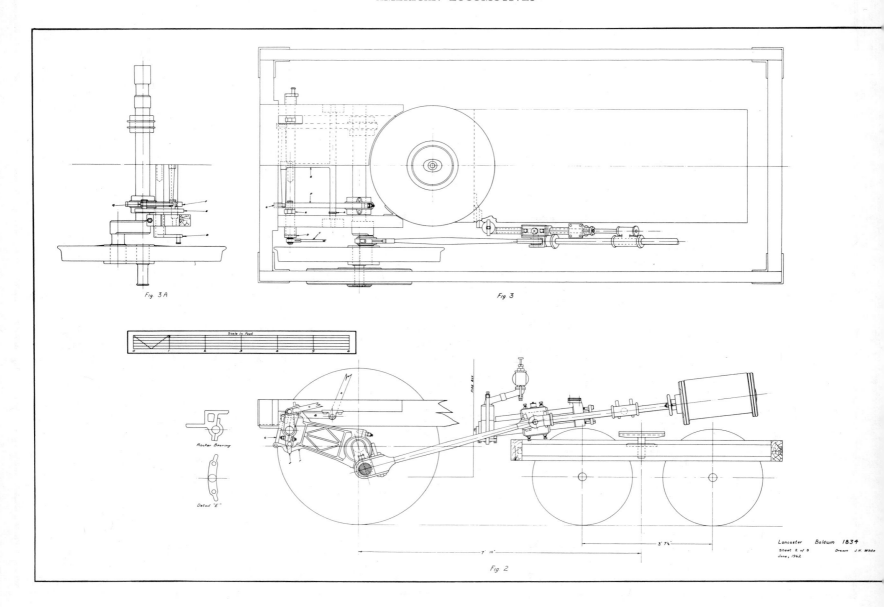

Fig. 111. *A reconstruction of the* Lancaster *showing details of the running gear.*

REPRESENTATIVE LOCOMOTIVES

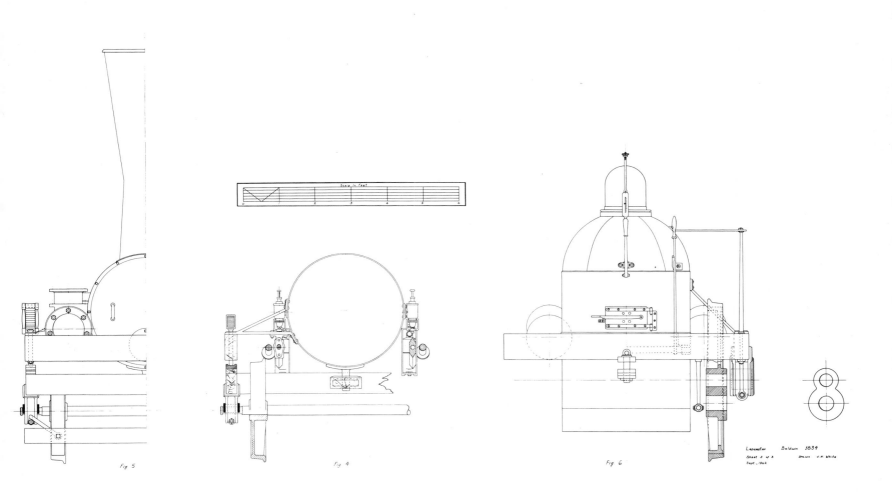

Fig. 112. *A reconstruction of the* Lancaster *showing front elevation, cross-section, rear elevation, and a detail of the crank boss.*

DUNHAM, CA. 1837

Henry R. Dunham was one of the many operators of small machine shops attracted to locomotive building during the nineteenth century. Little is known of Dunham's beginnings except that he was first listed in the 1834 New York City directory as a machinist. His rise was mercurial for by 1837 he opened a shop, known as the Archimedes Works, at 100 North Moore Street. Next door was the machine shop of his partner William Browning, who had started his shop in about 1830. Browning and Dunham conducted their firm under the name of H. R. Dunham and Company. The partnership continued at least as late as 1847. Dunham is last listed in the 1856 New York directory, which is probably the year he went out of business.

The earliest evidence that Dunham built locomotives was an advertisement appearing in the February 12, 1836, issue of the *American Railroad Journal*. Between 1836 and 1838 only about sixteen engines were produced. No record can be found of an engine built after 1838. Dunham locomotives were supplied to the Camden and Amboy, Michigan Central, New York and Harlem, and others. Curiously enough, out of this tiny production a Dunham engine may have survived. Although the identification is not positive, the *Mississippi*, presently on exhibit in the Museum of Science and Industry in Chicago, may well have been built by this obscure firm.

The drawings reproduced here are from Paul R. Hodge's book, *The Steam Engine,* published in 1840. They represent a Dunham locomotive of about 1837. Hodge came to this country from England sometime before 1837. He worked as a draftsman for Thomas Rogers making the drawings for the *Sandusky*. According to one account, Hodge's design was so poorly prepared that the firebox would not fit into the boiler shell. Hodge was reportedly fired for his incompetence.[1]

Hodge's own story of his employment with Rogers does not agree with the above account. Hodge does not mention design-

---

[1] An account of Hodge's association with Rogers is given in the *Railroad Gazette,* December 10, 1880, p. 655. The account, apparently based on statements made by William Swinburne, was extracted from a series of newspaper articles that appeared in the Paterson *Press* and later were published in the book *A History of Industrial Paterson*.

ing the *Sandusky* but claims to have designed two locomotives for the "New-Brunswick and Jersey City road" (New Jersey Railroad and Transportation Company). A drawing of one of these machines, the *Juno*, appears in *The Steam Engine*.[2] These two engines, actually named the *Arresseoli* and *Uncle Sam*, were built by Rogers in 1838 and 1839, respectively.

Whatever the precise circumstances of Hodge's relationship with Rogers, he was not a resident of New York until 1839–40 and may well have been employed by Rogers until that date.[3] It is possible he worked for both Rogers and Dunham (1837–39) on a consulting basis since neither was fully engaged in locomotive building during that period. It should be noted that Hodge settled in New York at the time Dunham dropped the locomotive business.

In 1840 Hodge's book *The Steam Engine* was published in New York. It was the first work to contain drawings of an American locomotive. The drawings of the Dunham engine enjoyed two reprintings. The *Practical Mechanic Magazine* reprinted the full set of drawings in its issue of January, 1845 (Volume 4). A shaded side and front elevation was published in the 1848 edition of James Renwick's book *The Steam Engine*.

Before returning to England in 1847, Hodge built a steam fire engine for use in New York City. This machine, built in 1841, was similar to the Dunham locomotive. The boilers and the over-all size of the machines, 7 or 8 tons, were nearly identical.

After returning to England Hodge published a second book, *Analytical Principles and Practical Application of the Expansive Steam Engine* (1849). The last reference found by the present author concerns a few remarks made by Hodge to the Institution of Civil Engineers of London in 1869.[4] No record can be found of his last years or death.

---
[2] P. R. Hodge, *The Steam Engine* . . . (New York, 1840), p. 243 and Plate XLVII.
[3] Hodge is first listed in the 1839–40 New York directory. No Paterson city directories were issued early enough to check his residence for 1836–39.

A study of the Dunham drawings clearly shows the marked similarity of the Dunham engine to the engines built by Baldwin. The only major differences lie in the crank axle and crosshead.

A full crank axle was used in imitation of Baldwin's half-crank, but it was carefully contrived so as not to infringe on Baldwin's patent. The crank was set in a heavy shoulder of the driving-wheel casting. The same design was used on early Rogers engines and is more evidence that Hodge designed for both Rogers and Dunham. Wrought-iron bands, about an inch square, loop around the driving-wheel hub and crank-axle boss to provide extra support.

The valve gear is identical to that of the *Lancaster*. Colburn noted that Dunham used a considerable lap of ½ inch, thus effecting a cutoff at ¾ stroke.[5] The valve travel was stated by Colburn to be 2 inches.

One last mechanical feature worth mentioning is the location of the feed-water check valve on the rear tube sheet under the boiler shell. The cold feed water was injected into the water space surrounding the firebox. Since this was the hottest part of the boiler, the cold water would cause a great turbulence within the boiler.

Hodge's description of the Dunham engine was little more than a listing of the parts (see below). He did mention that the usual steam pressure was 50 pounds per square inch and that the engine pulled a 220-ton train (locomotive and tender included) up a slope at 14 miles per hour. Hodge calculated that 77 horsepower was required to move the train at this speed.

Cylinders	10½″ × 16″
Wheels	54″ diameter
Truck wheels	32″ diameter
Boiler	37″ diameter
Tubes (136)	1½″ diameter

---
[4] Institution of Civil Engineers, *Proceedings*, Vol. 28 (1869), p. 402.
[5] Colburn, *Locomotive Engineering*, p. 59.

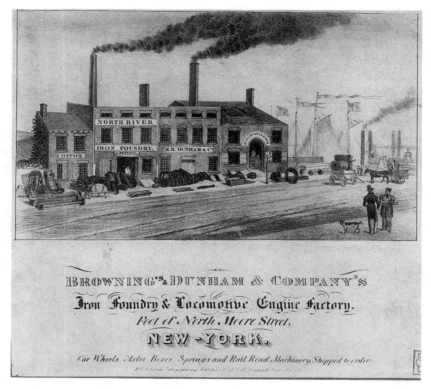

Fig. 113. An 1837 lithograph of H. R. Dunham's foundry.

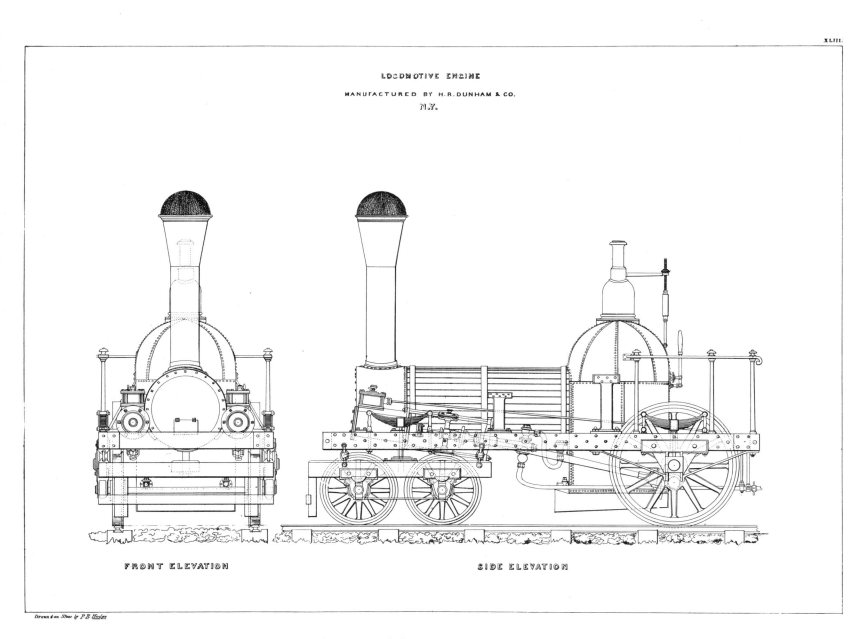

Fig. 114. Front and side elevation of Dunham's locomotive of about 1837.

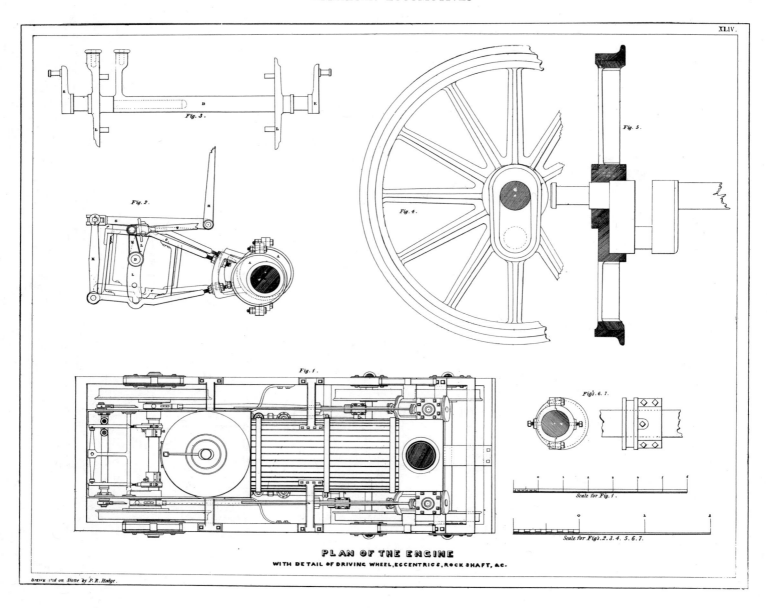

Fig. 115. General plan and details of Dunham's 1837 locomotive.

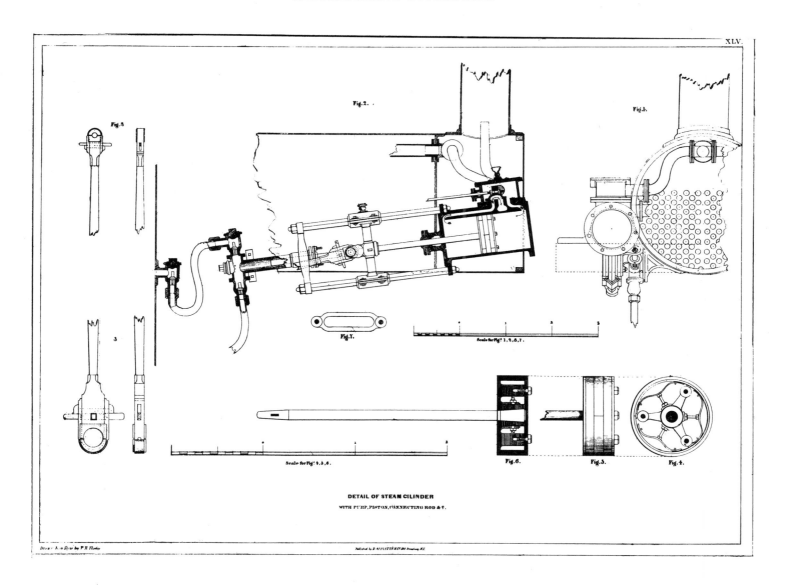

Fig. 116. *Cylinder details of Dunham's 1837 locomotive.*

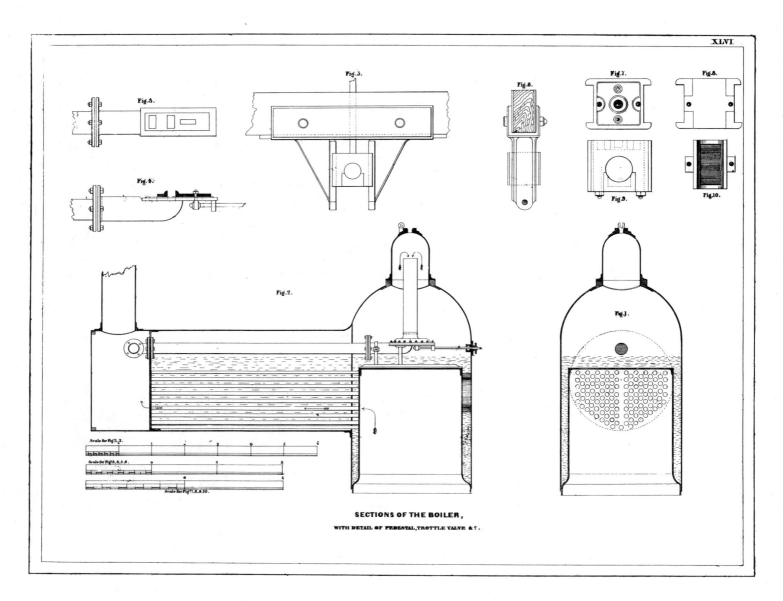

Fig. 117. Boiler and pedestal detail of Dunham's 1837 locomotive.

GOWAN AND MARX, 1839

The *Gowan and Marx*, one of the earliest 4–4–0's constructed, was built for slow-speed coal traffic. It won international acclaim for hauling a train forty times its own weight. Famous in its own time, it remains today one of the best-known early American locomotives.

The Philadelphia and Reading company ordered the *Gowan and Marx* in the summer of 1839 from Eastwick and Harrison of Philadelphia. The machine was named for a London banking firm. A short wheel base, a concentration of weight on the driving wheels, and a large firebox and boiler combined to produce a compact and powerful locomotive. The total weight was 11 tons with 9 tons on the drivers. The wheel base measured 10 feet; the drivers were set only 3 feet 8 inches apart. The engine developed an estimated 147 horse power at its usual operating speed of 8–10 miles per hour.[1]

The records differ on the precise cylinder and driving-wheel size. A report issued by G. A. Nicolls, superintendent of the Philadelphia and Reading, on February 24, 1840, stated that the cylinders had a 12⅜-inch bore and 16-inch stroke and that the drivers were 40 inches in diameter. Karl Ghega, an Austrian engineer who examined the railroads of the United States

Fig. 118. *Joseph Harrison, Jr. (1810–1874), a partner in the firm that built the* Gowan and Marx.

[1] Ghega, *Die Baltimore-Ohio Eisenbahn*.

in 1842, agreed with Nicolls on the cylinder dimensions but stated that the wheels were 42 inches in diameter. Writing in 1872, Joseph Harrison, the engine's designer, recalled a cylinder size of 12½ inches by 18 inches and wheels 42 inches in diameter.

A Bury-style boiler was used, but the firebox was oblong, thus providing a larger fire grate than did the usual round shape. This variety of Bury boiler was used by Eastwick and Harrison on other locomotives. The inside of the *Gowan and Marx* firebox measured about 48 inches long by 36 inches wide. The boiler shell was approximately 40 inches in diameter and contained 129 tubes, each 2 inches in diameter. The steam pressure varied between 80 and 130 pounds per square inch.

The feed-water pump was placed in the cylinder saddle parallel to the cylinder itself. The saddle, warmed by the steam passing to and from the cylinder, preheated the feed water and prevented the pump from freezing in the water. This clever design was not, to the best knowledge of the writer, used by other United States builders. One objection may have been the interference or cramping of the steam passages by placing the pump in the already crowded saddle casting. The pump was driven by the crosshead. The center of the pump and the crosshead guide were about 10 inches apart. This great distance undoubtedly caused severe racking of the crosshead and was unquestionably a constant source of trouble.

The valve gear, patented by Andrew M. Eastwick on July 21, 1835, was favored by Eastwick and Harrison. The valve ports were cast as a separate block and were shifted to reverse the engine. Only two fixed eccentrics were used. The beauty of this design was its simplicity, but it was effective only in forward motion. Because of the peculiar problems created by shifting the ports rather than the valve, the valve had no lead. A small lap was possible on the outside of the valve but not on the inside. This poorly conceived valve operated well enough in forward motion but permitted only slow speeds in reverse. The cutoff, as with other early valve gears, was not variable. The position of the eccentric is an open question. The placement of the rear driving axle under the firebox and the narrow clearances of the frame and boiler make it difficult to determine how an eccentric of sufficient size was attached. The Ghega drawing, Fig. 122, shows the eccentric rod (item 9) extending from inside the frame, but the clearances are not sufficient to permit this placement. The best compromise, but admittedly not entirely satisfactory, is a small eccentric under the firebox.

The Eastwick reversing block was used as late as 1861 on some 4–4–0's built by Harrison, Eastwick, and Winans for the Volga-Don Railway.[2] The addition of a cutoff valve was an important modification of Eastwick's earlier arrangement.

The most important mechanical feature of the *Gowan and Marx* is the equalizing lever. The device was basic to the success of the coupled locomotives and was used from its inception in 1837 until the end of steam locomotive construction. Andrew Eastwick originated the idea of equalizing in his patent of November 20, 1837 (No. 471), but was unable to develop a practical design. His partner, Joseph Harrison, perfected a workable design and secured a patent on April 24, 1838. The heavy cast-iron beam used on the *Gowan and Marx* is representative of the earliest type of equalizer used in the United States. The Peoples Railway No. 3 exhibited at the Franklin Institute has equalizing levers of this type and was consulted for the *Gowan and Marx* reconstruction (Fig. 126).

No data is available on the tender used for the *Gowan and Marx* except that it weighed 6 tons. It was undoubtedly a common four-wheel U tank tender of the period. A reconstruction of this type of tender appears in Fig. 125.

The reconstruction drawings of the *Gowan and Marx* shown in this volume as Figs. 123–125 were based primarily on the Ghega drawing. Another contemporary drawing was prepared in 1841 by Enoch Lewis, an employee of Eastwick & Harrison.[3] This drawing is preserved in the John B. Parson Collection at Columbia University. The Lewis and Ghega drawings agree exactly in general arrangement and boiler details but differ markedly in the running-gear detail. A riveted frame with massive plate pedestals is shown. The wheels are T-

---

[2] *The Engineer*, March 12, 1880, p. 199. In this article Ross Winans was incorrectly credited with building locomotives in Russia.

[3] The Lewis drawing was reproduced in *A Century of Reading Company Motive Power*, p. 12.

rather than oval-spoked. A double-hook valve gear is shown rather than the Eastwick arrangement. The truck has a bar frame with individual springs for each wheel in place of the "spring" truck shown in the Ghega drawing. In short, the Lewis drawing differs enough from the Ghega drawing to indicate a rebuilding of the *Gowan and Marx* between 1839 and 1841. A third contemporary drawing of the *Gowan and Marx* is in a French publication of 1843.[4] The engraving (see Fig. 121) agrees more closely with the Lewis drawing and the *Ontalaunee* (Fig. 120). The most notable mechanical variation from Ghega's drawing is the large spring used in place of the equalizing lever. Other minor variations will be seen in the frame, truck, valve gear, and feed pumps.

Having briefly described the *Gowan and Marx,* a few notes will be added concerning its history. The *Gowan and Marx* was the eighteenth locomotive completed by Eastwick and Harrison. It was used to pull the first train between Reading and Philadelphia on December 5, 1839. On February 20, 1840, it pulled a train of 101 four-wheel cars weighing 423 tons from Reading to Philadelphia. This remarkable performance established a permanent place in railroad history for the *Gowan and Marx.* It so enhanced the reputation of Eastwick and Harrison that they were invited to Russia to build locomotives for the Moscow and St. Petersburg Railway. Curiously, despite their reputation, Eastwick and Harrison were not large builders. Before leaving the United States in 1844, they had built fewer than 50 locomotives.

The *Gowan and Marx* was originally designed to burn hard coal, but did not prove a success with this fuel. Wood was used until January, 1855, when the engine was again converted to burn coal. The conversion was not entirely successful, for the 1857 Philadelphia and Reading annual report notes that the *Gowan and Marx* was again rebuilt in August, 1856.[5] Its weight was reported to be 13.8 tons, which indicated some major alterations. An extension of the firebox similar to that added to the Peoples Railway No. 3 (Fig. 126) would explain in part the additional 2.8 tons. This boxlike projection enlarged the grate area enough to permit successful coal-burning.

After twenty years and over 144,000 miles of service on the Reading, the *Gowan and Marx* was traded to the Baldwin Locomotive Works as partial payment for a new locomotive. The old engine was probably sold off to work out its last days in some obscure corner of industrial America. Its final disposition is unknown.

[4] *Annales des Points et Chanssee's* (2nd series; Paris, 1841–45), Plate 47.

[5] Annual report of the Philadelphia and Reading for 1856, table.

Fig. 119. An advertisement from the 1839 Philadelphia directory.

Fig. 120. The Ontalaunee, *built for the Philadelphia and Reading Railroad by the Newcastle Manufacturing Co. in 1843. Note its similarity to the* Gowan and Marx.

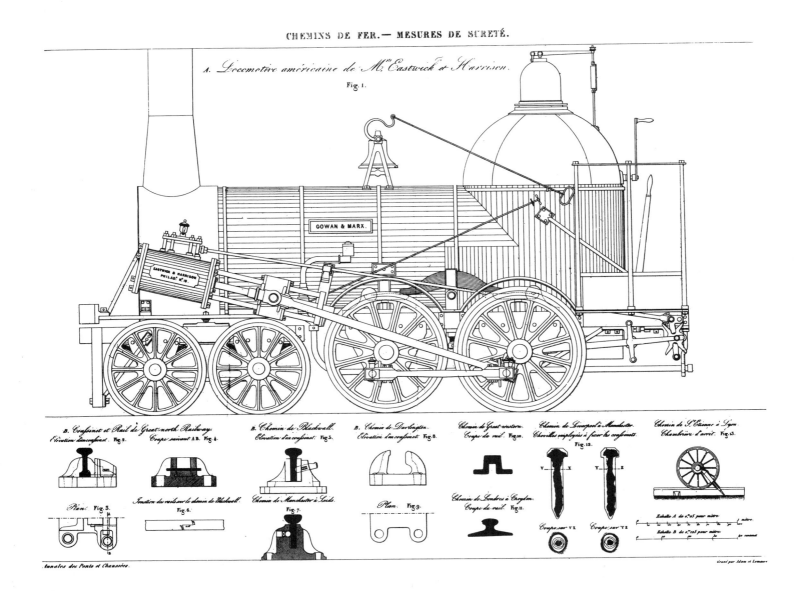

Fig. 121. A French engraving of the Gowan and Marx published in 1843.

AMERICAN LOCOMOTIVES

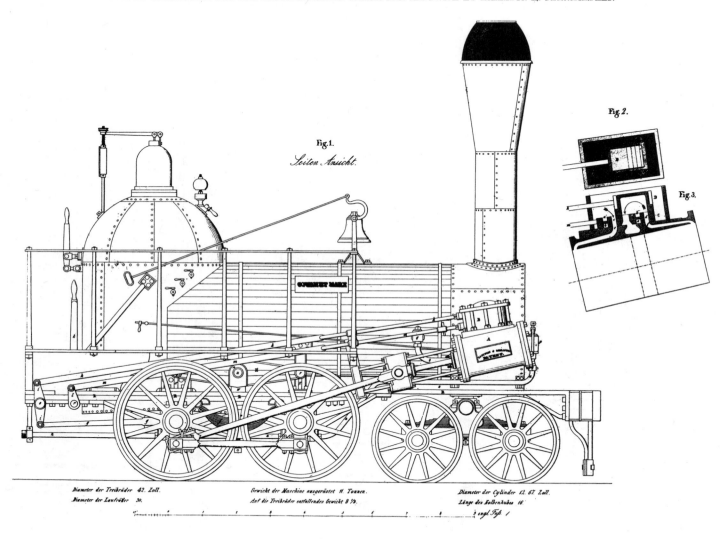

Fig. 122. *A side elevation of the* Gowan and Marx.

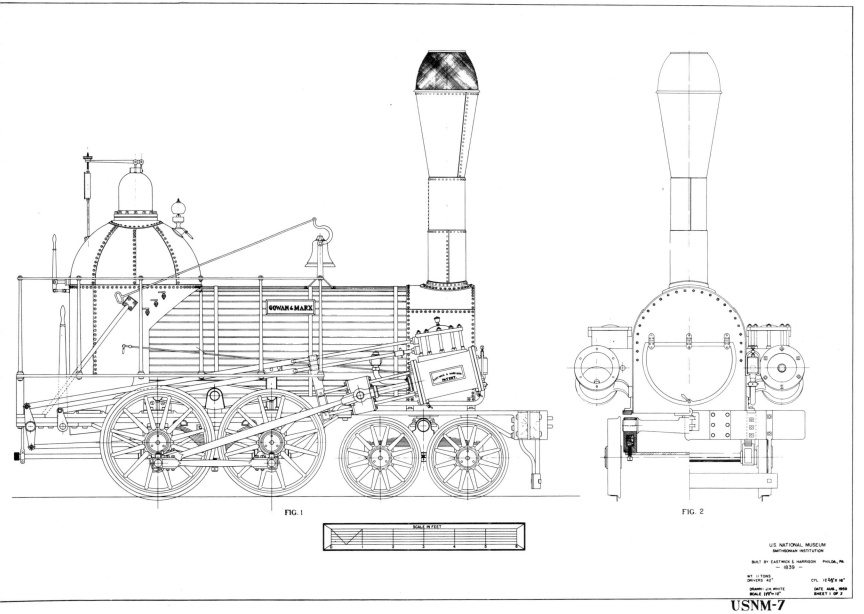

Fig. 123. *A reconstruction drawing of the Gowan and Marx showing side and front elevations.*

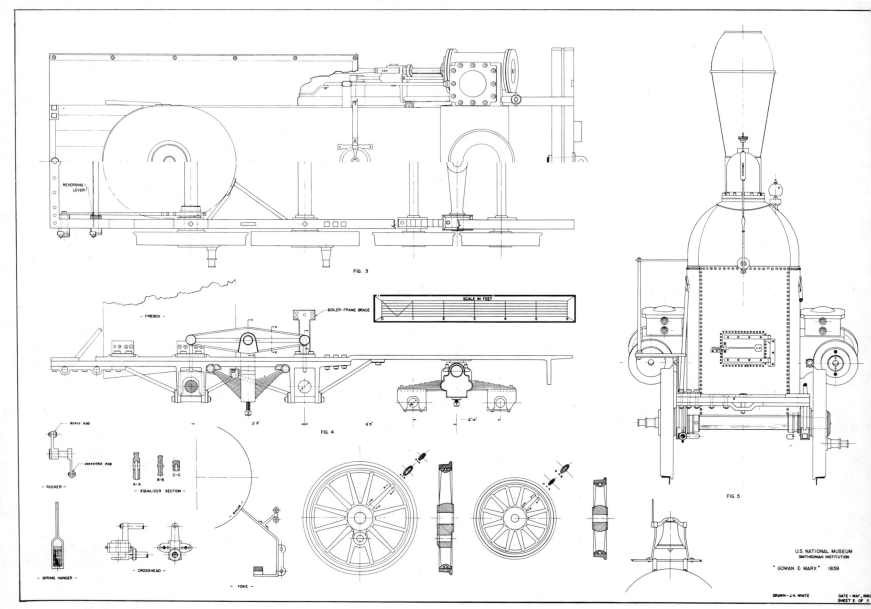

Fig. 124. A reconstruction drawing of the Gowan and Marx showing details of the engine.

REPRESENTATIVE LOCOMOTIVES

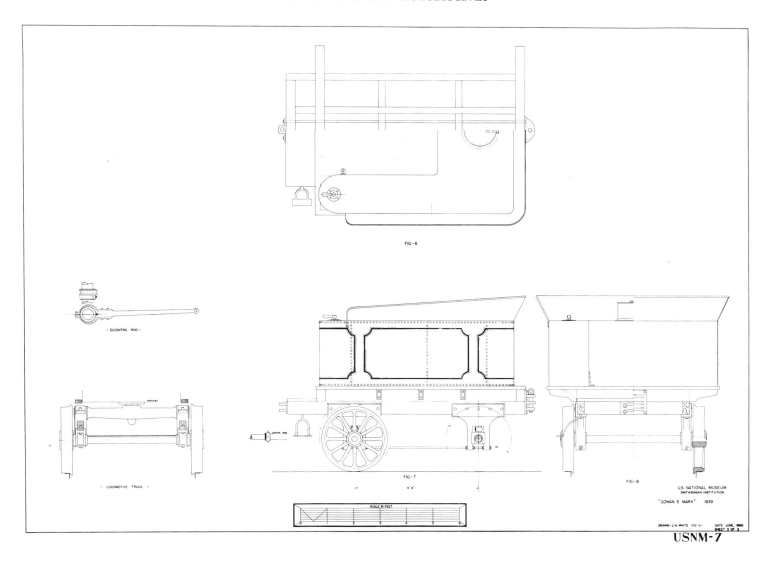

Fig. 125. A reconstruction drawing of the style of tender used on the Gowan and Marx.

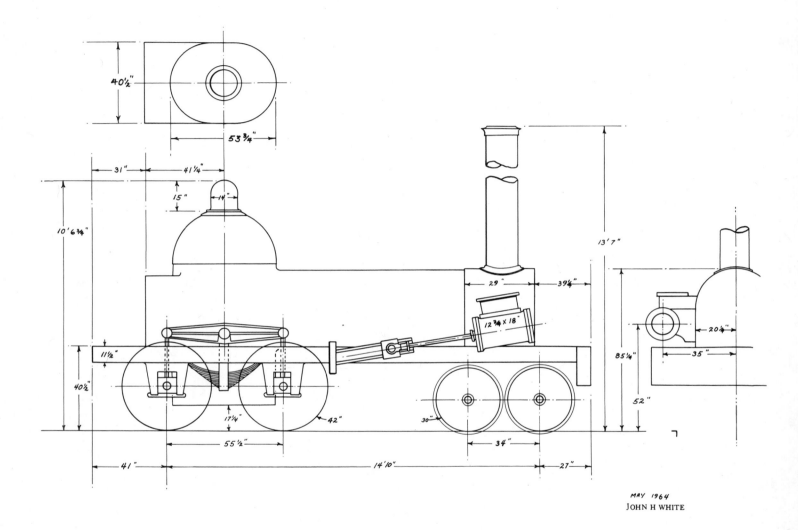

Fig. 126. *The Peoples Railway's No. 3, now exhibited by the Franklin Institute. The exact identity of this ex-Reading locomotive is unknown, but it was probably built by Eastwick and Harrison in about 1842. Note the firebox extension on the rear of the boiler and the cast-iron equalizing lever.*

WINANS' 4-4-0, 1843-49

Ross Winans, usually associated exclusively with the Camel locomotive, built at least five American-type locomotives. These can be accounted for as follows:

Atalanta	October, 1843	Rebuilt in 1853
Reindeer	December, 1845	Rebuilt in 1854
Rough and Ready	June, 1847	South Carolina Railroad
Juno	January, 1848	Rebuilt in 1856
Major Whistler	January, 1849	Northern Central Railroad

All were built for the Baltimore and Ohio except the *Rough and Ready* and the *Major Whistler*. The five engines are believed to have been built according to the same drawings, although there was a slight variation in the wheel and cylinder sizes.

Figures 127 and 128 are after drawings by the late C. B. Chaney showing his conception of the *Juno* as originally built and subsequently rebuilt. The unique stack shown in Fig. 127 is based on the widely published drawing of the *Delaware* built by Winans for the Philadelphia and Reading in 1847.

Figures 129 to 130–A are tracings from an old set of prints formerly owned by J. Snowden Bell. Bell presumably found these drawings while employed as a draftsman at the Baltimore and Ohio's Mt. Clare shops in the 1860's. The drawings are an excellent record of the early 4–4–0 but are a better record of the earliest type of eight-wheel tender. Also, notice in Fig. 129 the detail of the link motion. This indicates that interest in this form of valve motion was not entirely dead between James's experiments of 1832 and Rogers' reintroduction of the link in 1849.

The following dimensions are taken from the drawings:

Cylinders	14¼" diameter × 20"(?) stroke
Driving wheel	52"
Truck wheel	29"
Truck-wheel base	33"
Driving-wheel base	54"
Total wheel base	13' 6½"
Tubes	150 (2" diameter)
Tender	1,000 gal. capacity

Fig. 127. *A reconstruction drawing of the* Juno *showing the engine as originally constructed by Ross Winans for the Baltimore and Ohio Railroad in 1848.*

Fig. 128. *A reconstruction drawing of the* Juno *showing the engine as rebuilt by the Baltimore and Ohio Railroad in 1856.*

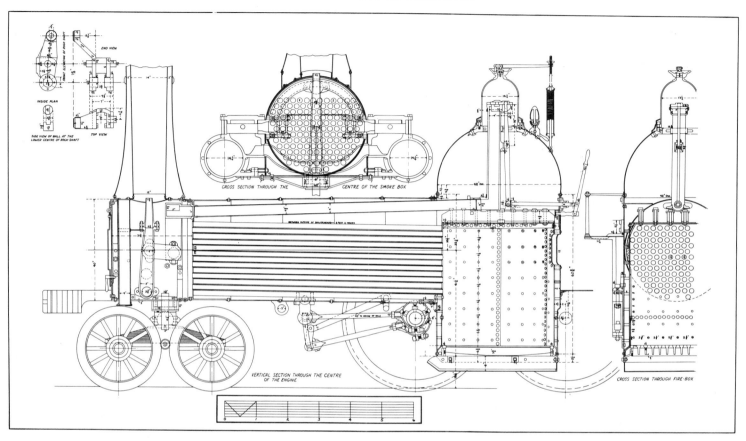

Fig. 129. Detail drawing of a 4-4-0 built by Ross Winans in about 1845.

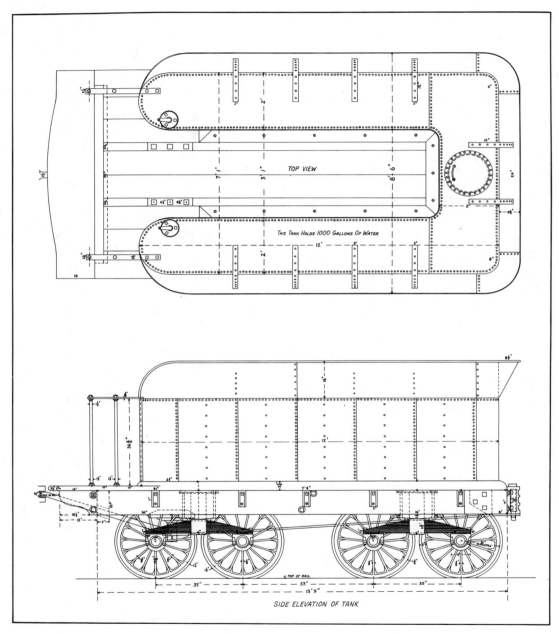

Fig. 130. Detail drawing of an eight-wheel tender built by Ross Winans in about 1845.

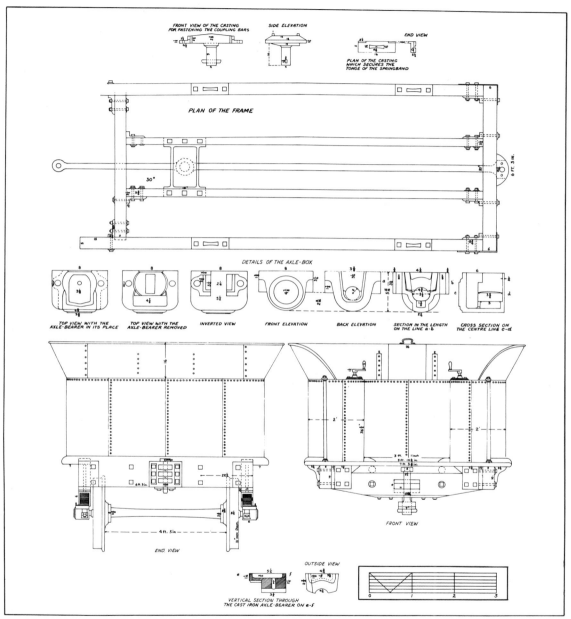

Fig. 130–A. Drawing of an eight-wheel tender built by Ross Winans in about 1845.

PHILADELPHIA, 1849

The *Philadelphia* was a six-wheel wood-burning freight locomotive rebuilt from the *Richmond*, which exploded in August, 1844, shortly after being placed in service on the Philadelphia and Reading Railroad.

The Philadelphia and Reading was built to move coal from the mines near Pottsville to the port of Philadelphia. The road was carefully laid so that the loaded trains could move downgrade while the empty cars were returned upgrade to the mines. Well-constructed track and easy grades permitted 500- to 600-ton trains at a time (the 1840's) when most United States roads rarely operated trains of half that weight. Substantial locomotives were required to handle the heavy trains of the Reading. The 0–6–0's were well suited for this slow but heavy traffic. Not only were new engines of this wheel arrangement purchased, but at least one 4–2–0 was rebuilt as an 0–6–0.[1]

The Norris Brothers of Philadelphia on August 14, 1844, delivered the *Richmond* to the Philadelphia and Reading. It was a standard design with Norris and was designated a "third-class six-wheel combined locomotive." The following description of the *Richmond* is taken from a contemporary account of Dionysius Lardner which is treated more fully later in this discussion.

Whole weight of engine, with wood and water, 36,925 lbs. Cylinders 14½ inches diameter, 20 inches stroke. The boiler contains 127 tubes 2 inches diameter, 4 tubes 1¾ inches diameter, in all 131 copper tubes 11 feet 8½ inches long. The waist of boiler is 40¾ inches diameter. The fire box 40½ inches wide, 40½ long, and 42½ inches in height, and presents a surface to the action of the fire of 48 square feet. Fire surface of the tubes 800 square feet.

The cylinders are attached to the outside of the boiler, and made fast to a frame extending the whole length of the boiler on each side, varying in thickness from 3 × 2½ inches to 4 × 2½. These frames are made fast to the boiler by lugs extending from the fire box end of boiler, and by braces extending from cylindrical part of boiler to the top of frame, secured by turned bolts of 1 inch in diameter.

The boiler, with its frame, cylinders, and all other machinery, rests upon an independent truck frame, constructed of the best fag-

---

[1] The *Neversink* was built by Baldwin in 1836; in 1846 it was rebuilt as an 0–6–0 by the Reading shops. It exploded on January 14, 1847.

gotted iron 4 × 3 inches. In this independent frame are placed the 6 driving wheels, 46 inches diameter; the power is applied direct from the cylinders to the centre pair of wheels, and from them communicated to the forward and back wheels by means of coupling rods. On the dome of the boiler, immediately in front of the engineman, is placed a safety valve of 2 inches in diameter, with lever 25 inches long, fulcrum 2⅝ inches, with a spring balance attached of 48 lbs. graduation. A second safety valve, of the same dimensions, is placed on the cylindrical part of boiler, not within reach of the engineman while standing on the platform.

The fire box is constructed of the best quality iron, 5–16 thick, the crown of which is supported by 8 cast iron stay bars, arched in the centre, and extending to each side of fire box; these bars are 4 inches deep by 1⅜ thick, and secured to the crown sheet by 1 inch bolts.

The new engine gave good service except for the minor problems of repair to a valve rod and sticking valves in the feed-water pumps. On one occasion it pulled a 760-ton train and was calculated to produce about 150 horse power at 10 miles per hour, its normal operating speed.

On September 2, 1844, the *Richmond* left Reading, Pennsylvania, with a train of eighty-eight empty coal cars. Joseph Ward, the engineer, had removed his oilcloth coat (necessary since there was no cab), for the evening's thunderstorm had passed. At about 8:40 P.M., two miles outside of Reading, the *Richmond* blew up. The violent explosion killed Ward and three other crewmen. The boiler was thrown nearly 250 feet.

The builders, at a loss to explain the frightful failure of a locomotive only 19 days in service, hired Dionysius Lardner to make a report.[2] Lardner, the author of several popular books on engineering, was not universally respected as a first-rate authority; his curious conclusion that the explosion was caused by lightning striking the locomotive, supports this opinion.

A second, more factual report was published by the Franklin Institute several months after the disaster.[3] The report dismissed lightning as the cause of the boiler explosion. Defective feed-water pumps and low water on the crown sheet were

[2] Dionysius Lardner, *Investigation of the Causes of the Explosion of the Locomotive Engine Richmond* (Philadelphia, 1844).
[3] *Journal of the Franklin Institute*, Vol. 9 (1845), pp. 16–31.

Fig. 131. *The* Richmond, *built by Norris in 1844 for the Philadelphia and Reading Railroad, was similar to the six-wheel locomotive shown in this advertisement.*

Fig. 132. James Millholland (1812–1875), a locomotive designer known for his development of successful anthracite-burning locomotives, rebuilt the Philadelphia in 1848–49.

blamed instead. The use of cast-iron crown bars was also condemned.

According to the 1845 annual report of the Philadelphia and Reading, the boiler and other salvageable parts of the *Richmond* were reconstructed by the Reading shops as the *Philadelphia* in October, 1844. The engine was remodeled by James Millholland in 1848 or 1849 as shown in Figs. 133–136.[4] These drawings are from Emil Reuter's *American Locomotives,* the earliest work published in the United States devoted exclusively to locomotives. Reuter, a draftsman employed by the Philadelphia and Reading, published only about half of the sixteen projected installments of this work.[5]

The *Philadelphia* of 1848–49 differed from the *Richmond* in its new frame, main-rod-to-rear rather than center drivers, a forward steam dome, and poppet- rather than slide-throttle valve. The wood-burning stack probably was not originally used on the *Richmond.* It appears to have been made after a patented design of James A. Cutting. Cutting obtained two patents for spark arrestors, No. 6559 of June 26, 1849, and No. 8077 of May 6, 1851, both of which enjoyed moderately wide use for wood-burning engines.[6] The most significant alteration made by Millholland was the balanced poppet valve throttle. This is the earliest known drawing of the poppet, which by about 1870 had been adopted as the standard style of throttle valve in this country.

The variable exhaust located in the smokebox and shown in Figs. 134 and 135 (front elevation) is patterned after Ross Winans' patent No. 1868, issued on November 26, 1840. This device was intended for coal-burning and apparently was part of Millholland's unsuccessful attempt to convert the engine from a wood-burner.

The V-hook valve gear was used. The fixed cutoff took its motion from the eccentric rods as shown in Fig. 133. The valve

---

[4] Sinclair, *Locomotive Engine,* p. 288, incorrectly states that the *Philadelphia* was a new engine built by Millholland in about 1849.
[5] *American Railroad Journal,* June 16, 1849, p. 373.
[6] In an article on the *Philadelphia* in *Railroad Gazette,* January 6, 1899, C. H. Caruthers states that a French and Baird stack was used. This is incorrect, for the Reuter drawing shows a stack markedly different from the well-known French and Baird design.

gear differed little from that used on the *Richmond* and other Norris locomotives of the period.

The cylinders were set at a steep angle (about 15 degrees) to the rear driving wheels. Because of the D-shaped smokebox no saddle was used; however, the cylinders were attached to the frame and smokebox by a "box" frame cast integral with the cylinders. This substantial arrangement certainly answered the repeated criticism aimed at weak cylinder attachments before the saddle was introduced in the early 1850's.

A simple canopy cab was used. It had metal sides, no windows, and was open at both ends. The cab is shown in Fig. 135 (rear elevation) but for some unknown reason is not shown in the side elevation, Fig. 133.

No sandbox appears in the Millholland-Reuter drawings, but the *Richmond* was known to have a sandbox immediately behind the stack.[7]

The *Philadelphia* was rebuilt in 1851 and again in April,

[7] *Journal of the Franklin Institute*, Vol. 9, p. 18.

1854, as noted by the annual reports of the company. Nothing is known of the 1851 rebuilding, but the weight of the engine was increased by nearly 2 tons in the 1854 rebuilding. The increase in weight may be explained by the addition of an enlarged firebox for coal-burning similar to that shown by Fig. 126.

The *Philadelphia* continued in service for another twenty years. It had, of course, been retired from main-line service before its scrapping in 1870.

Dimensions of the *Philadelphia*

Weight	18½ tons (only engine in working order)
	20¼ tons (as remodeled in 1848–49)
Cylinders	14½" × 20"
Wheels	46" in diameter
Steam pressure	120 lbs.
Heating surface	848 sq. ft.
Fuel	wood

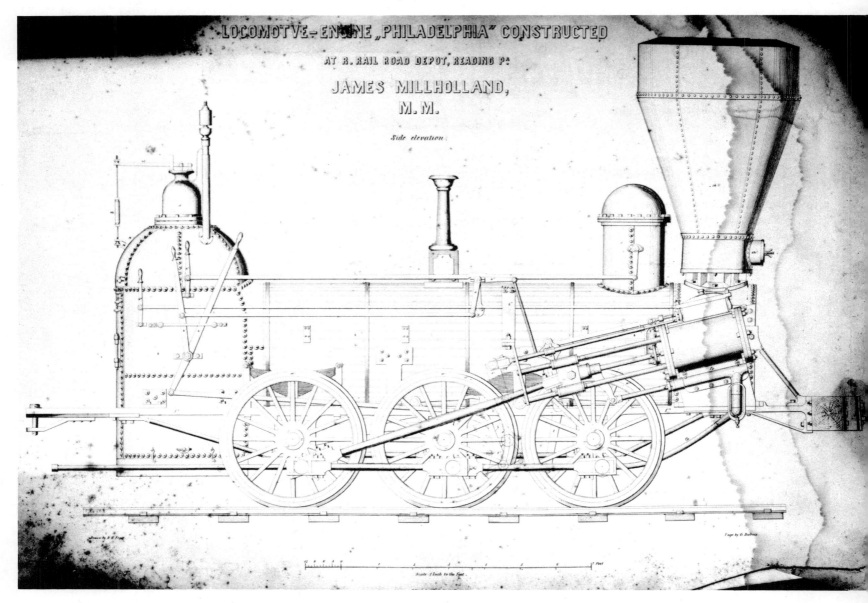

Fig. 133. The Philadelphia, *shown as remodeled by Millholland in 1848 or 1849.*

REPRESENTATIVE LOCOMOTIVES

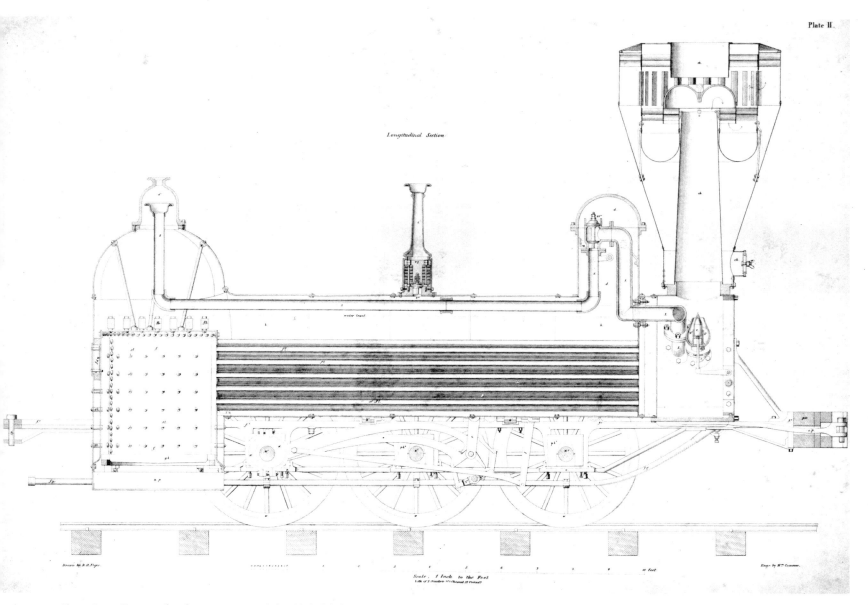

Fig. 134. Longitudinal cross-section of the Philadelphia.

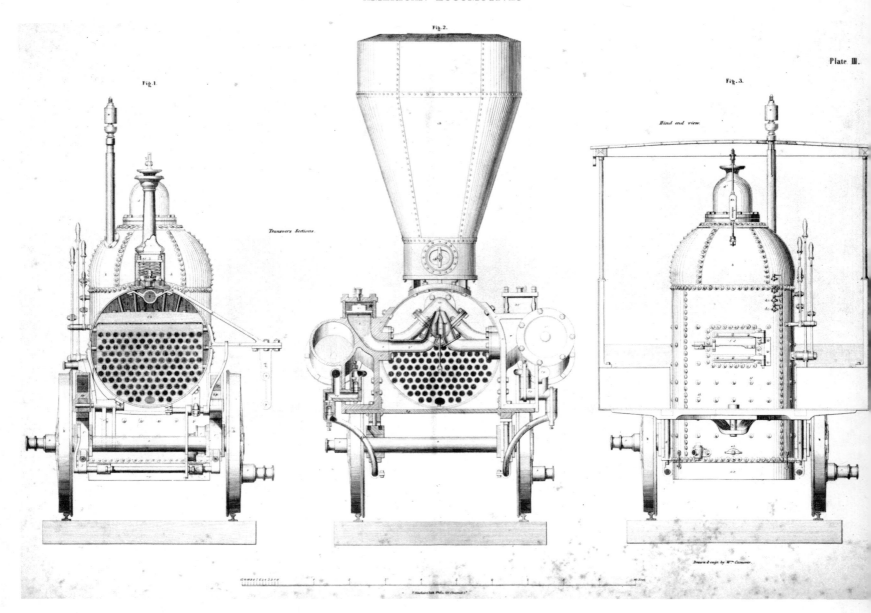

Fig. 135. Cross-section and elevation of the Philadelphia.

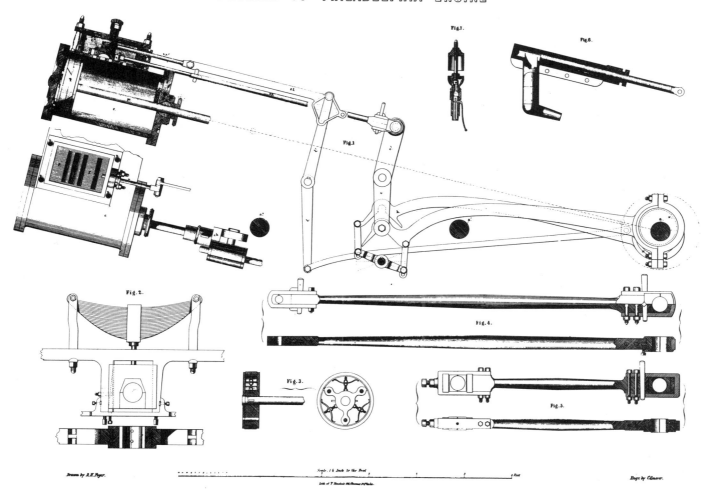

Fig. 136. Details of the Philadelphia.

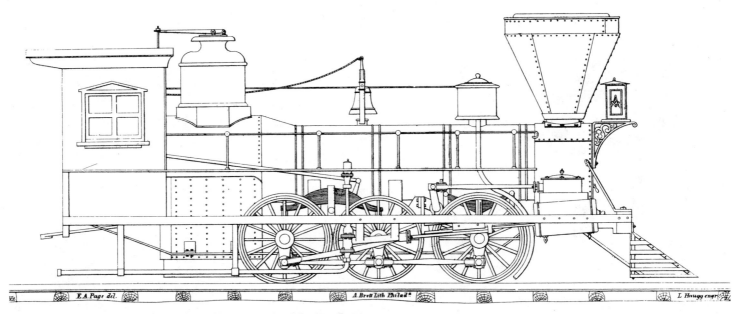

Fig. 137. *A Norris six-coupled freight locomotive of about 1855.*

COPIAPO, 1850

The *Copiapo*, built by Norris Brothers in 1850 and now preserved in Copiapo, Chile, is a unique example of early American locomotive engineering. It is the only surviving 4-4-0 of that period; it is the standard product of a major builder and, unlike most locomotives preserved in the United States, it has undergone little reconstruction. Whereas nearly all of about twenty pre-1860 locomotives existing in this country have been reframed, reboilered, or in other respects have lost their identity through reconstruction, the *Copiapo* remains a remarkably unspoiled specimen with such antique features as a Bury firebox, short wheel-base truck, round side rods, and an independent cutoff valve gear. Because of this combination of features and the rarity of engineering drawings for the period 1845–50, the *Copiapo* serves as the missing link between the earliest pioneers and the reformed American locomotives that emerged during the early and middle 1850's. That it is a standard wheel arrangement with a reasonably well-documented pedigree adds immeasurably to its value in our study of representative machines. Were it an odd wheel arrangement, the product of an obscure maker, or unidentified, as is the Peoples Railway's No. 3 in the Franklin Institute, it would be of little consequence to this study. Considering its obvious rarity and importance, it is surprisingly relatively unknown in this country.

The machine was constructed for the Copiapo Railroad, which was built to accommodate the tapping of rich silver and copper deposits near Copiapo in north-central Chile, a barren rocky region at the southern end of the Atacama Desert, about 500 miles north of Santiago. The line, often cited as the first in South America, runs fifty miles between the mining center and the small port of Caldera. It was conceived of by William Wheelwright (1798–1873) an energetic Yankee promoter who pioneered steam navigation in the Pacific and was involved in many Latin American enterprises.[1] Wheelwright employed Allan and Alexander Campbell of Albany, New York, to build the road. The contractors sailed for Chile with

---

[1] Wheelwright was the subject of an eulogistic biography by Juan B. Albérdi, *The Life and Industrial Labors of William Wheelwright in South America* (Boston, 1877), and was regarded as a benefactor rather than an exploiter.

"artizans and mechanics of every description, as well as locomotives and the rolling stock of the road."[2] Work began at the Caldera end of the line in March, 1850, and the road opened on Christmas Day of the next year.[3]

Built at a total cost of 1.4 million dollars, the line had a maximum grade of 60 feet per mile and made a total climb of 1,327 feet from the sea to Copiapo.[4]

The locomotive continued in service for forty years, traveling a total of about 190,000 miles.[5] The 1891 Chilean revolution, which included the shelling of Caldera and damage to the railroad by the insurgents, led to the conversion of the standard-gauge road to the meter gauge. The other railroads of the region were the meter-gauge variety; thus, the necessity to make major repairs presented a logical opportunity to regauge the Copiapo line. The *Copiapo* was retired at this time, but because of its historic value it was preserved. In 1894 it was exhibited at the Santiago Exposition of Mines. In 1901 it was sent to Buffalo, New York, as part of the Chilean exhibit at the Pan-American Exposition.[6] In 1929 the venerable machine was put under steam to pull a special train of delegates attending the South American Railway Congress in Santiago.[7] Reportedly, only a few minor repairs and new boiler tubes were needed to put the machine in good working condition. At this time, and presumably during previous years, the *Copiapo* was exhibited at the State Technical University in Santiago. In about 1945 the locomotive was returned to Copiapo for exhibit at the State School of Mines. Unfortunately, the tender disappeared at this time; no record of its disposition can be found. The machine is displayed in the open courtyard of the school. The school itself is only a few hundred feet from the railway.

Mechanically the engine exemplifies the usual construction practices of American builders for the period. It is not a specialized or non-standard machine built for export. A comparison of the *Copiapo*, the *Tioga* (Fig. 83), and the *Beaver* (Fig. 141) will show that in addition to the general arrangement it was in almost every detail a standard Norris product. The survival of this machine and a detailed inspection of its construction have greatly broadened our understanding of locomotive construction as practiced during these early years.

The boiler is on the Bury plan. The waist is 38 inches in diameter and contains 113 tubes. The precise diameter of the waist was difficult to determine because of the jacketing; a diameter of 36 inches has been stated by the Chilean Railways, but this is decidedly too small. The plate is in excellent condition; this and the long life of the boiler may be credited to the use of treated water. The water of the region has a high lime content; hence a distilling plant for pure boiler water was established during the first year of the railway's operation.[8] The small dome atop the Bury firebox is cast brass. It is fitted for a single safety valve; the valve, valve lever, and spring balance are presently missing. This and the throttle lever, also missing, were restored in the drawings (Figs. 142–144). Inside the cab are the remains of an early diaphragm steam gauge. It is uncertain that this gauge was original to the engine, but its general construction indicates it was an early addition. The latch on the fire door is a detail worthy of mention. Notice that it is spring-riveted at the far end and thus is under pressure to hold the door closed (Fig. 144).

The smokebox front is an ornamental iron casting and, as is clearly visible in Fig. 138, has the name of the maker and the

---

[2] *Ibid.*, p. 119.

[3] *Los Ferrocarriles de Chile* (Santiago, 1912), pp. 51–53. This same account states that the locomotive was landed at Caldera by the American frigate *Switzerland* in June, 1851. However, as previously noted, the Wheelwright biography states that the locomotive and rolling stock were brought with the contractors. The engine itself is dated 1850.

[4] *American Railroad Journal*, July 23, 1853, p. 476, notes the cost of the line.

[5] *Los Ferrocarriles de Chile*, p. 53.

[6] A photograph of the *Copiapo* appears on p. 28 of the official Chilean Government Catalogue for the Pan-American Exhibit. The engine is in the same condition as shown in the early view, Fig. 1, except for the addition of the headlight and a good coat of paint. The wooden cab and tender remain intact.

[7] *Baldwin Magazine*, October, 1940, pp. 11–13; a photo included in this article shows the engine in an open-roof shed. The iron cab, buffers, and headlight had been added. The tender and several primitive four-wheel cars were shown with the locomotive.

[8] Albérdi, *The Life of William Wheelwright*, p. 120.

date of construction in raised letters on its face. A heavy bed plate running between the frames forms the bottom of the smokebox and the main cross-brace for the frame. Two massive cast-iron legs projecting downward from the same casting support the truck's center pin. The smokestack appears a bit undersized and is very likely a late replacement.

The cylinders are cast separately and are mounted at a slight angle to the axles. The stroke of 26 inches seems unusually long; however, in an 1845 advertisement Norris offered engines with a 24-inch stroke as the standard size.[9] The mounting brackets and steam and exhaust pipes are cast integral with the cylinders. The valve box and cover are separate castings held in place by six bolts. No jacketing was used; there is evidence that the cylinder heads were brightly finished. The valve oil cups are missing but a threaded hole in the valve cover's center indicates their original presence.

The valve gear is an independent cutoff fitted with a riding valve. Six eccentrics are employed. To the author's knowledge, this is the only surviving example of the American cutoff valve gear; as a general form of valve gear it was at one time a very important feature of American locomotive design. Considering the long service of the *Copiapo* it is surprising that the clumsy cutoff gear was not replaced by link motion as it certainly would have been had the machine been in use on an American road. A description of the Norris independent cutoff is given in the valve-gear section (see p. 190).

The feed-water pump has an exceptionally long body to accommodate the long stroke of the engine. It is of cast iron construction; the air dome is of the same material. The check valves are not original and have a homemade look. They were probably applied during the 1929 repairs.

The pilot appears to be original; slender, round bar pilots were popular in the early 1850's. Unfortunately, the draw bar shown in Fig. 138 was removed after the engine's retirement. The buffers, hook coupler, and iron deck over the front of the engine are additions made after the Pan-American exhibit in 1901.

[9] This advertisement is reproduced in the *Philadelphia* section as Fig. 132.

The sandbox is of the old, true, box style. It was, before 1855, the most popular type of sandbox in use. The paneled sides are of cast iron, the ends of heavy sheet iron; the entire top is hinged as a cover.

The bell and whistle were reconstructed (see Fig. 142); however, the original stands of both are intact.

A very light one-piece bar frame with a top rail of only 2 inches by 4 inches is used. The forward section between the front pedestal and the cylinders is even lighter, measuring only 3¼ inches in width. As can be seen in Fig. 143, a single spring is used for the driving axles. The spring hangers are rigidly bolted to the equalizer, but a certain amount of motion is possible because of the ball-and-socket bearing at the top of the spring strap. This arrangement is shown by other Norris locomotive drawings dating from the late 1840's, but it appears to have been abandoned for the more conventional two-spring plan in the early 1850's. It might be well to mention that the rear cross-brace of the frame is a heavy iron casting that doubles as the cab deck. It runs across the locomotive and is bolted to the tops of each frame. The rear "snubbers," or stops for the equalizing lever, are cast integral with this casting, as are the heavy mounting brackets for the draw bar. Even the fulcrum for the reversing levers was a boss on this casting.

The driving wheels are made in the T spoke pattern. The spokes are very heavy in cross-section and have almost no taper. The double crank-pin boss was peculiar to Norris engines in this country, but a number of British builders followed the plan in the nineteenth century. The idea was that the crank pin could be pressed in the second hole if it worked loose from the first. This may have ensured double life for the wheel center, but it made a heavy, cumbersome wheel. The counterweights were bolted in place as shown. The wheel covers are of light sheet iron. A decorative border of polished sheet brass was attached with flush rivets. Note the similarity of this detail to the wheel covers of the *Beaver* (Fig. 141).

The leading truck has a moderately short wheel base of 42 inches. The 28-inch spoke wheels are identical in pattern to the driving-wheel centers, although it is doubtful that they are original, considering the engine's mileage. The general plan of the truck's frame is similar to the McQueen truck in that two pieces of boiler plate form the bolster (see Fig. 143). The axle-box pedestals are of cast iron while the top and bottom rails are wrought-iron bars. The forward weight of the engine is carried by the two cast-iron side bearings that chafe the underside of the frame. The size of the bearing surface is large and the truck's freedom in turning was surely restricted. Note that the side bearings are hollow so that the equalizing lever may pass through. It is possible that the equalizing lever was a later addition and that originally the spring was mounted in the hollow opening of the side bearing (above the wheels) as shown in the trucks of the *Licking,* Fig. 16, and the Baldwin 4–4–0 (*ca.* 1848), Fig. 26.

The mechanical drawings reproduced in this section were prepared after a first-hand inspection of the locomotive in August, 1965. The author is indebted to Oscar Barrios for some of the sketches and dimensions taken and to Carlos Arriagada, director of the School of Mines, as well as to other members of his staff for their thoughtful assistance.

Dimensions of the *Copiapo*

Cylinders	13″ bore by 26″ stroke
Wheels	60″
Heating surface	*ca.* 645 sq. ft.
Weight (engine)	19 tons
Weight (tender)	16 tons

REPRESENTATIVE LOCOMOTIVES

Fig. 138. The Copiapo, *built by Norris Brothers in 1850 for the Copiapo Railway, Chile. An early photograph taken at about the time of its retirement in 1891 shows wooden cab, tender, and other parts now missing.*

Fig. 139. The Copiapo *as presently exhibited at the School of Mines, Copiapo, Chile. The iron cab and headlight were added after the engine's retirement.*

AMERICAN LOCOMOTIVES

Fig. 140. The Copiapo *on exhibit in Copiapo, Chile. The buffers and hook coupler are later additions.*

Fig. 141. The Beaver, *built by Norris Brothers in 1849 for the Pittsburgh, Fort Wayne and Chicago Railroad. Note that this machine agrees with the Copiapo in almost every detail.*

REPRESENTATIVE LOCOMOTIVES

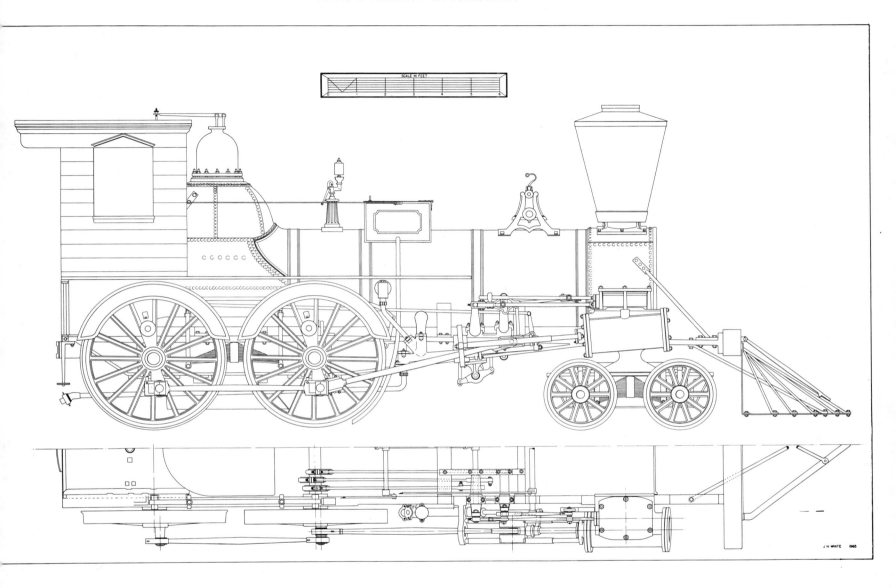

Fig. 142. Side elevation and plan view of the running gear of the Copiapo, 1850.

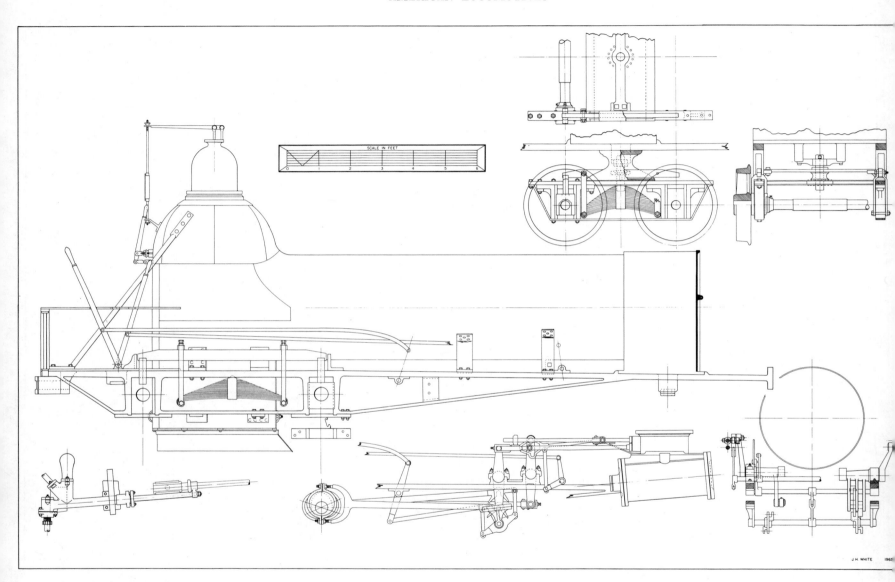

Fig. 143. Frame, truck, and valve-gear details of the *Copiapo*, 1850.

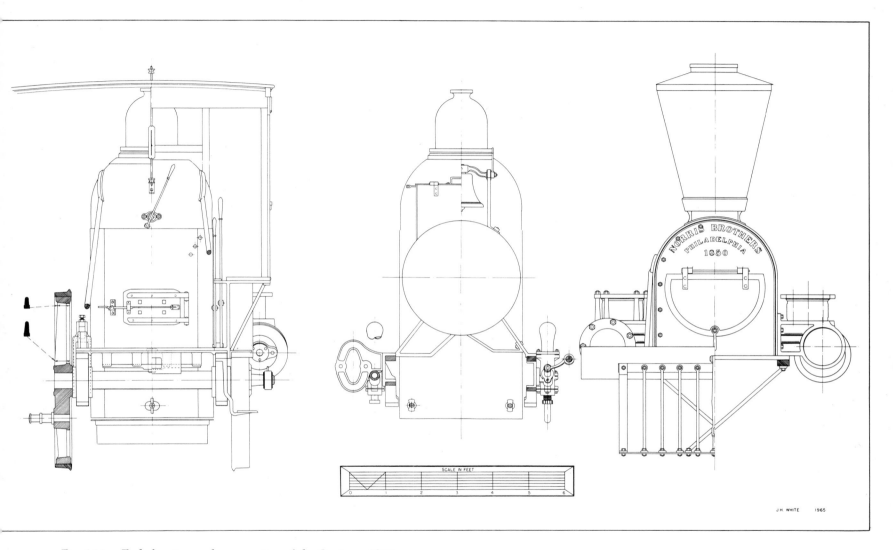

Fig. 144. End elevation and cross-section of the Copiapo, 1850.

FOUR FAST PASSENGER LOCOMOTIVES OF THE HUDSON RIVER RAILROAD

As early as 1832 a railroad was projected up the Hudson River from New York to Albany. The project failed to attract any support and few investors then or in subsequent years believed the public could be lured away from the swift steamers connecting these two principal eastern cities. The elegant Hudson River steamers were renowned for their comfortable cabins, luxurious appointments, and great speeds often as high as 25 miles per hour. Fares were low because of bitter competition. Add to all of this the magnificent scenery of the Hudson River Valley and what chance would a rattling, jolting train have to attract passengers?

Thus the project lay moribund until 1845 when some aggressive pushers in New York City agitated for a new charter. After incorporating the Hudson River Railroad in May, 1846, the promoters wisely selected John B. Jervis as their chief engineer. The task of building a railroad 143 miles up the east bank of the Hudson River to Albany would seem at first an ordinary task. But the Hudson River Railroad was to be no ordinary line. It was to be built for high speed service with heavy rail, sweeping curves (less than 2,800-foot radius), and easy grades. Cost per mile was $80,000, more than twice the national average. To win patronage away from the river, trains had to substantially better the time of the best steamers.

Express trains were scheduled to run at average speeds of almost 50 miles per hour, thus cutting several hours from traveling time between New York and Albany. The Hudson River Railroad was, when fully opened in October, 1851, the first high-speed line of any size to operate in the United States. It was more a model of British than American engineering.

Magnificent as the line was, it lost money steadily until the 1860's. Fares were kept low to compete with the steamers. Locomotives, cars, and roadbed repairs were large because of the high-speed operation. The Vanderbilt interests began buying into the line at about this time and saw to it that it made money. It was consolidated with the New York Central in 1869.

The first locomotives of the Hudson River Railroad were ordinary machines for the times, a good example being the *Champlain* which is treated in more detail in the next chapter.

The *Croton*, while specifically built for local trains, was also a conventional inside-connected passenger locomotive of the period. The management of the road soon became aware of the need for special locomotives to maintain its ambitious schedule. Initial selections were not favored with great success, however. One high-wheeled, single driving axle (probably a 4-2-2) was built by Thomas Rogers in 1849. In 1850 Baldwin followed with the *Susquehanna,* a 4-2-2 modeled after several other machines made by the same builder for the Vermont Central and Pennsylvania railroads. Both machines were deemed lacking in tractive power. The Rogers engine, the *Pacific,* was rebuilt as a 4-4-0. The fate of the *Susquehanna* is unknown but it was probably returned to its builder.

The failure of the "singles" prompted the Hudson River line to modify the faithful 4-4-0 locomotive for high-speed work. Accordingly, Walter McQueen, master mechanic of the Hudson River Railroad, prepared drawings for a large 4-4-0 with 78-inch wheels. Four of these express engines were completed by Wilmarth in time for the road's opening. They proved to be a great success and were followed by even larger engines of the same design, the *Columbia* and the *Rensselaer*. These and the other express locomotives of the Hudson River should not be viewed as freaks or oddities. Aside from their size they followed standard locomotive practice very closely and represent only an extension of orthodox design to meet an unusual speed requirement. The next few chapters deal with several high-speed engines of the Hudson River Railroad.

CHAMPLAIN, 1849

The *Champlain* was built for the Hudson River Railroad in December, 1849, by the Taunton Locomotive Manufacturing Company of Taunton, Massachusetts. It was a standard Taunton inside-connection engine and was probably of the largest size built by that concern. The general design is a slavish copy of the engines built by George S. Griggs, master mechanic of the Boston and Providence Railroad. Griggs was an early and positive advocate of inside connection; he developed a design in 1845 that was taken up by most New England builders as the *sine qua non* in locomotive perfection. Taunton was even more strongly influenced by this prevailing admiration for Griggs's design since the gentleman in question was a stockholder. In fact, the first Taunton engine completed in 1847 was built from castings and drawings supplied by Griggs.[1]

Several hundred locomotives of this style were built by Taunton until about 1857 when at last it adopted the more progressive outside connection, spread truck, and bar frame.

Even though the *Champlain* was built for ordinary service, it performed well, at least for a time, with the high-speed trains of the Hudson River Railroad. On one occasion, in the summer of 1850, when traffic was heaviest, the *Champlain* pulled an eleven-car passenger train weighing 120 tons at an average speed of 44 miles per hour. Between December, 1849, and March, 1851, the *Champlain* traveled 45,111 miles, averaged 2½ cents per mile for repairs, and consumed one cord of wood every 30 miles. Despite this very creditable performance the *Champlain* was for some unknown reason returned to its builder in 1852 where it was rebuilt and sold to the New York and Harlem Railroad as the *Seneca*.

Before beginning the mechanical description of the *Champlain* the reader is referred to Figs. 146–148, which were taken from the *Practical Mechanics Journal* (Glasgow).[2] Most of the particulars mentioned above are from the same journal.

The boiler is remarkably similar to the style developed by Robert Stephenson in 1830. It is of course larger in all dimensions, but the shape of the firebox wrapper of the boiler, rising

---

[1] *Locomotive Engineering*, July, 1892, p. 228.

[2] III (1850) and IV (1851).

just a few inches above the main shell of the boiler, duplicates the construction of the boiler used early in 1830 on Stephenson's Planet class. Notice also that the firebox and the shell of the boiler are bluntly joined with an angle-iron connection. Also in keeping with Stephenson's practice is the manhole (the low squatty dome) placed atop the firebox end of the boiler. The firebox measured 41½ inches long by 39⅜ inches wide on the inside. The boiler shell was 42 inches in diameter and contained 144 tubes that measured 1¾ inches in outside diameter by 11 feet 3 inches in length. The total heating surface was 824.43 square feet. The steam dome was near the forward end of the boiler. The dome was cylindrical in shape and was housed in an ornamental box after the Egyptian style.

The valve motion has a cutoff set at one-half stroke. The cutoff was driven by a crank attached to the front side-rod pin while the regular valve was driven by the orthodox plan of V hooks. Notice that the cutoff in this case is not a "riding cutoff" since the valve has its own seat. The lead of the main valve was 3/16 of an inch and its lap ⅝ of an inch. The steam ports were 1 inch by 14 inches; the exhaust port was 2 inches by 14 inches.

A riveted frame was used. The top rail was formed from two iron plates ⅝ of an inch thick by 5 inches deep with a 2-inch-square bar riveted between them. The cast-iron pedestals are riveted between the top-rail plates as well. A 1¼-inch bolt running between the two pedestals formed the bottom rail of the frame. A light, supplementary outside frame supporting the running board was made of 3-inch angle iron.

The driving wheels were of the common T spoke pattern. The leading truck wheels, 32 inches in diameter, were of Whitney's cast-iron corrugated plate pattern (see Fig. 149).

No drawings exist for the tender but it was on eight wheels and held 1,500 gallons.

Dimensions of the *Champlain*

Cylinders	15″ bore × 20″ stroke
Wheels	66″ diameter
Weight	23½ tons
Grate area	11.34 sq. ft.
Total heating surface	824.43 sq. ft.
Tubes (144)	1¾″ outside diameter × 11′13″ long

Fig. 145. The Gasconade, *built by Taunton in 1853 for the Pacific Railroad, Missouri. This engine, while lighter, was in general plan identical to the* Champlain.

REPRESENTATIVE LOCOMOTIVES

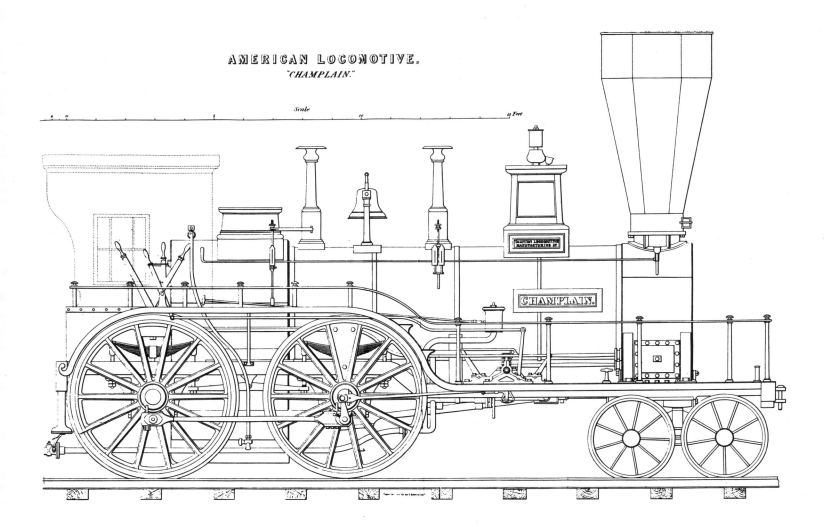

Fig. 146. The Champlain, *built by Taunton in 1849 for the Hudson River Railroad.*

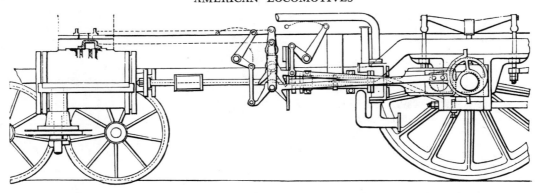

Fig. 147. *A detail of the Champlain's frame and valve gear.*

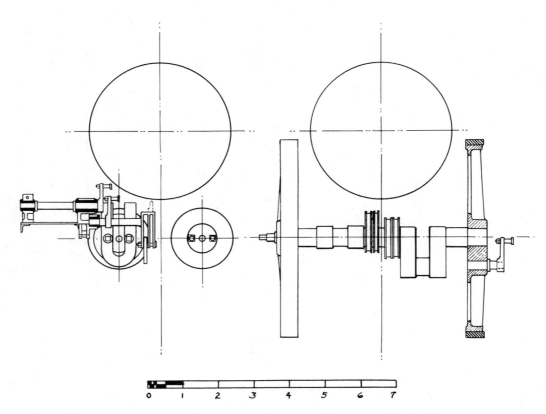

Fig. 148. *Two cross-section elevations of the* Champlain *showing crank axle, driving wheel, and yoke.*

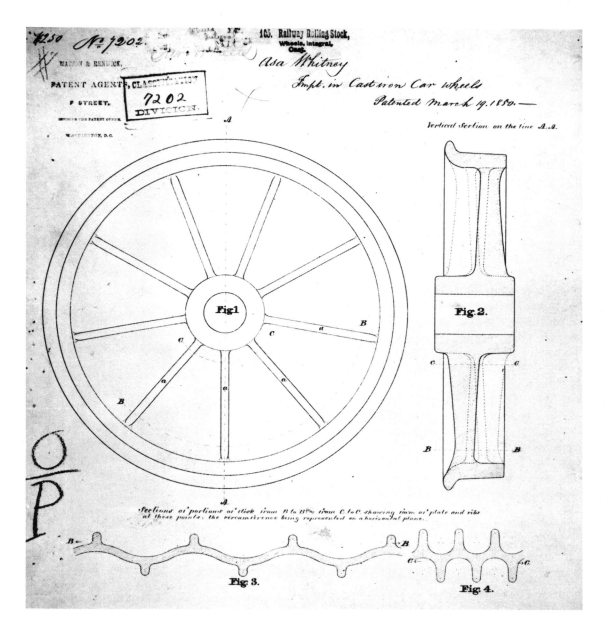

Fig. 149. Patent drawing (No. 7202, dated March 19, 1850) for the type of truck wheel used on the Champlain. Known as a corrugated wheel, it was patented and manufactured by Asa Whitney of Philadelphia.

CROTON, 1851

This small, inside-connected eight wheeler was built for light local trains. It was designed by Walter McQueen and built for the Hudson River Railroad by the Lowell Machine Shop in the summer of 1851. The *Croton* was Lowell's ninety-first locomotive and was one of six nearly identical engines built for the Hudson River Railroad between July, 1851, and November, 1852.[1]

The *Croton* made its first run on July 24, 1851, and proved to be quite a serviceable machine for local trains between New York City and Poughkeepsie. In August, 1851, it pulled a 90-ton train between these two cities; the train consisted of six well-filled passenger cars, one baggage car, and one sprinkler car. Despite the heavy load, the *Croton* made the run at its usual average speed of about 40 miles per hour.[2]

Fig. 150. *Walter McQueen (1817–1893), designer of the* Croton.

[1] Edwin R. Clark, in *Railway and Locomotive Historical Society Bulletin*, No. 7 (1924), p. 49, incorrectly includes the *Albany* in the Croton class of engines. He also lists a *West Point* which cannot be verified. According to the New York State Railroad Report, 1856, the six engines in this class were the *Croton, Spuyten Duyvil, Kinderhook, Matteawan, Sing Sing,* and *Peekskill.*

[2] Moné, *Outline of Mechanical Engineering;* the pages are not numbered in this book, but the above data are from the description accompanying the plates of the *Croton.*

The *Croton* became the New York Central's No. 10 sometime after the Hudson River Railroad merged with the Central in 1869. It continued in service until 1877. William Edson of the New York Central informed the author that a second No. 10 was constructed in 1877. He believes that parts of the *Croton* may have been salvaged for the second engine.

MECHANICAL DESCRIPTION

The boiler is a low-crown wagon top. The waist is only 37 inches in diameter but is quite ample for the small 12½-inch by 20-inch cylinder. The firebox is nearly square and quite deep. The crown sheet is 5/16 of an inch of wrought iron supported by seven crown bars measuring 2 inches thick by 3½ inches at the center. The rear tube sheet was 5/8 of an inch of copper while the front tube sheet was fabricated from ½ an inch of iron. One hundred and eighteen copper tubes measured 1¾ inches in outside diameter by 10 feet 4 inches in length. The total heating surface was 612.09 square feet.

The boiler is stayed by eight bolts, about 1 inch in diameter, which run its full length from the back sheet of the firebox to the front tube sheet. These bolts can be seen in Fig. 154 and are located between the dry pipe and the boiler tubes. The bolts are the only stays indicated by the drawings, despite a very large opening for the steam dome which would seem to call for more substantial staying. One very good design feature of the boiler is the gusset plate used to join the underside of the boiler shell to the firebox. This design permitted a freer circulation of water than the square angle-iron connection formerly used.

The throttle, feed pumps, and most other boiler fittings are typical of the period and are clearly shown by the accompanying drawings. It might be noted that two safety valves placed side-by-side atop the steam dome are not evident unless a careful study is made of the drawings.

A Radley and Hunter stack was used. Its construction is clearly shown in Figs. 154 and 155. The shape of the outer shell, which in this instance might be termed as an elongated diamond stack, illustrates one of the several outer-case designs used by Radley and Hunter.

The running gear is conventional, employing the usual bar frame, equalizer, and T-spoke wheels of the period. It is noteworthy, however, that the cab deck, a sheet of three-eighths of an inch of wrought iron, served also as a rear cross-brace for the frame. Note that the drawbar pin pocket is riveted directly to the deck plate.

The gallery, or running boards going around the engine, with its delicate handrail, is one of the most pleasing features of the engine. The gallery allowed the fireman easy and safe access to the cylinders for lubrication while the engine was moving.

The valve gear was designed by Walter McQueen and became known as Croton valve motion. It was of the riding cutoff style. The main or lower valve, set at about two-thirds cutoff, was driven by the four eccentrics on the front crank axle. V hooks were employed. The cutoff, the upper valve, was driven by the crank attached to the side-rod crank pin of the forward driving wheels. It was set at about one-half cutoff, could be worked only in forward motion, and was, of course, used only after the locomotive was well underway. The reversing lever is unfortunately not shown in the side elevation (Fig. 153), but it can be seen in the rear elevation (Fig. 155).

The leading truck is identical to a design perfected by McQueen as early as 1848 or 1849. See the drawing of the *Mohawk* reproduced in the section dealing with the 4-4-0 type. The side frame is a simple bar frame with a long bolt passing through a pipe spacer to form the bottom rail. The use of

⅜-inch iron plate for the cross-brace should remind the reader of the similar construction of the rear cross-brace of the engine frame. Notice that the *Columbia* and the *Superior* have the same style of truck frame. This distinctive truck frame was used at least as late as 1857, an example being the Breese-Kneeland locomotive on display in El Paso, Texas. The truck wheels were single plate on the corrugated pattern and were made after a design patented by Asa Whitney on March 19, 1850 (No. 7202); see Fig. 149 of the *Champlain* section.

The *Croton*'s tender was a rigid, six-wheel 1,300-gallon tank. Note that the two rear axles are equalized. While not shown on the drawings, Moné's description of the *Croton* noted that an "Indian rubber" hose was used.

Dimensions of the *Croton*

Cylinders	12½" base × 20" stroke
Wheels	66" diameter
Weight	16 tons
Grate area	11.08 sq. ft.
Total heating surface	612.09 sq. ft.
Tubes (118)	1¾" outside diameter × 10' 4" long

Fig. 151. The Essex of the Great Western Railway of Canada was nearly identical to the Croton. It was built in 1853 by the Lowell Shops.

AMERICAN LOCOMOTIVES

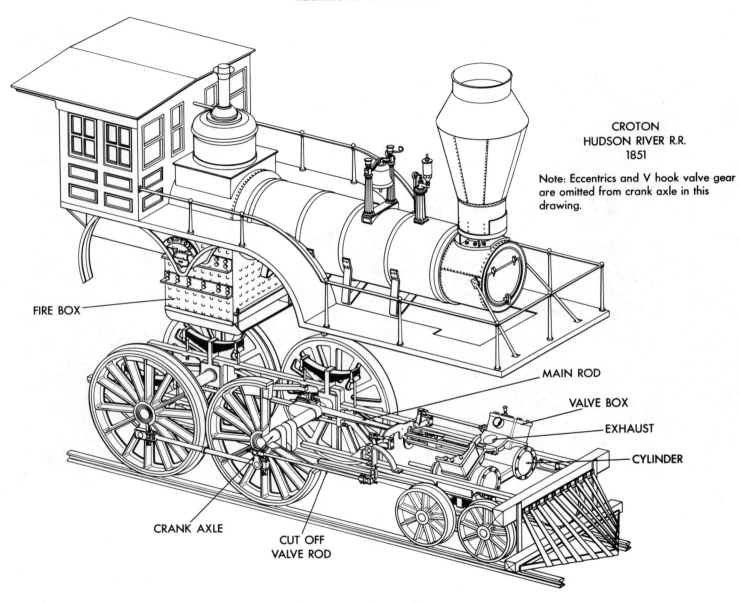

Fig. 152. A schematic drawing of the Croton *showing the principal parts of a typical mid-nineteenth-century inside-connected locomotive.*

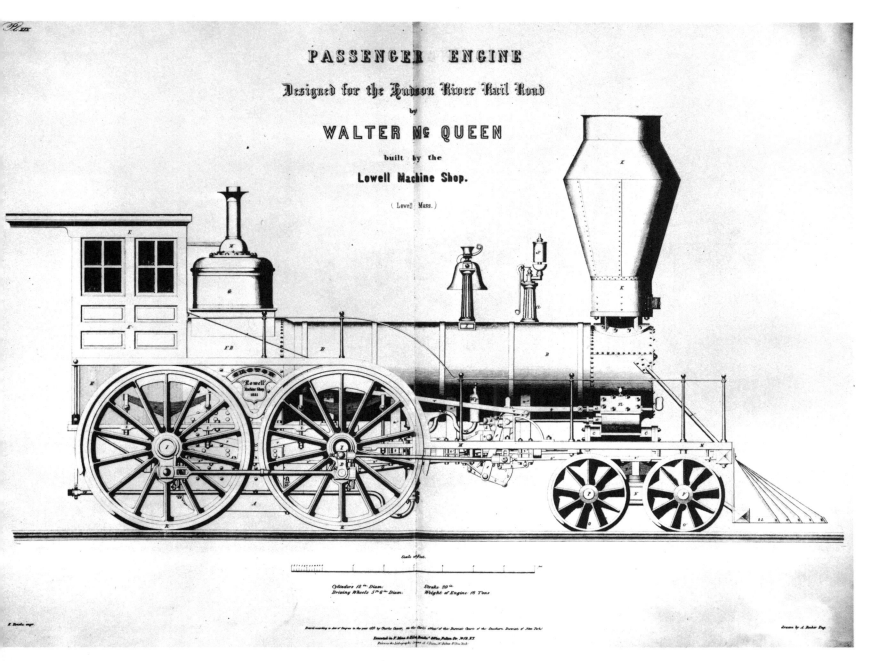

Fig. 153. Side elevation of the Croton. Built by the Lowell Shops in 1851 for the Hudson River Railroad, the engine was intended for light passenger service.

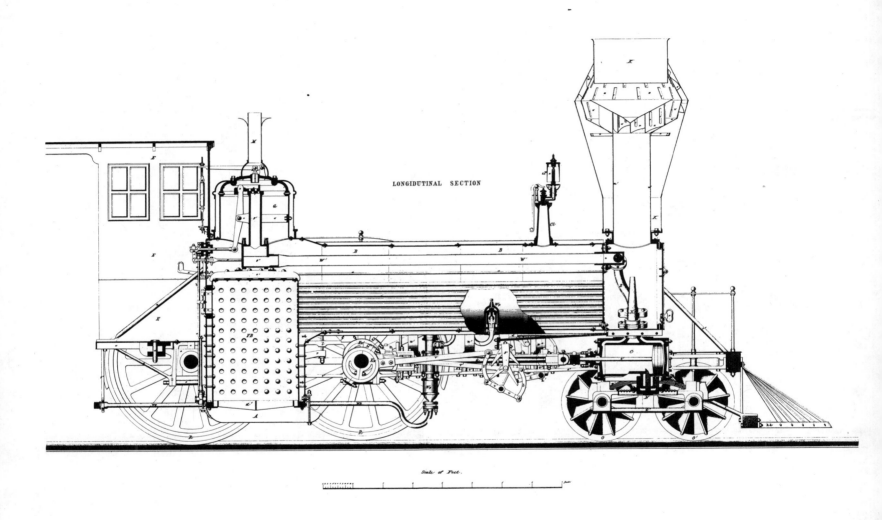

Fig. 154. A longitudinal section of the Croton.

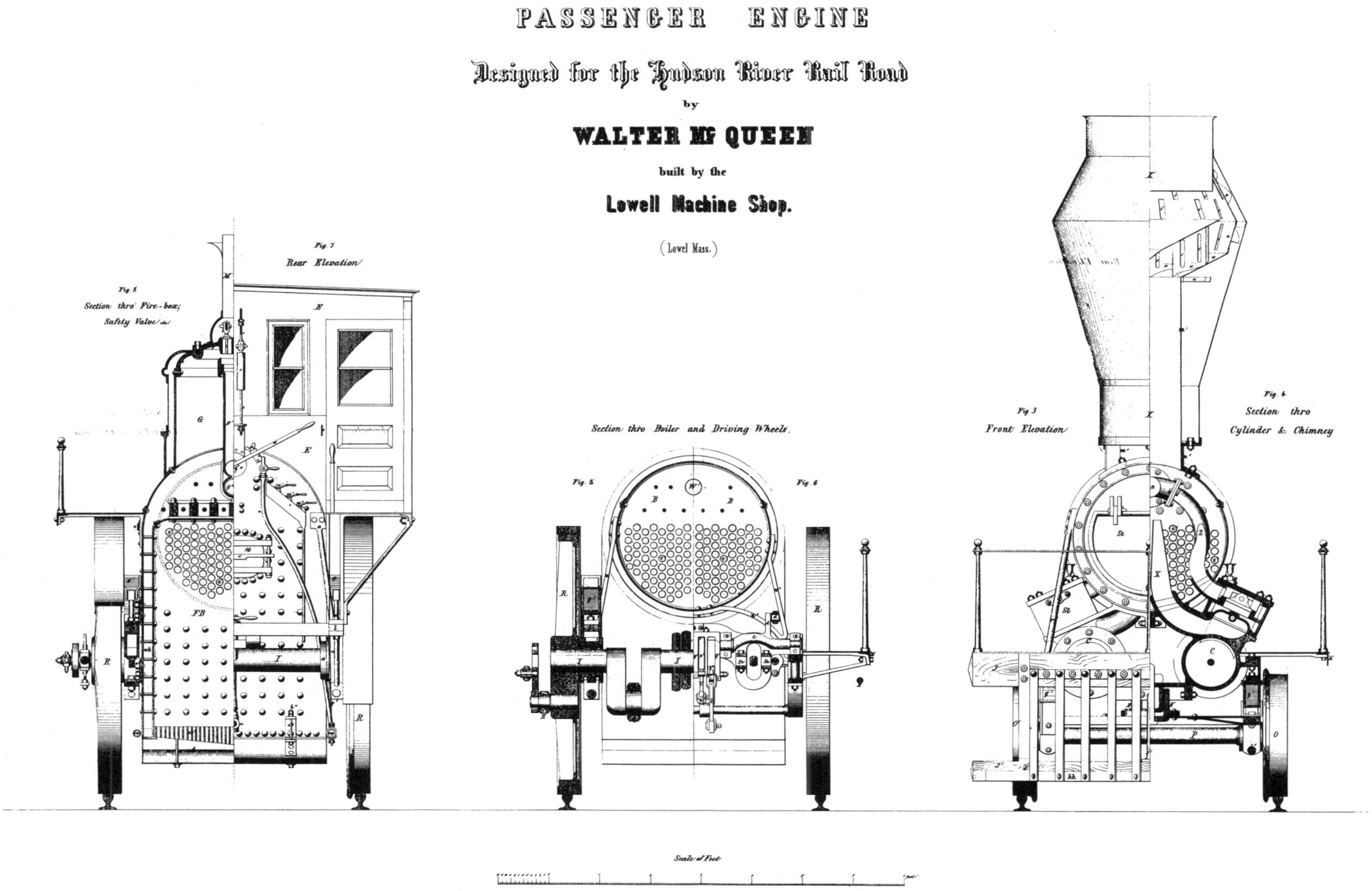

Fig. 155. Front and rear elevations with cross-section drawings of the Croton.

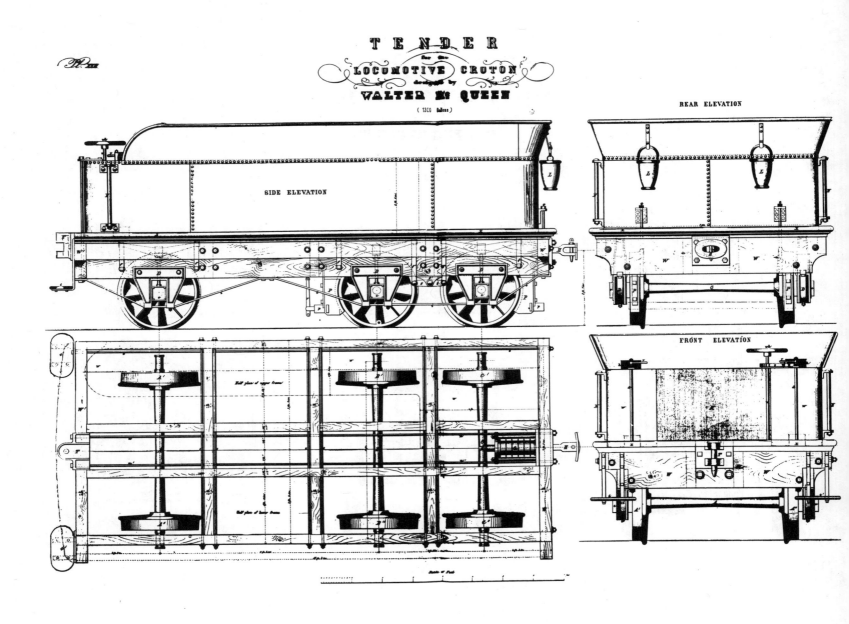

Fig. 156. The six-wheel tender of the Croton.

COLUMBIA, 1852

The *Columbia*, with its giant 7-foot drivers and lean boiler was obviously a racer, the result of the Hudson River's continuing search for faster and more powerful express locomotives. Indeed, there was probably no finer show of speed in the 1850's than the *Columbia* racing along the banks of the Hudson River at better than 50 miles per hour.

The *Columbia* was built by the Lowell Shops in July, 1852. Although somewhat larger, it was built after the design prepared by McQueen a year earlier for Hudson River's *New York*. Lowell also built the *Columbia*'s sister engine the *Rensselaer*.

The boiler, while considerably larger, was identical in design to the *Croton*'s boiler. The outside diameter was 42½ inches. It contained 140 tubes 2 inches in outside diameter and 12 feet long. The inside of the firebox measured only 44 inches wide by 48 inches long.

The cylinders were set level because of the high wheels. Notice also that it was *not* necessary to spread the truck for level cylinders. The *Columbia* would have been a steadier and safer engine with a spread truck. Since this improvement was gaining universal acceptance by the mid-1850's, it is likely that the *Columbia* received a long wheel-base truck not many years after entering service. The truck shown in this section has an inside and outside frame. This was for added safety, there being four extra bearings in case an axle broke. In addition, the outside truck frame gave wider support, thus preventing rocking at high speed.

The huge 84-inch driving wheels are at once the most impressive and frightening feature of the *Columbia*. They were of cast iron made in the old-fashioned T pattern. Although quite successful for the usual 60-inch wheel, it was hardly proper for the larger wheel meant for hard use in express service. While no account of disaster resulting from these wheels is at hand, it is noteworthy that three years earlier Winans' attempted use of 7-foot cast-iron wheels on his experimental *Carroll of Carrollton* failed after less than two months' service.

No drawings for the tender exist, but it was an eight wheeler, weighed 10 tons, and carried 2,300 gallons of water.

Dimensions of the *Columbia*

Cylinders	16½″ bore × 22″ stroke
Wheels	84″
Weight	27 tons
Grate area	12.8 sq. ft.
Total heating surface	914 sq. ft.
Tubes (140)	2″ outside diameter × 12′ long
Steam pressure	80 lbs.

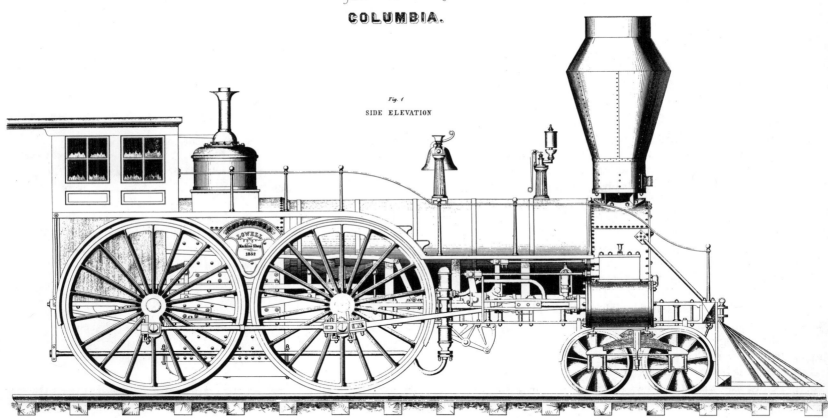

Fig. 157. *The Columbia, built in July, 1852, by the Lowell Machine Shop for express service on the Hudson River Railroad.*

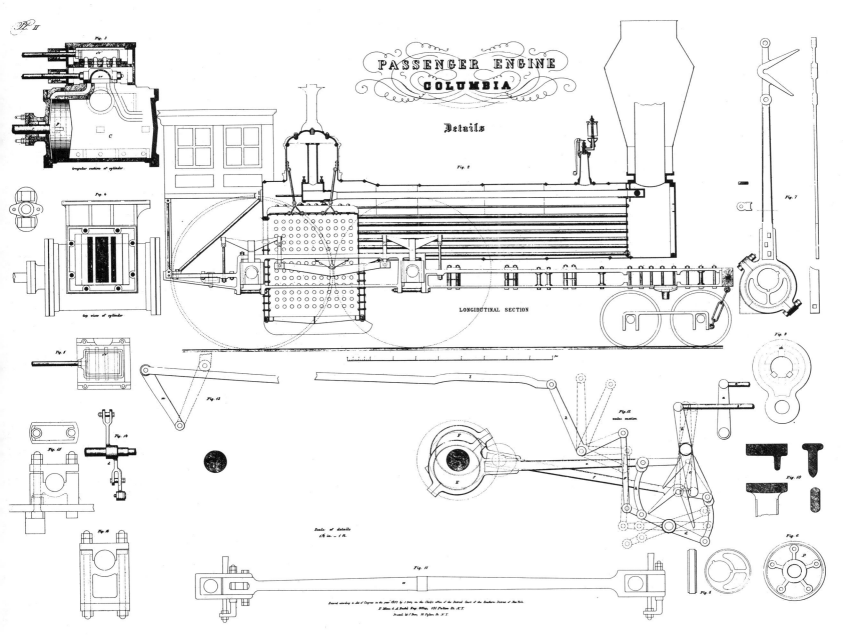

Fig. 158. Longitudinal section and details of the Columbia. While not shown, the cutoff was worked by a third eccentric on the front driving axle.

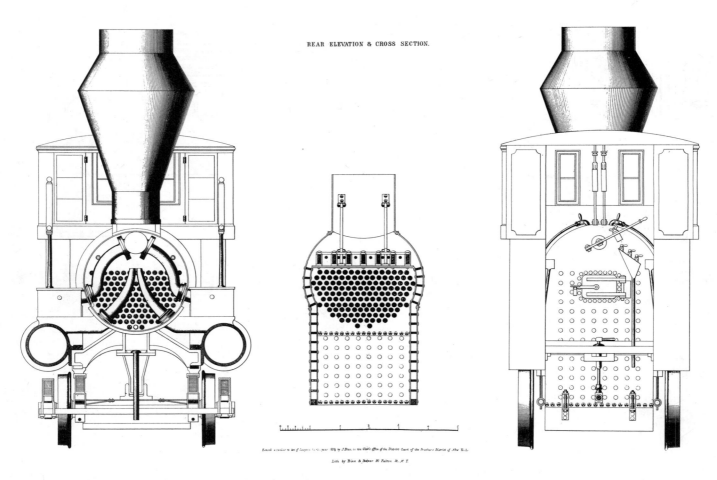

Fig. 159. End elevations and cross-section of the Columbia.

SUPERIOR, 1854

In the spring of 1854 the newly founded New York Locomotive Works built two express locomotives for the Hudson River Railroad. These engines, the *Superior* and the *Baltic,* were heavy and powerful machines for the times. Their excellent performance won an enviable reputation for their builder.

The *Superior* was the first locomotive constructed by Breese, Kneeland and Company, proprietors of the New York Locomotive Works located in Jersey City. It was placed in service in March, 1854, and embodied the several curious deviations from standard practice which set Breese-Kneeland apart from other builders in this country. The most notable and persistent variation was the slab frame. This odd design was used by the Jersey City works (successors to Breese-Kneeland) at least as late as 1860. Prior to the founding of the New York Locomotive Works, the superintendent of Breese-Kneeland, E. P. Gould, was superintendent of the Hudson River Railroad. Gould's former employment was of unquestioned assistance in obtaining orders from the Hudson River Railroad.

The boiler was straight, without the slightest suggestion of a wagon top or enlarged firebox wrapper. It was 46 inches in diameter and about as large as a boiler could be safely fabricated from 5/16 of an inch wrought-iron plate. The firebox was equally large, measuring 39 inches by 50½ inches on the inside. The rear tube sheet was of ¾-inch copper. There were 179 tubes 2 inches in outside diameter and 12 feet long. In all, this large boiler yielded 1,049 square feet of heating surface, quite a sizable record for the period.

Seven 5/8-of-an-inch-square rods running from the front tube sheet to the rear wall of the firebox, and several short rods at each end, stayed the boiler. Two steam domes were used to provide extra steam room since the boiler was straight. A dome at each end of the boiler also permitted, or so it was believed, a more uniform collection of steam.

The throttle is located in the smokebox, and, although the drawing is incomplete, it is evident that a slide valve was used.

Excellent as this boiler was in most respects, it had too many tubes, a defect commonly criticized in other Breese-Kneeland engines. The plan was to increase heating surface, but the placement of so many tubes in the boiler resulted in too small a space between the tubes. This weakened the tube sheets and produced a dangerous boiler.

The frame, as already noted, was of the slab pattern. It was 1¼ inches thick by 8 inches deep. While solid and neat in appearance, it was more difficult to forge and offered fewer places for attachment than the regular bar frame.

The driving wheels were of forged wrought iron rather than the usual cast iron. For large wheels, wrought iron was considered by some builders to be superior to cast iron. The spokes were unusually slender, producing a very pleasing, delicate wheel.

The leading truck was all but identical to the *Columbia*'s. It was side-bearing with the center pin acting only as a pintle.

The cylinder cocks with control linkage to the cab are one of the more interesting details and represent one of the earliest-known applications of this arrangement.

Stephenson link motion was used and another indication that Breese, Kneeland and Company was a progressive builder who wisely adopted the most advanced practice available. Two exhaust ports were used to reduce back pressure on the piston. This was an odd arrangement and was seldom used. The lap measured ⅞ of an inch and the lead ¼ of an inch.

Dimensions of the *Superior*

Cylinders	16″ bore × 22″ stroke
Wheels	78″ diameter
Weight	29 tons
Total heating surface	1,049 sq. ft.
Tubes (179)	2″ outside diameter × 12′ long

Fig. 160. *The New York Locomotive Works; an advertisement appearing in the Jersey City directory of 1854–55.*

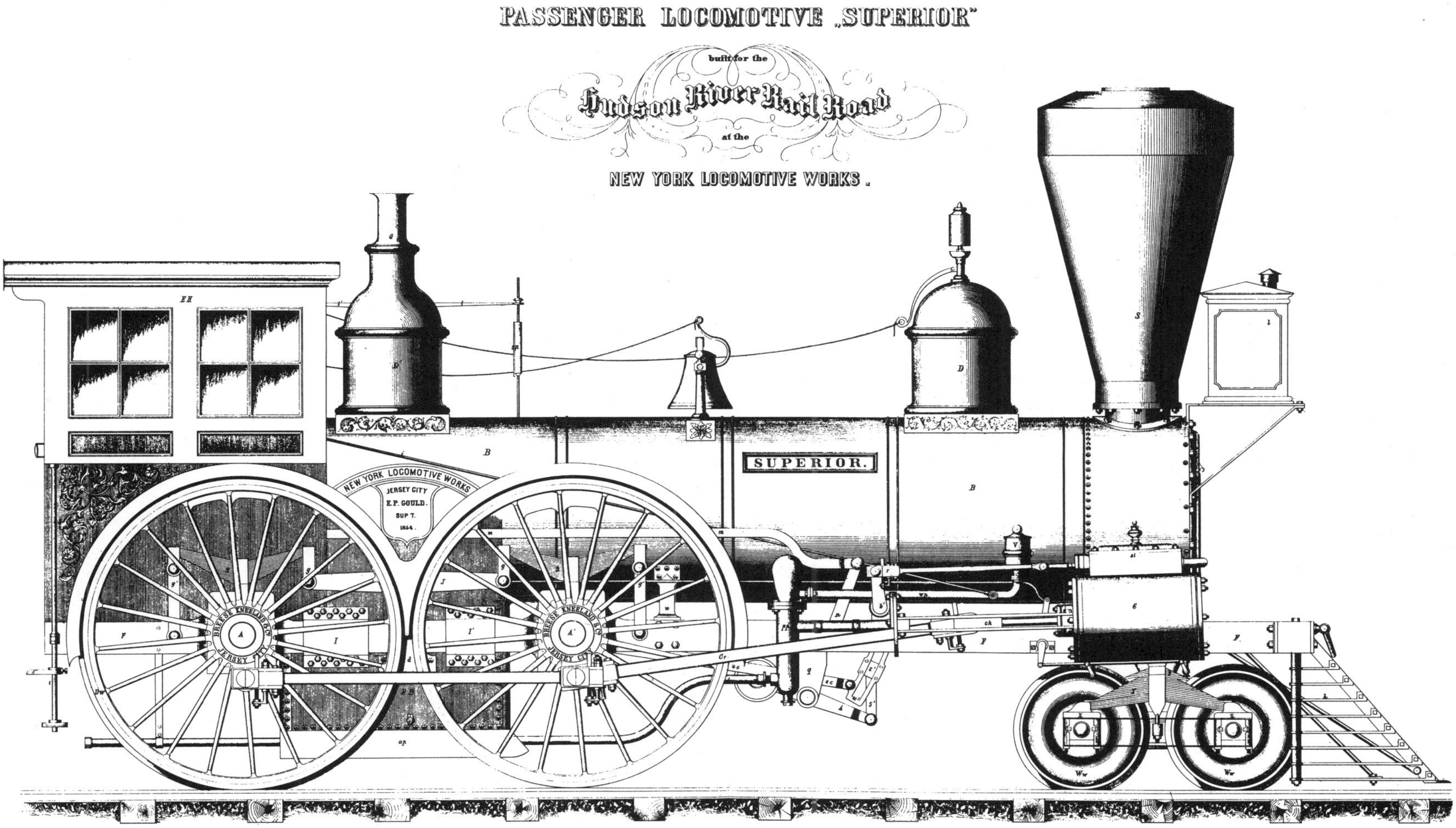

Fig. 161.   The Superior was the first locomotive built by Breese, Kneeland & Co. It was delivered to the Hudson River Railroad in March 1854.

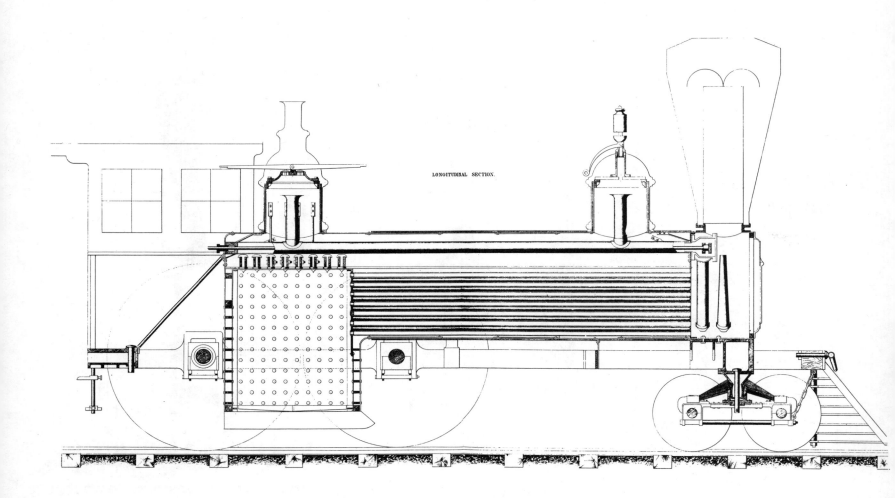

Fig. 162. Details of the Superior, 1854.

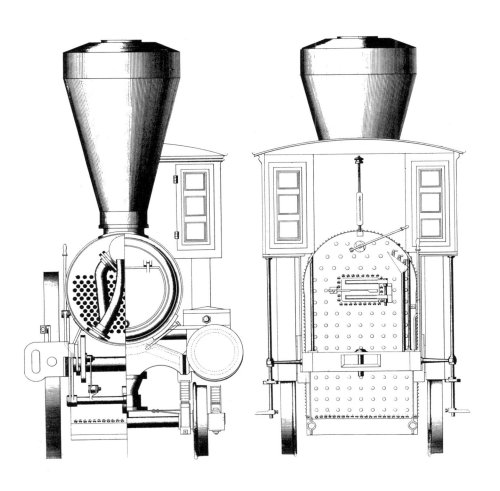 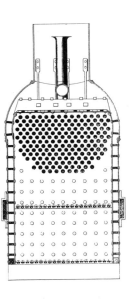

Fig. 163. Details of the Superior, 1854.

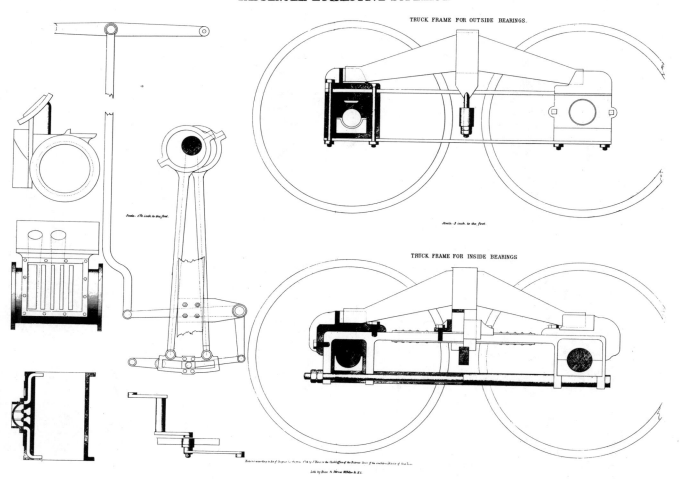

Fig. 164. Details of the Superior, 1854.

SUSQUEHANNA
A WINANS CAMEL, 1854

Ross Winans built from two to three hundred eight-wheel coal-burning locomotives known as Camels between 1848 and 1860. These ponderous, rough machines were built for slow freight service and maximum tractive effort. This was achieved by eight small coupled driving wheels, an ample heating surface, and large cylinders. Yet the over-all size of the engine was no larger than a good-size American type of the same period. The usual Camel was about 25 feet long and had a surprisingly short wheel base of 11 feet 3 inches. The weight was held to the safe limit of about 3 tons per wheel.

The Camel at once presents itself as an oddity among American locomotives because of its sloping firebox and its cab placed atop the boiler. But these are only the most obvious peculiarities; a careful study of the details reveals such variations as a plate frame, solid end rods, a cam-operated cutoff, chutes to feed the forward portion of the grates, stay bolts rather than crown bars, and an eccentric throttle. Zerah Colburn found Winans' design so queer that he described it as "alike peculiar, and in the strongest possible contrast with the proportions, arrangement and workmanship of the standard American engine."[1] In this work devoted to the history of standard practice why should so obvious an oddity as the Camel receive more than passing mention? Odd as they were, fully 200 Camels can be listed and as many as 300 may have been built.[2] The Baltimore and Ohio had 119 on their road and the Philadelphia and Reading and the Northern Central had half that many between them. With this many engines in service, the Camels cannot be dismissed as an experiment or mere freak, even though they exerted little influence on American locomotive practice.

The Camel was the first coal-burning locomotive to be built in large number. Here, then, lies its true significance. It represents Winans' final triumph after a fifteen-year search for

---

[1] Clark and Colburn, *Recent Practice,* p. 50.
[2] No complete list of Winans' engines exists. Larry Sagle accounted for nearly two hundred Camels in his history of Ross Winans but stated that his list was by no means complete; see *Railway and Locomotive Historical Society Bulletin,* No. 70 (1947). Colburn stated that "three hundred or more are running"; see p. 53 of Clark and Colburn, *Recent Practice.*

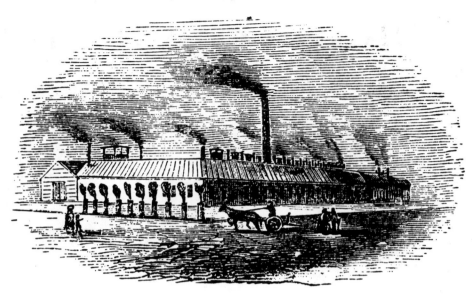

Fig. 165. Ross Winans' Locomotive Works was located in Baltimore adjacent to the Mt. Clare Shops of the Baltimore and Ohio Railroad.

a successful coal-burning locomotive.[3] Winans combined the following elements to achieve success. A large grate area was absolutely essential for proper combustion. Copper tubes were quickly eroded by the abrasive coal ash, so he used iron tubes. These he had used in his earliest vertical-boiler engines of the 1830's. The variable exhaust, although not essential, was useful in controlling the exhaust and thus in preventing the coal from being drawn off the grates when the engine was working hard.

The large grate was realized by placing the firebox *behind* the frame and making it as wide as the outside width of the frame. Since the firebox was not supported by the frame, several steps were taken to make it as light as possible. It was designed to slope down in order to decrease its over-all size, and there was no water space at its rear. With the cab and engineer placed over the boiler and the fireman on the tender, the weight of the cab, deck plate, and crew was removed from the firebox. As a curious aside on the subject of coal-burning, wood-burning Camels were built for the Erie in 1851.

The first Camel was built in June, 1848, for the Baltimore and Ohio. It was named the *Camel*, and, as with the *Consolidation*, all locomotives subsequently built on the same pattern were known as Camels. The engine probably took its name from the likeness of its large steam dome to the hump of a camel. A wag may have suggested a further resemblance in the contrary natures of the machine and beast! The *Camel* was a short-furnace engine, as were apparently all of the earliest Camels. The first detailed description for one of these machines was given by Colburn in his account of Winans' engine built in November, 1849, for tests on the Boston and Maine

[3] Colburn did not consider the Camel an efficient coal-burner. He stated that combustion was very poor, despite good coal. Much smoke was produced because of poor combustion (Clark and Colburn, *Recent Practice*, p. 69).

Fig. 166. The C. E. Detmold of the Cumberland and Pennsylvania Railroad was built by Winans in 1859. Scene: Wills Creek in about 1865.

Railroad.[4] The engine had 17-inch by 22-inch cylinders, 43-inch wheels and an 11-foot 3-inch wheel base. The boiler was of 5/16-inch iron plate, 41 inches in diameter. It contained 101 tubes 2½ inches in diameter and 13 feet long. The grates were 56½ inches long by 42½ inches wide. Colburn specifically mentioned a *step* in the back of the firebox, thus verifying that the slope-backed firebox was not used by Winans on the first Camels. In fact, the earliest surviving illustration showing a slope-backed firebox is an advertising line engraving issued in 1852 by Winans.[5] Returning to Colburn's account, the dome was described as large (41 inches in diameter, 51 inches high) and was placed on the forward end of the boiler. The total heating surface was estimated to be 926 square feet. The engine hauled a 76-car train (433 tons) up a grade of 47 feet per mile. It averaged a speed of 14 3/10 miles per hour on a trip from Boston to Great Falls, Massachusetts, a distance of 74 miles. Despite this very creditable performance the engine was rejected and sold to the Reading Railroad.

The second class of Camel was known as the medium furnace size. Its general arrangement was identical to the short-furnace class, but it had a larger firebox and 19-inch by 22-inch cylinders. It was the most common style of Camel and was first built in 1850 or 1852. The *Susquehanna* was a medium-furnace Camel and will be described in some detail later in this section.

The long furnace was Winans' final development of the Camel. Few engines of this class were built, and those received by the Baltimore and Ohio were found so unsatisfactory that they were rebuilt as medium-furnace engines. The firebox measured 8 feet 2 inches long by 42 inches wide inside. This enormous unsupported structure placed a severe strain on the

[4] Colburn, *The Locomotive Engine*, p. 110.
[5] Bell, *Early Motive Power*, Fig. 32.

Fig. 167. The Tuscarora of the Huntington and Broad Top Mountain Railroad. This medium-furnace Camel was built in about 1859. It exploded September 11, 1868, killing four crewmen.

connection between the firebox and the boiler waist. A long-furnace engine is pictured in Fig. 168.

All three classes followed the same general plan; all had 43-inch wheels and all except the earliest short-furnace engines had 19-inch by 22-inch cylinders. A group of Camels with 6-foot gauges was built for the Delaware, Lackawanna, and Western with 22-inch by 22-inch cylinders. The *Centipede*, an experimental 4–8–0 built by Winans in 1854, also had cylinders of this size.[6] The boiler size was not increased, however, and the large cylinders could not be properly supplied with steam even at low speeds.

[6] *Railway and Locomotive Historical Society Bulletin*, No. 109 (1963), pp. 11–15.

The Camel locomotive was not universally or even widely received with enthusiasm. Despite the continued interest in coal-burning locomotives and Winans' success in developing such an engine, the Camel was poorly conceived of and designed in several important respects. Only roads moving heavy coal trains, such as the Baltimore and Ohio and the Philadelphia and Reading, found the Camel well-tailored to their needs. Both roads operated slow trains and had ready access to cheap coal. The Boston and Maine, as already noted, found the Camel wanting. The Erie thought of purchasing a dozen, but, after testing the first two, canceled the order. The Cleveland and Pittsburgh purchased two Camels in 1852 but found them so unadaptable for general freight service that it set them aside after very limited use.

Fig. 168. The 199, the last Camel purchased by the Baltimore and Ohio Railroad, was built in 1860 and purchased in 1863 when the Baltimore and Ohio was desperate for locomotives because of the Civil War. It was a long-furnace engine.

Several defects have already been alluded to; these chiefly concern the boiler and firebox. The cantilevered weight of the firebox caused a leaky and troublesome joint between the boiler and firebox. The boiler waist was severely weakened by the huge dome opening. The dome itself was poorly stayed. Small wheels made the engine useful only for the slowest freight traffic. The Camel rarely functioned well at over 15 miles per hour and usually it was operated at less than 10 miles per hour. The boiler could not produce sufficient steam for the large cylinders except at very slow speeds.

Winans finished the engines very roughly. The forged side rods were not actually finished but were merely painted. While this hardly affected the engine's working, it illustrated Winans' disinterest in any refinements such as lagging the boiler and cylinders. For a cheaply finished engine, he demanded between $9,500 and $10,000, of which $750 was for patent fees.[7]

The Camel was the creation of an independent and original mind. Winans was a stubborn and proud person as well. His continual involvement in lawsuits, such as the eight-car patent case that dragged on for twenty years, illustrates a tenacious, if not cranky, personality. This turn of mind was most dramatically exhibited in the controversy with Henry Tyson, master mechanic of the Baltimore and Ohio, over the merits of the Camel and ten-wheel locomotives. Tyson was not the first Baltimore and Ohio mechanical officer to criticize the Camel. In 1855 Samuel L. Hayes, Tyson's predecessor, reportedly re-

[7] *American Railway Review,* January 16, 1862, p. 230.

fused acceptance of a new delivery of Camels. In the dispute that followed, Winans, a large stockholder in the Baltimore and Ohio, was successful in forcing Hayes's resignation. All of this was reported on July 19, 1856, by the partisan *Railroad Advocate*, a paper so unfriendly to Winans that he was termed an "impostor" on another occasion. Two years later Tyson was successful in defying Winans. The loss of the Baltimore and Ohio as a customer was a fatal blow to Winans' business; he closed his shop some time in 1862.[8] Hayward-Bartlett leased Winans' shops in 1863–64 and built several engines under the name of the Baltimore Locomotive Works before closing in 1867.

In spite of their faults, the Camel engines gave good service and some were in use for more than forty years. By the early 1890's most of the Camels had been retired. The Cumberland and Pennsylvania retired its last in 1896. The Baltimore and Ohio kept several wheezing around the Baltimore yards but scrapped its last Camel in 1898 (see Fig. 170).

THE SUSQUEHANNA

The *Susquehanna* was completed on March 31, 1854, and is shown in the present volume as it appeared in 1859 after slight modifications made by the Philadelphia and Reading Railroad (see Fig. 169). The drawing is the only complete plan for a Camel known to exist. It is from a German publication on American railways published in 1862 but it represents data gathered by the author three years earlier.[9] Other than the boiler jacket and a new crosshead, the engine differs little from a stock Camel.

We have already noticed the Camel's general mechanical arrangement. A more careful study of the *Susquehanna* will give a better understanding of Winans' design.

The boiler, as on all Camels, was straight and had a large forward steam dome and a sloping firebox. The waist was 46 inches in diameter. The grates measured 6 feet 8 inches long by 3 feet 6 inches wide. The coal chutes atop the firebox were for feeding coal to the forward end of the grates. Winans added the chutes believing that no fireman would heave coal to the front of the grates; it would be pointless to build so large a firebox if only the rear end of the firebox were to be fired. The fireman was to load a hopper on top of the two chutes. The coal was dumped from the hopper into the chutes by tripping the long lever seen at the center of the top of the firebox (Fig. 169). This arrangement, while sensible, was not popular with firemen, who objected to heaving coal first to the firing platform of the tender, which was level with the hopper, and then to the hopper itself. It is probable that few Camels were fired according to the builder's specification, that most operated only with a fire at the rear end of the grates.

The *Susquehanna* drawing shows the cast-iron rocking grates originally supplied by Winans. No complex linkage was employed. To shake the grates was simplicity itself. A jacking bar placed in the hole at the rear end of each grate bar tipped the grate from side to side, thus allowing the ash to fall through. Millholland later replaced these with water-tube grates since the intense heat of the anthracite fire rapidly burned out the cast-iron bars.

The boiler contained 101 iron tubes 2¼ inches in outside diameter and 14 feet long. Because the tubes were so long they were supported at the center by an iron plate. The plate was cut off at the top and bottom so as not to block the water's circulation through the boiler. The tubes were fitted loosely in holes bored through the plate.

---

[8] 1860 is commonly given as the date Winans' shop closed. However, the repair order books of Winans preserved by the Maryland Historical Society reveal that the shop was in operation through December, 1861. No orders are shown for 1862, but the shops reopened temporarily in 1863 to complete several Camels practically finished; these were sold to the Baltimore and Ohio, the Huntington and Broad Top Mountain, and the Cumberland and Pennsylvania railroads.

[9] A. Bendel, *Aufsätze Eisenbahnwesen in Nord-Amerika* (Berlin, 1862), pp. 49–50.

We have already noted that the boiler was weakened by a large steam dome, heavy unsupported firebox, two holes necessary for the coal chutes, and poor staying. Yet only two cases of explosions are known to the author. These were the *Minnesota* of the Philadelphia and Reading, which had a crown-sheet failure in 1851, and the Huntington and Broad Top Mountain Railroad's *Tuscarora*, which blew up in 1868 (see Fig. 167).

The throttle was set high in the dome and was of the common slide type. The valve was actuated by a lever, an eccentric, and the necessary linkage; it could be operated from either the right- or left-hand side.

The stack was again peculiar to Winans' products. The forward pipe, of the same diameter as the stack and parallel to it, was a hopper for cinders. It was emptied by a hatch hinged to the bottom. No cone was used; the smoke, exhaust steam, and light cinders passed out through screening fitted over the metal cowl atop the twin pipes. The rivets on the stack and the coal chutes shown in the Bendel drawing (Fig. 169) are oversized.

The feed pump was mounted on the side of the firebox, a singularly poor placement. Being attached to the thin outer wrapper of the firebox, it was a weak mounting and added weight to the rear of the locomotive, which was already out of balance because of the heavy firebox. In addition, it required a long pump rod from the crosshead. Actually, a long rod was not necessary, but it was Winans' way. The Baltimore and Ohio replaced Winans' arrangement with a half-stroke pump driven by a crank on the rear connecting rod pin. Note also that while the pump rod and valve rod appear to be one they are not, but are merely parallel. The air dome is probably an addition made by the Philadelphia and Reading; it was not used by Winans even though it had been accepted years before as a common improvement by other builders. The cold feed water was injected into the side of the firebox, the hottest part of the boiler. Winans was not alone in this poor practice, as we have noted in the discussion of several other engines.

The frame was made of two iron plates ⅝ of an inch thick set 6 inches apart and 20 inches deep at the pedestals. U-shaped castings served as guides for the wheel boxes. They were also used as spacers between the frame plates.

The springs served also as equalizers. They were originally mounted out of sight between the two plates of the frame but are shown in Bendel's drawings as remade to fit above the frame.

The large box on top of the boiler just in front of the cab is the sandbox. The sand pipe was omitted from the drawing.

The wheels were 43 inches in diameter, the standard Winans size. They were fitted with thick cast-iron tires. These were bolted to the wheels and could not be shrunk on as was the common attachment procedure for wrought-iron tires. Complaints were made that the cast-iron tires broke easily and did not have the adhesion qualities of wrought-iron tires. They were cheap and thus enjoyed a limited popularity for small wheels.

Winans was probably the earliest advocate of solid end rods. Since no wedges were used, Winans contended that inept crewmen had no opportunity to misalign the rods. He also correctly recognized that the solid rods were safer since there were no straps that could work loose. Winans' views, though much disputed at the time, were finally adopted as standard. Though right in principle, Winans made the bearings too narrow; they wore out quickly, thus requiring frequent replacement of the bushings. The rods were rough, unfinished forgings and, like the remainder of the engine, were painted a dreary green.[10]

The crosshead shown in the Bendel drawing is a replacement made by Millholland. The original was made of plate with a single square guide of the style best shown in Fig. 168.

The *Susquehanna* had dead-level cylinders with a 19-inch bore and 22-inch stroke. They were attached to the sheet-iron sides of the smokebox. Long bolts passed through the smokebox between the cylinders. The mounting was weak for such large cylinders and they were known to work loose. No insulation was used. The Philadelphia and Reading found that steam condensed so readily in cold weather that the cylinder cocks had to be left open.[11] No further comment on the inefficiency of such a system is necessary.

[10] *Railroad Advocate*, January 27, 1855, p. 4.
[11] *Railroad Advocate*, February 3, 1855, p. 3.

The Bendel drawing (Fig. 169–A) shows three detailed views of the *Susquehanna*'s valve motion. This valve motion was used on all the Camels and is largely based on a design patented on July 29, 1837 (No. 311). It exhibits Winans' usual capacity for novelty and is in many respects a clever plan. The drop-hook motion is conventional enough, but the massive cast-iron rocker and cam-driven cutoff are decidedly odd. It is noteworthy that only *one* slide valve is used for both the cutoff and regular strokes. The hooks engage the lower arm of the rocker and are lowered or raised on or off the rocker by cams fixed to the reversing shaft. This shaft is below the hooks. The cams are set so that only one hook may be engaged on the rocker at a time. The rocker, it might be added, is an example of the extensive and clever use of cast iron made on nineteenth-century American locomotives. The reversing shaft and the cams were rotated by a lever through a geared crescent. The gear teeth are not shown on the crescent in the Bendel drawing, but the pinion on the reversing shaft is fully illustrated on the front elevation of the valve-gear detail.

Winans made the same claims for simplicity and ruggedness of the valve gear as he did for all the design features of the Camel. Against these claims for superiority came equally valid complaints of clumsiness and inconvenience. It was difficult to line up the hooks, a defect common to all such motions. It was an awkward gear to reverse, again a characteristic of all hook motions.

The cam-driven cutoff with its intermittent exhaust (the valve was stationary at two points of the cam's rotation) was said to produce a poor draft for the fire. Winans defended the cutoff by noting its early action and rapid motion. He also claimed it had less back pressure than slower-moving valves driven by link motion. It was, in theory, more efficient than link motion, but its complexity and clumsiness precluded its general adoption.

The drawbar connection between the locomotive and tender was one of the weakest points of the Camel. In a conventional locomotive it was possible to make a simple, strong coupling by attaching the drawbar to the rear of the frame behind the firebox, but in the Camel it was necessary to pass under the firebox to reach the rear end of the frame. Winans' usual arrangement was to rivet a V-shaped piece of iron plate to the engine's frame. The vortex of the V faced the rear of the engine. A flat, iron drawbar was attached to this end of the V and passed through the ashpan to the tender. This arrangement, however, is not shown on the *Susquehanna*. In its place are two stout plates, 12 inches wide, riveted across the frame between the two rear axles. A round drawbar is pinned between these plates and then passes through the ashpan to the tender. Drawbars were reported to heat to a cherry-red color, being so close to the fire, and to pull apart when under a heavy load.[12]

The tender was on eight wheels and carried 5 tons of coal, 8½ tons of water. When fully loaded it weighed 23 tons, only 4 tons less than the locomotive. A "live" or spring truck was used; see Figs. 166 and 167 for a better understanding of the truck and tender.

A good deal has been said about the faults of the Camel, its odd design, and the general contempt heaped upon it by rival builders. Its builder, however, is redeemed, in part, by the service performed by these unique machines. An example of this good service is the *Susquehanna* as reported by Bendel. A train of 110 four-wheel coal cars was moved at 8 miles per hour from the mines to Philadelphia, a distance of 95 miles. Four and a half tons of coal at $2.50 per ton was consumed on the trip. The fuel cost came to an average of 11.74 cents per mile. This compared favorably with the fuel costs of a woodburner of the same period, which often ran as high as 25 cents per mile. Of course, this great difference was not entirely attributable to the efficiency of the Camel; an additional factor was the cheapness of coal on the Reading line.

Dimensions of the *Susquehanna*

Cylinders	19″ bore × 22″ stroke
Wheels	43″ diameter
Weight (engine only)	27 tons
Grate area	23½ sq. ft.
Total heating surface	*ca.* 1,000 sq. ft.
Tubes (101)	2¼″ diameter × 14′ long
Steam pressure	90 lbs. per sq. in.

[12] *American Railway Times*, January 31, 1861.

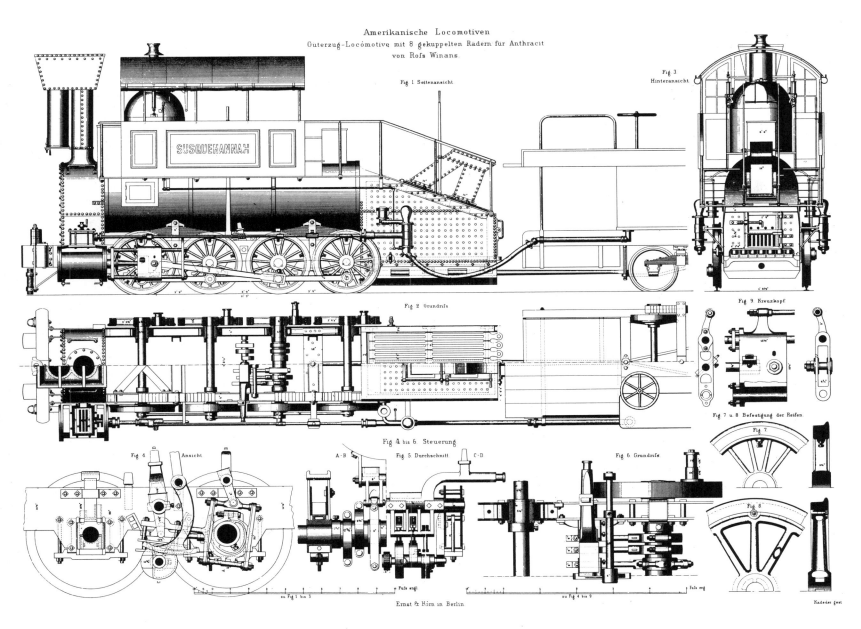

Fig. 169. The Susquehanna, built in 1854 for the Philadelphia and Reading Railroad, is shown in this drawing as it appeared in 1859 after slight modifications made in the road's shops.

AMERICAN LOCOMOTIVES

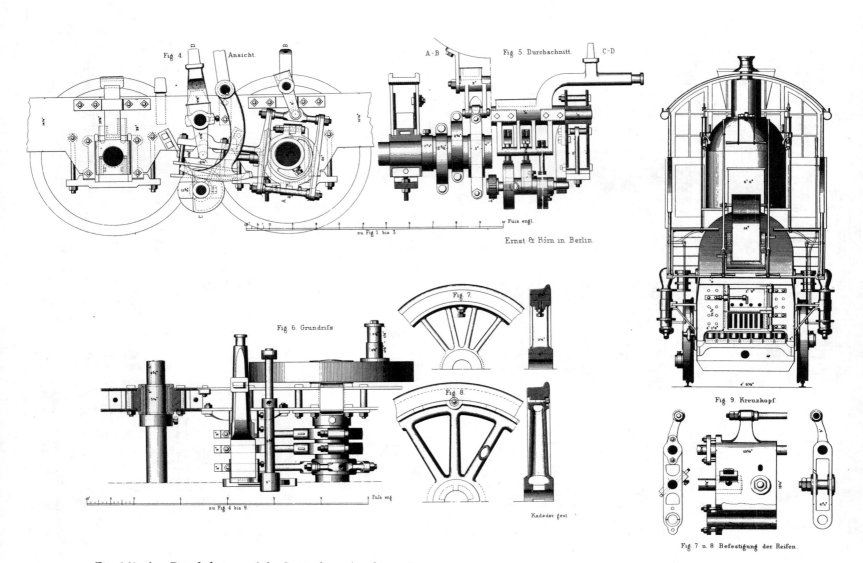

Fig. 169–A.   Detailed views of the Susquehanna's valve motion.

REPRESENTATIVE LOCOMOTIVES

Fig. 170. The last Camel *being scrapped at the Mt. Clare Shops, Baltimore, in 1898. The 143 was built in 1853. It is shown here as rebuilt and much patched over after fifty-five years of service.*

TALISMAN, 1856-57

The handsome set of drawings for this broad-gauge passenger locomotive are from Gustavus Weissenborn's *American Engineering* (1857). Weissenborn did not name the road that the *Talisman* was built for, and no record of a 6-foot gauge engine of this name can be found by the present author. Even so, these engravings are a remarkable record for a mid-nineteenth-century American locomotive.

The New Jersey Locomotive and Machine Company, builder of the *Talisman,* was a small establishment located in Paterson, New Jersey. Chartered in April, 1851, to succeed the machine shop established a few years earlier by William Swinburne, the new firm earned a good reputation for well-designed locomotives. Even though its production was limited to about two locomotives a month during its early years, the firm was fortunate in attracting such capable designers as John Brandt and Zerah Colburn. After building more than 300 engines, the business fell into hard times and was taken over by D. B. Grant in 1863 or 1864. Rarely enjoying continuous financial success, the plant was moved to Chicago in 1893 with the hope of capturing a larger share of western orders. This hope proved false and the company was dissolved soon thereafter.

The boiler of the *Talisman* is a typical Paterson wagon top of the mid-1850's. Two steam domes are used for more equal steam collection. The throttle is located in the smokebox. The deep firebox is a positive indication that the engine was a wood-burner.

A common bar frame was used. It should be noted that the top rail is not split, a practice employed by most progressive builders for ease of repair in case of a wreck. The rear of the frame is stiffened by a V-shaped truss in place of the usual back braces. When the back brace was attached to the boiler's backhead and the frame, it interfered with the free expansion of the boiler. The truss was not attached to the boiler and permitted it to expand freely when heated.

While the *Talisman* is an ordinary engine in its basic layout and closely copies Rogers' standard design, the valve gear shown in the drawings of this section (especially Fig. 174) is a decided departure from standard practice. This valve motion was devised in 1855 by H. Uhry and H. A. Luttgens, both employed by the New Jersey Locomotive and Machine Company. Uhry and Luttgens attempted to improve the quality of the steam distribution of the Stephenson link motion when working in advanced cutoff by introducing a wider opening of the steam ports and a delayed exhaust. The valve was driven by eccentrics and the cutoff was regulated by the links, but the speed of the valve was governed by a cam through a differential rocker. A patent was issued May 20, 1855 (No. 12564). According to the original plan two links and valves were to be used; however, no locomotives were built with a valve gear on this first plan. The next year Uhry and Luttgens modified their design by eliminating one link and valve. A second patent was issued on September 7, 1858 (No. 21455), for this improved version of the original design. The following roads received engines with this style of valve gear: The Central Railroad of New Jersey, the Saratoga and Whitehall, the Ohio and Mississippi, and the Iron Mountain Railroad.[1] The *Talisman*'s valve gear was based on the 1858 patent, but, as already noted, this motion was used by the New Jersey Works two years before the patent was granted.

The Uhry and Luttgens valve gear was little used. One difficulty experienced was cam slippage. A second objection was the additional cost of $500 over and above the cost of a conventional link motion. Finally, the valve gear failed to demonstrate any marked fuel economy when compared with the "unimproved" link motion. In later years, when Luttgens was chief draftsman for Rogers, the old valve motion, slightly modified, was given a final test on an engine built in 1866 for the Central Railroad of New Jersey.[2]

The cylinders were bolted to the sides of the smokebox. Some attempt was made to stiffen this weak plan with cross-braces and ample bolting. For some reason Rogers and the other Paterson builders were slow to adopt the cylinder saddle. While speaking of the cylinders, some mention should be made of the clever scheme for adjusting the piston rings. Without removing the cylinder head an adjustment nut could be reached by removing the ornamental acorn nut shown in one of the details of Fig. 174.

The tender was a standard eight-wheel tank of about 2,000 gallon capacity. The all-metal trucks were a distinct improvement over the wood-beam truck previously favored. Notice also that equalizers and three-point suspension were used to improve the tracking quality of this heavy vehicle.

The following table from Weissenborn's descriptions of the *Talisman* summarizes the important dimensions of that machine:

Diameter of cylinder	17	inches.
Length of stroke	22	"
Length of fire box	4 feet 5	"
Width " " "	3 " 11	"
Depth " " "	65	"
Length of flues	11 "	
Outside diameter of flues	2	"
Space between flues	¾	"
Area of grate	17 " 3	
Length of boiler, including fire and smoke box	18 " 1	"
Diameter of boiler	3 " 11	"
Diameter of boiler near smoke-box	3 " 9½	"
Diameter of driving-wheels	5 " 6	"
Face of driving-wheels	5¾	"
Copper flues	153 pieces	
Diameter of steam domes	21	"
Height of steam domes	22	"
Diameter of main steam-pipe placed in the interior of the boiler	5½	"
Diameter of steam-pipe near cylinder	4½	inches.
Diameter of plunger for feed-pumps	2	"
Stroke of plunger	22	"
Diameter of chimney	13½	"
Diameter of main driving-axle	6¼	"
Diameter of centre-pin on the main driving-wheel	4	"
Diameter of centre for the outside journal	3	"
Diameter of piston-rod	2¼	"
Depth of piston	5½	"
Diameter of valve-stem	1¼	"

Pressure in boiler, 100 to 120 lbs.
One cord of wood (dry pine) to run 60 miles
Length of fire-box for bituminous coal-burner, to produce the same quantity of steam, 5 feet 9 inches
Length of fire-box for anthracite coal-burner, 6 feet, with the addition of a combustion chamber, projecting some 24 inches into the barrel of the boiler

---
[1] *Master Mechanics Report*, 1890, pp. 72–77.
[2] Forney, *Locomotives and Locomotive Building*, p. 55.

AMERICAN LOCOMOTIVES

Fig. 171. The J. H. Devereux of the U.S. Military Railroad, built in March, 1862, by the New Jersey Locomotive and Machine Company. This machine is very similar to the Talisman, built five years earlier by the same manufacturer.

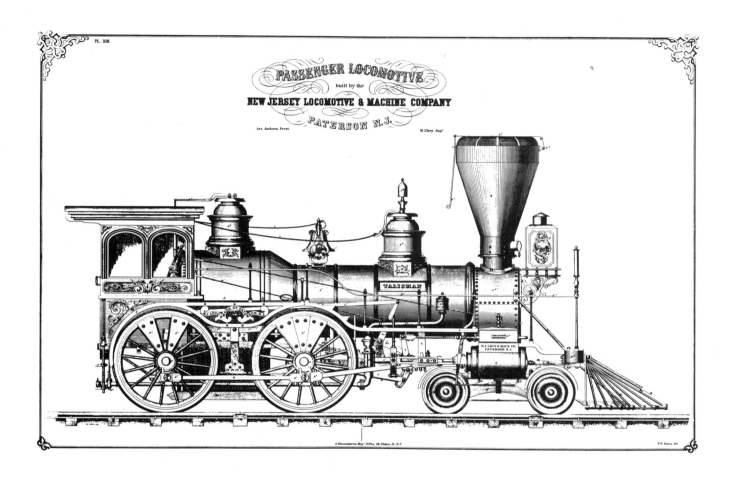

Fig. 172. The Talisman, *built in about 1857 by the New Jersey Locomotive and Machine Company.*

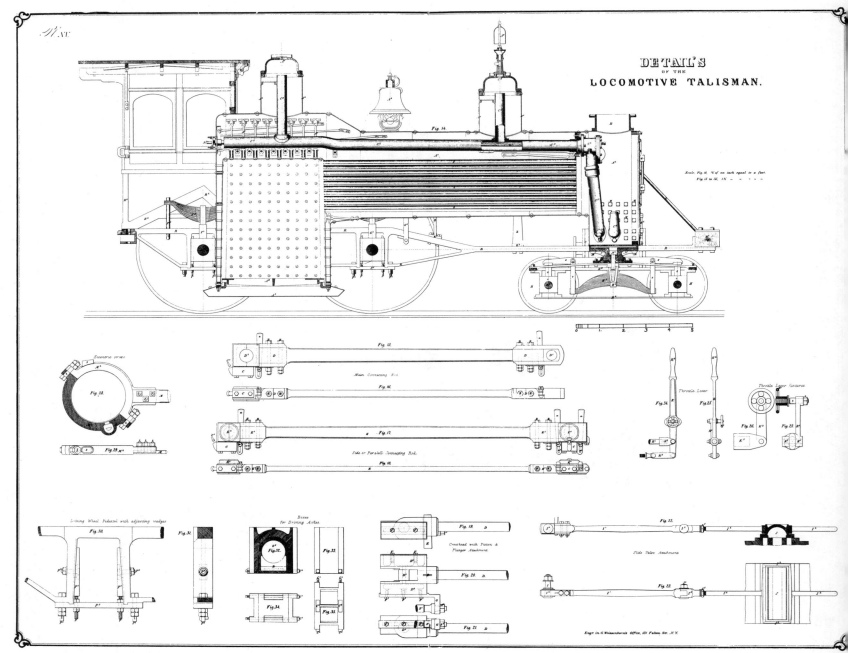

Fig. 173. Details of the Talisman, 1857.

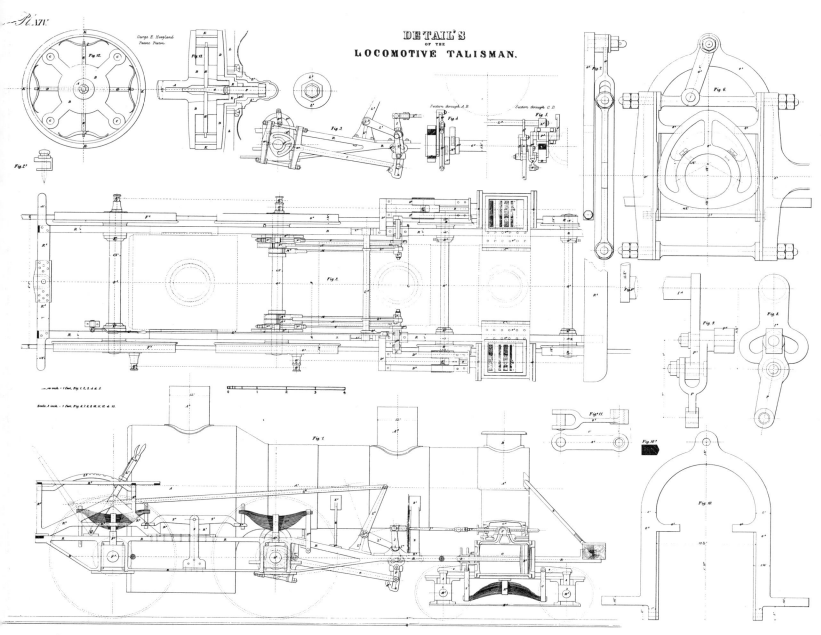

Fig. 174. Details of the Talisman, 1857.

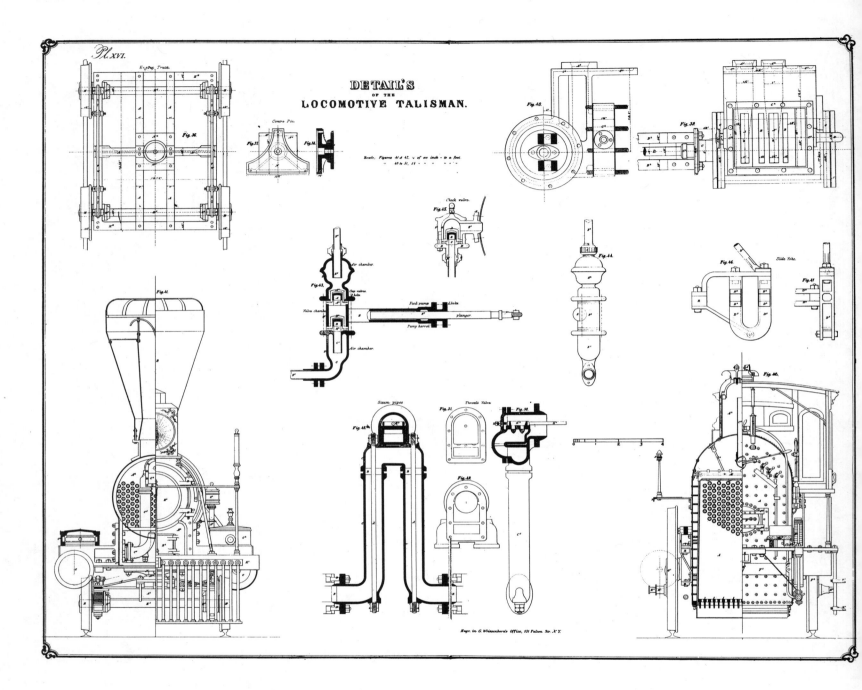

Fig. 175. Details of the Talisman, 1857.

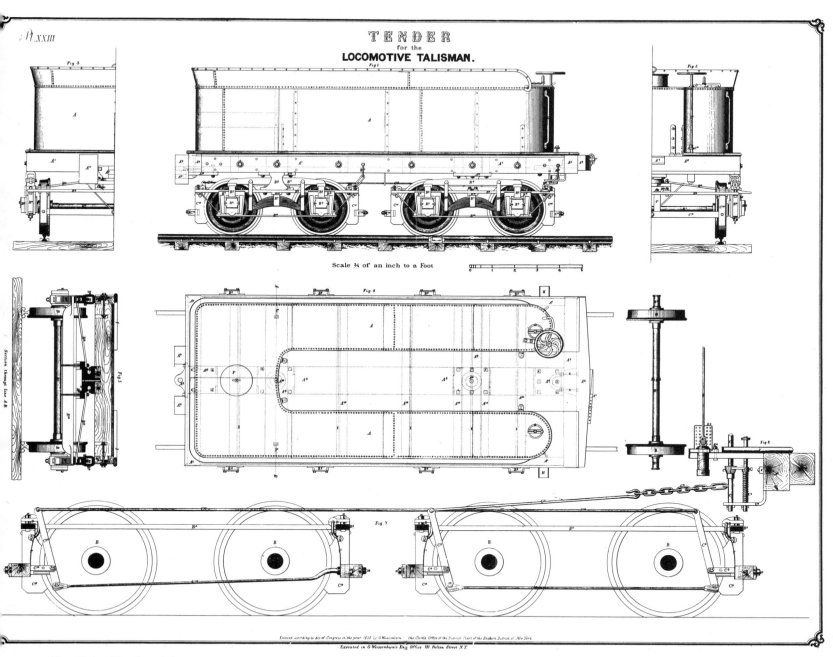

Fig. 176. *Tender of the Talisman, 1857.*

THE TYSON TEN WHEELER, 1857

The ten-wheel locomotive designed by Henry Tyson, master mechanic of the Baltimore and Ohio Railroad, is the earliest machine of the "modern" style for which we have anything in the way of complete drawings. This machine, built for both passenger and freight service, incorporates the various design reforms (spread truck, level cylinders, link motion) introduced by Thomas Rogers in the early 1850's.

Tyson intended these engines primarily for fast freight trains but planned them as well for passenger service if necessary. The drawings were completed in September, 1856, just a few months after Tyson's appointment as master mechanic of the Baltimore and Ohio. Tyson approached Ross Winans for the building of the first group of engines. Winans refused, offering instead to modify Tyson's design so that standard Camel locomotive parts could be used. Tyson's refusal of Winans' offer started the bitter controversy already mentioned in the chapter on the Camel locomotive and more fully discussed in Bell's *Early Motive Power of the Baltimore and Ohio Railroad*. In the end the Baltimore and Ohio management sided with Tyson and in doing so firmly committed the road to a progressive motive-power policy.

A. and W. Denmead and Sons of Baltimore built seven engines, numbered 222–228, after Tyson's design. Delivery was made between April and December of 1857. Two more, Nos. 229 and 230, were built by the Baltimore and Ohio at its Mt. Clare Shops in September, 1857. In the company's annual report for 1857, Tyson proudly reported that No. 230 was giving good service with high-speed cattle trains on the Washington branch. He also noted that No. 229 had hauled a 150-ton train up the 17-mile grade (116 feet per mile) near Piedmont. The engines were rated as being able to pull 800 tons at 20 miles per hour on level ground and 600 tons at 20 miles per hour up a 20-foot-per-mile grade.[1]

Tyson's ten wheelers were not entirely successful, however, for only the initial nine were built. Like many early ten wheelers, they were said to derail easily. Whether or not this was a well-founded criticism of Tyson's design is an open question. Tyson was in office less than three years and succeeding master mechanics apparently had no interest in perpetuating their predecessor's design.

By 1890 most of the Tysons were out of service. The last one, No. 230 (renumbered No. 294 in 1884) was retired in 1894.

[1] Douglas Galton, *Supplement to the Report on the Railways of the United States* (London, 1858), Plate 5.

REPRESENTATIVE LOCOMOTIVES

MECHANICAL DESCRIPTION

A good general description of the Tyson Ten Wheeler is given in the following specification prepared by Tyson in 1856:

The weight of each engine to be thirty tons (60,000 lbs.) with water in boiler.

The material used in their construction to be of the best quality, and the workmanship to be of the most perfect description.

The boiler to be plain wagon top; the cylinder part to be 48 inches at fire-box and 46 inches at smoke-arch, 14¼ feet long outside dimensions; the fire-box outside dimensions, at top 74½ × 48¾ inches; inside dimensions, 66 × 41 inches, and 57½ inches deep; the smoke-arch, 46⅞ inches diameter by 36½ inches long on outside; the tubes 125 in number, 2⅛ and 2⅜ inches outer diameters, length 14 feet 3½ inches. The inside of fire-box to be of copper, except the crown-sheet which is to be of iron; the crown is to be supported by vertical bolts ⅞ inches diameter, tapped through the outer shell of boiler and crown-sheet, with nut underneath and rivetted on ends. The iron in boiler and tubes to be made from charcoal blooms after the most approved mode of manufacture.

The driving wheels, six in number, 50 inches diameter, with hollow spokes and chilled cast-iron "slip tires;" the back drivers to be flanged, the middle and forward ones smooth, all to work in front of fire-box.

The cylinders, 18 inches diameter and 24 inches stroke; their position to be horizontal; the steam-valves to be operated by stationary links, and rockers.

The frame to be made of a solid bar of 3 × 4½ inch iron, to be flattened to 1¼ × 8 inches to pass the sides of the fire-box and extend to support the footboard and receive the draught-bar. The pedestals to be 15 inches long and 3 inches wide and forged solid to the frame, the top and sides of the frame are to be planed smooth; the feet of pedestals are to be planed to receive the brace. The inside and sides of pedestal jaws are to be furnished with wrought iron facings planed and accurately finished to receive the brass boxes.

The truck wheels to be (Bush & Lobdell's) 28 inches diameter and 68 inches apart from centre to centre; the frame to be made of rectangular bars of wrought iron, with cast iron pedestals properly fitted and secured by bolts and braced to receive the weight on side bearings; distance from centre of back driver to centre of truck pintel, 15 feet.

The engines to be furnished with cast-iron steam-pipes, and variable exhaust operated by a hollow plug. All bolts to be turned and fitted, the heads and nuts of bolts of parts that require frequent adjustment to be case-hardened, the dimensions and threads of those bolts to be made according to sample. The cylinder part of boiler to be covered with ¾ inch hemlock lagging and cased with Russia iron.

The tender: tank to be of 2,000 gallons capacity; the top, sides and bottom to be secured by 1½ × ¼ inch angle iron and ⅜ inch rivets, securely braced; the trucks will be similar to those used for the engines except that the wheels will be 50 inches apart from centre to centre.

These engines to be delivered to the company in the city of Baltimore, in perfect running order, with all tools and appurtenances necessary for their operation, subject to the approval of the company's Master of Machinery.

The payments to be made in cash after each Machine has been delivered and successfully operated on the road for thirty days, satisfactory guarantee being given that the company will be reimbursed for any damage from accident that may arise from defect in material or imperfect construction during the first six months of their service.[2]

The specification describes most elements of the engine in sufficient detail, but a few additional comments are necessary to emphasize some of the unusual mechanical features. The absence of crown bars is the most remarkable feature of the locomotive. The boiler depends entirely on stay bolts to support the crown sheet and is the earliest example of a conventional locomotive boiler so-constructed. It was very likely inspired by Winans' practice, but, unlike Winans, Tyson did not use a slope-backed firebox.

The engine was intended for coal-burning and correspondingly has a large firebox measuring 41 inches by 66 inches inside. Note that a variable exhaust after Winans' plan is shown in Fig. 183.

The feed pump is again patterned after Winans'. It is a full-stroke pump driven by a long rod attached to the crosshead. These pumps were later altered to half-stroke pumps driven by a crank on the rear side-rod pin. Tyson avoided at

[2] *Papers Relative to the Recent Contract for Motive Power by the Baltimore & Ohio Railroad Co . . .* (Baltimore, 1857).

Fig. 177. A Tyson Ten Wheeler with the Artists Excursion Train on the Baltimore and Ohio Railroad west of Piedmont, Va., in June, 1858. Note the polished cast-iron cylinder heads.

Fig. 178. The Baltimore and Ohio Railroad's No. 230 at Keyser, W. Va., in 1862. This Tyson ten wheeler was built by the Baltimore and Ohio at Mt. Clare in 1857. The 230 was renumbered the 294 in 1884. It was the last Tyson Ten Wheeler in service and was retired in 1894.

least one defect of Winans' usual plan by injecting the water ahead of the firebox.

The frame closely copies Samual Hayes's (Tyson's predecessor) 1853 design for a ten-wheel Camel passenger locomotive. The forward part of the frame has a conventional bar pattern. However, the rear of the top rail is made deep and thin, thus permitting a wide firebox. The combination spring-equalizer rig is also copied from Hayes's 1853 design.

A modern, wide spread truck (Fig. 184) with a 68-inch wheel base was used. The general plan is identical to the much-admired Mason truck, which is distinguished by rocker side bearings and individual, non-equalized springs.

The valve gear shown in Fig. 185 is a stationary link or Gooch motion. Note that the eccentrics are set well inside the frame, thus necessitating a long upper rocker arm not much different from Winans' design. The stationary link was at first the most popular style but was quickly succeeded by the shifting link; by about 1860 the former had all but disappeared from the American scene.

We are fortunate that Douglas Galton's report included tender drawings equally as elaborate as those of the locomotive. Commonly, no drawings of the tender were reproduced, the thought being that it was too simple a vehicle to deserve the treatment afforded the locomotive. The tender on eight wheels had a capacity of 2,000 gallons. The truck was *inside-bearing*, built on the "live" pattern with large leaf springs serving as the side frames. A careful study of Fig. 177 shows the inside-bearing truck where only the wheels are visible.

Dimensions of the Tyson Ten Wheeler	
Cylinders	18″ × 24″
Wheels	50″
Weight	30 tons

AMERICAN LOCOMOTIVES

Fig. 179. The 227, built by Denmead in December, 1857, is shown as rebuilt with an extended smokebox, cap stack, new cab, half-stroke pump, and outside, wooden-beam tender trucks. The scene is probably Martinsburg, W. Va., in about 1880. The 227 was renumbered the 292 in January, 1884.

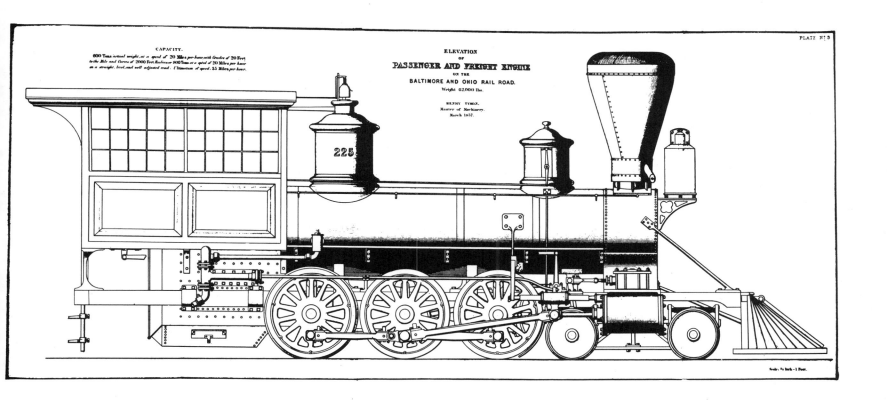

Fig. 180.   *The Tyson Ten Wheeler No. 225, built by Denmead in August, 1857.*

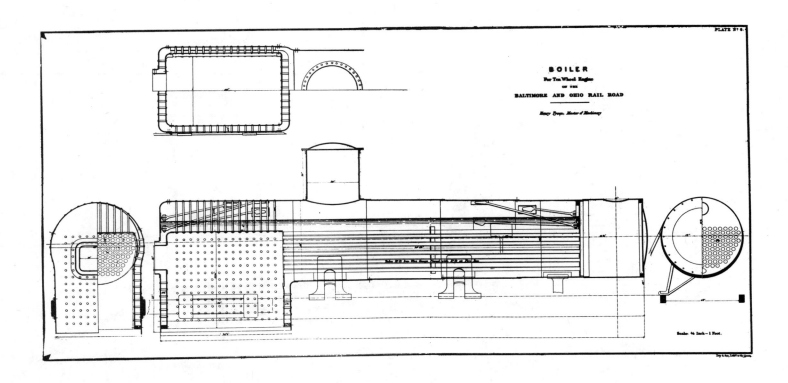

Fig. 181. Detail of Tyson's ten-wheel boiler. Note that crown bars are not used to support the crown sheet.

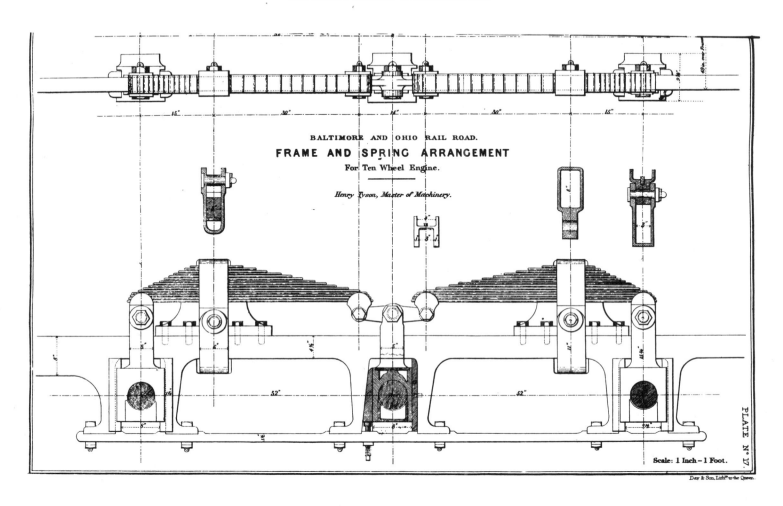

Fig. 182. Detail of the frame for the Tyson Ten Wheeler.

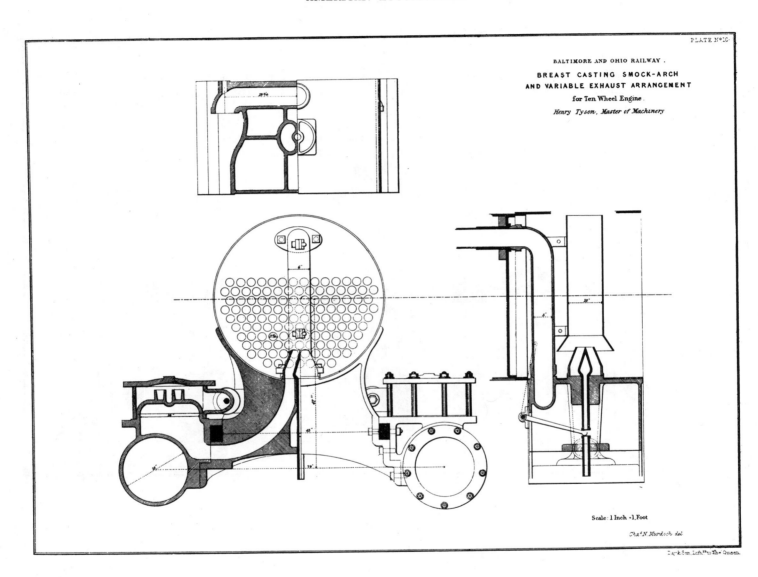

Fig. 183. Cylinder saddle and variable exhaust for the Tyson Ten Wheeler.

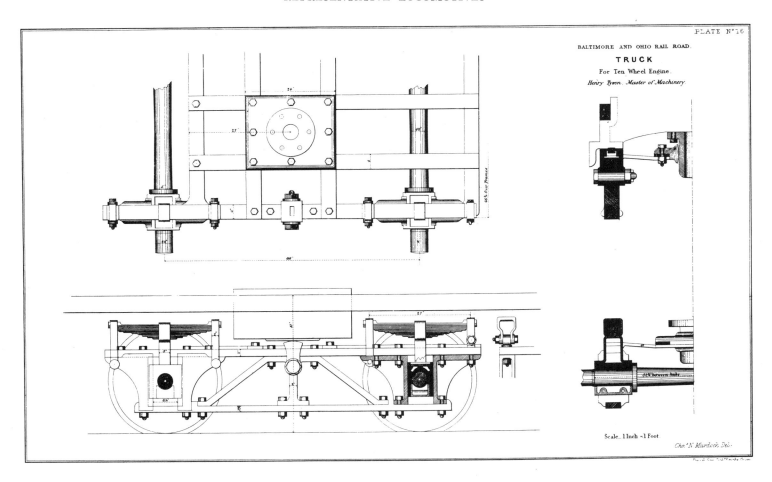

Fig. 184. Truck detail for the Tyson Ten Wheeler. Note the similarity of this truck to Mason's design.

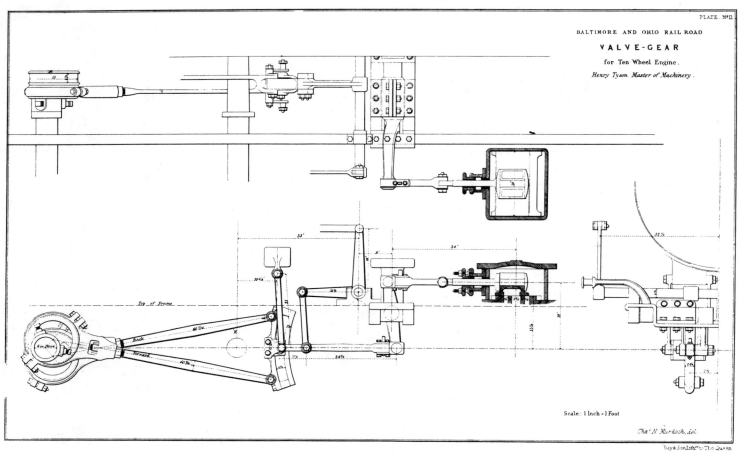

Fig. 185. *Valve-gear detail for the Tyson Ten Wheeler. A stationary link was employed.*

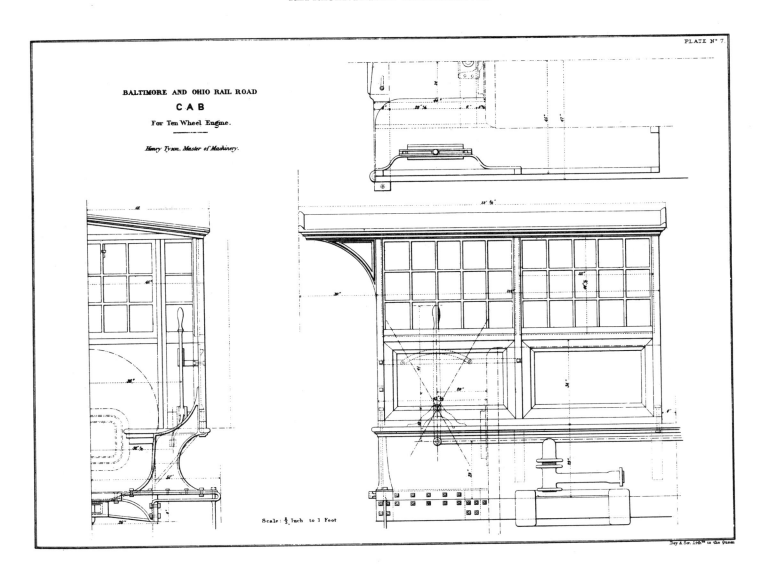

Fig. 186. Cab and rear-frame detail for the Tyson Ten Wheeler. The cast-iron cab bracket is like that used by Mason.

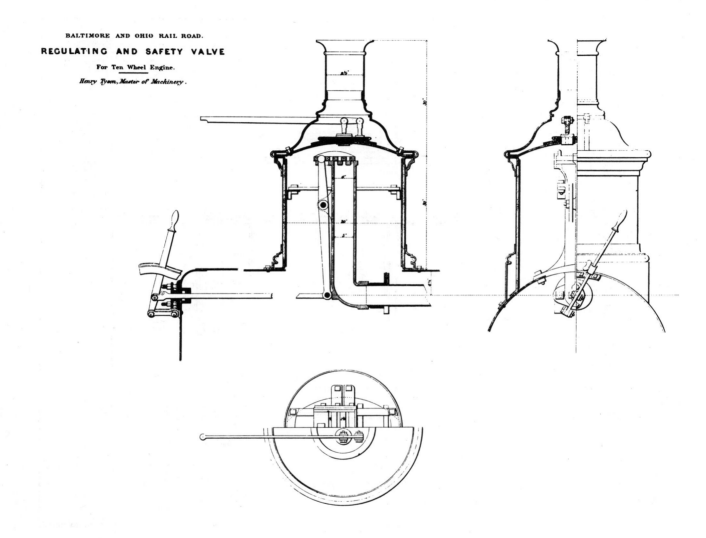

Fig. 187. Throttle and safety-valve detail for the Tyson Ten Wheeler. The throttle is a common slide valve. The spring balances for the safety valves are not shown.

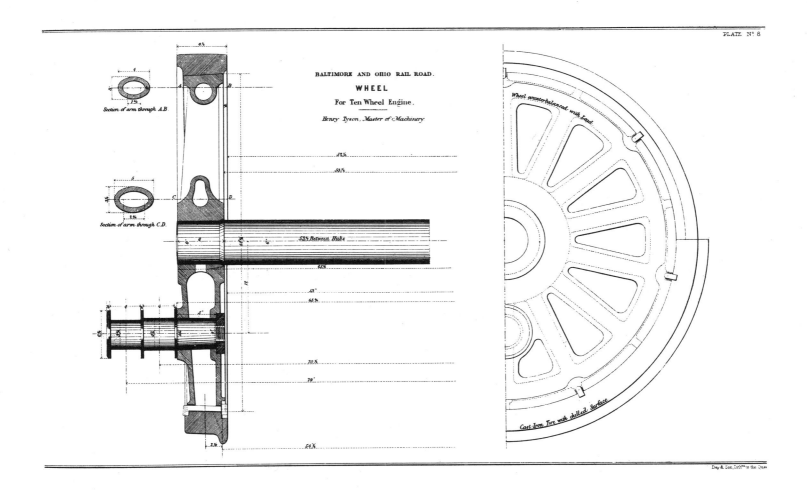

Fig. 188. Driving-wheel detail for the Tyson Ten Wheeler. Notice the heavy cast-iron tire.

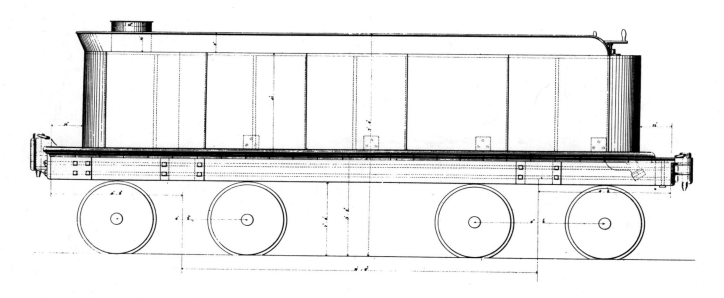

Fig. 189. Side elevation of the tender for the Tyson Ten Wheeler. The tank held 2,000 gallons of water.

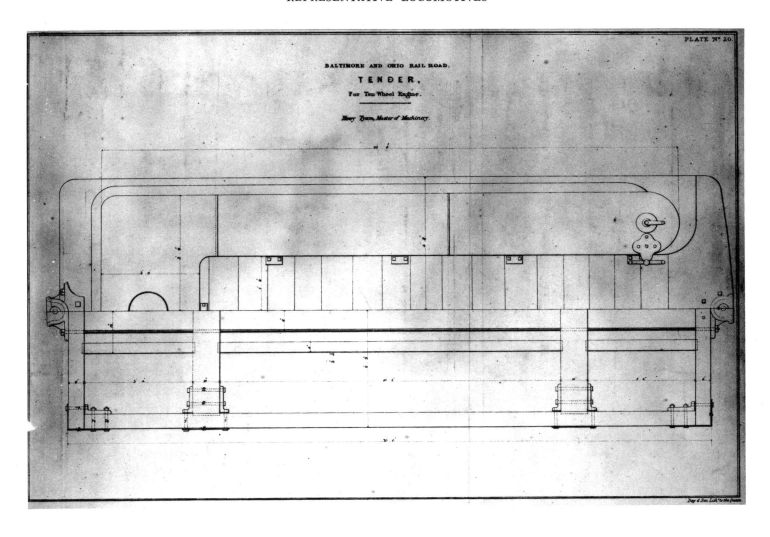

Fig. 190. Plan view of tender tank and frame for the Tyson Ten Wheeler.

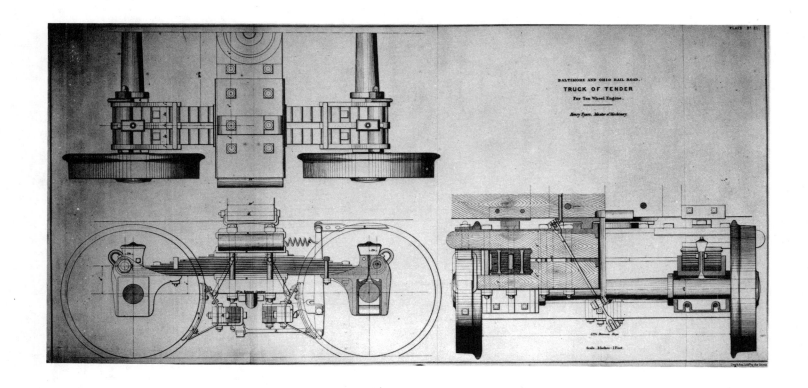

Fig. 190–A. Tender truck of the Tyson Ten Wheeler. Notice that the truck has inside bearings and is built on the "live" pattern.

PHANTOM, 1857

The *Phantom* was an elegant wood-burning passenger locomotive built in February, 1857, for the Toledo and Illinois Railroad by William Mason of Taunton, Massachusetts. The *Phantom* was Mason's fifty-ninth locomotive. The Toledo and Illinois, later part of the Wabash system, received nine similar engines from the same builder between 1855 and 1857. One of these machines, the *Boreas,* was in service for about thirty-four years, a fact that speaks well for this class of engine.[1]

Mason is commonly credited with radical and far-reaching advances in locomotive design. One popular account proclaims Mason as the "father of the American type locomotive." Much of this acclaim undoubtedly stems from Mason's own statements.

In 1882 Mason claimed credit for the cylinder saddle, among other improvements.[2] In an interview with Angus Sinclair he took credit for the split frame and level cylinders and indirectly suggested that hollow-spoke wheels were his idea.[3] Actually, these improvements had been developed by other builders, notably Rogers, several years earlier. Both of these statements were made late in Mason's life and can very likely be credited to the unfortunate clouding of a brilliant mind turned egocentric by advanced years.

[1] Bell, *Early Motive Power*, p. 47.
[2] *Railroad Gazette*, October 27, 1882, p. 656.
[3] Sinclair, *Locomotive Engine*, p. 186.

Fig. 191. William Mason (1808–1883), locomotive and textile machinery builder of Taunton, Mass. This portrait shows Mason at about forty-five years of age.

Mason's chief contribution to locomotive construction was *symmetrical design* rather than basic mechanical reform. His engines were characterized by neat, simple lines, pleasing proportions, and careful treatment of small details. Mason's original intention was to build a good-looking locomotive, as was clearly indicated in the following statement written just before his first engine was completed:

He is now laying the foundation of a large building, to be devoted to locomotive-building. We should not be at all surprised should he do as much for the branch of mechanics as he has for cotton machinery. He is yet in the very prime and vigor of his faculties, and he cannot do anything in a humdrum, routine way. His mind and taste will be strongly concentrated upon the subject before him, and he will, most assuredly, bring forth new things constantly. One thing is certain; we shall, at least, have some better-*looking* locomotives—some in which the lines of beauty will be more prominent. We want them, of course, strong workers, but we want them also good lookers. We have too much respect and admiration for the *iron horse* to allow his legs to be all straightened out and his lines all angularized. As good taste has been successfully at work on cotton machinery, we shall hope to see something soon on the rails that does not look exactly like a "cooking stove on wheels.[4]

Mason achieved a beautifully symmetrical design by carefully arranging the basic external parts of the locomotive. The steam dome was placed exactly between the driving wheels in line with the center of the equalizing lever. The smokestack, smokebox, cylinders, and truck were all placed on the same vertical center line. The sandbox was placed between the steam dome and smokestack. The awkward, outside-frame running board was replaced by a simple, straight running board inconspicuously supported by small brackets attached to the boiler shell. The handrail and straight boiler, together with the running board, all combined to produce a strong, simple horizontal line. These complemented the carefully arranged vertical lines of the steam dome and stack. The rich classical mouldings of the dome, sandbox, and cab, plus other tasteful details, produced a pleasing effect. By the mid- and late 1850's other builders, particularly the New England firms, were producing locomotives on the same handsome lines as Mason's. Mason should then be credited as a gifted *stylist* rather than a mechanical innovator as far as locomotive history is concerned.

All of this should in no way suggest that Mason did not build a fine mechanical product. The precision of design was carried over into manufacturing. All important parts were machine-fitted rather than hand-chipped and filed. Holes were reamed and bolts snugly fitted. All assemblies were carefully aligned and checked for free, smooth running.

Despite the fine reputation enjoyed by Mason, the production of his shop was not particularly large nor was the works ever devoted entirely to locomotive-building. Between 1853 and the end of production in 1890, only 754 locomotives were built.

MECHANICAL DESCRIPTION

Mechanically, the *Phantom* incorporated the best features of the American locomotive that had been developed by the mid-1850's. These included level cylinders, spread truck, link motion, and the bar frame. The general arrangement and details are probably more familiar and certainly less awkward than the earlier machines described earlier in this work.

The boiler was straight and had a deep, narrow wood-burning firebox. A perforated dry pipe is shown in Fig. 193. The existence of the steam dome may therefore be questioned since the perforated dry pipe was developed expressly to eliminate the dome. Possibly the dome was added for more steam room or because the builder would not have his design ruined for the lack of a dome. A slide-type throttle valve is located in the smokebox. Note the steam gauge shown in Fig. 194. A common bonnet stack was used.

A shifting-link Stephenson valve gear was used. Notice that the lifting shaft is above rather than below the links as was the

---

[4] John Livingston, *Portraits of Eminent Americans* (New York, 1853), p. 18.

common practice of the times. A coil spring attached to the reversing lever by a link chain was used to counterbalance the valve gear. This curious arrangement was favored by Mason for many years. It was also used by Wilson Eddy on many of the locomotives he built for the Boston and Albany.

The conventional bar frame needs no description. However, the heavy cast-iron bed plate below the cab deck is worthy of mention. It served both as rear frame support and drawbar pocket.

The leading truck had rocker side bearings and individual non-equalized springs. The rockers are not shown in Fig. 193, but the reader is referred to the Eddy Clock frame drawing (Fig. 64) in the introductory section for this detail. Mason favored this style of truck for some years, but it was not as safe as the center-bearing, equalized trucks used by the Paterson builders. The truck wheels are spoked and were advocated by the builder for looks rather than utility. Mason thought that plate wheels looked like "cheeses" and would not use them on leading or tender trucks.

Only the side elevation of the tender was included in the engravings of the *Phantom* published in Clark and Colburn's *Recent Practice in the Locomotive Engine* (see Fig. 195). Fortunately, J. Snowden Bell prepared a sketch in 1863 of a tender from one of the engines supplied to the Baltimore and Ohio by Mason in 1856–57. This sketch, redrawn by the present author, provided basic dimensions for a suitable reconstruction. Mason continued to build tenders with frames *inside* the truck wheel until the early 1870's. The apparent reason for this design was the style of truck used; it included four individual springs. This arrangement, inside frame and unequalized trucks, must have produced an unstable and hard-riding tender. Because of the narrow tender frame and the outside bearings required for both trucks, it was not possible to achieve the three-point suspension so desirable for any railway vehicle.

Dimensions of the *Phantom*

Cylinders	15″ × 22″
Wheels	66″ diameter
Weight	22 tons (empty)
Tubes (136)	2″ outside diameter × 11′ 3″ long

Fig. 192. The Baltimore and Ohio Railroad's No. 232, built by Mason in June, 1857. Scene is probably Cumberland, Md., in 1858.

REPRESENTATIVE LOCOMOTIVES

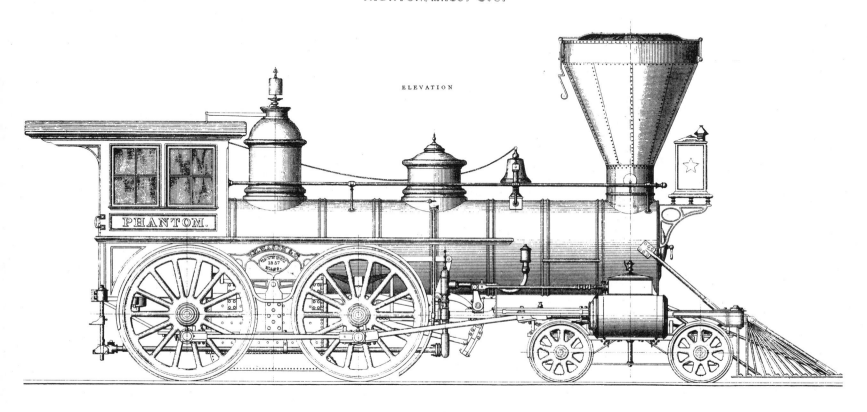

Fig. 193. Elevation of the Phantom, built by Mason in 1857 for the Toledo and Illinois Railroad.

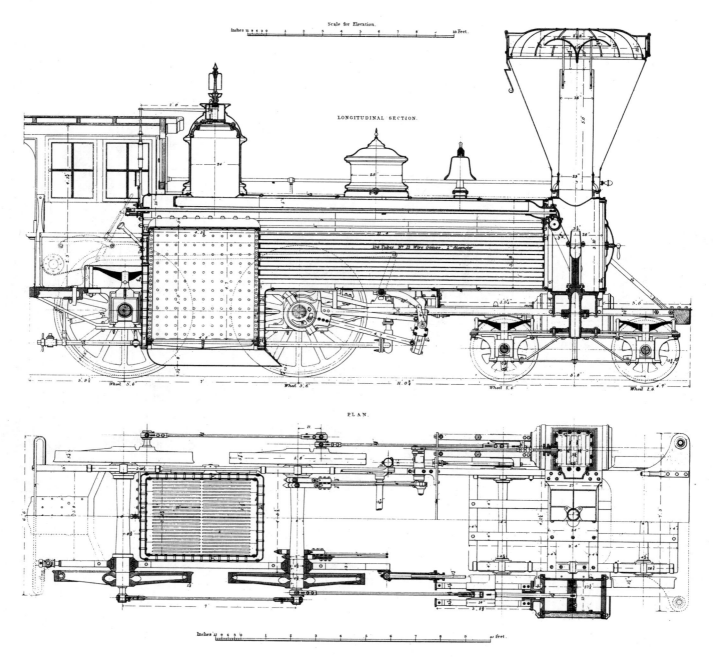

Fig. 193–A. Longitudinal section and plan view of the Phantom.

REPRESENTATIVE LOCOMOTIVES

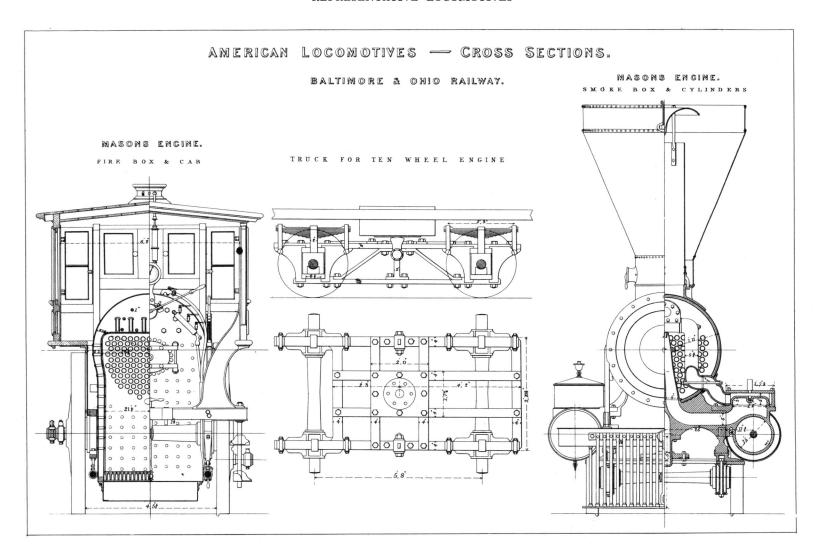

Fig. 194. End elevation and sections of the Phantom. The truck detail is for a Tyson Ten Wheeler and should not be considered as part of the Phantom drawing.

TENDERS FOR AMERICAN LOCOMOTIVES.

MASON'S TENDER

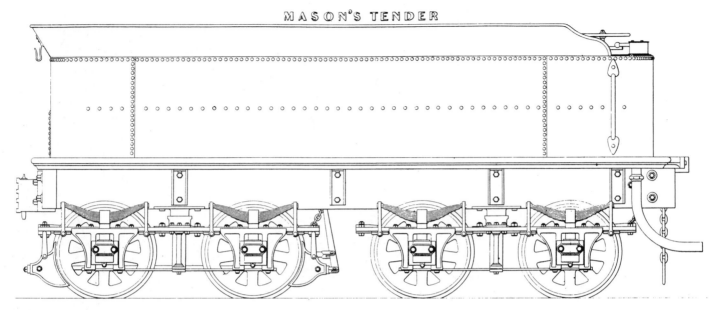

Fig. 195. Side elevation of the Phantom's tender.

Fig. 196. Tender built by Mason in May, 1862, for the U.S. Military Railroad. This tender is nearly identical to that of the Phantom.

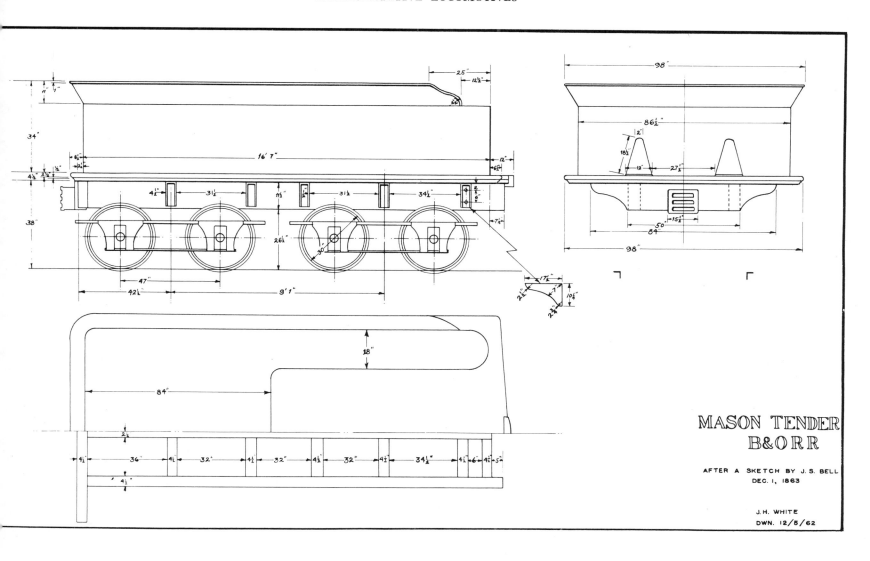

Fig. 197. A Mason tender of 1856 or 1857 built by Mason for the Baltimore and Ohio Railroad and sketched in 1863 by J. S. Bell.

SOUTHPORT, 1857

The *Southport*, a 6-foot-gauge, wood-burning passenger locomotive, was built in 1857 by Danforth, Cooke and Company, a noted locomotive builder of Paterson, New Jersey. The engine was reportedly built for the Ohio and Mississippi Railway, a broad-gauge road, but was refused before delivery.[1] The builders then sold the engine to the Delaware, Lackawanna and Western Railroad in February, 1858, for $11,500. In 1865 the *Southport* was renamed the *W. E. Dodge* and again in 1876 it was renamed the *Sam Sloan*. It was last used on the Morris and Essex division of the Delaware, Lackawanna and Western and was scrapped in 1912.

The *Southport* was a typical eight wheeler of the period. Its mechanical features are so standard that further comment is almost unnecessary. The front-end throttle, however, is worthy of notice, as is the leading truck. The double-bearing truck enjoyed a limited popularity at this time, particularly on passenger locomotives where the extra bearing was considered a safety feature should an axle break. The Kite safety bearing, an iron frame strapped loosely around the axle, was another precautionary device used on the *Southport*'s truck. Introduced in the 1830's, it was commonly used on passenger cars but was rarely used on locomotives; the present example is the only one known by the author.

[1] *Railroad Gazette,* July 18, 1902, p. 567. In an article on early Delaware, Lackawanna and Western locomotives, Herbert L. Walker contends that the *Southport* was not accepted by the Ohio and Mississippi because the road had not opened. However, the entire line from Cincinnati to St. Louis opened in June, 1857, and certain portions of the line had been in operation since 1853.

Dimensions of the *Southport*

Cylinders	17″ × 22″
Wheels	66″ diameter
Tubes (150)	2″ diameter × 11′ 6″ long
Grate area	18.38 sq. ft.

REPRESENTATIVE LOCOMOTIVES

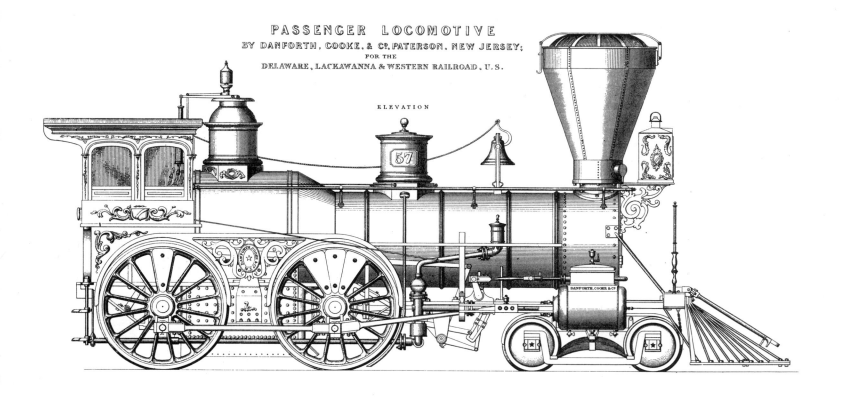

Fig. 198. *The Southport, built in 1857 by Danforth, Cooke & Co., was used for passenger service on the Delaware, Lackawanna and Western Railroad.*

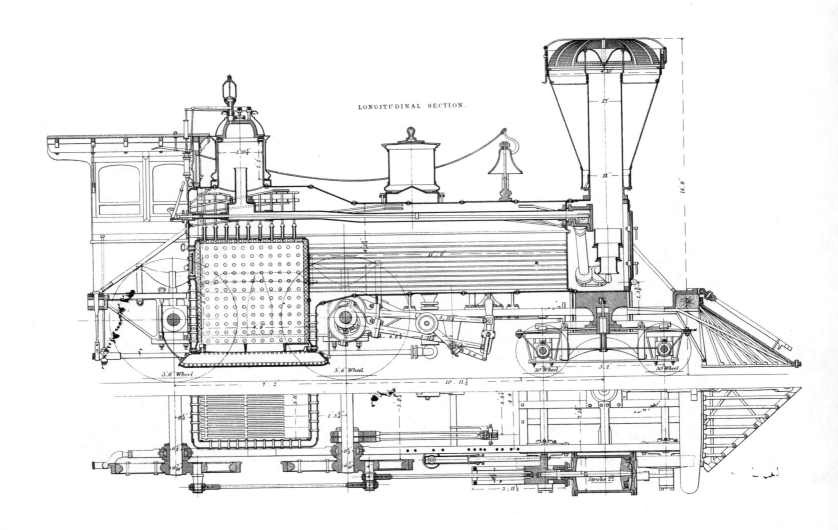

Fig. 198–A. *Longitudinal section of the* Southport.

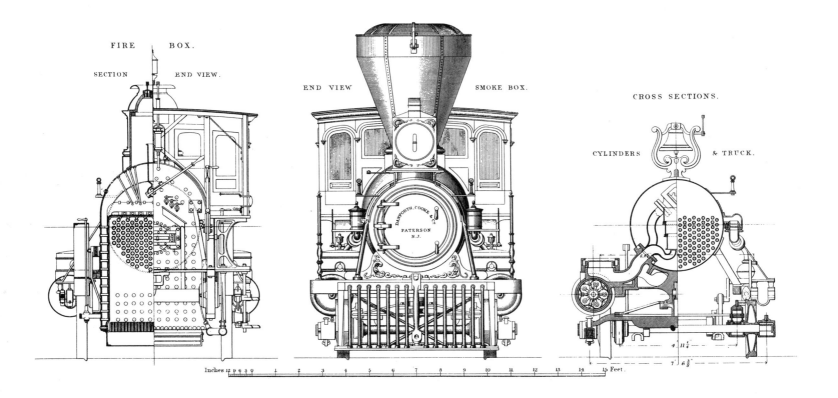

Fig. 198–B. *End view of the* Southport.

THE FLEXIBLE-BEAM-TRUCK LOCOMOTIVE, 1857

While many early United States railroads were built and equipped primarily for passenger traffic, the rapid development of a lucrative freight business in the early 1840's brought with it a demand for more powerful locomotives. Conventional six- and eight-coupled freight engines were capable of moving heavy trains but had difficulty in passing around sharp curves. When forced through tight radii, such engines, with their long, rigid wheel bases, would turn rails over or suffer rapid flange wear. American locomotive builders soon remedied this problem.

In 1842 M. W. Baldwin developed a freight locomotive specifically for poorly built roads needing powerful but flexible locomotives. It was known as a flexible-beam-truck engine. All wheels were coupled, yet the drivers were not closely coupled together like the early short-wheel-base 0–6–0's (see the *Philadelphia*); they could be spread out to afford a smooth-running engine. In Baldwin's plan the two front driving wheels were carried by a "truck" that was arranged so that the axles could slide laterally when on a curve but would remain parallel to the rigid, conventionally mounted rear driving axle.

Since the axles were always parallel the rods did not bind. The movement of the front driving wheels was not great. Even on a curve as extreme as 150 feet in radius, the wheels' movement was only 3 inches out of line from the rigid rear wheels.

The scheme was ingenious, its details well worked out, and, unlike many variants of conventional practice, the flexible-beam-truck effectively solved a practical operating problem. It was so enthusiastically received that for the first years after its introduction Baldwin's production was all but entirely devoted to engines of this pattern. During the years of its production, 1842–66, about three hundred were built.

The eulogistic memorial to Baldwin's life claims that the idea of a flexible-beam locomotive "occurred to him suddenly in the depths of the night." This account continues with a romantic story, so typical of nineteenth-century industrial biographies, of Baldwin's dramatic, last-minute triumph over his creditors, the sheriff, and foreclosure, all occasioned by his timely invention of the night before.[1] Actually, Baldwin had

[1] *Memorial of M. W. Baldwin* (Philadelphia, 1867).

been working for some time to perfect an engine with greater tractive power than the 4–2–0 without resorting to coupled driving wheels. In 1839 he perfected a plan to power the leading wheels of a 4–2–0 by means of gearing. In August, 1841, an engine was built on this plan, and, though it performed well, it was not duplicated. Its complexity over conventional engines of the same capacity and the tendency of its gears to strip brought no further orders. Only one demonstrator engine was built. The geared engine was not a complete loss for it prompted Baldwin to develop the flexible-beam truck. The idea of powering all the wheels of a 4–2–0, accomplished in the geared locomotive, was carried over to the flexible-beam engine. In the geared engine the front "drivers" were small truck wheels carried on a conventional truck, which when on a curve was no longer at right angles to the boiler's axis. Therefore, since the axles were not parallel, side rods could not be used. The power was transmitted by broad-faced gears. Baldwin's solution to this problem of transmission was to design a truck that could swivel while maintaining parallel axles. The principle of the design can be illustrated with a parallel ruler where the pivoting arms represent the driving-wheel axles. Thus, when the engine is on a straight track, the truck is a rectangle, while on a curve it is a parallelogram. Since all axles remain at right angles to the boiler's center line the axle centers remain constant and the wheels can be connected with rods. The truck beams were connected only by the axles, which in turn were held by cylindrical boxes and thus were able to accommodate any lateral movement. The truck beam itself was attached to the main frame by a ball joint and could vibrate laterally and vertically. These features were patented by Baldwin on August 25, 1842 (No. 2759).

The first flexible-beam engine was rebuilt from a 4–2–0 built by Baldwin in 1837 as the *Tennessee* for the Georgia Railroad and Banking Company. The *Tennessee*, rebuilt as an 0–6–0, was completed on December 8, 1842, and was assigned a new construction number, No. 180. One wonders how many other engines listed by the builder as new machines might in fact have been old engines made over? While no drawings of the *Tennessee* exist, we are fortunate in having a drawing of one of the first flexible-truck engines built, the *John C. Calhoun* of the Wilmington and Raleigh Railroad.[2] This drawing, reproduced as Fig. 199, shows an 0–6–0 with the rear driving wheels set behind the firebox. The same wheel arrangement was also offered with all drivers in front of the firebox, thus affording greater flexibility for sharp curves but a less-steady engine on straight track. About 120 six-wheel flexible beams were constructed by Baldwin. The majority of these were built before 1855.

An odd variety of the six-wheel flexible beam was built for passenger work. It might be termed a 2–4–0 since a small leading wheel was used in place of the front drivers. The first engine of this type was built in October, 1843, for the Erie and Kalamazoo Railroad. Several engines of this plan were also built for the Württemberg Railway in 1845. One of these machines is illustrated in the introductory section of the present volume (see Fig. 82). In all, about 21 six-wheel flexible-beam passenger engines were constructed.

Baldwin considered the flexible-beam patent to be of considerable value and refused to sell more than licensing rights to produce engines of this design. No United States manufacturer other than Baldwin built flexible-beam locomotives. Some flexible beams were built in Germany by the Esslinger Works in about 1850.[3] These eight-wheel engines varied from Baldwin's design in that both the front and rear pairs of axles were carried in a truck, not merely the front pair. The truck beams or frames were outside the wheels. Cranks were mounted on the axle ends. The number of these machines built cannot be determined, nor is it known if Esslinger made any patent agreement with Baldwin.

The eight-wheel flexible-beam engines, with two trucks and two rigid axles, were mentioned in the 1842 patent specification, and Baldwin offered to build such an engine in a prospec-

---

[2] The *John C. Calhoun* was the Baldwin Works's No. 181, dated November, 1843. It is not the second flexible-beam locomotive, as might be imagined, for the construction numbers were not assigned in chronological order; for example, No. 185 was delivered in October, 1843.

[3] Max Mayer, *Esslinger Lokomotiven: Wagen und Bergbahnen* (Berlin, 1924).

tus of 1844, yet none appeared until 1846. The *Atlas,* built in 1846 for the Philadelphia and Reading, is said to have been the first eight-wheel flexible beam. It was accordingly the first of some 150 machines built by Baldwin during the next twenty years. In 1850, two 36½-ton eight-wheel flexible beams were built for the six-foot-gauge Erie. Remarkable though it may seem, engines approaching this size were successfully worked around curves of a 230-foot radius on the Don Pedro II Railway in Brazil.[4] As successful as the flexible beam was in moving tonnage trains over roads with sharp curves, it was adept only in slow traffic. As road beds improved and the demand for faster freight service increased, the purchase of flexible beams fell off. Between 1842 and 1854, flexible beams accounted for 200 of the 450 locomotives built by Baldwin. After that time, and until the end of production in 1866, they accounted for only about 110 of the 930 engines produced. Aside from improved roads, the introduction of the 4–6–0 and later the 2–6–0 can account in part for the flexible beam's demise. Like all 0–6–0's and 0–8–0's the front wheels were overloaded. This was particularly true of flexible beams since large, heavy cylinders were the rule. Another mechanical failing of the flexible beam was the placement of the cylinders. Set forward and high to clear the front drivers, the cylinders were weakly attached to the engine.

The official Baldwin history is clearly wrong in stating that the last flexible-beam engine for domestic service was built in 1859.[5] A number of machines were built after that date for the Mine Hill Railroad, the Iron Railroad, and the Pennsylvania Railroad. The last flexible beam was very likely the *Santiago De Cuba* shown in Fig. 208. Cuba imported a number of flexible beams and had one in service as late as 1911.

Figures 203–205 illustrate the final development of the powerful eight-wheel flexible-beam locomotive. These drawings are from Clark and Colburn's *Recent Practice in the Locomotive Engine*. While the authors did not definitely identify the engine, the wheel and cylinder size indicate that it must be one of the following 25-ton Class E engines:[6]

Const. No.	779	Little Schuylkill R.R.	*Schuylkill*	Aug. 24, 1857
"	" 787	Penna. Railroad	No. 44	Oct. 15, 1857
"	" 788	North Penna. R.R.	*Carbon*	Oct. 21, 1857
"	" 789	Penna. Railroad	No. 45	Nov., 1857
"	" 793	Penna. Railroad	No. 128	"
"	" 794	Penna. Railroad	No. 129	"

We have the history of one of these engines, the Pennsylvania Railroad's No. 45, *The Northumberland*. It operated as built until 1864 when it was rebuilt at Altoona. It emerged as a curious eight-wheel engine (see Fig. 206). The flexible-beam truck was moved forward about two feet, apparently in an attempt to distribute the weight more evenly. As already mentioned, the front wheels of the flexible beams were overloaded. The front drivers were replaced by smaller leading wheels. This reduced traction but was necessary in order to clear the cylinders. The driving-wheel size was increased from 43 inches to 48 inches. New crossheads and guides were added. Link motion replaced the variable cutoff and hook motion. Finally, the feed pumps were replaced by injectors. The engine probably was not successful in its rebuilt form for it was scrapped in December, 1869.

Returning to Figs. 203–205 we find an eight-wheel connected engine nearly identical in size to a Winans' Camel and intended for the same service, slow freight. The large grate area (18.6 square feet) and the iron firebox and tubes are unmistakable features of a coal-burning locomotive. The variable exhaust, controlled by the handwheel on the left side of the cab (Fig. 204), and the steam jet visible in the smokebox are other features common to coal-burning engines.

The ratchet throttle is an oddity and no other examples of this arrangement are known to the author. Aside from the throttle the details of this engine are ordinary. The throttle

---

[4] *Engineering,* May 15, 1868, p. 467.
[5] *History of the Baldwin Locomotive Works,* p. 57.

[6] Data from the original specifications of the Baldwin Locomotive Works, courtesy of the De Golyer Foundation Library, Dallas, Texas.

valve is a common slide valve, the stack a straightforward bonnet design.

Baldwin's variable cutoff, patented in 1853, is shown in Fig. 203. It has already been discussed in the introductory section on valve gears. The main valves are operated by V hook motion. Baldwin continued to use this old-fashioned form of valve gear, including the variable exhaust, as late as 1860.

The flexible-beam truck is clearly shown in Figs. 203–205. Each beam consists of two half-inch wrought-iron beams riveted through a cast-iron center. The cast center does not entirely occupy the space between the plates but is made that way so that the spring and its straps fit neatly between the plates as well. The composite wrought- and cast-iron beams were introduced in about 1848 and replaced the earlier all-cast-iron beam. The old-style beams, shown in Figs. 199–201, were cast solid with heavy strengthening ribs. As engine weights increased, however, these beams proved inadequate and were replaced by the more complex, but stronger, composite beams.

Dimensions of the
25-Ton Class E Flexible-Beam-Truck Locomotive

Cylinders	18″ × 20″
Wheels	43″ diameter
Weight	31¾ tons
Tubes (110)	2⅛″ outside diameter × 11′ 6″ long

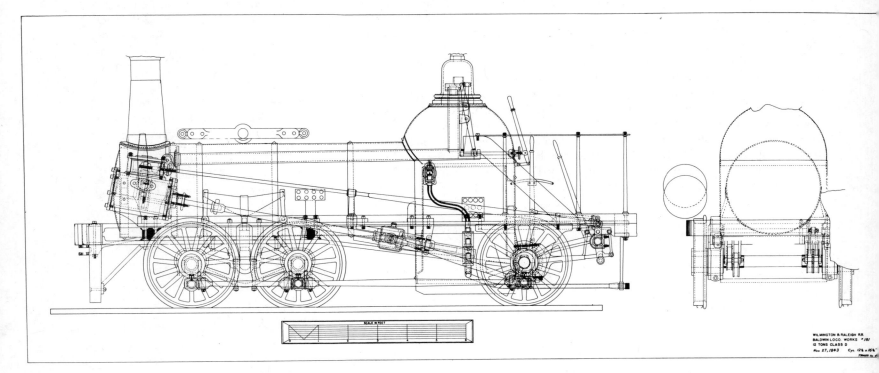

Fig. 199. One of the first flexible-beam-truck locomotives, built by Baldwin in November, 1843, for the Wilmington and Raleigh Railroad. Named the John C. Calhoun, it had 12½-in. by 16½-in. cylinders, 36-in. wheels, and weighed 12 tons. Note the plan-view detail of the truck beam above the boiler and the combination crosshead guide and feed-water pump.

Fig. 200. A flexible-beam-truck locomotive of 1844.

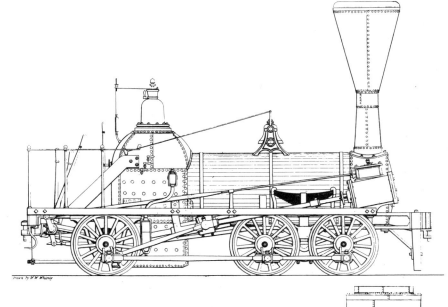

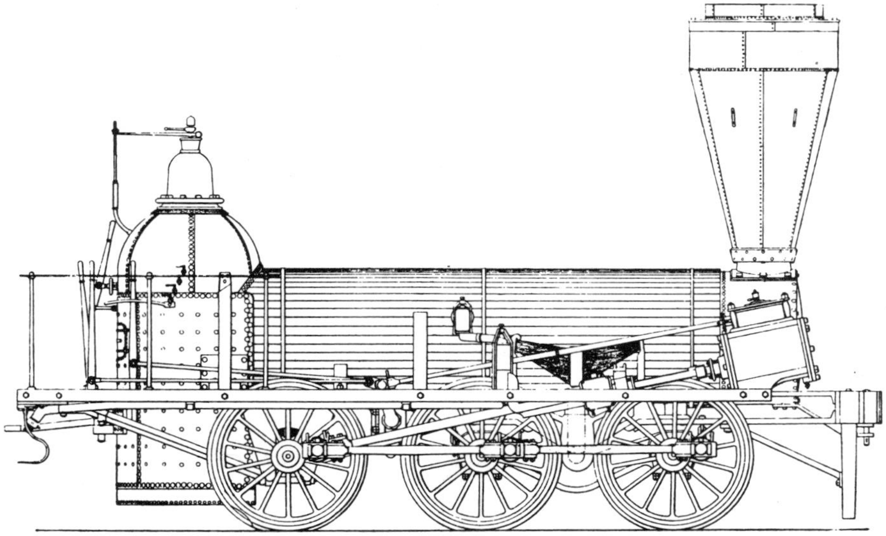

Fig. 201. A flexible-beam-truck locomotive of about 1845.

Fig. 202. A flexible-beam-truck locomotive of about 1848. Note the wrought-iron truck beam.

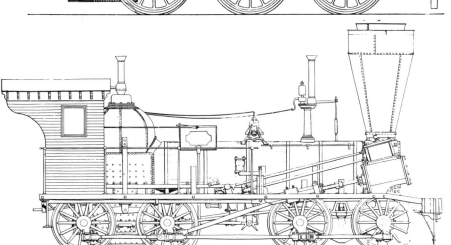

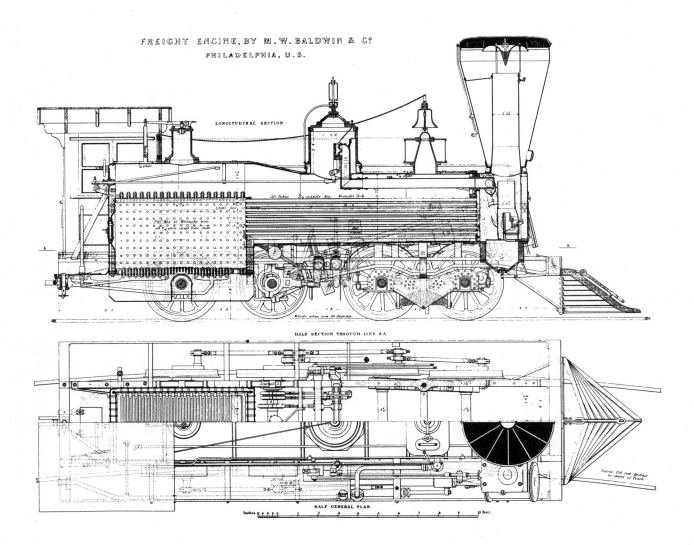

Fig. 203. *A flexible-beam-truck locomotive of 1857.*

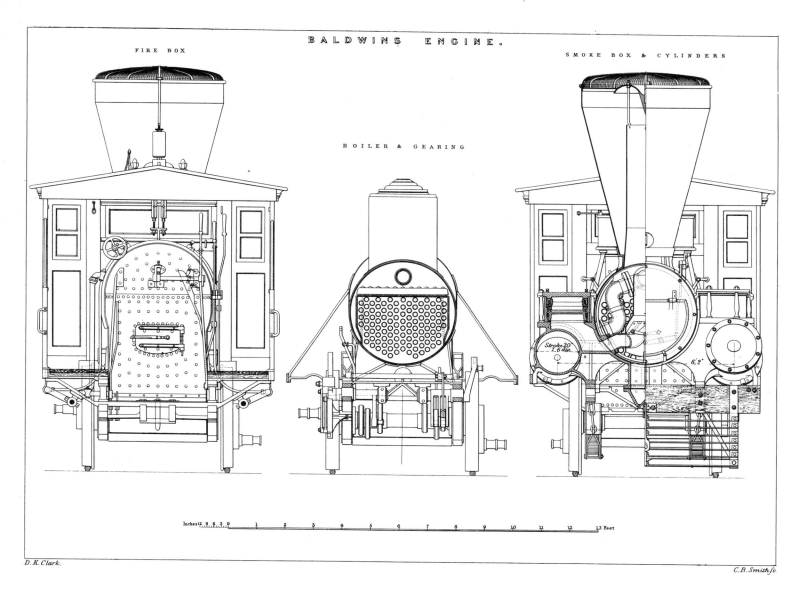

Fig. 204. Front and end elevations of a flexible-beam-truck locomotive of 1857.

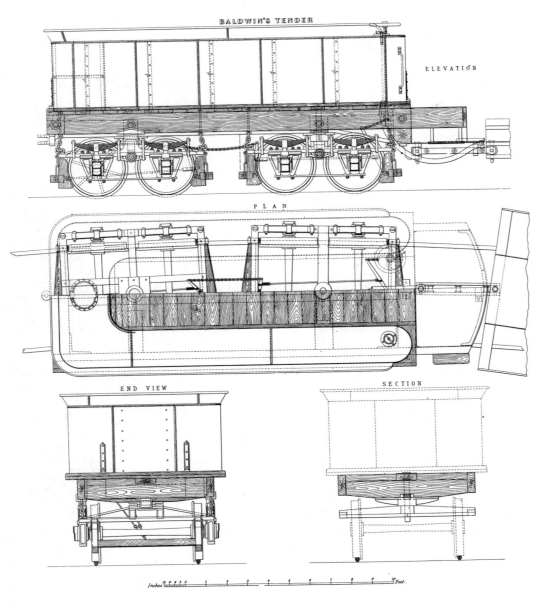

Fig. 205. Tender drawings for a flexible-beam-truck locomotive of 1857.

REPRESENTATIVE LOCOMOTIVES

Fig. 206. The Pennsylvania Railroad's The Northumberland, shown as rebuilt in 1864, was built in 1857 after the drawings shown in Figs. 203 and 204.

Fig. 207. A flexible-beam truck of about 1855–60.

Fig. 208. The Santiago De Cuba, *built by Baldwin in May, 1866. This is thought to have been the last flexible-beam-truck locomotive constructed.*

A ROGERS MOGUL OF 1863

This well-proportioned coal-burning Mogul locomotive was built by the Rogers Locomotive and Machine Works for the New Jersey Railroad and Transportation Company, now an important segment of the Pennsylvania Railroad's main line between Philadelphia and New York. The engine was intended for passenger, freight, and fast-freight service, but because of its relatively large drivers it could be used for heavy passenger trains as well. Rogers delivered two Moguls, Nos. 35 and 36, to the New Jersey road in October and November of 1863. They were so successful that a third machine of the same pattern was ordered from the Rogers plant in February, 1864. Numbered 39, it was not delivered until January, 1865, because of the priority of government orders for the Civil War.

Late in 1871 the Pennsylvania Railroad leased the New Jersey Railroad and Transportation Company (previously acquired by the United Railroads of New Jersey) and acquired the above-mentioned Moguls with the property. The engines were renumbered as 735, 736, and 739 respectively. The 735 and 736 were sold in October, 1879. The 739, well worn after seventeen years of use, was scrapped in October, 1880.

Before going on to the mechanical description of these engines, it would be well to comment on the popular misconception that these three early Moguls were the first 2–6–0's built in the United States. This error probably originated from a careless reading of Forney's Rogers history, which stated only that the firm built its *first Mogul* in 1863.[1] Several examples of Moguls built prior to 1863 by other builders have been given in the introductory section of this work.

The New Jersey Railroad and Transportation Company's, Nos. 35, 36, and 39 were modeled very closely on Moguls built a year or so earlier by the Cooke and the New Jersey Locomotive and Machine companies for the Erie and Central of New Jersey railroads. Since these two works were also in Paterson, Rogers undoubtedly had the opportunity to examine the products of its competitors. All three builders used the Millholland boiler. However, Rogers used a single steam dome in place of Millholland's customary two domes. The size of the cylinders and wheels and the weight of all these engines were nearly identical.

[1] Forney, *Locomotives and Locomotive Building*, p. 20.

Rogers used a massive four-bar crosshead guide because of the large cylinder size. Two half-stroke pumps, driven by a crank on the rear driver, were used because of the forward placement of the front driving wheels. There was not sufficient space for a regular crosshead pump.

The New Jersey Railroad and Transportation Company's Nos. 35 and 36 were fitted with Bissell two-wheel leading trucks as were other early Moguls. The poor tracking qualities of the Bissell pony truck led W. S. Hudson to equalize the front driving axle and the pony-truck axle. This arrangement is believed first used on the No. 39. Hudson's design was so successful that it became the standard plan for all locomotives with single-axle trucks subsequently built in this country.

The tender drawing (Fig. 215) is from Weissenborn's *American Locomotive Engineering* and shows the standard ten-wheel arrangement used by the New Jersey Railroad and Transportation Company. Notice the use of rubber blocks for springs. The neighboring Camden and Amboy Railroad also used the unusual combination of four- and six-wheel trucks on its tenders.

The present author based reconstruction drawings of the 36 (Figs. 211–214) on the two builder's photographs, Figs. 209 and 210, and the engraving in Forney's history of the Rogers Works.[2] A comparison of the Forney engraving and the photographs revealed several minor differences in the details of the frame and the placement of the dome sandbox and bell; these were accordingly adjusted in the reconstruction.

Dimensions of the New Jersey Railroad and Transportation Company's No. 36

Cylinders	17" bore × 22" stroke
Wheels	54" diameter
Total engine weight	36½ tons

[2] *Ibid.*

Fig. 209. Builder's photograph of the New Jersey Railroad and Transportation Company's No. 36, built in October, 1863, by Rogers Locomotive and Machine Works of Paterson, N. J.

Fig. 210. *The New Jersey Railroad and Transportation Company's No. 36 in service; shown with Hudson's coal-burning stack.*

Fig. 211. *Reconstruction drawing of the New Jersey Railroad and Transportation Company's No. 36, side elevation.*

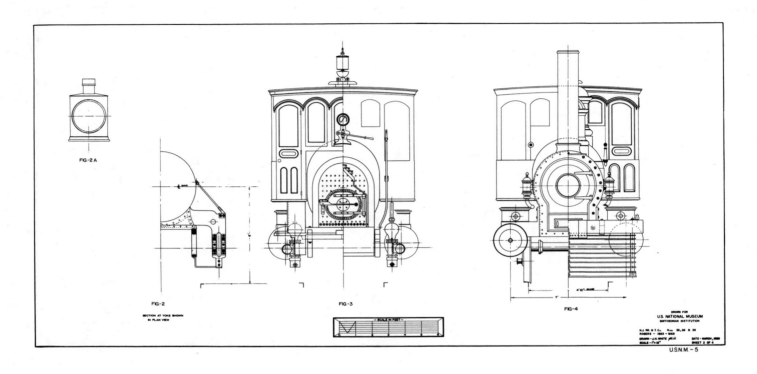

Fig. 212. Reconstruction drawing of the New Jersey Railroad and Transportation Company's No. 36, details.

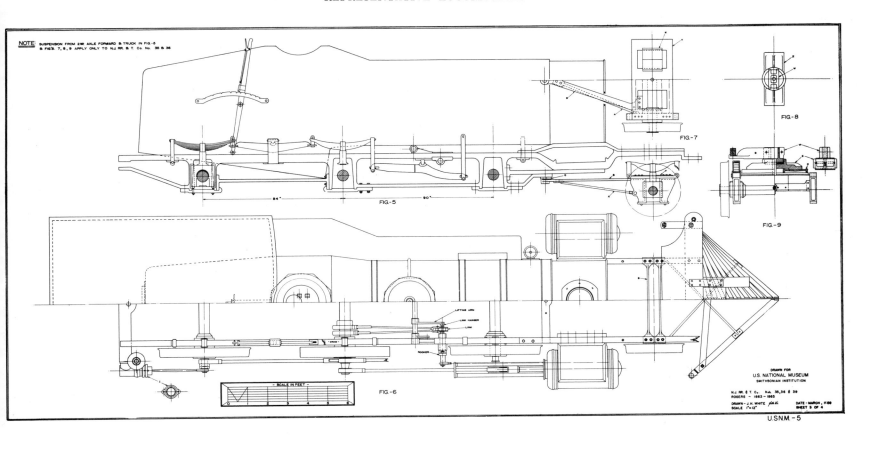

Fig. 213. Reconstruction drawing of the New Jersey Railroad and Transportation Company's No. 36, details.

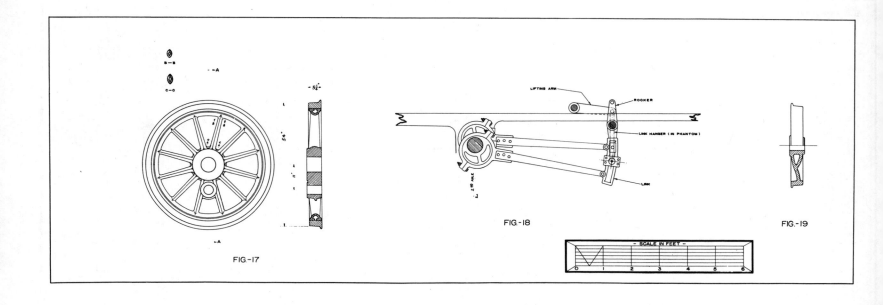

Fig. 214. *Reconstruction drawing of the New Jersey Railroad and Transportation Company's No. 36, details.*

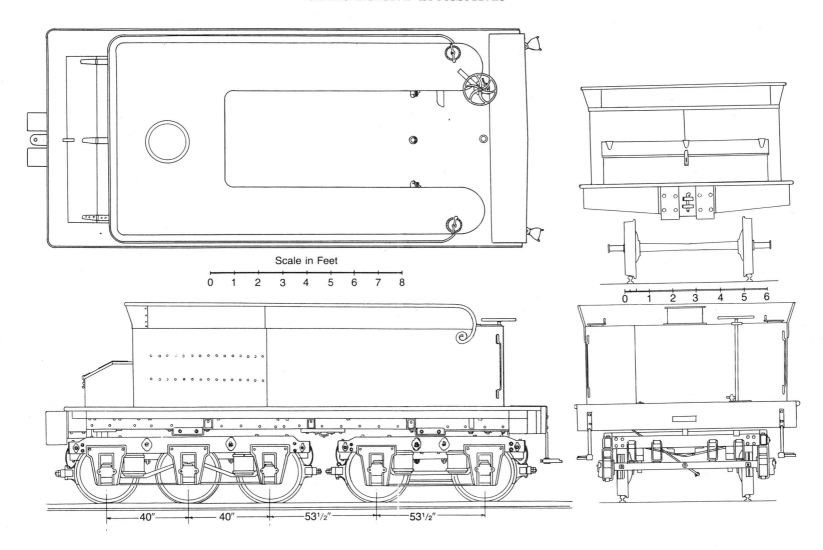

Fig. 215. Standard ten-wheel tender of the New Jersey Railroad and Transportation Company. This style of tender was used on the 36.

THE ERIE MOGUL NO. 254, 1865

Until the early 1860's the Erie Railway was content to move its ponderous broad-gauge freight trains with ordinary eight- and ten-wheel engines. The great expense of operating short trains with light locomotives caused the road to adopt the powerful Mogul for freight service. Accordingly, ten 2–6–0's were ordered from Danforth, Cooke and Company late in 1862. The new machines were such a success that several other orders followed and the Erie soon became one of the first roads in the United States to operate Moguls in large numbers.

The subject of the present discussion, the No. 254, was part of this great fleet and, though delivered in March, 1865, was identical to the first order filled by Cooke in 1862. The Erie's 250 class was described in the British publication *Engineering* dated May 11, 1866. Much of the information given here, together with Figs. 218 and 219, is from that account.

Aside from its wheel arrangement, the most notable feature of the 254 is its boiler. It was built on Millholland's plan for anthracite-burning. The boiler's general plan is very closely patterned on the modified wagon top used on Millholland's *Hiawatha* (1859). Notice the large firebox, the absence of crown bars, and the water-tube grate. The five lower grate bars were not water tubes since they were fairly well insulated from the direct action of the fire by the upper 16 water tubes. The boiler's waist was 46 inches in diameter and cannot be considered too ample for cylinders 17 inches by 22 inches. There were 175 iron tubes measuring 2 inches in diameter and 11 feet long. The total heating surface was 1,255 square feet. The two steam domes, double-poppet throttle valve, and the small vertical throttle lever with a ratchet are directly copied from Millholland's design. Here again the Erie showed good sense by not only purchasing the most advanced type of coal-burning locomotive but also adopting the most progressive design available.

Three safety valves were used. Two ordinary lever valves were placed on the rear steam dome. The spring holding these valves closed was so arranged that the crew could manually open the valve if necessary. The third valve, located atop the forward dome, was a pop valve held shut by the leaf springs clearly shown in Fig. 219.

The boiler was fed by a crosshead pump on the right side

REPRESENTATIVE LOCOMOTIVES

and an injector on the left side. The injector was attached to the frame between the rear and middle driving wheel. The placement of the pump and injector is shown in Fig. 218. Because injectors were considered unreliable it was the usual practice at this time to have two pumps.

The running gear and valve motion were standard and require no further comment. It should be noted that steel tires were employed. The suspension and truck are worthy of comment in that they illustrate the failing of most early Moguls. Notice that the driving axles are sprung together. The leading truck, constructed on the Bissell plan, but more interesting because of its early use of swing links instead of inclined planes, is not connected to the driver suspension. Because of this, the truck was at times called upon to bear too great or too small a portion of the front-end weight. Too much weight broke the springs; too little weight might cause the truck to derail. This condition was corrected by connecting the truck and the front driving axle with an equalizer and thus more evenly distributing the varying weights caused by rough track to two axles. This scheme, which brought success to the Mogul locomotive, was developed and patented by William S. Hudson in 1864.

Dimensions of the Erie Railway's No. 254

Cylinders	17″ bore × 22″ stroke
Wheels	54″ diameter
Engine weight	35½ tons
Tender weight	20 tons

Fig. 216. Mogul freight locomotive No. 254, built in 1865 for the Erie Railway by Danforth, Cooke & Co.

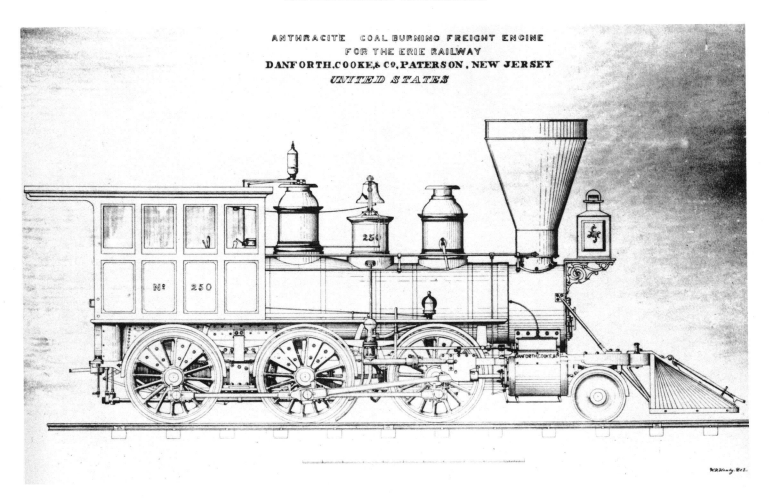

Fig. 217. Side elevation of a Mogul freight locomotive built for the Erie Railway by Danforth, Cooke & Co.

AMERICAN LOCOMOTIVES

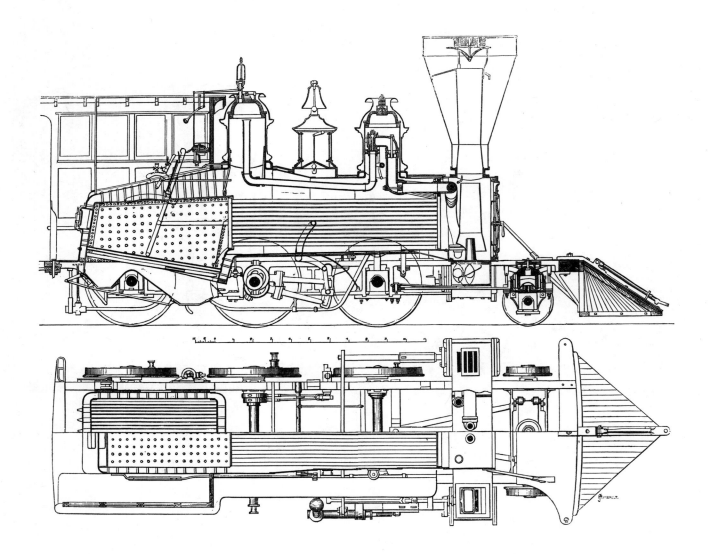

Fig. 218. Longitudinal section and plan view of a Mogul freight locomotive built by Danforth, Cooke & Co. for the Erie Railway.

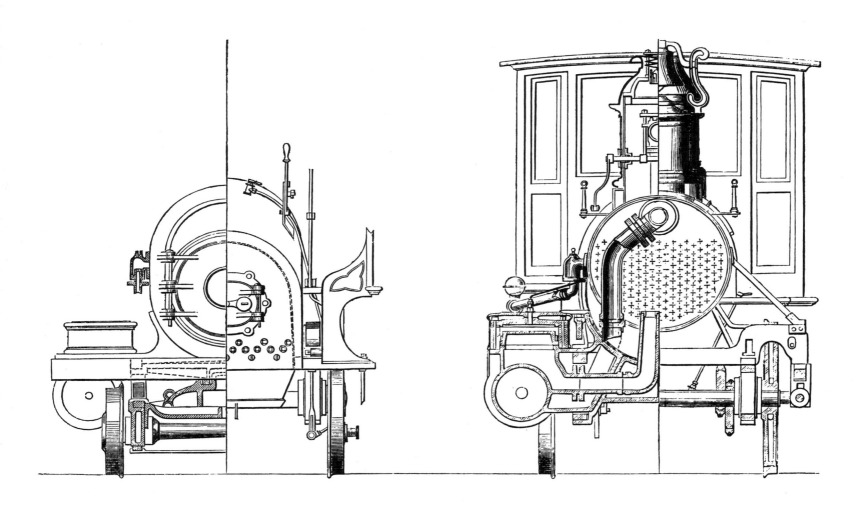

Fig. 219. Front elevation and cross-section of a Mogul freight locomotive built by Danforth, Cooke & Co. for the Erie Railway.

THE ROGERS 4-4-0, CA. 1865

During the early 1850's Rogers did more than any other builder to reform the American locomotive by introducing such basic improvements as the wagon-top boiler, spread truck, and link motion. Unfortunately, no working drawings exist for Rogers engines during this period of rapid development. Drawings do exist, however, for a Rogers eight wheeler of about 1865, but they show no marked improvement over any engine built by the same manufacturer ten years earlier. The drawings, shown here as Figs. 221–223, are from *Locomotive Engineering and The Mechanism of Railways* by Colburn. This comprehensive publication dealt mainly with British locomotives; only the Rogers drawings represented American practice.

The most advanced American design is shown by the coal-burning firebox with rocking grates and a large combustion chamber. However, other features of the Rogers 4-4-0 are not representative of the most progressive United States designers. The cylinders are inclined, very slightly, of course, but still not level. The smokebox is an old-fashioned D-shaped affair. No cylinder saddle is used, the cylinders being bolted to the sides of the smokebox. The leading truck has inclined planes for a centering device rather than swing links which were a recognized improvement over the original Bissell design. The reversing shaft is under the links while the best placement, adopted earlier by other builders, was above the links. Most of these obsolete features are rather trivial and should not be taken as a condemnation of this well-proportioned machine. The point is that Rogers, a pioneer in locomotive modernization, adopted a fixed design and soon found rival builders taking the lead in locomotive improvement.

Dimensions of the Rogers 4–4–0

Cylinders	16″ bore × 24″ stroke
Wheels	68″ diameter
Total heating surface	787 sq. ft.
Grate area	13 sq. ft.

Fig. 220. *The New Jersey Railroad and Transportation Company's No. 8, built by Rogers in 1865. Notice the similarity of this engine and the drawings in Figs. 221 and 222.*

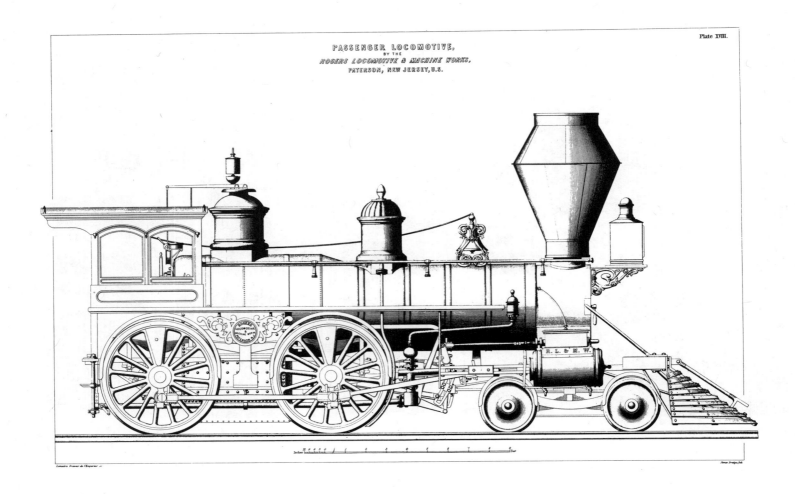

Fig. 221. A coal-burning passenger locomotive built by Rogers in about 1865.

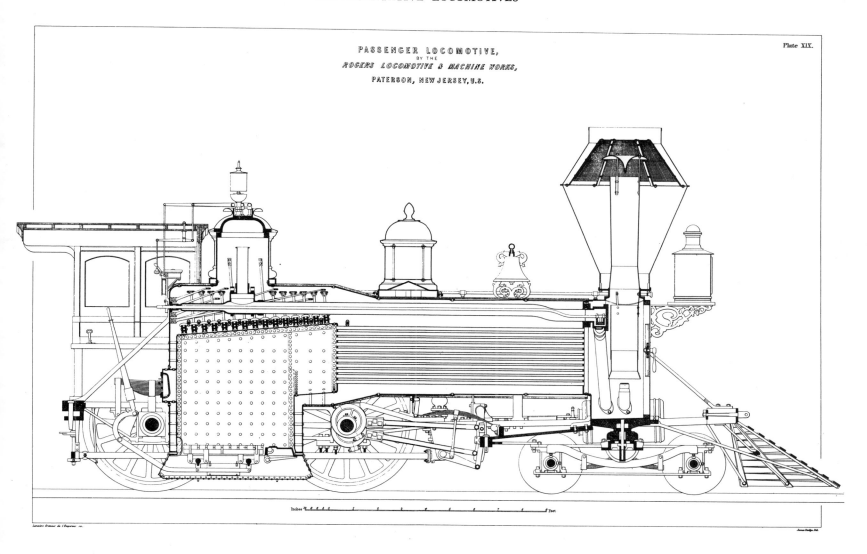

Fig. 222. *A coal-burning passenger locomotive built by Rogers in about 1865.*

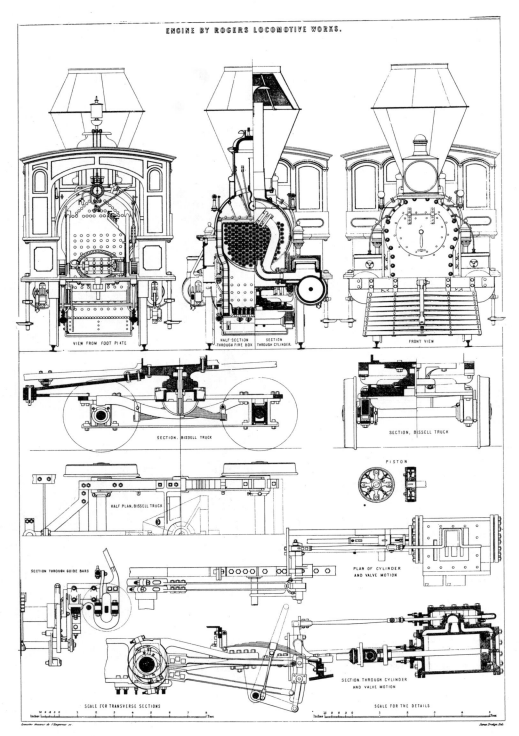

Fig. 223. Details of a coal-burning passenger locomotive built by Rogers in about 1865.

CONSOLIDATION, 1866

The designer of the *Consolidation*, Alexander Mitchell, probably had little idea that he had created what was to become the most popular and widely built freight locomotive in the United States. Mitchell perfected plans for the engine late in 1865, intending it to work long trains of anthracite over the steep grades of the newly opened Lehigh and Mahanoy Railroad. One of these grades, rising 133 feet per mile for 2½ miles, was such a severe obstacle that the road's heaviest power, 4–6–0's and 2–6–0's, could move only short, light trains over it. Recognizing the need for more expedient and efficient movement of this traffic, Mitchell suggested the construction of a high capacity eight-wheel connected locomotive with a two-wheel leading truck for better distribution of front-end weight. The truck would also permit greater road speed and less chance of track damage than an ordinary 0–8–0. Mitchell completed a drawing of his proposed "super" freight locomotive in January, 1866.[1] The drawing was first submitted to Baldwin but the

[1] *Railroad Gazette*, December 28, 1877, p. 571, published a brief history of the first 2-8-0 based on the comments of Mitchell, Blakslee, Henszey, and others connected with its construction.

Fig. 224. *Alexander Mitchell (1832–1908), master mechanic of the Mahanoy Division of the Lehigh Valley Railroad and designer of the* Consolidation.

design was not approved by Matthew Baird, a partner in the firm, who suggested instead one of the Baldwin old-fashioned flexible-beam locomotives. Baldwin finally agreed to build the engine but would neither guarantee its success nor name a firm price. William S. Hudson of the Rogers Works declined to bid for the engine. The design was finally shown to Grant, who quoted $19,500.

For reasons unknown to the author, the contract was not given to Grant, possibly because of slow delivery or a lack of confidence in the builder's ability to meet the purchaser's expectations. Baldwin was given the order, probably after redrafting its earlier, disinterested proposal and promising instead to produce a successful locomotive. William P. Henszey, of Baldwin's engineering department, worked with Mitchell in revising and detailing the design.

The story of the *Consolidation*'s construction is told in the Baldwin letters now preserved by the Historical Society of Pennsylvania. The correspondence passed between James I. Blakslee, superintendent of the Lehigh and Mahanoy, and George Burnham of the Baldwin Locomotive Works. The first letter, dated April 9, 1866, requested that an eight-wheel engine be built, "to new drawings provided you." The letter went on to say that Mitchell wanted to make a few changes. On June 1, 1866, Blakslee inquired how the work was progressing on the "Consolidation No. 14" and said: "I wish we had her now, I hope you will do your best on her, and that when finished she may out pull any Engine, ever built." Burnham replied: "We mean to make her all you anticipate or desire as [to] power and endurance—a strong and healthy specimen of a locomotive that it will be pleasant to see about." Burnham expected some delay in completing the engine for he was awaiting the arrival of steel tires from Germany. During the time that these letters were written the Lehigh and Mahanoy consolidated with the Lehigh Valley Railroad and the new engine was named in honor of the merger. The *Consolidation* was listed as the new road's No. 63 in place of the Lehigh and Mahanoy's No. 14.

On or about July 20 the engine was put under steam and tested at the Baldwin Works. Some adjustments were required and the engine was not shipped until August 17, 1866. On that date Burnham sent the following bill:

Aug. 17, 1866
Lehigh & Mahanoy R.R. Co.
   August 17. For one 34 ton 10 wheel E. Locomotive with 8 wheels. Tender complete, with Steel fire box & steel tires.
"Consolidation" $19,000.00
   5% war tax   950.00
                                        $19,950.00

James I. Blakslee, Esq.
Supt. L & M R.R.
Mauch Chunk, Pa.
Dear Sir:
   Above please find bill for Engine Consolidation $19,950.00 including tax. Johnson left with the machine this morning and expects to be at Mauch Chunk tomorrow afternoon.
                                       Yours truly,
                                       M. W. Baldwin & Co.
                                       G. Burnham

The initial tests were a great success as noted by the following letter of Blakslee:

Mauch Chunk
Aug. 22, 1866
Messrs. M. W. Baldwin & Co.
Gents,
   I would like to have some of you come up on Thursday evening to see the "Consolidation" work, on Friday. . . . *She is a perfect success,* (Mr. Baird will not undertake to make us believe that the old 8 wheel connected Engine is the best hereafter) I am satisfied she can out pull any machine ever built of her weight. She has plenty of speed and will make any curve that your 10 wheelers dare make. There . . . [are] a few small matters Mr. Mitchell will find fault with & point out when you come up—you may calculate on orders for more. This letter is not for the public.
                                       Truly yours,
                                       James I. Blakslee
                                       Superintendent

During these tests the engine was carefully weighed at Weatherly, Pennsylvania.[2] The total engine weight was 85,792

---

[2] The engine was weighed on August 26, 1866. This data is from the Baldwin specification of that date, courtesy of the De Golyer Foundation Library, Dallas, Texas.

pounds with only 9,632 pounds on the leading truck. The new engine proved quite an attraction. Blakslee, in his final letter concerning the engine (September 18, 1866) reported: "We have a good many visitors to see her and all seem highly pleased with her."

No account of the *Consolidation* at work can be found, but several reports are available for the performance of sister engines on the Lehigh Valley Railroad.[3] On grades of 96 feet per mile the Lehigh Valley reported moving 290-ton trains with "Consolidations"; on 146-foot grades "Consolidations" pulled 182-ton trains. This ability to move tonnage was not that impressive, for a common 0–8–0 could do the same work. The remarkable facility of the Consolidation to move around sharp curves at respectable speeds without derailing or causing severe flange wear was its true merit. The Baldwin Works claimed in its 1872 catalog that 2–8–0's could pass around 400-foot curves. The ordinary eight-wheel connected engine could do this safely only at slow speeds.

Only a few other facts are known about the *Consolidation*'s service history. In May, 1875, it received a new boiler. It continued in operation until 1886 when it was retired after completing nearly 377,000 miles of service.

MECHANICAL DESCRIPTION

The *Consolidation*'s general appearance seems to deny that it was a locomotive of the 1860's. The sturdy, plain lines, eight coupled wheels, and the straight stack gave it the appearance of a far more modern machine. Yet, aside from its great size and wheel arrangement, it was built according to orthodox designs and standards of the time. The boiler and possibly the cylinders were oversized, but none of the other parts should have offered any unusual manufacturing problems. The Baldwin was a large and well-equipped shop. For these reasons the author has often questioned the reported difficulty experienced by Baldwin in building the *Consolidation*. It might be added that Millholland built an even larger engine, a giant 0–12–0 pusher, at the Reading shops three years earlier.

Mr. W. E. Lehr, formerly superintendent of motive power for the Lehigh Valley Railroad, kindly supplied the following data on the *Consolidation*. This information was taken from the old motive-power ledger presently preserved by the Lehigh Valley Railroad at Sayre, Pennsylvania:

[3] Silas Seymour, *A Review of the Theory of Narrow Gauges* (New York, 1871), p. 55; *Baldwin Locomotive Works Catalogue*, 1881, p. 115; *Railroad Gazette*, June 22, 1872, p. 263.

Locomotive Name	*Consolidation*
" number	63
Built by	Baldwin
Placed on road	July 1866 [date of completion; placed on road August, 1866]
Kind of fuel	Coal
Description of engine	8-wheel connected, pony truck
No. pairs of drivers	4
Diameter of drivers	48″
Diameter of cylinders	20″
Length of stroke	24″
Weight of engine with fuel and water	85,720 lbs.
Weight on drivers with fuel and water	75,160 lbs. [see note 2 for weight taken August, 1866]
Grate area	3,780 sq. in.
Length of combustion chamber	18″
Number of flues	173
Length of flues	142″
Outside diameter of flues	2¼″

In February, 1893, *Locomotive Engineering* presented a slightly different set of measurements for the *Consolidation*.

These measurements were based on the original drawings at the Baldwin Locomotive Works which, unfortunately, are no longer in existence.

Boiler waist	50" diameter
Firebox	108" long by 34" wide
Combustion chamber	16¾" long
Tubes (179)	2" diameter by 143" long
Cylinders	20" by 24"
Driving wheels	48½" diameter[4]
Driving axle journal (main)	6½" × 8"
Driving axle journal (others)	6" × 8"
Tender	2,000 gal. capacity

[4] *Railway and Locomotive Historical Society Bulletin*, No. 42 (1936), p. 35, lists wheel diameter as 48¾ inches.

The *Consolidation* was intended to burn anthracite coal; accordingly, the firebox was quite large, 25 square feet, and a combustion chamber was employed. A variable exhaust and water grates completed the coal-burning equipment. Other mechanical features that distinguished the *Consolidation* were a poppet throttle valve, an injector (on the left side only), and all-iron tender trucks.

The reconstruction drawings of the *Consolidation* (Figs. 227–230) are based on an elevation drawing by W. J. McCarroll[5] and the two builder's photographs shown as Figs. 225 and 226, together with the specifications noted in this discussion.

[5] *Baldwin Magazine*, July, 1925, p. 8.

Fig. 225. The Consolidation, *built by the Baldwin Locomotive Works in 1866 for the Lehigh and Mahanoy Railroad. Scene: the Baldwin works.*

Fig. 226. *The Consolidation. Scene: the Baldwin works.*

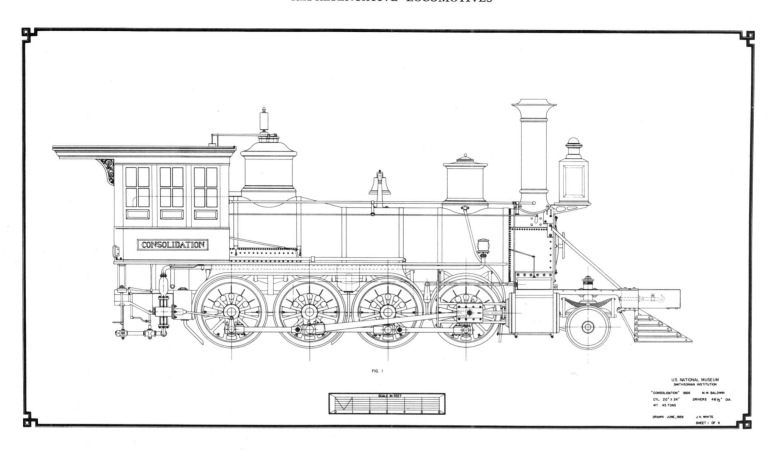

Fig. 227. Reconstruction drawing of the Consolidation, 1866.

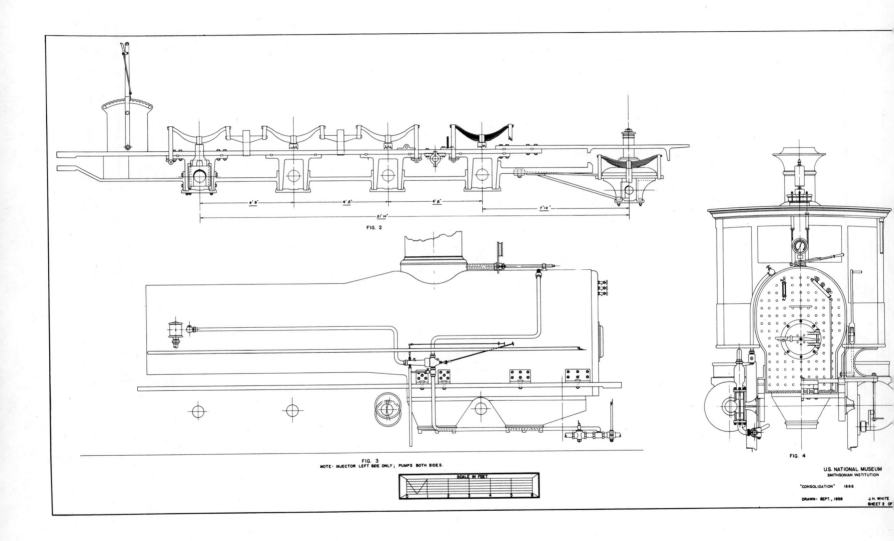

Fig. 228. Reconstruction drawing of the Consolidation showing frame and boiler details.

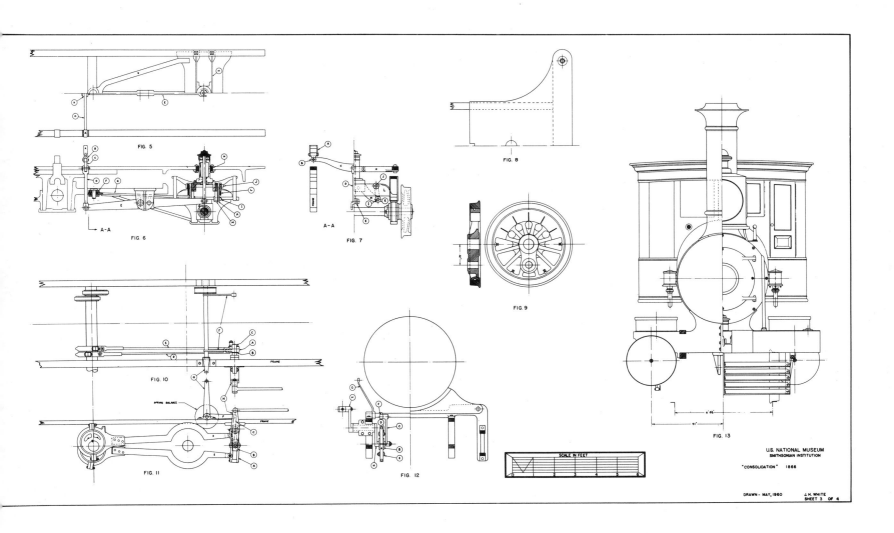

Fig. 229. Reconstruction drawing of the Consolidation showing truck and wheel details.

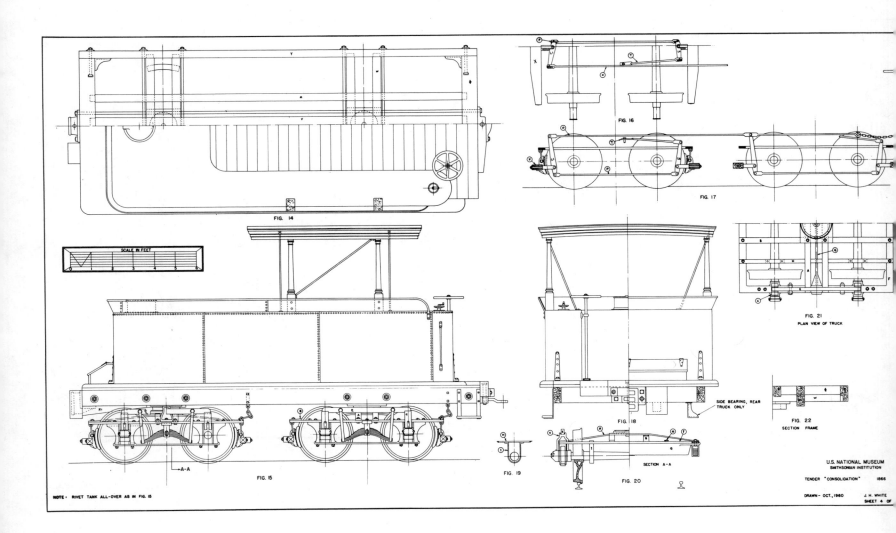

Fig. 230. Reconstruction drawing of the Consolidation's 2,000-gallon tender tank.

THE BALDWIN
TEN WHEELER, 1870

The Baldwin Ten Wheeler is the final selection in our study of representative American locomotives. The machine is typical of locomotive engineering practice as it stood in 1870. It clearly shows how little the basic design and details of engines changed from the beginnings of "modern" American locomotive practice in 1855.

Figures 233 and 234 are taken from the original erecting drawings of two ten-wheel freight engines built for the Danville, Hazleton and Wilkes Barre Railroad. This road was absorbed by the Pennsylvania Railroad in 1900. Baldwin delivered the No. 4 on November 28, 1870 (Construction No. 2295), and the No. 3 on the following day (Construction No. 2296). Both machines were class 24½D anthracite-coal-burning freight locomotives. The boiler was 46 inches in diameter and contained 130 tubes 2 inches in diameter and 152 inches long. The long, shallow firebox (34½ inches wide, 93¼ inches long) was set above the rear driving axle. It should be noted that the crown sheet sloped down toward the rear of the boiler. This was deemed necessary on long-furnace, hard-coal engines to keep water on the rear of the crown sheet when the engine moved downhill. No precautionary measure was required for the front of the crown sheet on an ascending grade because the water level did not vary so greatly near the center of the boiler.

Other noteworthy boiler features included a short (5-inch) combustion chamber, water grates, and a variable exhaust. Two crosshead feed-water pumps and one injector (on the left side) supplied water to the boiler. A balanced poppet throttle valve was employed and was located, as usual, in the steam dome.

The 1873 Baldwin catalog gives several other particulars for 24½ D-class ten-wheel freight locomotives. This was the lightest class of 4–6–0's offered by Baldwin at the time. Weight on the drivers was given as 51,000 pounds. The engine's capacity was rated as follows:

This class of engine was also available for soft-coal roads. Figs. 231 and 232 show such a machine. Notice that the firebox is short and deep with the rear driver behind, rather than under, the firebox. It will also be seen from Figs. 231 and 232 that 24½D-class engines could be obtained with straight boilers. Two steam domes were employed for added steam room.

Load in Addition to Engine and Tender	
On level	1,230 gross tons
20-ft. grade	570 " "
40 " "	360 " "
60 " "	260 " "
80 " "	195 " "
100 " "	155 " "

Dimensions of the Danville, Hazleton and Wilkes Barre Railroad's Nos. 3 and 4	
Cylinders	16" by 24"
Wheels	54¾"
Total heating surface	969 sq. ft.
Total engine weight	33½ tons

REPRESENTATIVE LOCOMOTIVES

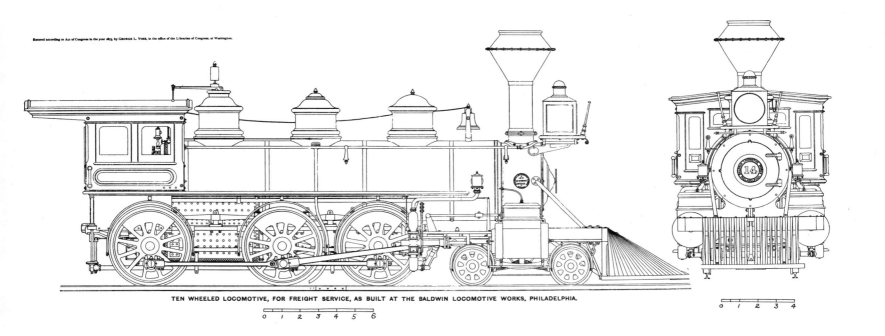

Fig. 231.   A Baldwin ten-wheel freight locomotive of 1871.

Fig. 232. *A soft-coal-burning, class 24½D locomotive built by Baldwin for the Evansville, Hendersonville and Nashville Railroad in May, 1870.*

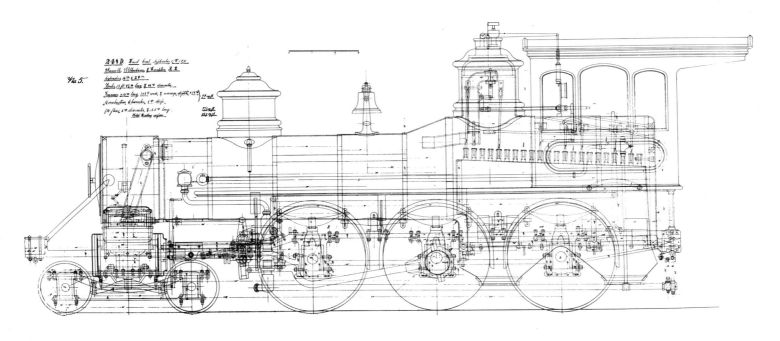

Fig. 233. *An anthracite-coal-burning, class 24½D locomotive built by Baldwin for the Danville, Hazleton and Wilkes Barre Railroad in November, 1870.*

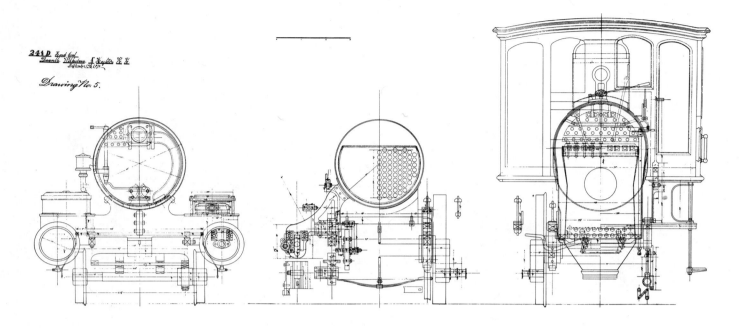

Fig. 234. *An anthracite-coal-burning, class 24½D locomotive built by Baldwin for the Danville, Hazleton and Wilkes Barre Railroad in November, 1870.*

SUMMARY

"I have endeavored to put together in these engines all the good things and get rid of all the bad things experience has developed up to this time. . . ."
George Washington Whistler, 1843[1]

The early development of the American steam locomotive was directed by practical mechanics who depended on experience and pragmatic methods to solve design problems. The main objective of these mechanics was to produce a cheap, practical machine and to increase its power without unduly increasing its complexity or weight.

Low first cost was eminently important to the capital-poor American lines. Economy was achieved mainly by the use of simple, elementary designs unencumbered by auxiliary apparatus. Costs were further reduced by the generous use of cheap materials, particularly the extensive use of cast iron. As late as 1880 it was reported that nearly one-third of the material used in locomotive construction was cast iron.[2]

It was not only railroad promoters and officials who were interested in cheap locomotives. Commercial locomotive builders showed a reluctance to produce machines that were more complex or expensive than those offered by their competitors. They resisted scrapping patterns and successful designs for costly innovations unless forced to do so by the specific demands of the purchaser. For example, Colburn considered the New England builders as merely "manufacturers" who held a particularly complacent attitude toward design reforms. Finally, many builders ignored the regular use of recognized improvements, such as balanced slide valves, because they "couldn't be seen" and would probably go unappreciated by the unknowing purchaser, who would object to the greater cost of the machine as compared to an "equal one" offered by another maker.

Economical operation and fuel economy comprised a never-ending dialogue among railroad officials in the nineteenth century. Yet, despite the talk and projected schemes for saving fuel, it is the author's opinion that locomotive designers always considered fuel economy as a decidedly secondary considera-

[1] From the letter of G. W. Whistler to J. G. Swift, November 3, 1843, in Patten Papers, New York Public Library.
[2] Bureau of the Census, *Tenth Census*, II, 46.

tion compared to performance. This unspoken indifference to fuel economy was encouraged by the abundance of cheap fuel in the United States. The constant rejection of improvements such as feed-water heaters and superheaters as needless complications indicated the prevalence of this attitude prior to 1900.

Once a practical design had been perfected for steam locomotives, a reluctance to radically alter it naturally developed. The basic arrangement introduced in 1829 by the *Rocket* was never abandoned. Despite the attempts of many gifted engineers, no plan was offered that surpassed the old arrangement in performance, dependability, simplicity, or low first cost. The history of steam locomotive development has been one of refinement rather than revolution.

Several fumbling attempts were made in the early 1830's by Colonel Long, Childs, and other American mechanics to devise a new plan for locomotives distinct from the British pattern. It was soon apparent, however, that the basic British design was sound and that it could be readily *adapted* for our use. The fire-tube boiler and engine parts were copied directly from imported British engines; however, the weight of such boilers was greatly reduced by using very thin boiler iron. Greater power was realized with higher steam pressure. American locomotives commonly carried higher boiler pressures than were considered safe in Europe. It was a simple, if not risky, method of increasing horsepower without increasing weight. A new, flexible running gear was developed by American builders. Its major features—the truck, equalizing lever, and bar frame—can be traced to earlier British designers, but these elements were combined and adapted to the extent that they soon distinguished the American locomotive from its British counterpart.

These developments took place in the early and mid-1830's. The end product was the light, simple, easy-riding 4–2–0. This wheel arrangement came into favor rapidly, but its insufficient traction, caused by its having only one pair of driving wheels, proved to be its undoing. The 4–4–0, introduced in 1837, was favored after 1840 and was to remain the dominant style of locomotive on American railroads for the next thirty-five to forty years.

The 1840's witnessed the steady growth and development of the 4–4–0. The boiler was lengthened considerably and with it the locomotive's wheel base. The stumpy, compact appearance of the small locomotives of the preceding decade disappeared. The diameter of the boiler was kept small to hold wheel loading within the limited capacity of light tracks. Total engine weight rarely exceeded 20 tons and was more generally held at 15 tons. The engines of this period were awkward in appearance, poorly proportioned, and often cluttered with outside supplementary frames and complex cutoff valve gears. Few basic or long-lasting improvements were introduced during these years.

The early 1850's saw the perfection of the 4–4–0. The spread truck, level cylinders, wagon-top boiler, and link motion were all adopted in these years. The appearance of the machine was notably improved in this period. Much of the engine's clutter was removed. The basic outline was simplified and better proportioned. A clean-cut, balanced design of remarkable grace was perfected by the mid-1850's. Elaborate brass trim and decoration were adopted. Bury-dome boilers, cutoff gears, outside frames, and inside connections, all disappeared in this period. A serious interest in coal-burning developed late in the decade.

In all, the developments of the 1850's marked the culmination of the work of many pioneer locomotive designers who found the "reformed" 4–4–0 so satisfactory that it was left virtually unchanged for the next twenty-five years.

The 1860's showed continued contentment with the 4–4–0 although several new important wheel arrangements were introduced (the 2–6–0 and 2–8–0). Construction practices and design showed little improvement. The more widespread use of steel and the substitution of coal as fuel were the most important general reforms. Increased size along orthodox designs took place. In general, it was a quiet time with few startling changes.

The 1870's and 1880's did not appear much different from the preceding decade. The most important happening was the decline of the 4–4–0 and its relegation to passenger service only on main-line roads.

REPRESENTATIVE LOCOMOTIVES

In the 1890's things again began to stir. The 4–4–0, even when built in sizes up to 60 tons, was proving to be incapable of handling main-line passenger trains. Not only were newer wheel arrangements finding more favor, but the need for larger boilers and fireboxes was becoming urgent. Grate area and heating surfaces were doubled and tripled. Locomotives double in weight compared to those of twenty years earlier were now commonly accepted. The old limits of 3–4 tons per wheel were forgotten as heavier bridges and rails were installed to carry the increasing traffic funneled onto railroads by America's burgeoning industries. The light, slim-boilered locomotive with its high smokestack and long, elegant cowcatcher vanished overnight.

MORE REPRESENTATIVE LOCOMOTIVES

IV

The following essays describe an additional selection of steam locomotives once commonly found on American railroads in the nineteenth century. The selection was dictated more by the availability of information than by the importance of the individual locomotives portrayed. All are, however, more or less conventional engines, and all represent the standard engineering practices of their day.

LA JUNTA, A CUBAN 4–2–2, 1843

La Junta is a survivor. It survived tropical storms, revolutions, insurrections, and a Communist takeover. It escaped any major rebuilding and survives very much as built. While the scrappers dismembered all of its contemporaries, La Junta alone survived. Some insist that this is because La Junta was the first locomotive on the island—it is called No. 1, after all. But in fact, although La Junta was No. 1 on the Matanzas Railroad, it arrived in Cuba some six or seven years after the first engine reached the Caribbean's largest island.[1] However, I would not argue with the motives of its preservers.

Just why the engine was built with a set of trailing wheels rather than a second pair of driving wheels is unknown. There is no obvious advantage to this wheel arrangement, other than the fact that it relieves some of the weight from the main drivers and thus preserves the track. But the disadvantage is that the adhesion of the locomotive is somewhat lessened by diverting weight to the trailing wheels. Nevertheless, Thomas Rogers produced about ten engines on this plan during the 1840's—not a large number, to be sure, but enough to suggest that it was a reasonably successful design.

This is a small locomotive, as locomotives go. It weighs an estimated 12 to 13 tons. The engine's overall length, including the lengthy cowcatcher, is only 22 feet. The open cab and four-wheel tender underscores its lack of mass. This standard-gauge engine is a mere pigmy when compared to most narrow-gauge engines dating from a century ago.

If you revere the antique, then an examination of La Junta is like viewing a sacred relic. All of its basic features are of the ancient order. The **D**-shaped smokebox, inclined cylinders, short wheelbase, leading truck, Bury boiler, **V**-hook valve gear, and square sandbox were all obsolete by about 1855. But they survive on this machine, despite the fact that it was not retired until about 1900. Why wasn't it modernized? This same little engine in the United States would have undergone a systematic series of changes if it had remained in service for even one-half of La Junta's service life.

Straightaway it would have become a 4–4–0. The **V** hooks would have been replaced by link motion. A spread truck and

---

[1] Trains, September, 1984, pp. 21–28, contains an article on La Junta.

level cylinders would have come next, and very soon it would have looked like a nice little eight wheeler, outshopped in around 1870. Just why these improvements were not made in Cuba is uncertain. Perhaps it was because of the engine's size. It was so small—why invest capital in a machine too little to do much useful work? Perhaps it was just poverty, the greatest of all forces preserving antiquities. Or perhaps it was the Latin American dedication to tradition and keeping things as they were first made. As usual, we have more questions than answers.

Mention has been made of some of the old-fashioned features of this relic, but the massive cast-iron rocker is surely one of the most interesting elements of the engine. Most valve gear rockers are fairly light wrought-iron or steel fabrications, but La Junta's rocker is a sizable chunk of common cast iron. It is best illustrated in the cross-section drawing (Fig. 236).

The tender is a good-sized four wheeler measuring 11 feet 8 inches over the ends of the frame. The tank is 29 inches high and 10 feet 8 inches long. It had a capacity estimated at 600 to 700 gallons. The inside bearings are an interesting feature of early tender design, and one that, as existing illustrations suggest, may not have been that uncommon. In this book there are six examples of inside-bearing tenders (Figs. 25, 66, 90, 138, 145 and 190-A). Surely outside bearings predominate, but some designers felt that inside bearings were worthy of wider use. The wooden brake shoes are another emblem of railroading in the nineteenth century, when cheapness and practicability were synonymous.

Dimensions of La Junta

Estimated weight	12 to 13 tons (engine only)
Cylinders	11″ bore × 18″ stroke
Driving wheels	54″
Boiler	36″ in diameter, 85 tubes
Price	$6,700

Fig. 235. La Junta, 1843, *built by Thomas Rogers in Paterson, New Jersey, for service in Cuba. The engine is on exhibit in Havana.*

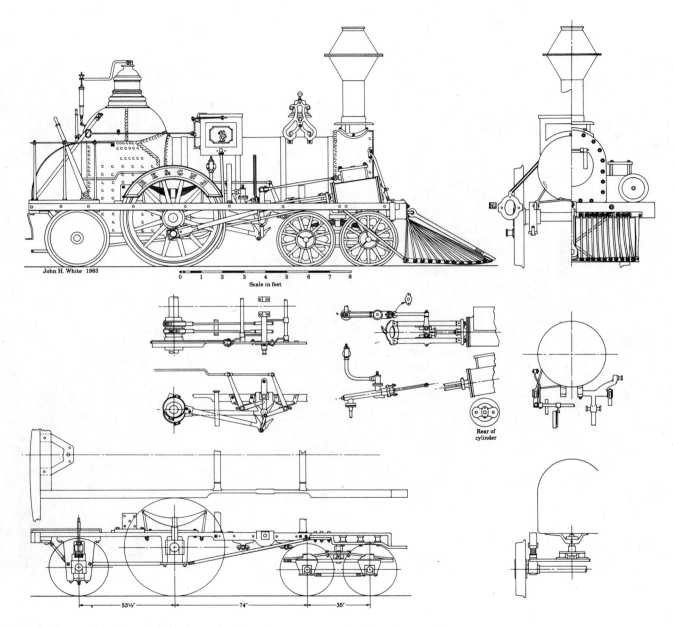

Fig. 236. La Junta, *drawings made by the author from measurements taken from the original machine while it was on display at Lenin Park, Havana, Cuba, in 1983.*

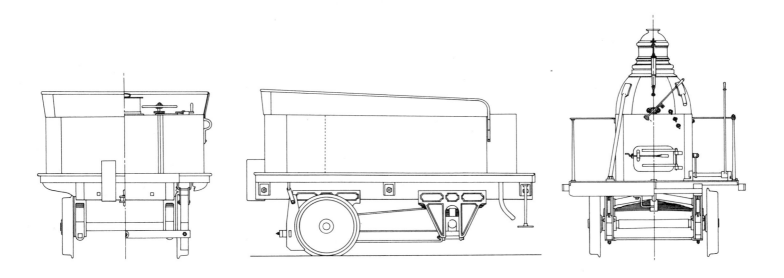

Fig. 237. More drawings of La Junta *and its tender, now preserved in Havana, Cuba.*

THE STRETCH PLANET
A HINKLEY 4-4-0, 1845

The drawings shown here are not for a specific locomotive; rather, they represent a general design or pattern of engine produced by Hinkley and Drury, of Boston, Massachusetts, between roughly 1843 and 1855.[1] The smallest had cylinders only 12 inches by 18 inches and surely weighed no more than 15 tons. The largest had 15-inch by 20-inch cylinders and weighed around 20 tons. During this decade of the Hinkley stretch Planet, about five hundred such engines left Hinkley and Drury's Harrison Street works. Other than a modest growth in size, about the only notable change was the relocation of the steam dome from the top of the firebox to the forward end of the boiler. By the early 1850's, newfangled add-ons such as cabs, headlights, and cowcatchers were becoming more common, but they hardly altered the basic design. Most of Hinkley's early production went to New England railroads, where inside-cylinder 4-4-0's were the norm until around 1855. Some of the stretch Planets were produced for service outside of the region; engines of this pattern were produced for the Macon and Western, the Michigan Central, the Philadelphia and Reading, and the Richmond and Danville railroads.

Normally, historians of technology study the innovators, those dedicated to change and improvement. These were the risk takers, the daring mechanics who were on the cutting edge of change. Clearly none of this was true of Messrs. Hinkley and Drury. These conservative mechanics and their design staff were very satisfied with the tested and proven. Why change things when you have a perfectly workable plan? Being original and forward-looking required time, experiment, retooling, and often, large-scale investment. As every New Englander understood, money was to be saved and not spent, especially not on idle experimentation. Such a mindset might sound timid, unprogressive, or even lazy, but there was a good argument for staying with a design proven by years of dependable performance. After all, the basic Stephensonian plan for the steam locomotive was never changed despite numerous efforts to revolutionize it.

[1] *Railroad History*, Spring, 1980, contains a lengthy article on the Hinkley Locomotive Works, starting on p. 27.

There is no direct evidence about how or why the Hinkley stretch Planet came into being. Locomotive designers did not write memoirs, nor did they leave notes or published accounts regarding their design work. Yet their plans tell us something about their thinking and allow us to speculate on the how and why of the design process. What inspired John Souther as he sat before his drafting table and that blank sheet of paper? Souther was a pattern maker and a practical mechanic, but not an academic engineer given to theory and mathematics. To a man like Souther, design was most likely a matter of example and experience. The smart designer stayed with a proven plan that evolved from established proportions and dimensions. These provided elementary formulas that could be contracted or expanded as the needs of the job required, just as the common knowledge that a 2-inch by 12-inch plank would safely span a 10-foot gap could be extrapolated to determine that a 2-inch by 14-inch plank would do the same for a 12-foot span. It was all just horse sense.

Now, every locomotive man in New England knew that the Planet engines put out by the Locks and Canals Machine Shop in Lowell were dandy little engines. They were surprisingly powerful for their size and had proven wonderfully dependable over the years (see Fig. 1). To be fair, they did have a few drawbacks. They were a mite rigid, but nothing a leading truck would not improve. At 10 tons they were rather undersized, so why not make them bigger? Locks and Canals had boldly copied Stephenson's 1830 design without pretending to do otherwise. Yes, they had added a bell, a whistle, a cowcatcher, and a wood-burning smokestack, but these were add-ons only.

And so what did Souther do to fill up his blank drafting paper? I really don't know, but from the design itself (Fig. 238) I think that he simply elongated the Planet. It became the stretch Planet. The boiler grew by 2 feet. The diameter of 36 inches remained unchanged. The non-swiveling single-axle lead wheels were replaced by a four-wheel truck with 26-inch-diameter wheels. A second pair of drivers was placed behind the firebox.

But the other features of the original Stephenson design were simply copied: the inside cylinders and crank axle, the D-shaped smokebox, the low-arched firebox, and the wooden outside frame—timber and riveted plate clad on sandwich frame and a forged-iron inside frame. The driver diameter was maintained at 60 inches. The wheelbase of a Planet was 5 feet; the wheelbase of the Hinkley 4–4–0 between the driving axle was the same.

Souther changed nothing that could be saved. The result was a stretched-out duplicate of a design only a little more than a decade old. The table below helps to illustrate the similarities of the two styles of locomotives.

	Hinkley 4–4–0, 1845	Stephenson 2–2–0, 1830's
Cylinder diameter	13"	11"
Cylinder stroke	20"	16"
Driver diameter	60" (4)	60" (2)
Boiler diameter	37"	36"
Tube length	9' 6"	6' 8"
Tubes (number)	88	129
Tubes OD	2"	1.625"
Grate length	30"	N/A
Grate width	39"	N/A
Firebox depth	36"	N/A
Tube surface	438 sq.ft.	365.5 sq.ft.
Firebox surface	39.3 sq.ft.	42.16 sq.ft.
Grate area	8.1 sq.ft.	6.5 sq.ft.
Total heat surface	N/A	408 sq.ft.
Weight (engine only)	15 tons	8 tons
Steam pressure (estimated)	100 psi	50 psi
Truck wheel diameter	27" (4)	36" (2)
Tender	6 wheels	4 wheels
Tender tank (estimated)	1,200 gal.	500 gal.
Tender weight (fuel and water) (estimated)	12 tons	5 tons

The basic document for our reconstruction of the stretch Planet is a lithograph issued by Hinkley in around 1845 as an advertisement for the current design (Fig. 238). It does not show a particular locomotive, as is the case with some lithographs of the period; rather, it shows a generic plan followed for a series of machines produced by this early New England machine shop. From this print other views of the 4–4–0 were derived. As in all reconstructions, a certain amount of speculation is necessary, and some elements of the original design cannot be interpreted with absolute assurance. Even so, I think that the several drawings reproduced here are reasonably accurate renderings of a design created almost 150 years ago. Those having better information are encouraged to come forward and enlighten us as to the true particulars of Hinkley and Drury's original creation.

The simple, direct lines of the engine are obvious from the drawings, but a few features might be noted. The inside sandwich-style frame has large cast-iron pedestals on the inside—note the ogee shape away from the driving boxes. The end jaws appear to be part of the top rail forging. The outside running board with the railing all around allows the crew easy access to the machinery, even while the train is under way, for oiling and inspection. The valve box protrudes from the side of the smokebox—an improvement on Stephenson's original plan, which buried the valve boxes inside the smokebox in a most inaccessible manner. The valve gear looks more complex than it actually is because, instead of using rockers and lifting links, the eccentric rods and the D or V hooks were raised or lowered by cams. The cams, which look like the letter D placed with its curved side down, are worked by half crescent gears through rods and levers going back to the deck plate. This odd plan can best be seen in Fig. 65. A surviving example of the Hinkley valve gear can be found on the *Lion* at the Maine State Museum.

The feed-water pumps are placed at the very rear of the engine. They are of the half-stroke style and are worked by the motion of engine through small cranks fastened to the forward crank pin. The plunger of the pump is so large in diameter that it is self-supporting and does not require a crosshead.

There are few decorative features on this locomotive. It exhibits the usual Yankee penchant for restraint and economy. However, the sides of the sandbox have cast in relief a fine reproduction of the Great Seal of the Commonwealth of Massachusetts.

The six-wheel tender shown in Figs. 242 and 243 may strike many as a peculiar novelty, although in fact it was something of a standard plan. Zerah Colburn describes a tender of this exact design in his 1851 book *The Locomotive Engine* (pp. 148–150). He notes that the tank was 12 feet long, 7 feet 1 inch wide, and 35 inches high, with a capacity of 1,398.5 gallons. The running gear and frame have been reconstructed from several photographs of Hinkley engines of the period—the best one of these is reproduced here as Fig. 242. Note the odd spacing of the wheels, which was intended to provide better support for the heavy rear end of the vehicle. I have no exact information on the style of springs used, but I assume that they were India rubber cylinders.

Fig. 238. A lithograph of an inside-cylinder 4–4–0 issued by Hinkley and Drury, locomotive builders of Boston, in about 1845.

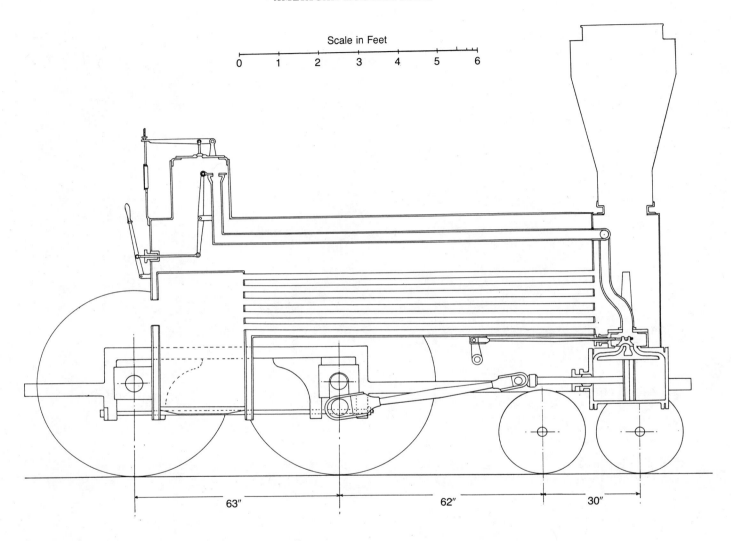

Fig. 239. A longitudinal cross-section representing a Hinkley locomotive of the same design as shown in the previous illustration.

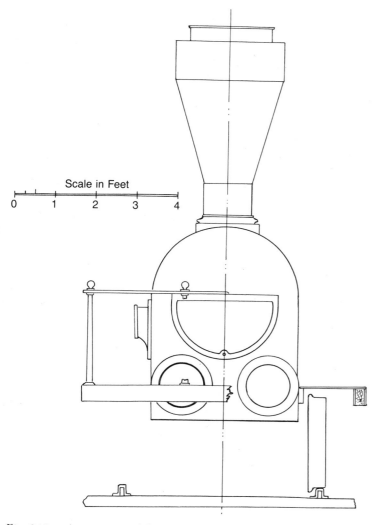

Fig. 240. A reconstructed front elevation of a Hinkley 4–4–0 of about 1845.

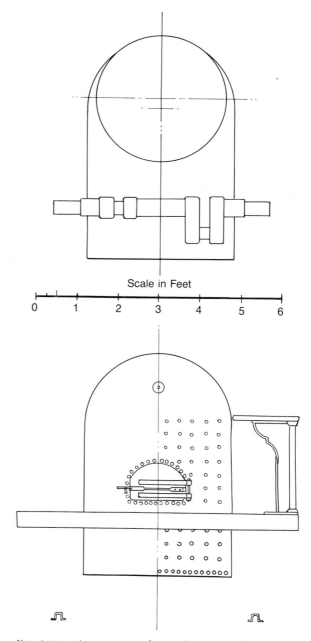

Fig. 241. A reconstructed rear elevation and a detail of the crank axle of a Hinkley 4–4–0 of about 1845.

Fig. 242. The Fitchburg Railroad's Brattleboro, 1844, an outside-connected Hinkley eight wheeler. This view, copied from a Daguerreotype of 1846, is especially valuable for its depiction of Hinkley's standard six-wheel tender.

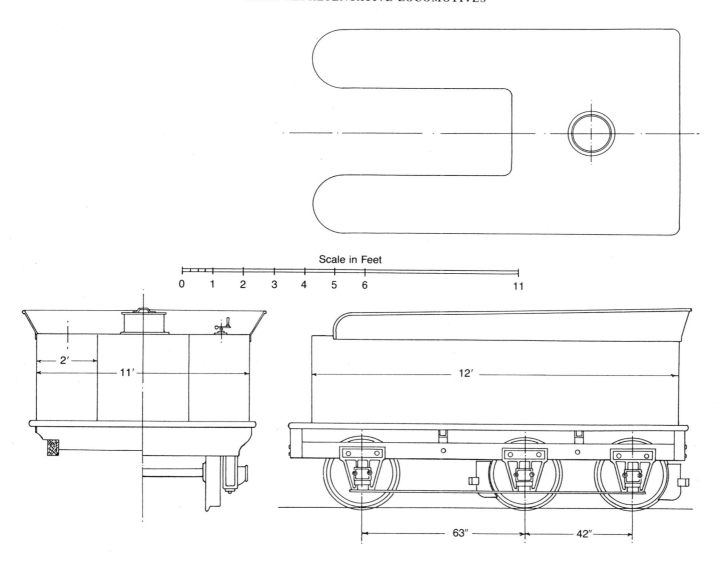

Fig. 243. *A reconstruction of a Hinkley six-wheel tender of about 1845. The tank held 1,398 gallons of water.*

ELEPHANT, 1849: THE FIRST IN THE WEST

So far as history records, the first steam locomotive in the far western United States was a high-wheeled eight wheeler named the *Elephant*. It was built in South Boston at the Globe Works, of which John Souther was the proprietor.[1] Presumably the engine was built for a railroad running out of Norfolk, Virginia, that, like most southern railroads, had a track gauge of 5 feet. Whether the engine actually worked on this line or remained in storage at the maker's shop is uncertain, but in July, 1850, the *Elephant*, along with an Otis steam excavator, also manufactured by Souther, and a number of tip cars, plus 150 tons of rail, was sent to James Cunningham of San Francisco. Cunningham used the equipment to level sand dunes along the San Francisco waterfront. The goal was usable land for real estate development, but before any work was accomplished the city authorities prohibited use of the engine. No doubt it was felt that a steam railway operation, even so limited a one as this, posed a nuisance and fire hazard in a frontier city hastily built without much regard for a fire code. And so the *Elephant* was placed in storage.

Just why a contractor would select for slow-speed construction work a locomotive designed for express main-line service is not easily explained. Most likely, Souther, rather than custom-building a new engine designed for switching dirt cars, offered Cunningham the high-wheeled *Elephant* at a cheap rate. Surely it was more than big enough for the job. But the bargain that sat in storage was worthless unless it could be put to work elsewhere. To find a buyer in California at this time was not easy, however, for there were no railroads anywhere on the Pacific coast. Finally, in September, 1855, the low-mileage *Elephant* was dusted off and sent upriver to the state capital for use on California's premiere railroad, the Sacramento Valley Railroad. The *Elephant*, quickly renamed after the railroad's president, C. K. Garrison, was placed in regular service late in November. It was the third locomotive on the line, having been preceded by two Hinkley engines.

---

[1] My account of the *Elephant* is based on an unpublished article by Wendell W. Huffman of Carson City, Nev. Huffman's account is based on extensive research in the newspapers of the day, together with an examination of the secondary literature, such as Gilbert Kniess, *Bonanza Railroads* (Stanford, Calif., 1941).

MORE REPRESENTATIVE LOCOMOTIVES

The *Garrison* (formerly the *Elephant*) performed well for its new owner, averaging 24,000 miles per year. It was said to be the most economical engine on the line, but, like all inside-connected engines, it had a propensity for crank-axle failures. In the summer of 1856, with less than a year of running, it required new axles and driving wheel tires. It broke an axle in 1862 and jumped the track the next year, but without major damage. When the Big Four took over the Sacramento Valley in 1865 to make it a branch of the Transcontinental Railroad's western end, it was necessary to change the track gauge from 5 feet to standard gauge. The *Garrison* went into the shop with its sisters for regauging.

More drastic surgery awaited the West Coast's pioneer locomotive. Early in 1869 it was taken to the Folsom repair shop for a complete rebuilding. The twenty-year-old *Garrison* received a new boiler, and the hook-motion valve gear was replaced by link motion. The old-fashioned riding cutoff and outside frame and gallery railing were removed. The obsolete inside cylinder and crank-axle arrangement were retained, however. It is likely that the cab, cowcatcher, and other details were replaced or modernized.

It has been contended that new, larger, 71-inch driving wheels were installed during this rebuilding, but if this was true why would T-spoke wheels have been used? They are right for 1849 but well out of date for 1869. It might also be noted that the rebuilders were not too quick to discard all vestiges of Souther's handiwork. The curious round valve box covers, the large forward steam dome, and the very slender round main rods were retained. After many months of work, the engine was returned to service in November, 1869. In deference to its history, the engine was renumbered as 1 and rechristened the *Pioneer*. Yet the honor of being No. 1 hardly made the *Pioneer* a sacred cow. It was retired in 1879, because by then it was considered too small and obsolete for general use on a main-line railroad.

When the occasional need for a light engine arose, the *Pioneer* was steamed up. Efforts were even made to find a buyer, but not even the most impecunious short line seemed ready to acquire such an old-fashioned crank-axle engine. It was a museum piece and little more. There was no last-minute reprieve from the governor, no just-in-time appearance of a curator from the state museum—no one, in fact, who might have saved the old relic came forward, and so the *Pioneer* was scrapped late in 1886. The business of preserving antique locomotives was just beginning, and many a fine specimen was lost as a consequence.

If the history of this engine appears sketchy and uncertain, so too is our picture of its appearance. There are actually only a few relevant illustrations that are of sufficient quality to help us create a mechanical reconstruction of the *Elephant*. The best is the often-reproduced photograph of the engine taken in around 1877 on Front Street in Sacramento. It is surely a clear image, but it shows the machine in a much rebuilt condition and near the end of its service career. There is also a lithograph of the *Washington* issued in about 1853 by John Souther as an advertising giveaway to railway officials to promote sales of engines. It is safe to assume that the print is as accurate as it is handsome because it was drawn by Zehra Colburn, a leading locomotive authority, who was at the time a draftsman and designer employed by Souther.[2] While most lithographs are purely pictorial, Colburn drew in some phantom lines to indicate some of the parts hidden by the fender and outside frame. This is visible only on the original print, which is drawn at a scale of 1 inch to the foot and so reveals details lost on reduced reproductions. Another relevant picture is a photograph of the *Roanoke*, which was built in 1854 for the Virginia and Tennessee Railroad (Fig. 245). At the time, Souther was managing the locomotive shop of the Tredegar Iron Works, so this engine is a product not of South Boston but of Richmond, Virginia, yet it is clearly a duplicate of the machines produced in Massachusetts. When comparing the lithograph of the *Washington* with this photograph, it is difficult to find many differences except for the obvious add-ons of a headlight and sandbox. The reader might also refer to the *Reindeer* photograph, Fig. 78, for another example of a globe engine of the 1850's.

---

[2] The *Washington* lithograph is reproduced in color on the cover of my *Early American Locomotives* (New York, 1972).

The drawings shown here were prepared for a large-scale display model and are hardly adequate for the construction of a full-sized locomotive; however, they are useful for explaining the general arrangement. The similarity between the products of Souther and those of Hinkley should be obvious from a simple comparison of the stretch Planet, as shown in previous pages, with the *Elephant*. The boilers, the frames, the feed pumps and just about all other design features of the two locomotives are identical, which is as it should be, because John Souther's plan was followed for both.

As a young man, Souther worked as a pattern maker for Holmes Hinkley. When Hinkley entered the locomotive business, Souther made the patterns and took over the design duties as well. In 1846 he left Hinkley to open his own machine shop. Within two years he entered the locomotive trade, and to no one's surprise he remained loyal to the design he had perfected for his former employer. In time, however, certain differences appeared. These may have been a conscious effort to demonstrate that Souther was not simply producing knock-offs of the old standard Boston Locomotive Works design.

Souther abandoned the ancient order of the D-shaped smokebox for a cylinder saddle. The valve boxes—tall and square-shaped—stood up at an angle, just like those used by the Lowell Machine Shop (see Fig. 155). We will never know whether Souther copied this scheme or came upon it independently, but he did devise a method to give it a new look. The inclined valve boxes were shrouded in spun brass jackets. These cylindrical cannisters give Souther engines a distinctive appearance that helped set Souther's product apart from that of his former employer and now rival, the Boston Locomotive Works.

For its time, the *Elephant* was a large locomotive, and while all early 4–4–0's tend to look alike to modern eyes, they were sometimes very different in size. In comparing the Hinkley 4–4–0 stretch Planet and the *Elephant,* we find that the Souther engine has a wheelbase 3 feet longer than that of the Hinkley and that its height, to the top of the smokestack, is nearly 3 feet greater.

Dimensions of the *Elephant*

Cylinders	15" bore × 20" stoke
Wheels	71" (some sources claim 66")
Weight	25 tons (estimate)

Colburn offers the following dimensions for a typical Souther engine in his 1851 book *The Locomotive Engine* (p. 54):

Diameter inside of boiler	42"
Length of tubes	10½'
Number of tubes	135
Outside diameter of tubes	1¾"
Length of grate	37"
Width of grate	37½"
Depth of firebox	53"
Tube surface	649.42 sq.ft.
Firebox	60.8 sq.ft.
Area of grate	9.63 sq.ft.
Size of steam ports	9⅞" × 1"
Size of exhaust ports	9⅞" × 1½"

MORE REPRESENTATIVE LOCOMOTIVES

Fig. 244. The Elephant, *built by John Souther of Boston in 1849, was the first locomotive in California. This reconstruction was drawn in 1993 by the author.*

AMERICAN LOCOMOTIVES

Fig. 245. The Roanoke *was built in 1854 by John Souther for the Virginia and Tennessee Railroad. It had 15-in. by 22-in. cylinders and 54-in. drivers.*

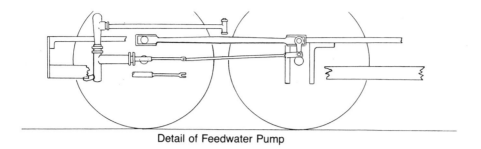

Detail of Feedwater Pump

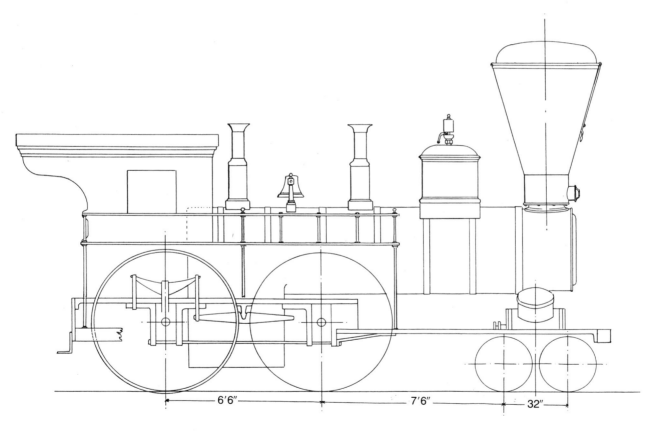

Fig. 246. A reconstructed side elevation of the Elephant. *The upper drawing shows details of the pump.*

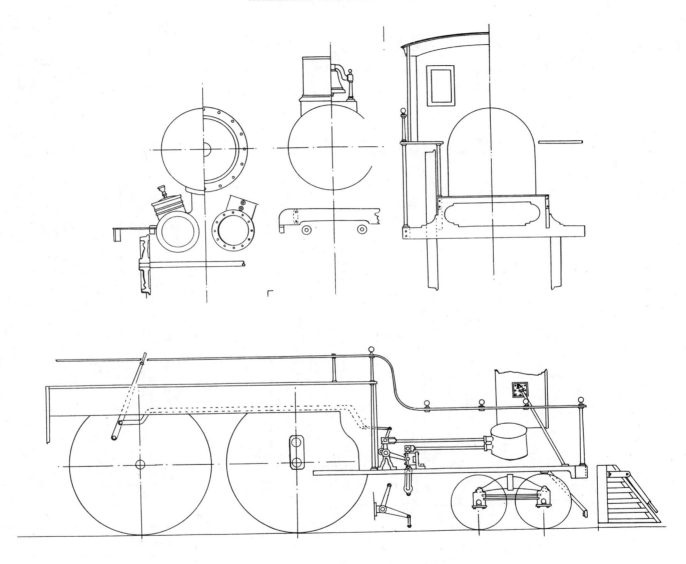

Fig. 247. A reconstruction of the Elephant. Many of its details are conjectural. For example, the rear beam of the frame is largely imaginary, but it does agree with the scant information available in side views of Souther engines of the period.

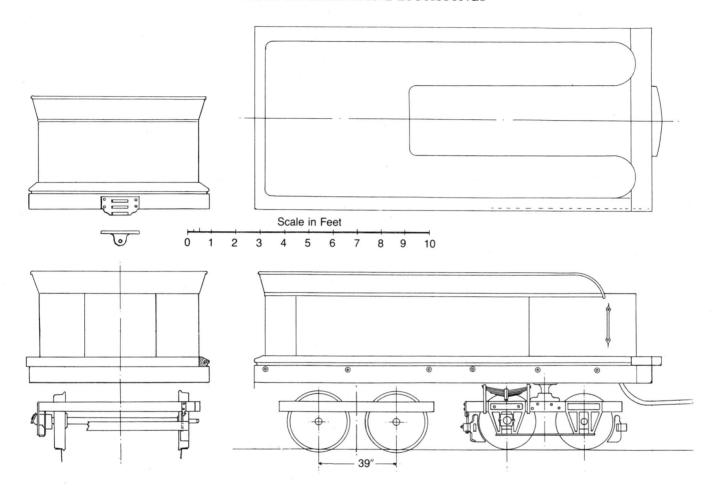

Fig. 248. The type of eight-wheel tender used by the Elephant, *shown in a reconstructed drawing by the author.*

SANDUSKY, A TEN-WHEEL FREIGHT ENGINE, 1851

Low-drivered ten wheelers were seen as the correct style of locomotive for heavy freight service around mid-century. Just about every American locomotive builder produced engines on this pattern. Fortunately, a drawing for one such machine was found in the Portland Company collection at the Maine Historical Society. Two engines, the *Sandusky* and the *Huntsville*, were constructed on this plan for the Mad River and Lake Erie Railroad. Railroad building was reaching a frenzy at this time in Ohio, as the railroad network reached out into what were then considered the western states. By 1860 Ohio had almost 3,000 miles of railroad, giving it the greatest track mileage of any state in the union. The Mad River line opened its first segment in 1838 and had ambitions to build tracks from the Lake Erie port of Sandusky as far south as the Ohio River. This goal was never reached, and the Mad River went through several mergers and name changes before being folded into the Cleveland, Cincinnati, Chicago, and St. Louis Railway in 1890. Ohio's pioneer railway disappeared as a corporation in 1858, but its lasting influence was in establishing 4 feet 10 inches as the standard railway gauge in Ohio.

The *Sandusky* of 1851 was the second locomotive on the Mad River line to bear this name. The original, a 4–2–0 built by Rogers in 1837, disappeared from the roster in 1850. Its successor was a long, rangy ten wheeler, depicted here by a rather sketchy outline drawing reproduced as Fig. 249. The long cylinder stroke and small wheels are a sure indication that this locomotive was meant for tractive power but not speed. The boiler is long but small in diameter, only 42 inches. There are many antique features obvious even in this elementary drawing. Among these are the inclined cylinders set at 10 degrees, the short-wheelbase truck, the riding cutoff or double valves, and the round side rods. The outside frame is used, but rather than the composite wood-metal affair still employed by conservative designers, Portland adopted an all-metal slab, measuring ¾ inch thick by 4¾ inches deep. Note that the lead drivers are blind, in keeping with standard practice of the time, when just about every locomotive running in the nation did so with flangeless leading driving wheels.[1] Some ten wheelers, such as

[1] *Railroad History*, No. 172, Spring, 1995, p. 63.

the Baltimore and Ohio's Hayes 4–6–0's, had both the lead and center pair of drivers fitted with flangeless tires.

Once again referring to the Portland drawing, one wonders why the draftsman spent such effort in depicting the adjustable key arrangement for the connecting-rod brasses. Was it a new feature or a special requirement of the purchaser? There is, of course, no answer, nor can I explain the tapered axle of the center pair of drivers. It is modeled on a car-wheel axle but its design is unique, or at least very uncommon, for locomotive driving axles. The idea behind making the axle smaller in diameter at its center was to allow a degree of flexibility, so that the axle would bend slightly rather than break if strong side thrusts were encountered. This scheme, universal for car axles, was, however, not adopted by locomotive designers. Again, we know what happened, but not why it happened.

There are no photographs or other pictorial images of the *Sandusky* or its sister, or in fact of any other Portland-built 4–6–0 of this period. To gain an idea of what the complete engine might have looked like, we can refer to other ten wheelers of the period. Aside from its spread truck, the *Wilmore*, built in 1856 for the Pennsylvania Railroad by Smith and Perkins of Alexandria, Virginia, was strikingly similar to the *Sandusky* (Fig. 250). The *Wilmore* was somewhat longer, but feature for feature—from the inclined cylinders to the half-stroke, rear-mounted feed pump, the two engines are near duplicates. Curiously, the *Wilmore* was sent out to Ohio in 1857 on loan to a Pennsylvania Railroad affiliate, the Steubenville and Indiana. In the next year, the *Wilmore* went to work on a much larger Pennsylvania Railroad affiliate (and in later years a major element in Pennsylvania's Lines West), the Pittsburgh, Columbus, and Cincinnati Railroad. It is possible that these featherweight ten wheelers may have passed by one another during journeys across the Buckeye State more than a century ago.

Another 4–6–0 of the period which is, in general, on the plan and scale of the *Sandusky* is shown here in Fig. 251. It is redrawn from the Baldwin advertising lithograph reproduced in Fig. 109. It very likely depicts Baldwin's first style of 4–6–0, which was introduced in 1852. The same features found in the *Sandusky*—from the outside frame to the double valves and the inclined cylinders—were used in this machine. The clumsy Matthew-style smokestack hardly adds to the engine's appearance, and some critics might even question the look of the iron hen-coop cowcatcher. This engine certainly seems to prove the contention that freight engines were not meant to be beautiful. Placing the rear drivers at the extreme end of the frame precluded the use of a rear style of feed pump that was used on the *Sandusky*. It is assumed that Baldwin's patented combine feed pump and crosshead guide were used on this engine.

A photograph of a Baldwin 4–6–0 that was completed at the end of 1855 offers a final picture of the ten wheeler as it appeared around the middle of the nineteenth century. This engine, the Pennsylvania Railroad's No. 118, started out very much as a *Sandusky* look-alike (Fig. 252). However, by the time this photograph was made, in the late 1860's, the No. 118 had been modernized. The double valves, outside frame, and feed pumps had already been removed. It is likely that the truck wheelbase had been lengthened and that the cylinders had been made more level. Yet the lightness of the frame, the dainty driving-wheel centers, and the round side rods help us to understand the appearance of first-generation ten wheelers.

Dimensions of the *Sandusky*	
Cylinders	14″ × 22″
Wheels	44″
Weight	20 tons (estimated)

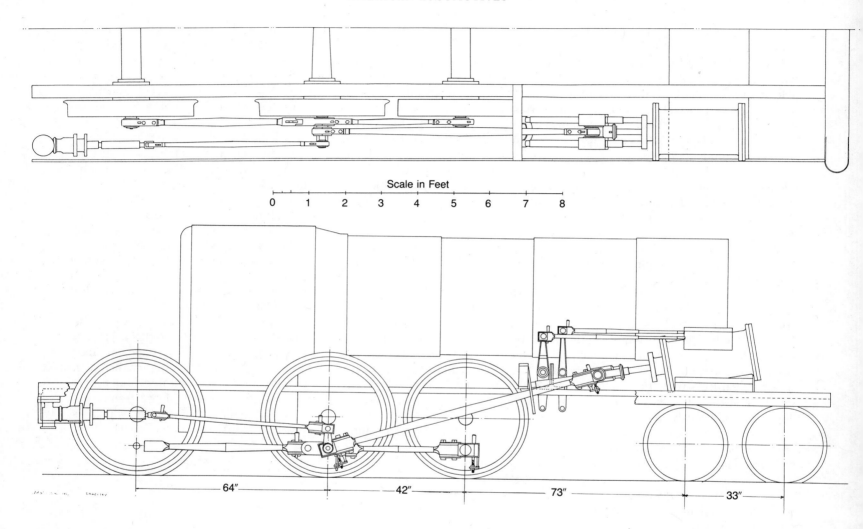

Fig. 249. The Sandusky, built by the Portland Company in 1851 for the Mad River and Lake Erie Railroad. This drawing was traced from the original in the collection of the Maine Historical Society.

MORE REPRESENTATIVE LOCOMOTIVES

Fig. 250. The Wilmore, *built in 1856 by Smith and Perkins of Alexandria, Virginia, for the Pennsylvania Railroad.*

Fig. 251. M. W. Baldwin produced his first ten wheelers in 1852 on a plan very similar to the one shown here. These machines had 18-in. by 22-in. cylinders and 44-in. drivers.

Fig. 252. *A Baldwin 4–6–0 built in 1855 for the Pennsylvania Railroad. It is shown here about ten years later in a somewhat modified condition.*

LADY ELGIN AND TWO LESSER SISTERS, 1852

Canada made a brave start in 1836 in opening its first steam railway; however, the vast northern land made little progress in expanding its system. By 1850, when the United States was building rail lines in a brisk manner, Canada had only 64 miles of track. During the next decade, considerably more energy was expended on railway construction, and British North America could boast of almost 1,900 miles of line. One of the region's pioneering rail lines was the Ontario, Simcoe, and Huron, which set out in October, 1851, to connect Toronto with Lake Simcoe.[1] This project was considered so important that the governor general, Lord Elgin (James Bruce, 1811–63), attended the ground-breaking ceremony. His wife, Lady Elgin, turned the first spade of earth to symbolize the beginning of construction. The first locomotive on the line, which was delivered some months later, was named in her honor. This machine, depicted here by Figs. 253 and 254, was completed at the Portland Company's works in June, 1852. Sending a locomotive to central Canada at this time was no easy matter, for the region was nearly devoid of railways, and there was no rail connection.

In addition, because the new locomotive was designed for the Ontario, Simcoe, and Huron's track gauge of 5 feet 6 inches, shipping it over most U.S. railroads was difficult. The engine was partially disassembled and sent by rail to Oswego, New York. There the bits and pieces were put aboard a schooner and sent across Lake Ontario to Toronto. The *Lady Elgin* shuffled back and forth over a short segment of line, but by 1855 the railroad extended to Collingwood on Georgian Bay, some 90 miles north of Toronto. A quarter century passed, and the *Lady Elgin* and most of its broad-gauge compatriots were at the end of their economic service life. They were too small and too obsolete for further use. When the last segment on what was then called the Northern Railway was converted to standard gauge in 1881, the *Lady Elgin* was retired.

In March, 1852, Portland completed for predecessors of the Grand Trunk Railway two locomotives, the *Danville* and the *Consuelo*, which were slightly smaller versions of the *Lady*

---

[1] The history of the *Lady Elgin* is drawn largely from the *Railway and Locomotive Historical Society Bulletin*, Nos. 56, 85, and 147.

Elgin. The broad-gauge railway between Montreal, Quebec, and Portland, Maine, was built in two sections; the Canadian end was called the St. Lawrence and Atlantic, and the U.S. segment, the Atlantic and St. Lawrence. Any identity problem caused by these dual corporate titles was quickly resolved, however, for in 1854, about a year after the dual railway opened, it was taken over by the Grand Trunk.

The *Danville* was taken over by the Grand Trunk and was designated as No. 111. It was sold in 1866 to the Portland and Oxford Central. The *Consuelo* was delivered to contractors for use in building the St. Lawrence and Atlantic. It remained in service on that line until at least 1858, and is so listed in the report on Canadian railways by Samuel Keefer, inspector of railways for Canada, that was published in that year. How much longer it remained on the St. Lawrence line is uncertain, as is its ultimate disposition.

After a casual inspection of the two drawings reproduced here as Figs. 253 and 255, one is tempted to conclude that these engines are duplicates. In a general sense this is surely true, for the pattern of both designs is the same, and they are so alike that they are probably from the same designer. In the original drawings, both elevations were drawn on the same sheet of paper. They overlapped, in fact, and the tender frames are drawn in at the ends so that they, too, presented a confused image. I traced and separated them for clarity. The differences between the two designs are surely marginal; the driving-wheel centers were made ¼ inch shorter on the *Lady Elgin* than on the other two engines. The *Lady Elgin*'s cylinders were 1 inch larger in the bore, and its boiler was 2 inches greater in diameter. It is true that the *Lady Elgin* had fifteen more tubes, double valves, and a forward half-crank rather than a crosshead-style pump, but in the name of common sense and efficiency, why not make such very similar engines exactly alike? There was every reason to do so and almost no reason not to do so. About the only plausible explanation was that purchasers required these differences. What else could explain the *Danville*'s having 119 tubes and *Consuelo*'s having 117?

Both engines were built from the same drawing. This aversion to standard construction seems to defy explanation. Surely it would require a twisted mind to justify varying the wheelbase by ¼ inch.

Regarding the external appearance of these three engines, we are fortunate to have a photograph of the *Lady Elgin* and a Portland lithograph dated 1852 of an engine named *Forest State*. No engine of this name is known to have been built by Portland; however, the exact mechanical nature of the rendering suggests that it is more than a casual sketch. From the lithograph it appears that these engines were delivered without cabs, headlights, and cowcatchers. Engines that never ran at night might fare well without a lamp, but there were few tracks in the land not visited by cattle, deer, or hogs, and it is hard to imagine the sturdiest crew member in Canada or New England who would consider a cab a luxury. The *Lady Elgin* had all of these fittings by the time the late in-service photograph shown in Fig. 254 was made. It appears to have lost its double valves for link motion but is otherwise very much a locomotive of the 1850's.

Dimensions of the *Lady Elgin*	
Cylinders	14″ bore × 20″ stroke
Driving wheels	60″
Tubes (132)	10′ 4¼″ long
Heating surface	684 sq. ft.
Weight	49,500 lb.
Tender	8 wheels, 1,738-gal. tank

Dimensions of the *Danville* and the *Consuelo*	
Cylinders	13″ bore × 20″ stroke
Driving wheels	60″
Tubes	10′ 10¼″ long × 1¾″ diameter
Heating surface	630 sq. ft.
Weight	40,000 lb.
Tender	8 wheels, 1,321-gal. tank

AMERICAN LOCOMOTIVES

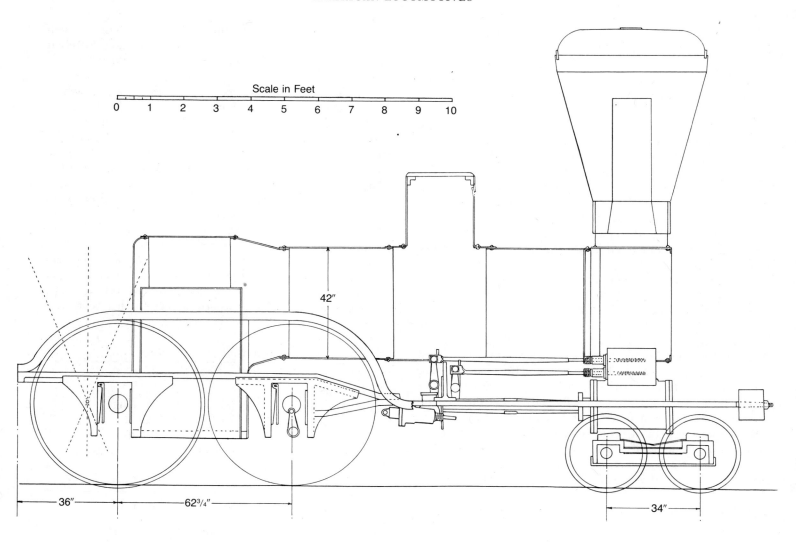

Fig. 253. The Lady Elgin, built in 1852 by the Portland Company for service on the first railroad in what is now Ontario, Canada. It was built for a track gauge of 5 ft. 6 in.

Fig. 254. The Lady Elgin *as it appeared late in its service life on the Grand Trunk Railway. The engine was retired at about the time the Grand Trunk converted the last of its broad-gauge lines to standard gauge in 1881.*

AMERICAN LOCOMOTIVES

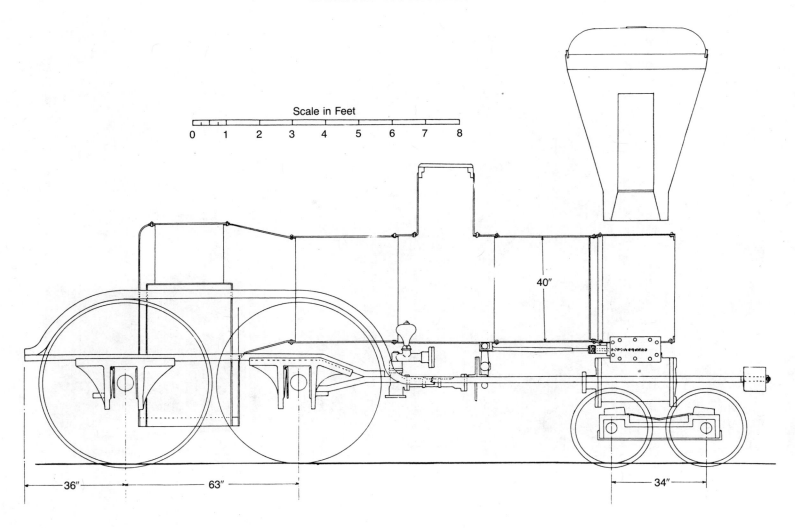

Fig. 255. In 1852 the Portland Company built two locomotives, the Danville *and the* Consuelo, *for service in Canada from this drawing.*

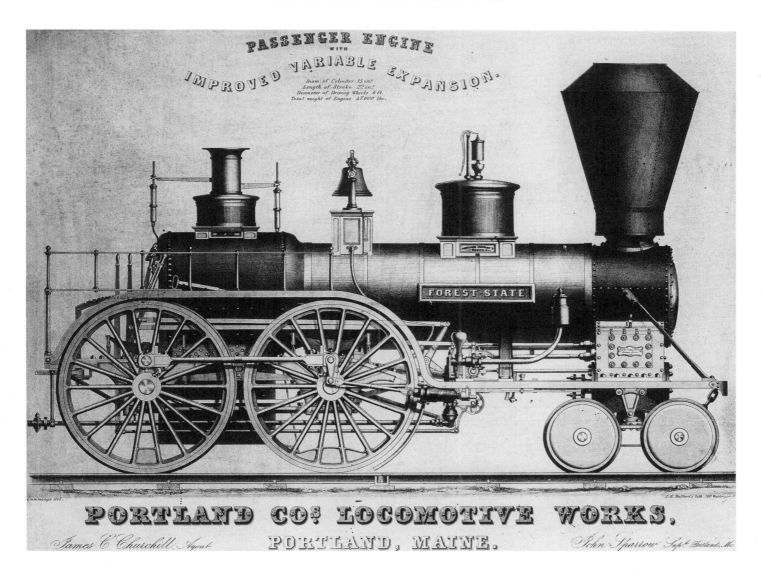

Fig. 256. The Forest State. *This lithograph is contemporaneous with the three Portland engines depicted above and is surely a good representation of their original appearance.*

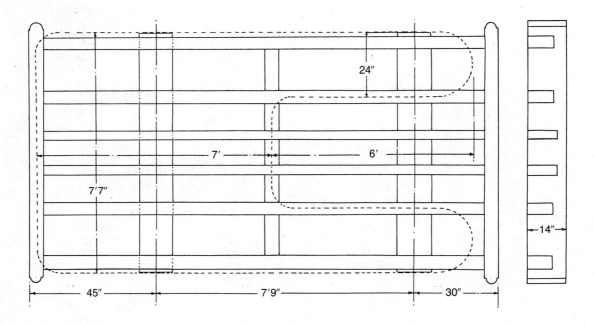

Fig. 257. The tender floor plan for the Danville *and the* Consuelo, *shown in a drawing traced from the original, which is preserved by the Maine Historical Society.*

LISLE, A HINKLEY 0-4-0, 1853

When I began gathering material for this book in about 1960, I chanced upon a side elevation of the *Lisle* in the collections of the Boston Public Library. I assumed that the engine was an early four-wheel switcher and put a copy of the large-scale drawing aside for future study. Many years later, I found the *Lisle*'s story to be far more complex than its simple profile might suggest. It was one of twelve engines built in 1853 and 1854 for the Syracuse, Binghamton, and New York Railroad. This 6-foot-gauge line later became part of the Delaware, Lackawanna, and Western.[1] At 22 tons, the *Lisle* was as heavy as most 4-4-0's of its day. Its cylinders were equally large, and although the 48-inch drivers appear undersized, they could surely have propelled a freight train along the line at 10 miles per hour without difficulty. For these reasons, it is likely that the *Lisle* was built as a road engine and was not designed for switching service. The eight-wheel, 1,000-gallon tender once again suggests over-the-road operations. Typically, switchers during this early period would more likely have had a four-wheel tender—perfectly adequate for short runs around yard trackage. But, of course, there is no way to determine exactly how this or any other engine was employed on a day-to-day basis.

At about the same time the *Lisle* was built, Hinkley issued a lithograph of an inside-cylinder 4-4-0 with the number 490, shown in raised numerals on the base of the steam dome; this graphic is reproduced in the first volume of Thomas T. Taber's invaluable three-volume history of the Lackawanna. The 0-4-0 *Lisle* also carried the construction number 490. The lithograph carries no name, so it is possible that it was issued to represent a generic design and the number was applied without reference to a specific engine. Such a fragmentary and confused record surely hampers the work of the locomotive historian.

Before proceeding with the history of the *Lisle*, I shall digress briefly to discuss the unusual plate style of driving wheels on this engine. They were made much like a double-plate car

---

[1] Thomas T. Taber, *The Delaware, Lackawanna and Western Railroad in the Nineteenth Century*, (Muncy, Pa., 1977), pp. 384-90; and *Railway and Locomotive Historical Society Bulletin*, No. 72. See also *Railroad History*, No. 142, for data on Hinkley.

wheel that was cast hollow with a thin wall on either side—except, of course, that they were larger in diameter than the typical 33-inch car wheel. It is not entirely clear whether the tread and flange were part of a one-piece casting or whether a wrought-iron tire was pressed on. If the wheel was all one piece, then it would be a one-wear wheel, and the entire unit would require replacement after the chilled surface of the tread or flange wore away. Some industrial locomotives, especially those with very small drivers, had plate wheels. But we can offer only a few examples of other main-line engines so equipped, and these are based on photographic evidence. A Norris 4–2–0, the *William Penn,* dating from the 1830's, was rebuilt in 1865 at Lancaster, Pennsylvania, for service in California—it had plate driving wheels. In a more modern instance, Hinkley produced three 0–6–0's with 42-inch disc drivers in 1863 for the Atlantic and Great Western.

Returning to the *Lisle*'s drivers—double bosses for the crank pins were reminiscent of the same scheme used for the *Copiapo*'s driving-wheel centers (Fig. 142). The two smaller circles on either side of the crank pin bosses are holes for the inner sand core and are a necessary part of the casting process.

In 1867, the *Lisle* was thoroughly rebuilt with outside cylinders. It had already received a link-motion valve sometime before that year. Just nine years later, the railroad converted to standard gauge. It is possible that the 1853 *Lisle* was retired at this time and replaced by a new engine of the same name or road number 3. We are once again uncertain as to what happened, but there is a record that the No. 3 was remodeled in 1893 as a 49-ton 0–6–0. If the engine that underwent this transformation began as a 22-ton 0–4–0, it is difficult to believe that much of the 1853 engine was incorporated into the new and vastly heavier No. 3.

There is one more intriguing puzzle about the little *Lisle,* and again one that possibly cannot ever be answered: I wonder whether this design was the basis for four 0–4–0's built in 1853 and 1854 for the Cleveland, Columbus, and Cincinnati Railroad by the Cuyahoga Steam Furnace Company of Cleveland, Ohio. They favor the Hinkley design in just about every way. They have inside cylinders, 13 inches by 20 inches; 48-inch drivers; and a straight boiler. Was the similarity of design features a mere coincidence, or had the draftsman at Cuyahoga ever peeked over the fence at any number of Poney engines that Hinkley sent to railroads in the central states?

Dimensions of the *Lisle*

Cylinders	13" bore × 20" stroke (The Syracuse, Binghamton, and New York 15" × 20".)
Wheels	48" in diameter
Tubes (94)	copper, 2" × 10' 6"
Cost	$6,000

MORE REPRESENTATIVE LOCOMOTIVES

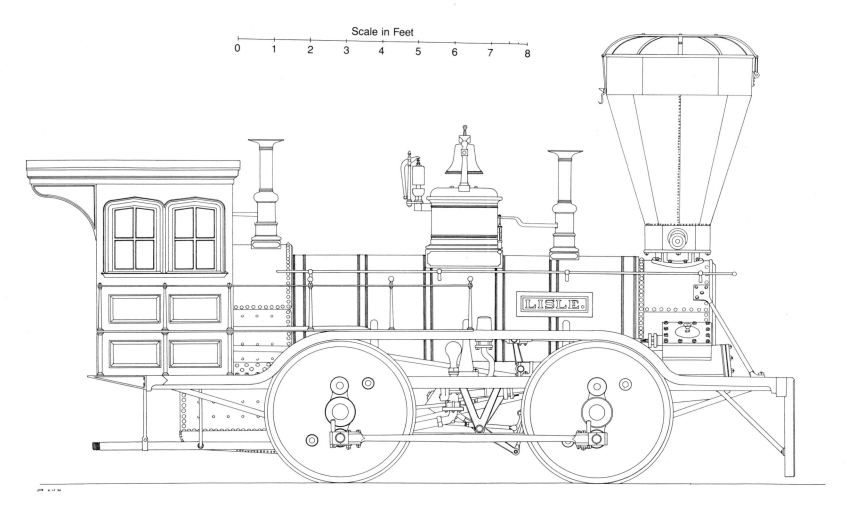

Fig. 258. The Lisle, *built in 1853 for the Syracuse and Binghamton Railroad. This 22-tonner was most likely designed for slow-speed freight service.*

STOREY, A WESTERN MOGUL, 1869

Few railroads are better documented than the Virginia and Truckee, a rich short line opened in 1869 to exploit Nevada's huge silver deposits near Virginia City. Much of its paperwork is preserved by the Nevada State Railroad Museum, in Carson City, Nevada, and a dozen or more books have been written to record its history.[1] There can be no lament, in this case, about the paucity of information available about its locomotives.

By the late 1860's, the better silver ore was exhausted, and it was no longer economical to process the ore at the steam-powered stamping mills located near the mines in Virginia City. The ore could be crushed far more cheaply in water-powered mills along the Carson River—15 miles away, with a drop in elevation of 1,600 feet. A railroad was needed to connect the mines and the mills, and so in early 1869 surveys began for the Virginia and Truckee Railroad. It would be an isolated short line with no connection to the national system, at least as first built. This meant that rails and rolling stock had to be transported overland to the construction site. The closest rail connection was Reno, some 30 miles to the north of Carson City, the Virginia and Truckee's western terminal. In September, 1869, as the railroad neared completion, teams of oxen—thirty-six per locomotive—pulled the first three engines overland to Carson City. Extra-wide treads were fitted over the regular tires, and the main and side rods were disconnected, but still, it was a matter of brute force and determination to drag the *Storey* and its two smaller sisters to their destination.

In November, 1869, the railroad opened, and the *Storey*, designated as the Virginia and Truckee's No. 3, was busy hauling trains over the line. It was surely a railroad of curves and steep grades. The line's serpentine route made it necessary to traverse 24 miles of track to travel 15 air miles, and most grades were 2 percent. Fortunately, the heavy ore traffic was downhill, while the lighter, empty ore cars, firewood, and

---

[1] The books on the Virginia and Truckee include Gilbert Kneiss, *Bonanza Railroads* (Stanford, Calif., 1941); Lucius Beebe and Charles Clegg's small volume *Virginia and Truckee: A Story of Virginia City* (Berkeley, Calif., 1949), which went through many printings and several revisions; and more recently, Ted Wurm and Harre Demoro's *The Silver Short Line* (Glendale, Calif., 1984).

mine-prop traffic went uphill. The *Storey* managed to handle trains that weighed 70 tons, exclusive of the weight of the cars.

The promotors of the railroad were western businessmen, with a natural self-interest in boosting the local economy. The Bank of California, an important investor, at the same time wanted to see Pacific Coast industries such as the Union Iron Works in San Francisco flourish, and thus locomotives No. 1 through No. 3 were ordered from this firm. The argument was repeatedly made that local machinists could produce a product equal to any made in the East while saving the delay and shipping costs inherent in imported machinery. The same logic had been used decades earlier by those wanting to establish locomotive plants in Nashville, Chicago, and other locations that did not, in the end, produce more than a handful of locomotives. Between 1865 and 1881, the Union Iron Works built about thirty locomotives, or fewer than were built in one month's work at Baldwin's plant in Philadelphia. As a specialist, Baldwin could simply produce better engines, more cheaply than could a general machine shop such as Union, which produced a variety of products ranging from mining machinery to marine engines. Union was paid $16,500 for the *Storey*, while Baldwin sold a comparable 2–6–0 to the Virginia and Truckee in 1869 for $15,000.

When it comes to quality, a comparison is more difficult to render, but all of the Union-built engines on the Virginia and Truckee were retired or sold at an early date. It is true that the *Storey* ran as late as 1920 for its subsequent owners, and thus vindicated its original owners to a degree, at least. Yet, for reasons not now apparent, the Virginia and Truckee became a loyal patron of Baldwin and seemed to find no problem with using a supplier situated on the other side of the continent.

The British technical journal *Engineering* featured the *Storey* as a centerfold drawing in its March 10, 1871, issue (Fig. 259). On a nearby page, a cross-section drawing and a brief description appeared. The plan is credited to one of Union's partners, Irving M. Scott. Scott had learned the machinery business in Baltimore and worked for a time for Murray and Hazelhurst, machinists of that port city who built a large variety of products, including a small number of locomotives. It seems clear from the engraving reproduced here that Scott sought to break no new ground. The general arrangement and details of the engine are conventional and well executed. If this machine incorporated any deficiencies, it was not in its conception but more likely in its execution, for bad workmanship or poor materials can ruin the best plan. Scott used a Bissell truck, link motion, crosshead pumps, and other standard and proven locomotive design features of the time.

There are few reminders of this doomed effort to establish a locomotive shop in the far West—a few engravings, drawings, and photographs, but little more. Yet at the time of this writing—October, 1996—a full-sized reproduction of the *Storey*'s smaller sister the *Lyon,* Virginia and Truckee's No. 1, is under way at the Strasburg Rail Road's repair shop in eastern Pennsylvania. This is a private effort, and the work goes forward slowly as finances permit, one part at a time. This is to be a full-sized operating replica. When will we see the *Lyon* under steam? It will be a few more years, surely, but consider how pleased Irving Scott and the others associated with this early Mogul would be to see their handiwork recreated in such a faithful fashion.

Dimensions of the *Storey*	
Cylinders	16″ bore × 24″ stroke
Wheels	48″ diameter
Weight	30 tons (engine only)
Tubes (160)	2″ diameter × 10′ long
Tender tank	40″ high × 16′ long, 2,200 gal. capacity
Tender wheels	26″ diameter (spoked)

Fig. 259. *The Virginia and Truckee Railroad's No. 3, the Storey, built in 1869 by the Union Iron Works of San Francisco.*

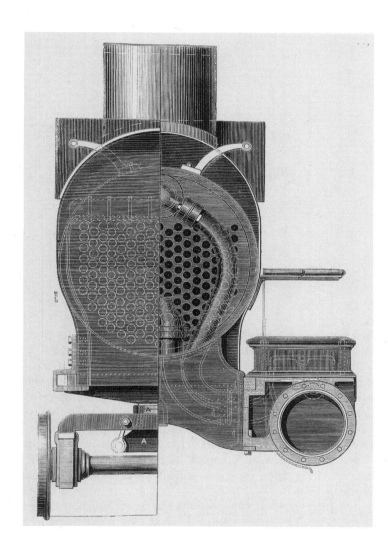

Fig. 260. A cross-section engraving of the Storey *as depicted in the British technical journal* Engineering.

PERKINS TEN WHEELER, 1871

Thatcher Perkins (1812–83) was one of those master mechanics who every few years seemed ready to move on.[1] He first rose to prominence in 1847, when he was appointed master of machinery for the Baltimore and Ohio Railroad. This lasted barely eight years. Then Perkins hired out to the Central Ohio Railroad; after a brief stay, he went to the Tredegar Iron Works and then back to the Baltimore and Ohio. In 1865, the newly formed Pittsburgh Locomotive Works attracted his attention, and he worked as superintendent for that company. Earlier, in the 1850's, he had been a partner in the locomotive-building firm of Smith and Perkins, in Alexandria, Virginia. In 1868 Perkins made his final career change, going to Louisville, Kentucky, as master mechanic for the Louisville and Nashville Railroad. His lifelong penchant for locomotive building soon found expression at the repair shops, where a number of freight and passenger engines were produced to his designs.

Among these machines were five 4–6–0 freight engines built in 1871 and 1872 to the design shown in Figs. 261 and 262.

Perkins was no stranger to the 4–6–0, for he had built a number of these engines for the Baltimore and Ohio at the Mt. Clare shops, starting in 1863. One of these engines is preserved at the B&O Railroad Museum in Baltimore, Maryland. This group of engines was designed for passenger service. They had larger drivers and a shorter wheelbase and were altogether more compact and symmetrical than the low-wheeled, rangy version designed for the Louisville and Nashville.

The plan shown here represents the conventional practices of the time. The boldly moulded cap stack gives the engine a distinctive look but not entirely a new one, for Perkins had used the design while still with the Baltimore and Ohio. There are a few mechanical items worthy of note. The boiler is shown without its lagging. The small scotch yoke surrounding the rocker's upper bearing is intended to relieve the up-and-down pressure that the valve stem and its packing gland normally

---

[1] *Railroad History*, No. 169, Autumn, 1993, pp. 54–68; *Engineer*, February 16, 1872, p. 119.

experienced at this point. The link is facing backwards, and although the drawing is not clear on this point, Perkins may have used a Gooch valve motion. The piston of the feed-water pump is made after the now-familiar automotive style and thus precludes the need for a crosshead or guide. This engine was built for a track gauge of 5 feet, as were most other Southern locomotives of the time.

Dimensions of the Perkins Ten Wheeler	
Cylinders	18″ bore × 24″ stroke
Wheels	53″ in diameter
Total driving wheel base	13′ 10″
Heating surface	1,062 sq. ft.
Tubes (150)	1⅞″ OD × 12′ 4″
Weight	40¼ tons

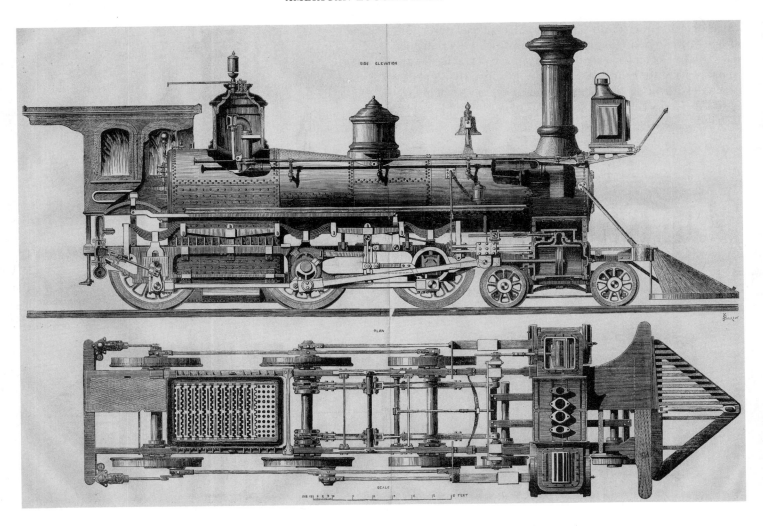

Fig. 261. *The Louisville and Nashville Railroad built several ten-wheel freight engines at its Louisville shops to the design of Thatcher Perkins in 1871.*

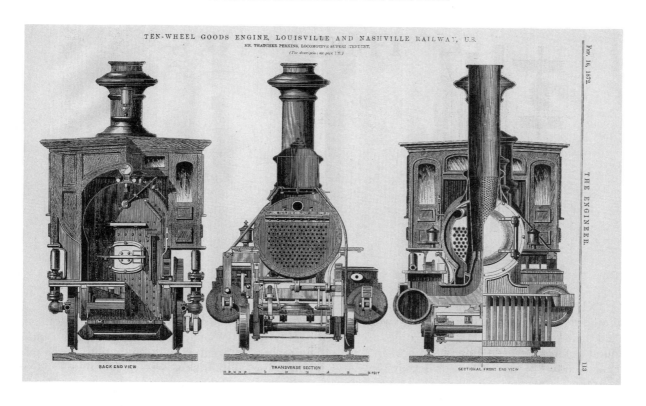

Fig. 262. *End and cross-section drawings of Perkins ten wheelers are shown in these engravings.*

SOME FOUR-WHEEL SWITCHERS OF THE 1870'S

In the beginning of the railroad era there were no switchers; road engines did their own shunting. Horses were also used to move cars around at terminals or industrial sidings. In time, obsolete road engines were re-employed to drill cars. Many 4–2–0's worked out their last years in such service. This was also the fate of the Baltimore and Ohio's Grasshopper class—in time they were remodeled into tank engines, and then were downgraded a second time to shop switchers. But the exact who, when, and where concerning the first new American yard locomotives remains a mystery. During the early 1850's a number of small 0–4–0's were delivered, bearing such names as *Poney, Pusher, Hoosier Poney,* and *Mule.* These names suggest to me a small, low-wheeled engine meant for switching cars. The first switcher on the Cleveland, Columbus, and Cincinnati was named for Myron Dow, a teamster who had previously switched cars in the Cleveland yard with a single horse.[1] This engine, built by the Cuyahoga Steam Furnace Company in 1853, was also known as *Poney No. 1.* During the next decade, photographs of switchers appear in enough numbers to suggest that such engines were commonplace. A fine example of a Civil War–vintage switcher is the picture of the Union Pacific Railroad's No. 3, a Grant 0–4–0T, reproduced in Barry R. Combs' *Westward to Promontory* (1969), p. 18.

By 1870, the four-wheel switcher was a well-defined style of locomotive that was found in increasing numbers on major railroads as freight traffic accelerated. With the growth of interchange freight car service, marshaling yards grew from meager end-of-the-division operations to huge complexes covering 500 acres of land. A special style of engine was needed to break down and build up arriving and departing trains. By 1890, the switcher, which had once been a rarity, had become commonplace. Switchers made up 28 percent of the American locomotive fleet, according to Arthur M. Wellington, editor of *Engineering News.* They generally ran at between 6 and 8 miles an hour, but because of the many starts and stops their operating and repair costs were about the same as those of a full-sized road engine.

The four-wheel switchers of the 1870's took on a charac-

---

[1] Alvin F. Harlow, *The Road of the Century* (New York, 1947), p. 346.

teristic pattern that prevailed for many years afterwards. Among their notable features were a straight boiler, with the steam dome centered over the drivers, and two sandboxes, one for each direction of travel. Having two sandboxes also ensured a good supply of sand—a necessity with so much starting and stopping. A backup headlight was a common but not universal accessory.

The four-wheel switcher as it appeared in 1870–80 is represented in Figs. 263 and 264. The *Suffolk* was built in 1871 for Manson and Company by the Rhode Island Locomotive Works. The firm had been active in locomotive business for just five years. Its plant was the former Burnside Rifle Company, a facility not well suited for locomotive construction. Despite this handicap, the Providence builder continued producing railway engines until 1908. I have been unable to identify Manson and Company, but I believe that it was somehow connected with the Merchants and Miners Transportation Company, a steamship line operating between Norfolk, Baltimore, and Providence. The *Suffolk* had two sister engines, the *Norfolk* and the *Useful*. All were used around the dockside terminals of the Merchants and Miners Transportation Company. The last-named engine became the property of the Boston and Providence Railroad and was not scrapped until 1904.

Just eight years after the *Suffolk* entered service, the Chicago, Burlington, and Quincy Railroad's Aurora Shops produced a very similar engine, the No. 335 (Fig. 264). The No. 335 had its bell on a more stable perch on top of the boiler, rather than above the forward sandbox. The backup headlight and slope-back tender improved the engineer's vision for backup moves. Feed water was supplied by an injector on the left side (shown in the drawing) and a feed-water pump mounted on the right side under the cab deck. The engine had a plain black finish relieved only by yellow lettering, according to the brief description accompanying the drawing in the February, 1880, *National Car Builder*. This contemporary notice states that the No. 335 was a Class F switcher, but recent historians claim that the No. 335 and all Chicago, Burlington, and Quincy Class F locomotives were 0–6–0's.[2] Apparently Class F was eliminated as a designation in the 1904 renumbering of the railroad. It is also recorded that the No. 335 was replaced in 1896 by a new 0–6–0, and this may be the basis for the confusion.

Dimensions of the *Suffolk*

Cylinders	14″ bore × 24″ stroke
Wheels	48″ in diameter

Dimensions of the Chicago, Burlington, and Quincy Railroad's No. 335

Cylinders	15″ bore × 22″ stroke
Wheels	44″ in diameter
Tubes (99)	2″ diameter × 11′1″ long
Weight (engine only)	25 tons
Tender tank	1,760 gal. capacity
Tender weight	18 tons (with coal and water)

[2] B. G. Corbin, *Steam Locomotives of the Burlington Route* (1960), pp. 43, 263; on early locomotives of the Chicago, Burlington, and Quincy, see *Railway and Locomotive History*, special bulletin 43A, July, 1937, p. 31.

Fig. 263. The Suffolk, built by Rhode Island in 1871, a prototypical four-wheel switcher *of its day.*

MORE REPRESENTATIVE LOCOMOTIVES

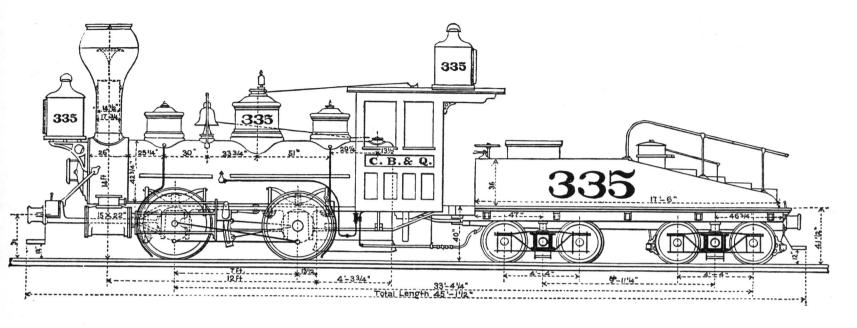

Fig. 264. The Burlington Lines built some of its own power, most of it in the main repair shops in Quincy, Illinois. The No. 335 was produced in 1879.

MARMORA, AN EDDY CLOCK, 1876

Wilson Eddy, master mechanic of the Boston and Albany Railroad, espoused many novel locomotive design features. He incorporated most of them in 135 locomotives built at the Springfield, Massachusetts, repair shops of the Boston and Albany between 1851 and 1881.[1] Eddy was an innovator and an independent thinker, but he cannot truly be called a leader, for he had relatively few followers. Most of his ideas were, in fact, rejected by his fellow locomotive designers. Eddy advocated straight boilers, without a steam dome, while just about everyone else remained steadfastly loyal to wagon-top boilers with steam domes. Those who accepted straight boilers generally insisted on two steam domes. Eddy believed in small steam ports, while the industry universally favored generous ports and steam passages. Eddy bolted the side rails directly to the firebox—much to the horror of his contemporaries. Expansion joints were built into the cylinder saddle. He remained loyal to the large safety-valve escape pipes, or "cannons," and the square style of sandbox long after both were considered obsolete. In defense of the safety-valve cannons, it must be said that they were something of a necessity on a boiler bereft of a steam dome. The valves had to be high above the water level in the boiler, so as to vent only steam and not hot water. It was also important to keep this discharge above the cab roof, in order to protect the crew. There wasn't much choice other than to put the valves on some type of pipe or column; hence the survival of the cannons.

Eddy placed the feed pumps under the cab deck and drove them from eccentrics mounted on the rear driving axle. This arrangement was also considered odd and was only occasionally used elsewhere. The fact that the several schemes mentioned here were not popular much beyond Springfield hardly condemns them as wrong or defective. Eddy's engines were not dubbed "Clocks" for nothing. They ran as smoothly and steadily as a fine mantle clock. They were dependable and economical; it's just that the majority of engine builders were not about to copy all of Eddy's odd notions.

---

[1] Many of Eddy's design ideas are scattered throughout the original text of this book; see the index for exact page numbers, and refer to Appendix A for an outline of Eddy's life.

It should be understood that Eddy was not always outside of the mainstream. He was an early user of the spread truck, balanced valves, front-end throttles, and injectors. He promoted the idea of large heating surfaces and managed to enlarge firebox width with his peculiar style of frame (see Fig. 64). He introduced oil cups inside the cab. A long tube ran the length of the engine to the valve box and cylinders. This convenient arrangement did much to assure proper lubrication of these vital parts. It also freed crewmen from the necessity of making a hazardous trip out over the running boards should more oil be needed while the engine was under way.

Eddy's peculiar breed of locomotive proved itself when tested against more conventional motive power. In the test of an Eddy Clock and a Rhode Island 2–6–0 conducted in 1876 on the Boston and Albany, the Clock showed a fuel saving of around 25 percent while doing the same work as its rival. It is also evident that a major railroad such as the Boston and Albany could not long indulge an eccentric locomotive superintendent unless his engines were up to the job. Moving freight and passengers through the Berkshires required a stable of dependable locomotives. Eddy spent forty years on the Boston and Albany and was, during most of that time, the head of the motive power department.

Engine building began in a small way in Springfield with Eddy's first engine, the *Addison Gilmore,* in 1851. Two years later, the directors of the Western Railroad, as the Boston and Albany was named before 1867, gave Eddy specific permission to construct new engines at the Springfield repair shops, as fill-in work. It was seen as a way to keep repair crews busy during slack times. By the early 1870's, Springfield was turning out a dozen or more new engines per year, but even at this rate most of the railroad's motive power came from regular suppliers. The earliest Clocks weighed 20 to 25 tons.

As the demand for more powerful locomotives grew, most railroads began to adopt Moguls, ten wheelers, or Consolidations, but Wilson Eddy did not. He simply built larger and larger 4–4–0's. By the late 1860's, Clocks were constructed in the 29-ton range. By the middle of the next decade, 33½ tons was reached. The cylinders in the Clocks were now 18 inches by 26 inches—dimensions larger than those found in many Moguls and ten wheelers. By the late 1870's, near the end of Eddy's career, 40-ton Clocks were in production.

In November 1880, Eddy retired. One of his sons stayed on at the shops and in a fine gesture of loyalty attempted to continue the old order. The last Clock was outshopped during the next year. The new superintendent of motive power, Arthur B. Underhill, apparently felt that the age of the Clocks, if not of the 4–4–0 itself, was long since over. In the summer of 1890, about forty Clocks, presumably all but the newest or best remaining in service, were sold at scrap metal prices to a speculator.[2] The new owner hoped to sell them handily down south for $1,000 apiece after performing some light repairs and adding a fresh coat of paint. This plan proved a failure, and so the entire lot was cut up for old iron late in the same year.

A few Clocks stayed on the Boston and Albany for several more years, doing light work on branch lines or doing work-train service. One of the last on the property, the No. 39, or *Marmora,* was assigned to the lowest type of service: it was used as a stationary boiler at the Worcester, Massachusetts, station. Here it stood, lonely and forlorn, all complete but in a most decrepit state, to make steam to heat coaches so that passengers might not enter cold cars. In about 1904, Purdue University began collecting locomotives for a proposed railway museum in Lafayette, Indiana. The seedy relic at Worcester came to Purdue's attention. An exchange of letters brought the desired result, and soon the *Marmora* was creaking its way westward in a slow procession of way freight trains. The proposed museum plan collapsed when its chief sponsor, Professor William F. M. Goss, left Lafayette for a position elsewhere. The collection remained, however, until just after the end of World War II, when it was given to the National Museum of Transport in St. Louis, Missouri. The preservation of the last Eddy Clock and its sister relics happened more by accident than by plan. Their survival to modern times is actually almost miraculous in light of the meager public support available for the preservation of technological relics.

[2] *Locomotive Engineering,* December, 1890, p. 224.

The exceptional longitudinal cross-section drawing shown here as Fig. 265 is typical of the excellent drawings produced by Richard K. Anderson, Jr. The original drawing was based on measurements taken from the original on exhibit in St. Louis in 1989. Plans to complete this drawing, as well as drawings of the other engines in the Museum of Transport's collection, have stopped because of a lack of funds. The production of plans of historic ships has been under way since the 1930's and continues today as an ongoing project of the National Park Service. An expansion of the program to include early railroad rolling stock will not happen without substantial political support. If you want this expansion to happen, write to your congressional representatives.

Dimensions of the *Marmora*

Cylinders	18″ bore × 26″ stroke
Wheels	64″ in diameter (60″ as built)
Weight	34 tons
Steam pressure	130 psi
Tender tank	2,500 gal.

MORE REPRESENTATIVE LOCOMOTIVES

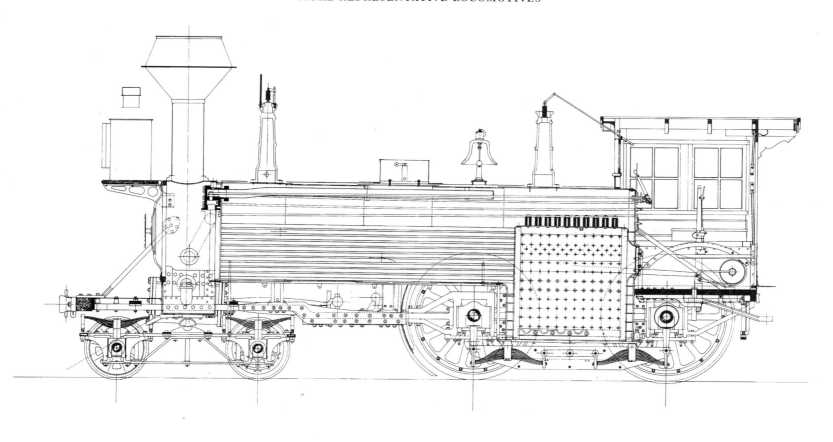

Fig. 265. The Marmora, *produced at the Boston and Albany Railroad's Springfield, Massachusetts, shops in 1876. It is preserved today at the National Museum of Transport.*

Fig. 266. The Barnes, *built in 1877,* a near duplicate of the Marmora. *Its drivers were 66 in. in diameter, and the tender tank was also larger, with a capacity of 3,000 gallons. This engine was sold or scrapped in 1893 or 1894.*

FORNEY ELEVATED RAILWAY LOCOMOTIVE, 1885

Steam locomotive designers realized from the beginning of the railway era that one size does not fit all needs. New engines were almost always designed to do a very specific job. Each class of service seemed to require specially tailored engines to do the job right, from high-speed express locomotives to sluggish dockside switchers. One of the more strenuous and specialized jobs that steam locomotives performed during the nineteenth century was on the New York Elevated railway system.[1] The New York Elevated was one of the busiest and most profitable railway operations in the nation. It ran 3,600 trains a day, with headways of only forty seconds during busy times of the day. Thus, millions of commuters rattled high above the streets of Manhattan on a slender iron trestlework that ran the length of the island. The weakness of the structure was also the inherent weakness of the elevated system. It could carry only very small locomotives, which in turn could propel only short trains (three to five cars); therefore, a larger number of trains had to be operated to move the traffic. Indeed, the first New York elevated line, which opened in 1868, was a cable-car line not intended for anything as heavy as a steam locomotive, even a small one.

The cable-car system proved a failure, and the successor company turned to steam locomotives built on a very diminutive scale. The little 4-tonners were right for the structure but too small to handle the trains. The bridgework was beefed up for slightly larger engines, and as new lines were built, a more substantial trestlework was erected. Even so, the structure was relatively light, and only undersized rolling stock could be employed.

Around 1880, the managers of the New York Elevated decided that the Forney style of locomotive was ideal for their needs. Forney engines were compact and agile and ran equally well in forward and in reverse. Concerns that the driver flanges would wear unduly when the engine was run with the driving wheels in front proved groundless. The first Forney that appeared over the streets of Gotham in 1878 weighed just under 15 tons and had 10-inch by 14-inch cylinders and 38-inch drivers. Over the next fifteen years, the Forneys were

---

[1] *Railroad History*, No. 162, Spring, 1990, pp. 20–79.

cautiously enlarged until 24 tons was reached. A few 27-tonners ran on the Suburban line. The larger Forneys consumed around 45 pounds of coal and 26 gallons of water per mile. The average speed was only 15 miles per hour, but it should be understood that the trains stopped every few blocks, so that a train had hardly started up before it was time to apply the brakes. In the later years of steam operations, when express tracks were installed, the little Forneys raced along at a respectable 45 miles per hour.

Mechanically, the elevated engines were conventional in their basic arrangement, but they did incorporate a few peculiar features. They were most notable for what they excluded, such as a bell and cowcatcher. The normal sandbox was replaced by two rectangular containers mounted under the running boards. In place of the usual large, boxy headlight was a small lamp that would have looked more at home in Britain. The engines were painted red and thus would have been noticeable even among the brightly painted engines common on main-line railroads. Some Elevated engines had metal cabs, which were surely a rarity in the United States at the time. Even more novel was the application of brakes on the locomotive. Almost no early U.S. locomotive had any form of brake other than a manual tender brake. But the Elevated applied vacuum brakes to the truck and driving wheels of its engines. The quick stops so characteristic of these engines' operation required power brakes on all wheels.

There were other, less obvious, apparatuses attached to these engines to adapt them to the city. An engine rattling through the countryside might cause few problems by belching out smoke and steam, or dribbling scalding water, ash, hot embers, oil, and grease along the track. But such leavings would never do in an urban environment. What about pedestrians' clothing, storefront awnings, fine carriages, or the apple lady's spread of fruit? All of these were along the streets under the tracks of the Elevated, and even the best-behaved locomotive was occasionally guilty of a certain amount of dripping and dropping. It proved necessary to put the locomotive in diapers. The ashpan was made watertight, for it was also called upon to retain the overflow from the injectors and the blowoff from the cylinder cocks. Drip pans were carefully fitted under the cylinders and crossheads to catch any oil drippings. To minimize smoke and ash, a very high grade of hard coal was burned. It was far more costly than ordinary grades of coal, but it did much to guarantee a clear stack.

The drawings shown here as Figs. 267 and 268 are engravings from the *American Machinist* of February 28, 1885, based on drawings prepared by the Manhattan Railway's engineering staff for the railway's Class F locomotive. This design was very successful, and it was copied by the Brooklyn Elevated for its first fifteen engines. These were built by Rhode Island Locomotive Works in the same year that the engravings were published. Several mechanical features might be noted. The eccentrics are mounted on the forward axle because of the lack of space between the ashpan and the rear driving axle. One of the vacuum brake diaphragms can be seen behind the valve gear link. The second diaphragm is forward of the truck, below the cab deck. The firebox is mounted on top of the frame rather than between it, to gain a few extra inches of width.

When the Pittsburgh Locomotive Works delivered the final batch of new elevated locomotives in 1894, time was already running out for steam operations. Electric power had taken over the transit industry. Within a few years it would vanquish the old-fashioned horse cars. In 1897 the Chicago elevated began to adopt electric traction and there were efforts under way to adopt it for main-line railways. The New York system came to accept the inevitable end of the spunky little red Forneys. The electrification of the system began in 1900 and was completed in April, 1903. More than three hundred Forneys were suddenly surplus property. Some were nearly new—but no matter: they, too, went off to the locomotive graveyard. They were sold off one or two at a time to short-line and industrial users around the world.

Dimensions of the Brooklyn Forney	
Cylinders	11″ bore × 16″ stroke
Wheels	42″ in diameter
Weight	22 tons
Heating surface	304 sq. ft.
Tubes (124)	1½″ diameter × 68 7/16″ long

MORE REPRESENTATIVE LOCOMOTIVES

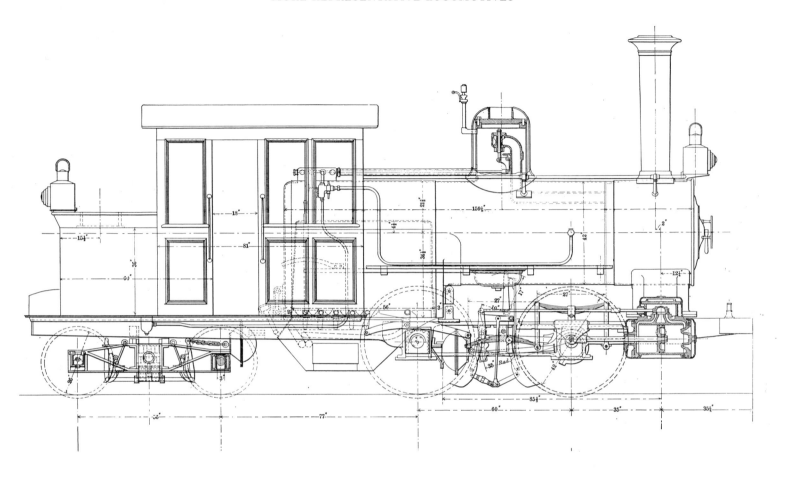

Fig. 267. *The Brooklyn Elevated Railway's first set of engines were copied from those serving on the New York Elevated lines. Rhode Island built these engines in 1885.*

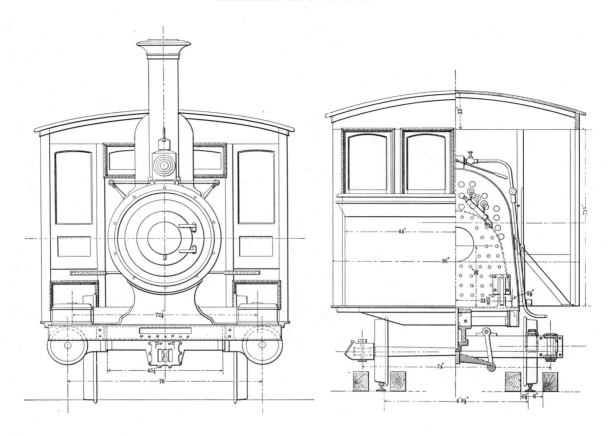

Fig. 268. End elevations of the Brooklyn Elevated's first group of locomotives.

COMMENTS AND NEW INFORMATION

V

COMMENTS AND NEW INFORMATION

The following pages are devoted to corrections and new information accumulated since the publication of the first edition of this book in 1968. In addition, I have reinterpreted some matters that were discussed in the original text but now seem to call for more emphasis or a simpler explanation. The true purpose of the cowcatcher is one example. I also discuss the work of other scholars active in the area of early American railroads and their locomotives.

References to the original text are made by page number and column; *RC* indicates the righthand column, *LC* the lefthand column. If both columns are referred to, only the page number is given.

Page xxv, RC

Since 1968, the early records of the Portland Company, Portland, Maine, have become available. The most valuable parts of this collection are a number of original drawings from the 1850's. Several are reproduced later in these pages. The Portland Company material is held by the Maine Historical Society, Portland, Maine. The Baldwin specifications are now available on microfilm from the DeGolyer Library, Dallas, Texas (see p. 577).

Page 4, LC

Hundreds of new books on American railroad history have appeared, and a few have dealt with the economics of the early lines. Most significant among these studies are two that I recommend highly: Robert W. Fogel's *Railroads and American Economic Growth* (1964) and Albert Fishlow's *American Railroads and the Transformation of the Ante-Bellum Economy* (1965). Readers seeking data on construction and operating costs would benefit from consulting a much earlier work by Arthur M. Wellington, *The Economic Theory of the Location of Railways*, 5th revised edition (1893).

Page 5, LC

Most "freak locomotives" were dead-end experiments. A few, however, had a long-term or perhaps a delayed influence on main-line locomotive designs. An early example is the articulated double enders designed by Horatio Allen for the South

Carolina Railroad. The first of these eight wheelers, named the *South Carolina,* entered service in February, 1832. The West Point Foundry of New York City delivered three sister engines the next year. Many readers will recognize the side elevation in Fig. 269; however, not all will be prepared for the bizarre design features revealed in plan view and front elevation. The boiler was made on a radical plan with two barrels at each end. A common firebox at the center provided a perch for the hapless engineer. The single steam dome was offset to one side of the engineer's platform. A single cylinder at each end mounted at the bottom of the smokebox was another highly original, if not so commendable, feature of Allen's design. Notice the very generously sized lubricator attached to the front cylinder head. The single crank axle was used in later years by the Philadelphia, Wilmington, and Baltimore Railroad on a series of three-cylinder engines and later yet by Alco and Baldwin on three-cylinder engines of the 1920's.

Allen's pioneering plan was for a very light style of engine which had a long wheelbase to spread the load for a weak track that could not sustain heavy axle loading. The double enders stretched out for 22 feet, yet weighed an estimated 8 to 10 tons. The driving wheels were 60 inches in diameter, while the front, or leading, wheels were 36 inches in diameter. The very long wheelbase apparent in these drawings was in fact no problem, for each of the wheel sets was attached to a separate underframe that was free to turn or twist below the boiler. There was no center pin—the weight of the boiler was supported by side bearings fitted with rollers. The main rods had ball joints at both ends to accommodate the undulations of the undercarriage. The design was surely original and imaginative, and Allen deserves much praise for producing such an ingenious plan. But, as George Stephenson was given to say, perhaps the plan suffered from a bit too much inventiveness. The double enders proved a colossal failure. Most spent more time in the shop than out on the road. All were retired by 1838. Officials of the South Carolina Railroad praised the design and steadfastly defended the honor of Horatio Allen. The failure was blamed on poor materials and workmanship, yet if these were the sole reasons for the failure, why wasn't a second batch of Allen double enders ordered from a more reputable builder? If the articulateds were such a noble creation, why did the South Carolina Railroad rely, for its subsequent motive power, on conventional, non-articulated designs? It would appear that the managers in Charleston had been once burned to become twice shy.[1]

Page 8, RC

The *Baltimore* (Fig. 1) was a faithful copy of Stephenson's Planet-class locomotive, but it incorporated several obvious modifications to make it more suitable for service in North America. A pilot, or cowcatcher, was added to protect the engine from trespassing livestock. Track brooms were fastened to the front bumper beam to sweep the track clear of mud or snow. A bell on top of the steam dome, and a whistle just behind the dome, allowed the engineer to sound an alarm as the engine ran along the line. The small 2–2–0's built at Lowell moved some sizable trains. In 1837 a 10-ton Lowell Planet pulled a forty-nine car train weighing 271 tons between Boston and Woburn (10 miles) in just 51 minutes.[2]

Most 2–2–0's surely paid back their owners by long years of good service, and some were rebuilt and enlarged as 4–4–0's or 4–2–2's. The *Baltimore* was remodeled in 1849 as a 4–4–0 at the Baltimore and Susquehanna's Bolton shops. It continued in regular service until 1865, when it was retired or sold. In the mid-1840's the New York, Providence, and Boston (the Stonington Line) began to rebuild its Lowell 2–2–0's into 4–2–2's.[3] The boiler was lengthened by 3 feet, and the cylinders grew from 11 inches by 16 inches to 13 inches by 16 inches. The weight was boosted to 12 tons. In other instances, the first-generation products of Lowell were sold off, pretty much as built, for secondary careers as contractors' engines. Two Planets went to the Marietta and Cincinnati Railroad in 1850 and were retained for light duty for another seven years.

[1] Rob Shorland-Ball (ed.), *Common Roots—Separate Branches* (London, 1994), pp. 80–83.
[2] *National Car Builder,* June, 1894, p. 89.
[3] Angus Sinclair, *Development of the Locomotive Engine* (New York, 1907), p. 178.

COMMENTS AND NEW INFORMATION

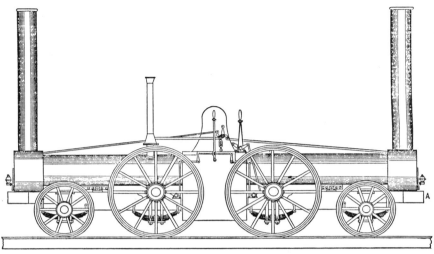

—LOCOMOTIVE FOR THE SOUTH CAROLINA RAILROAD.
DESIGNED BY HORATIO ALLEN, IN 1830 AND 1831.

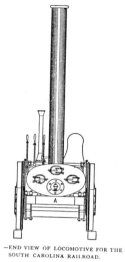

—END VIEW OF LOCOMOTIVE FOR THE SOUTH CAROLINA RAILROAD.

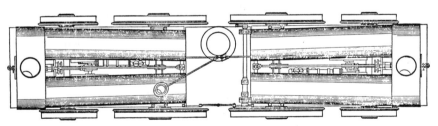

—PLAN OF LOCOMOTIVE FOR THE SOUTH CAROLINA RAILROAD.

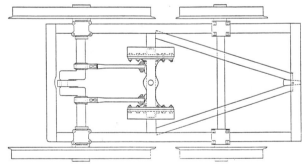

.—PLAN OF TRUCK OF LOCOMOTIVE FOR THE SOUTH CAROLINA RAILROAD.

Fig. 269. *The South Carolina,* one of Allen's double enders, built in 1832 by the West Point Foundry for the South Carolina Railroad.

British imports were recycled in a similar fashion, and a few had remarkably long careers, despite the fact that they seemed unsuited for service on American railways. It is also true that within a decade after delivery most would be considered undersized, yet they were suitable for switching or branch-line service. Braithwaite and Company of London delivered several 0–4–0's to the Philadelphia and Reading Railroad in 1838 and 1840.[4] The first of the lot, named the *Rocket,* ran in passenger service until 1845, when it was demoted to the roadway department as a ballast-train engine. In 1865 it was assigned to the Richmond Coal Wharves near Philadelphia as a yard switcher. When the *Rocket* was finally retired in 1879, after forty-one years of service and a total mileage of 310,164, the tiny machine had reached the relic stage. It is on exhibit at the Franklin Institute. A sister of the *Rocket,* named the *Spitfire,* ran on the Reading until 1849. It was sold the following year to a small Pennsylvania short line, the Leggetts Gap Railroad. At that time it was rebuilt for 6-foot gauge. It was eventually acquired by the Delaware, Lackawanna, and Western Railroad, which in turn sold the British four wheeler to the Spencer Coal Company. The coal company, intent on squeezing the last mile from its elderly switcher, did not scrap the *Spitfire* until around 1885. And even then, the boiler was salvaged for stationary use.

Page 12, RC

The Stephenson-built *Davy Crockett* mentioned here is shown in a longitudinal cross-section on page 169. *Railway and Locomotive Historical Society Bulletin,* No. 101, October, 1959, p. 63, lists for this engine a number of dimensions, which include cylinders, 9 inches by 14 inches; driving wheels, 53 inches in diameter; truck wheels, 32 inches in diameter; and boiler, 30 inches in diameter. There were 66 tubes, 1⅝ inches in diameter by 6 feet ¾ inch long. By scaling the drawing we find that the truck wheelbase is quite long for the period, at 4 feet. The driving axle and the rear truck wheels are on 6-foot 2-inch centers.

Page 13, RC

America's premiere locomotive, the *Best Friend of Charleston,* deserves more attention than it was given in the original text. This very small and presumably simple machine was technically more interesting than might be apparent from a casual look. With more study, we find some innovative design features. The *Best Friend* is a tank engine. A large, square tank was mounted at the front end under the frame and cylinders. This tank served as the water reservoir and also helped to counterbalance the weight of the boiler at the far end of the frame. The water supply arrangement suggests that exhaust steam was used to preheat the feed water.

The engineer sat over the tank between the cylinders, thus making the *Best Friend* a cab-in-front. The feed-water pumps are near the center of the engine and are worked by the crosshead via bell cranks. The boiler was an original piece of work patented by the engine's sponsor, Ezra L. Miller, in June, 1830. Unfortunately, the patent is lost, but the boiler has been described as a "teat-style" boiler, or what became known as the Porcupine, an elementary form of water-tube boiler. The fire door appears to be at the front of the boiler, placing the fireman between the pumps and the boiler itself. The wooden frame and wood center wheels were obsolete, or nearly so, even in 1830, but otherwise the diminutive little steam car incorporated many advanced ideas. The *Best Friend* was delivered in October, 1830, and was tested during the next month. However, it blew up in June, 1831, reportedly because the fireman, weary of the sound of escaping steam, tied the safety valve down. The original boiler was destroyed, but the running gear remained intact, and the engine was rebuilt with a conventional fire-tube boiler. Renamed the *Phoenix,* it was one of the more dependable engines on the South Carolina Railroad's roster. Some of the *Best Friend*'s leading dimensions are cylinders, 6 inches by 16 inches; driving wheels, 56 inches; and wheelbase, 66 inches. The boiler at its widest dimension was 44 inches in diameter. The crank axle was 4 inches in diameter.

---

[4] *Railroad Gazette,* June 13, 1879, p. 329, and June, 1902, p. 506. See also Thomas T. Taber, *The Delaware, Lackawanna, and Western Railroad in the Nineteenth Century* (Muncy, Pa., 1977), p. 151.

Exactly who designed the *Best Friend* is uncertain. The engine was ordered and paid for by Ezra Miller, a director of the railroad and an advocate of steam power. The boiler was one of his designs, and he may have taken part in designing other elements. Christian E. Detmold, a German civil engineer and for a time a surveyor on the South Carolina Railroad, was likely the chief architect of the machine. Then, too, Adam Hall, a Scottish machinist who was the superintendent of the West Point Foundry, very likely played a role in the engine's design. The machine shop where the *Best Friend* was built was located in Manhattan, on Beach Street, not far from the present entrance to the Holland Tunnel. The machine shop closed around 1839, and the West Point Foundry consolidated its operation upriver at Cold Springs, New York.

The engine was completed on June 18, 1830, and arrived by sailing ship in Charleston Harbor late in October. During a test on November 2, one of the wooden driving wheels failed, but all was put right, and the *Best Friend* entered regular service on Christmas Day. The engine was sold to the railroad at this time by Miller. It pulled trains of four or five small cars at speeds up to 21 miles per hour. Running alone, it ran easily at 30 and 35 miles per hour.

Engravings of the side elevation and plan, based on original drawings preserved by the Engineering Societies Library in New York City, are reproduced in Angus Sinclair's *Development of the Locomotive Engine* (1907), pp. 53 and 54. As might be expected, William Brown devotes considerable space to this engine in his *History of the First Locomotive in America*, revised edition (1874). The story of West Point and other builders of steam locomotives is given in my *Short History of American Locomotive Builders* (1982).

In addition to my small book, readers seeking information on American locomotive builders are urged to consult a series of articles that appeared in *Railroad History* (formerly the *Railway and Locomotive Historical Society Bulletin*) starting around 1970. Here will be found histories of Portland, Rogers, Norris, and several other smaller firms. *The Railroad History Index 1921–1984* (1985), a convenient index compiled by T. T. Taber, lists these and many other useful articles. John Brown's *Baldwin Locomotive Works, 1831–1915* (1995) treats the story of America's largest locomotive builder in considerable depth.

Page 25

Standard design in the nineteenth century may be more accurately described as the process of following a standard pattern; it differs from standard design as it is understood in the modern sense, which implies interchangeable parts. During a given time period, major and minor components were made on uniform plans, but they were not necessarily exactly alike. Hinkley, for example, produced frames and axle boxes, and even boilers, all after a basic design. These parts look very much alike, but they might vary in size according to the design of the individual engines. Duplicate engines were produced, of course, for larger railroads tended to order engines in batches of two, four, eight, and so forth. Sometimes, such purchases were spread out over several years, and so a limited consistency was possible, at least in a few instances. The Reading, for example, adopted a low-drivered type of 4–6–0 engine called the Gunboat class for freight service starting in 1863. During the next decade, 134 of these machines were built by the railroad's own repair shops, while others were produced by Baldwin and Lancaster.[5] Yet the Gunboats were not completely uniform: driving-wheel diameters, cylinder dimensions, and even wheelbases varied from batch to batch. Assembly drawings for a late-model Gunboat are shown in Fig. 30. J. C. Davis, master of machinery on the Baltimore and Ohio, built about seventy Camel-back 4–6–0's, also for freight service, between 1871 and 1873.[6] All were built to one size at the railroad's Mt. Clare shops in Baltimore. These engines had an old-fashioned look even in the 1870's, for they were an enlarged version of a design produced some twenty years earlier by Davis's predecessor, Samuel J. Hayes. Even so, the Davis Camels performed well, and many ran for thirty years.

[5] *James Millholland and Early Railroad Engineering*, U.S. National Museum Bulletin 252 (Washington, D.C., 1967); *A Century of Reading Company Motive Power* (Philadelphia, 1941).
[6] William D. Edson, *Steam Locomotives of the Baltimore and Ohio* (Potomac, Md., 1992).

As the Chicago and North Western expanded its lines westward into Iowa, it required new engines of a light pattern for construction work and initial freight service. New roads tended to have lightly built track, and so the heavier locomotives used on the established main stem of the North Western could not be used. The presiding master mechanic of the Chicago and North Western, George W. Cushing, designed a 30-ton 4–4–0 for Iowa service in 1865.[7] The first in the series, named *Missouri*, was produced at the railroad's Chicago Avenue repair shops. The Chicago shops built several identical engines over the next three years. Their leading dimensions were: cylinders, 15 inches by 24 inches; drivers, 54 inches; and boilers, 46 inches in diameter, with 154 steel tubes, 2 inches in diameter by 11 feet long. The interior sheets of the firebox were copper, the boiler was made from 5/16-inch wrought iron, and the steam pressure was 130 pounds per square inch.

The tires were made from Vickers steel. The eight-wheel tender had a capacity of 1,800 gallons of water and 5 tons of coal. The boiler had a very slight wagon top and two steam domes, and the sandbox was near the center of the boiler. A Hunter-style smokestack was used. These machines proved so satisfactory that Cushing ordered another fifty from Baldwin and Hinkley, with Baldwin receiving the greater share of these contracts. According to Cushing's own statement, all were to be built from the same specifications and drawings. One would assume from this assertion that they were all of a uniform plan. This amounted to standardization on a fairly large scale, at least for the 1860's, with sixty-some engines all alike. But were they all alike? The evidence is hardly definitive, but what evidence does exist suggests that a fairly large number of variations crept into the "standard" design. The Baldwin engines had 56-inch rather than 54-inch drivers. They had ten fewer fire tubes than did their Chicago shops sisters. A comparison of the original *Missouri* drawing reproduced in Cushing's article of June, 1900, and the Baldwin photo of the *Crawford* (Figs. 270 and 271, respectively) reveal many more differences. The basic arrangement was similar and perhaps identical, but just about every specific component, from the pilot beam to the counterweights on the drivers, was different. Standard design was rather loosely interpreted at this time.

Long before the time of Cushing's program to make the Chicago and North Western roster more rational, efforts were under way in England to achieve a similar end. In 1840, Daniel Gooch, mechanical chief of the Great Western Railway, had a series of sixty-two standard locomotives produced by seven different builders in a two-year period.[8] This is the earliest instance I can find of the implementation of a standard locomotive design on such a large scale. But one size did not fit all needs. The Great Western needed a smaller series of passenger engines for lighter and slower trains. Specialized freight engines were needed as well, and soon the Great Western had a variety of locomotives on its property, and not just the standard design.

Large railroads came to understand the need for special engines suited to specific needs. Express trains need powerful locomotives capable of high speeds. Lesser engines could be used for secondary passenger trains, locals, and accommodations. In the area of freight service, at least two classes were needed to handle fast and slow merchandise trains. If coal was a major item of business, special, slow-speed "drag" engines were needed. Major grades along the line meant pusher engines to get big trains over the hump. Large and small switchers were needed to service yards and industrial sidings. If the railroad operated branch lines, another set of special-purpose engines was needed for light rail and bridges. A large suburban traffic would require a fleet of double enders. And so, at the least, trunk lines would need a dozen different styles of locomotive. Standardization was very difficult with such a variable fleet.

To a degree, major components such as boilers might be of the same design for a few classes of engines. Smaller components such as feed-water pumps or bell stands might be made uniform for just about all of the different types, but in general

---

[7] *Railway Master Mechanics Magazine*, June, 1900, p. 301; *Railway and Locomotive Historical Society Bulletin*, No. 47-A, October, 1938, passim.

[8] E. L. Ahrons, *British Steam Railroad Locomotive* (London, 1927), p. 47.

Fig. 270. The Chicago and North Western's *Missouri*, 1865, the first of a series of standard 4–4–0's.

Fig. 271. The Chicago and North Western's *Crawford*, 1867, a Baldwin version of Cushing's standard 4–4–0.

it proved to be difficult to achieve true standardization. The nature of engineering progress was fundamentally against it. Mechanical design cannot be frozen for longer than a few years. New ideas are born, and locomotive design changes year by year, so that within a decade or two plans that originally seemed advanced have become obsolete. The human propensity to tinker, change, and improve nullified the most carefully constructed standardization program.

This can be explained in part by individuals' differing preferences. Every master mechanic considered himself as good as, if not better than, his fellows. When George Nixon succeeds Henry Briggs as master mechanic on the Wabash, he is ready to put things in order. Now, Henry was a fair mechanic, but those pop valves he favored are no good. Why did he pick that awful make? The cylinder saddles on the F class are far too light; no wonder we have so many in the shop. The netting in the spark arrestors is placed too low; that must be fixed as well. In fact, we really need a new style of express engine. I will get to work on that first thing. And so, a year or so after George replaced Henry, just about everything that Henry did has been undone. Once again the cause of standardization has been derailed. Some railroads seemed to acquire a new master mechanic every three or four years, and so these derailments were rather frequent.

There are many other probable causes for the failure to achieve much standardization in locomotive design, but I will mention just one more: the policy—or should we say, lack of a policy—that governed the acquiring of new locomotives. Long-range planning was more a matter of good intentions than of actions. When times were dull, railroad managers just limped along trying to hold down expenses, and no new engines were ordered. Then, when traffic began to swell again, came the belated rush to acquire new locomotives. Let us say that most engines on the road were Baldwins and it would make sense to order more engines from the Philadelphia builder. But Baldwin was overbooked and could not promise deliveries until next year, too late to meet current traffic needs. The railroad was forced to take engines from Mason, or Rogers, or whoever could promise fast delivery. Sometimes bad credit forced a railroad to go to a variety of engine builders. Whatever the cause of the diversity, it brought disorder and high costs to the motive power department. Jacob Johann (1830–1913), master mechanic of the Wabash, had eighty-two engines under his care, and they were a very mixed lot. In 1873 Johann had to keep in stock 26 different styles of driving-wheel springs, 15 kinds of crossheads, 13 types of driving boxes, 17 sizes of pilots, 8 types of grates, 8 styles of tender trucks, and 3 sizes of tender water hoses.[9] This plethora of parts was eventually cut down by the practice of rebuilding engines as they came in for major repairs, so that a far smaller number of parts would satisfy the needs of most engines. By 1878 only two styles of driving-wheel springs were needed; the same was true for pilots. Only one style of tender truck and tender hose was now needed.

This may have been small progress, but it was this kind of petty reform that lowered costs, speeded repairs, and made railroad operations more efficient and businesslike.

Page 28
Nowhere in my discussion of the locomotive business was the delivery of new engines mentioned. Shipping locomotives was a troublesome matter because of the fragmented nature of the American railroad network in its early years. The word *network* suggests a uniform and well-connected system, which did not exist much before 1880.[10] Railroads terminated at opposite sides of a city with no physical connection between them, meaning that all shipments, locomotives included, must be portaged across town. Few major rivers were bridged, so crossing a river meant another slow, costly ferrying operation. There were also the gauge differences that confounded the easy transfer of locomotives from one line to another. This problem was overcome to a degree by the Kasson Locomotive Express Company of Buffalo in 1852. The engine or engines in transit were partially disassembled and loaded on trucks or

[9] *Railroad Gazette*, December 6, 1878, p. 591.
[10] George R. Taylor and Irene D. Neu, *The American Railroad Network, 1861–1890* (Cambridge, Mass., 1956). See also "Moving and Losing Locomotives," *Inland Seas*, Spring, 1996, pp. 56–62.

heavy-duty flatcars; the smaller pieces went into boxcars. The system worked fairly well for rail lines of similar gauges, such as 4 feet 8½ inches, 4 feet 9 inches, and 4 feet 9½ inches. It was costly. In 1853 Kasson charged $190 to move an engine from Cleveland to Chicago. Lake transit for the same engine from Buffalo to Cleveland cost $112. These amounts appear modest by today's standards, but in the nineteenth century the U.S. dollar was a hard currency, and a single dollar was a day's pay for a laborer. It might also be noted that the cost of shipping the locomotive to Buffalo is not reflected in these figures.

Delivery time was another factor in finding more efficient ways to move an engine from the builder's shop to the purchaser. Locomotives moving from the East to the Midwest often required a month of transit time. It could take longer. In 1851 two engines went from Taunton, Massachusetts, to Terre Haute, Indiana, via rail to Buffalo and then by lake ship to Toledo, Ohio. The final leg of the journey was on canal boats. In all, it required seven weeks to complete the trip. The sea route was slower yet. The first locomotive for the Little Miami Railroad was finished by Rogers in April, 1841. The 11-ton 4–4–0 went by sea to New Orleans and then by riverboat to Cincinnati, arriving in July. Baldwin experienced a slower and surely a more uncertain shipping record early in 1852 when it sent the broad-gauge *Opelousas* to New Orleans. The sailing ship that was carrying the locomotive sank off Key West, Florida, but in shallow enough water that the cargo was later recovered, and the *Opelousas* arrived at the Algiers, Louisiana, wharf in September, 1853, about eighteen months after shipment.[11] There were, however, a number of engines that went down, never to be seen again. The *American Railroad Journal*, November 5, 1853, claims that fourteen locomotives were lost in the Great Lakes during the spring and autumn months. This brief notice did not offer any specifics as to which engines were lost or where the sinkings occurred. It is intriguing to speculate on what relics, if any, are resting on the muddy bottom of Lake Erie.

[11] The *Opelousous* story was kindly provided by T. T. Taber III of Muncy, Pa.

If moving locomotives to the interior of the country was a chore, moving them to the far West was a trial. Before the Transcontinental got under way in a serious fashion, there was not much demand for railway engines in California or any other of the Pacific states and territories. There were a few short lines, but in truth only a handful of Iron Horses could be found west of the Rockies in 1860. As the Central Pacific pushed eastward, there arose a serious demand for motive power. There was, of course, no through rail line, nor was there a series of connecting lakes and canals. The only way westward was the sea lanes, and they offered a long sail around South America—14,000 miles of whitecaps. Under favorable weather conditions, including favorable winds, a typical square-rigger could hope to cover about 100 or so miles in a day. But weather conditions varied, and a trip from New York to San Francisco generally required 150 days—five months at sea. Such service was fine if you were not in a hurry. But if you were, too bad—you could wait for the railroad to open in 1869. The Central Pacific shipped more than 150 locomotives on the long loop route around South America.

It is uncertain whether the machines went as deck cargo or whether they were disassembled and the parts went below decks. Stowing locomotive boilers would have been difficult in light of the low decks and small hatches typical of the time. Wendell Huffman, an expert on Central Pacific history, is convinced that such maneuvers were possible, but I think that the evidence on this subject is surely incomplete.

In 1868, a motive power crisis developed on the Central Pacific, which was in a race with the Union Pacific to build as much new line as possible.[12] More locomotives were desperately needed if the Central Pacific was to keep up with its rival, and they were needed quickly. Five-month delivery times were unacceptable. Officials in the West appealed to their eastern representative C. P. Huntington—a major if not the dominant

[12] Wendell W. Huffman, railroad historian of Carson City, Nev., shared his information on Panama route shipments of Central Pacific locomotives in a most complete and generous manner. His data came largely from the Pacific Railway Commission Report of 1887, and various California newspapers of the 1860's.

partner of the Central Pacific—to send more locomotives with dispatch. The scheme that was developed was a novel one—send the engines via the Panama route, and hence cut the journey to a more manageable length. The Pacific Mail Steamship Company was none too enthusiastic about transporting such heavy objects, but for a price they agreed to accommodate Huntington. The company's steamers carried the engines from New York to Panama, whence they were carried over the isthmian railway and reloaded on ships to continue on to San Francisco. The shipping charges per locomotive came to about $3,500. Payment in gold was demanded. At the same time, the Pacific Mail company required that the engines go below decks—just why they could not travel as deck cargo is not explained, but it might involve something as elementary as floor loadings. To go below decks, the boilers had to be cut in two. The tubes were removed and crated separately. This defeated some of the time saving involved in going the shorter route, for everything that was taken apart had to be reassembled at the California end of the trip.

In general, the engines sent via the Panama route were thoroughly disassembled. They resembled more a pile of boxes than a complete steam locomotive. When it came time to make the crossing over the Panama Railroad, there was no need to worry about the gauge difference between the Central Pacific's 4-foot 8½-inch gauge and the Panama's 5-foot gauge. The boxes were pushed aboard boxcars or flatcars and whisked overland from Colon to Panama City. At the western end of the run, just 50 miles from the Atlantic side of the isthmus, the boxes and loose bits were piled into the hold of another steamer, which soon departed for San Francisco, where the entire lot of locomotive parts was put on the dock for transshipment by Sacramento River schooner to the railroad's main shops at Sacramento. Here the arduous reassembly operation took place. Each engine arrived in about sixty separate crates or pieces. The name of the engine was to have been painted on each item, but we cannot be sure that this was always done or that, even if it was done, the script could still be read after so much buffeting about had taken place. It was a puzzle for the workers in the shop, and in at least one instance they took parts of several similar engines to make one operable machine. The sister locomotives were assembled later as the parts were received or found. Even with the express Panama route, it required an average of 108 days to get the parts to California and assembled in Sacramento. The time saving was significant, no doubt, by comparison with shipment via the sea route, but was it worth all of the extra cost and hassle? The answer was an obvious one—no more engines went via Panama. Few more engines went by sea either, for the Transcontinental Railroad was nearly finished.

Page 34, LC

E. L. Miller is almost always mentioned in early histories of American locomotives, but he is rarely identified other than as a citizen of Charleston. Actually, he was not a Southerner: he came from New England, having been born in Connecticut on August 29, 1784. He migrated to Charleston in about 1826 and so impressed the local business establishment that he was made a director of the South Carolina Railroad. His involvement with the pioneer American locomotive *Best Friend* is related elsewhere in this volume. Miller's patent traction increaser of 1834 was purchased by M. W. Baldwin. About a year later, Miller's restless nature prompted him to return north to Brooklyn, New York, and then to move on to Illinois. Miller's business affairs faltered, and he became so despondent that he ended his life at a Newark hotel in March, 1847. For more on Miller, see *Railroad History,* No. 150, Spring, 1984, pp. 115–17.

Page 50, Fig. 20

There was no Utica and Syracuse Railway: the actual corporate title was the Syracuse and Utica. It was established in 1836 and opened in 1839. Frank Stevens's book *Beginnings of the New York Central Railroad* (1926), p. 164, offers more details on the measurements of this machine.

Page 52, Fig. 22

The Baltimore and Susquehanna Railroad's annual report for 1849 notes the *Watson*'s weight as 26½ tons. The boiler pressure was listed as 95 pounds per square inch.

Page 59, LC

Colburn mentions a test run of the *New Hampshire* in his *Locomotive Engine* (1851), p. 112. The 74-mile trip between Boston and Great Falls was completed with a sixty-one-car train weighing 391 tons. The maximum grade encountered was 47 feet per mile. Snow covered the tracks much of the way, yet the stalwart *New Hampshire*'s average speed was just over 14 miles per hour. The engine consumed 3.4 cords of wood and 3,734 gallons of water.

The first generation of 4–6–0's included some notably ugly locomotives. Among the most ungainly were several inside-connected ten wheelers produced in 1848 by Rogers for the broad-gauge Erie Railway. Fig. 272 shows an engraving based on a photograph of the *Yates* taken by Zehra Colburn in 1851. Note the outside frame, the cranks for the side rods, and the inside connection of the main rods. This 30-ton engine had 18-inch by 20-inch cylinders.

Page 65

Wellington, in *Economic Theory of the Location of Railways* (p. 16), suggested that the best thing, other than steel rails, to appear on American railroads was the 2–8–0 locomotive. It was the superpower engine of the nineteenth century, a powerful brute that could empty a full yard and yet effortlessly sweep around curves without kinking a single rail. Octave Chanute, consulting engineer to the Erie, shared Wellington's admiration for the Consolidation type—a test engine produced by Cooke late in 1877 performed well on the Jefferson Branch, which had the heaviest grades on the Erie system. On the main line, the new 2–8–0's outperformed the older 4–4–0 and 4–6–0 freight engines as well.

A Consolidation could handle forty freight cars, while the eight wheelers were limited to sixteen, and the ten wheelers were limited to eighteen to twenty. Longer trains and fewer engines meant a cut in the work force, which led to a revolt among the train crews. In time, peace was made with the workers, although sixteen crews were eliminated by the big engines. As many as two thousand cars a day were handled by fifteen Consolidations, which had replaced thirty smaller engines. Erie Railroad workers came to call the Consolidations "iron mountains" because of their size.[13]

Fig. 272. The *Yates*, built in 1848 by Rogers for the New York and Erie Railway. The engraving is based on a contemporary photograph, apparently now lost.

Page 66

Readers seeking more information on the *Monster* and other Camden and Amboy locomotives should refer to my *John Bull: 150 Years a Locomotive* (1981).

The focus of this book is standard main-line engines; curiosities and experimental and very specialized locomotives were downplayed on purpose to give more space to what might be described as mainstream developments. Limited space was, however, given to switching locomotives and to one special-purpose type of locomotive, the Forney, in Part IV, above.

Page 71–80

The discussion of operations and repair could be expanded, for a great deal of material has come to my attention since the completion of the first edition of this book nearly thirty years ago. Rather than attempt a summary of all of this information, I will simply direct readers to the major sources. My own book

[13] *Railroad Gazette*, October 17, 1879, p. 553.

The American Railroad Freight Car (1993)—see especially the first chapter—contains much useful data on train speed and operations.

Edward B. Dorsey's *English and American Railroads Compared* (1887) contains hundreds of references to operations, repairs, and costs. On p. 6, for example, Dorsey notes that engines on the Hudson River Railroad averaged 39,948 miles per year, with repairs costing 2.7 cents per mile. On p. 7 he reports a typical coal train on the Lehigh Valley as having one hundred cars (weighing 340 tons). The 2–8–0's that were used to pull such trains burned 3¾ tons of coal a day.

Arthur M. Wellington's *Economic Theory of the Location of Railways,* 5th edition (1893), is a vast compendium of operating facts and costs. Locomotive data is especially rich on pp. 399–484. The weight of parts is given: the boiler and water equal one-third of the total engine weight. On another page, Wellington notes that locomotive axles have a life of 300,000 miles. He then notes that the Pennsylvania Railroad performed general repairs after eighteen to twenty months of service.

If one searches carefully enough, one can even find cost data for oil-burning headlights. In February, 1865, the Lackawanna and Bloomsburg Railroad published a table that chronicles the cost of lamp oil for each of its twenty engines. The most economical registered 29 cents per month—apparently it rarely ventured out after dark; the most costly guzzled $10.69 worth of oil per month. This piece of operating cost trivia appeared in *Railway and Locomotive Historical Society Bulletin,* No. 47, September, 1938, opposite p. 58.

Fleet management became a difficult problem as the railroad system expanded. Keeping track of a few dozen engines was easy, but when that grew to hundreds, the master mechanic's lot was not a happy one. For efficient deployment of the motive power, it was necessary to know each engine's condition, location, and type of service. Recording such data in a ledger book was fine for the bean counters, but master mechanics were not much given to abstraction. They wanted a more hands-on system of locomotive accounting. An article in the August, 1879, *Boston Herald* stated that the Pennsylvania Railroad used a system of painted metal discs—one disc for each engine. The discs were placed on hooks on a board divided into sections representing the various divisions of the railroad. An engine in perfect condition received a white disc; engines in a lesser state received colored discs whose shade reflected the engines' state of decrepitude. A black disc meant that the engine was ready for the scrapper. If an engine required more than $3,000 in repairs, the Pennsylvania Railroad considered it cheaper to put the money into a new engine.

The New York Central had an even more elaborate system, created by William Buchanan, its locomotive superintendent. Buchanan devised a huge pegboard, 4 feet high by 13 feet long, as an index of the Central's fleet. The board was divided into sections for each division of the railroad, from the Harlem to the Western. A sturdy iron peg represented each engine. The class type (e.g., *A, B,* or *C*) and the road number (e.g., 85 or 92) were painted on the end of the peg. The type of service was represented by a letter: *P* for passenger, *F* for freight, *W* for work. A star on the right side meant that the engine had an air pump. A star on the left side meant that it also had driver brakes. The cylinder size was inscribed on the side of the peg, while the wheel diameter was given on the small end of the peg. Condition was noted in nine columns, ranging from "good" to "sold" or "scrapped." Buchanan's board was described in the *Railroad Gazette,* February 8, 1884.

Pages 83–86

Some valuable publications on the use of wood as a fuel have appeared since the first edition of this book appeared in 1968. I especially recommend Arthur H. Cole's "The Mystery of Fuel Wood Marketing in the US," *Business History Review,* Autumn, 1970. Other useful publications include Albert Fishlow's *American Railroads and the Transformation of the Ante-Bellum Economy* (1965) and the 1880 U.S. Census special volume on transportation, containing information on both wood and coal as locomotive fuels. For a more general approach to wood and the railroad industry, I recommend Brooke Hindle (ed.), *Material Culture of the Wooden Age* (1981).

While the wood-burning era is long past, it lives on in curious ways. The huge firewood preserve of the former South

Carolina Railroad was eventually taken over by the Southern Railway, which leased the South Carolina Railroad in 1899. The new owners used the woodlands as a tree farm to teach local farmers how to grow trees. The farm was later abandoned and allowed to grow wild until William D. Brosnan became president of the Southern in 1962. He turned it into an executive retreat for the entertainment of railroad officials and large shippers, and so it remains today. I am indebted to Albert Eggerton, former public relations director of the Southern Railway, for this story.

Wood-burning steam engines were in commercial service as recently as April, 1995, according to an article published at that time in *Railfan Magazine*. A small railroad in Paraguay is portrayed in this illustrated feature article. Two of the wood-burning locomotives discussed were built as recently as 1953.

Page 93

Most discussions of locomotive boilers say much about fuel consumption but rather little about water usage. Railway engines were surely thirsty and wasteful consumers, for all of the water consumed was blown out of the stack as so much waste material. Engines of the 1850's used roughly 50 gallons of water per mile. By 1890 the consumption was up to 80 gallons per mile. The cost per mile was figured at one cent. On busy lines, water tanks were spaced at 5- or 6-mile intervals. On lines with average traffic, a water tank every 10 miles was sufficient. Modern steam locomotives, such as the Nickel Plate Road's 1949 Lima-class 2–8–4's, were comparatively moderate water consumers. Engines of this class used only 100 gallons per mile, or just twice the quantity used by a 25-ton machine of the 1850's.

There is a tendency to dismiss early locomotives as very simple, if not primitive, machines. I suppose that they are simple by comparison with the space shuttle, yet I find that early railway engine design is more ingenious and complex than might at first glance be imagined. The many small parts that were skillfully fabricated to make a strong, unified vessel such as the boiler convince me that the pioneer locomotive designers and builders were men of considerable talent. A drawing found in the Portland Company papers at the Maine Historical Society illustrates this point clearly. The engine in question is a 66-inch-gauge 4–4–0 of the 1850's, but it might be any engine of the period, for the design represented by this cross-section is generic and not experimental. Readers are referred to Fig. 273.

One last note on the subject of boilers and fireboxes: There seems to be a belief—at least among model builders—that when we speak of copper fireboxes, we mean that the entire structure was copper. This is not so: only the interior plates were made of copper. The exterior shell was made of common wrought iron; hence, the copper surface was not visible.

Page 100

Sheet copper, commonly used for high-quality roofing and downspouts, was used more rarely for locomotive jackets. A fine little 0–4–0 preserved at the National Railway Museum in York, England, has a copper jacket over its Bury firebox. This fine relic of the 1840's is known as "Old Coppernob." On this side of the Atlantic, about the only railway to employ copper boiler jackets was the Cincinnati, Hamilton, and Dayton Railroad. The *American Railway Review* for October 4, 1860, said that the *J. W. Ellis*'s boiler, cylinder, and steam chests were copper lagged.

Page 101, LC

A hint to restorers: an automotive finish that closely resembles Russia iron is Dulux Enamel Charcoal Metallic No. 4980-DX. Perhaps one day some enterprising firm will once again begin the manufacture of genuine Russia iron.

Page 114

It would be possible to fill a book with descriptions of the various smokestacks devised by nineteenth-century mechanics, but that is not the purpose of this volume. However, some fresh information was uncovered that seems worthy of inclusion in these pages. The Congdon stack was popular on the Union Pacific and some of its subsidiary lines. It was devised by Isaac H. Congdon, who was born in Granville, Massachusetts,

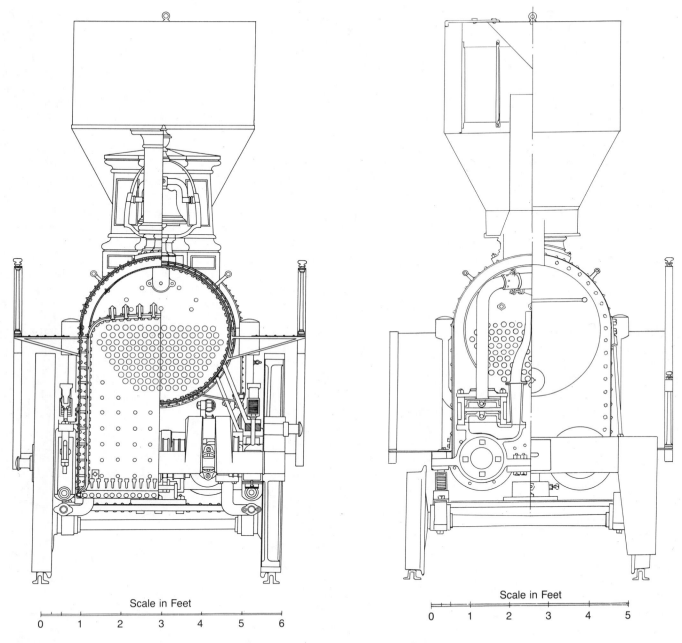

Fig. 273. The Portland Company prepared drawings for a 66-inch-gauge 4–4–0 in about 1855 which included this beautifully detailed cross-section through the firebox.

in June, 1833. He went west in 1851 to become a machinist with the Cleveland, Columbus, and Cincinnati Railroad. In about 1866 he became master mechanic with the Union Pacific, and he remained in that position until 1885. He died on August 21, 1899. The Congdon stack had an unmistakeable appearance and was notable for its great mass. The lower section looked like a conventional bonnet stack crowned by a very high upper course that tapered back ever so slightly toward the top ring. The last known example was preserved with the Utah and Northern's No. 7 at the University of Idaho. Unfortunately, the No. 7 was sacrificed to the World War II scrap drive, and thus the last example of Congdon's ingenuity was destroyed.

The end of wood burning did not end the spark problem on American railways. Coal burners were capable of ejecting a considerable firestorm of hot embers. This was true because the engines were worked so hard. Small engines pulling big trains could only do so if they were driven hard: everything was pushed to the maximum. In a desperate effort to keep up steam and move the train over the line, fuel was piled into the firebox faster than it could be efficiently burned. According to Wellington, such dense firing caused 20 percent of the fuel to go up the smokestack.[14] Some of the expelled pieces were sizable; in one instance, a lump of partially burned coal traveled over the train and dropped inside an open car window. It measured 1¼ inches in diameter.

Page 128

New devices such as the injector were accepted skeptically. In a letter dated January 11, 1868, W. S. Hudson, the veteran superintendent of the Rogers Locomotive Works, wrote to an official of the Union Pacific Railroad.[15] Injectors steal power from the boiler, according to Hudson. Boilers, in fact, steam more easily when using pumps. Injectors therefore should only be used as a backup if the engine is stationary for a long time.

In conclusion, Hudson concedes that injectors are a good thing in the unlikely event the pumps are disabled. A copy of this letter was kindly supplied to me by Jim Wilke of Los Angeles, California.

Page 134, RC

The inventor of the steam gauge manufactured stills in Paris, France. In 1849, Bourdon noticed that the tube of one of his still coils tended to unwind when flattened. Upon further examination, he discovered that the extent of the unwinding was directly related to the pressure inside the flattened tube. He connected one end of the tube to a needle, and the steam gauge was born. Rights to the invention were purchased in 1854 by the American Steam Gauge Company of Boston. This story is related in *Asher and Adams' Pictorial Album of American Industry* (reprint, 1976), p. 126.

Page 167

There was an almost universal agreement among locomotive men that leading trucks were essential to successful operation of the engine over America's uneven and serpentine tracks. Leading wheels steered the locomotive over and around the rough spots; without them, derailments would be common. This opinion was a "given" in basic locomotive design and is well documented in the literature. Moreover, the preference for leading wheels is documented by the vast majority of road engines built for U.S. service. Yet there was one holdout in the industry.

It was not a very large railroad, but it ran more trains per day than most trunk lines and was one of the most profitable steam railroads in the world. I am referring to the New York Elevated.[16] The members of its motive power department championed 0–4–4T Forney engines and said that they ran best with drivers facing forward. They claimed that derailments were rare and that flange wear on the drivers' tires was not excessive. In an even greater affront to the conventional wisdom,

---

[14] Arthur M. Wellington, *The Economic Theory of the Location of Railways*, 5th rev. ed. (New York, 1893), p. 449.
[15] Levi O. Leonard Collection, University of Nebraska, Lincoln.

[16] *Railroad History*, No. 162, Spring, 1990, pp. 20–58.

they removed leading wheels from the few tank engines on the roster, saying that they ran better without them. The courageous stand of the Elevated line won no converts, and the Elevated remained an isolated island of anti–leading truck sentiment in a vast ocean of railway men committed to the opposite opinion.

Page 184

In hindsight, the air brake is usually praised as one of the great inventions of its time. It was a device that vastly improved railway safety and saved the lives of hundreds of trainmen and passengers each year. This sanguine view of the atmospheric brake was not shared by certain members of the railway engineering world. Distinguished members of that fraternity, such as Wilson Eddy, perceived the air brake as a downright nuisance and a maintenance nightmare. See my *American Railroad Freight Car,* pp. 539–41, for more on this subject.

Page 204

Just when the first oil cup came into being is a matter of speculation, but we can find some very early examples in the drawings of the West Point Foundry locomotives—for example, the *Best Friend* (see Sinclair, *Development of the Locomotive Engine,* p. 53) and the *South Carolina,* shown here as Fig. 269.

Page 211

There is an erroneous perception that the purpose of the cowcatcher was to save the cows. The primary purpose was actually to save the locomotive and its train of cars. Livestock, once under the wheels, could derail the locomotive's truck wheels and so cause a costly accident. The job of the cowcatcher, or pilot, was to push the carcass off to one side and so clear the track. In some cases this could be done with little or no harm to the animal, and so much the better, but railroad officials preferred to pay a farmer for the loss of his animal than to suffer the far greater loss of a wrecked train. In 1849 the Little Miami Railroad noted, in its annual report, payment of $2,500 in claims for lost cattle—plans were under way to fence the entire right-of-way. Farmers did not feel that it was their responsibility to keep their beasts confined; they clearly favored the laissez-faire notion of an open range. Most railroads, however, pushed ahead with track-fencing programs. The Erie had both sides of its entire line fenced by 1862.[17] Claims for dead cattle kept rising along the Union Pacific, so that by 1891 they had reached $221,000.[18] Wire fencing, at $375 per mile, seemed a good bargain.

The traditional cowcatcher was made with vertical bars running from the front bumper beam to a triangular bottom frame placed just above the rails. It was produced in either wood or iron. There were at least two variants to this basic design. One, which might best be described as the drooping-mustache style, had vertical bars just like a conventional cowcatcher, but only the center bars reached the bumper beam. It was in effect a double V, as can be seen in the photograph of the *New Hampshire* reproduced here as Fig. 274. Note that support rods outfitted with adjusting nuts were mounted between the second and third bar. The so-called hen-coop style of cowcatcher is illustrated by Fig. 275, the Cleveland and Toledo's *Ottawa.* The horizontal slats apparently suggested the wooden cages in which, at the time, farmers took chickens to market. There is much more to this picture than the cowcatcher. The general lightness of engines of the period is well illustrated here—the rods look like twigs. The engine looks new, but notice the dent in the smokestack and the mismatched truck wheels. The lantern placed on a stand on the top of the boiler, just ahead of the forward steam dome, is believed to be a train indicator lamp—that is, a white lamp for an extra or a green lamp for a second section.

Page 215

Restorers and researchers occasionally write requesting a source for details of engine components such as whistles. A very good and obvious source is Gustavus Weissenborn's *Ameri-*

---

[17] Henry M. Flint, *Railroads of the United States* (1868), p. 183.
[18] Maury Klein, *The Union Pacific Railroad* (New York, 1887), vol. 1, p. 500.

Fig. 274. *The Cheshire Railroad's* New Hampshire, *built in 1847 by the Hinkley Locomotive Works. Front-end views of engines of the period are scarce. Hence, I picked this picture to best represent the double-V style of cowcatcher.*

Fig. 275. The Amoskeag Manufacturing Company produced this light eight wheeler for the Cleveland and Toledo Railroad in 1853. It was later operated on the Lake Shore and Michigan Southern.

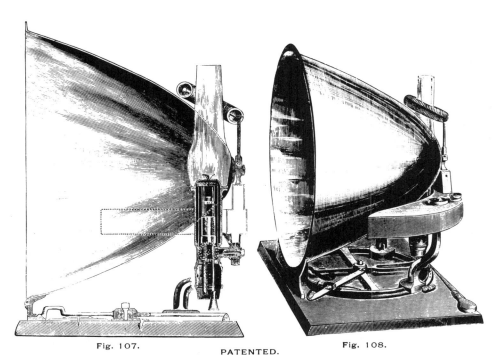

Fig. 276. *Interior hardware of an oil-burning headlamp from Adams and Westlake catalog of 1887.*

can Locomotive Engineering and Railway Mechanism (1871). The reprint makes these detailed engravings available to anyone willing to make a reasonable effort to find a copy (see the Bibliography, below). Less accessible, but also valuable, is James Dredge's 1879 folio-sized volume *The Pennsylvania Railroad*.

Page 216

The interior construction of oil-burning headlights seems to hold a particular fascination for many readers of previous editions of this book. Many seem to overlook the cutaway drawing of the Grant locomotive used as the endpapers in the original hardbound book, and the current edition. It is sketchy, I will agree, and so I am including a more detailed drawing copied from the 1887 catalog of a railway car lamp manufacturer, Adams and Westlake of Chicago. In most respects, this interior lamp hardware is typical for the period (Fig. 276). The deep reflector, the horseshoe-shaped fuel tank, and the long burner with a cylindrical wick are standard practice. The rotating table is less common, but it is surely a good feature, for it allows the lamp to be serviced from a side door rather than a front door. The lamp can be swung around to clean the reflector or the wick. The wick can be adjusted by the long rod with a flat finger-knob shown in the right end of the base in the right-hand side of Fig. 276. The spring around the upper end of the glass chimney is to keep the chimney from vibrating against the opening in the reflector when the engine is under way.

Sometimes it is possible to identify the headlight maker.

Sometimes the maker's name is stenciled on the lamp itself. There is a fine engraving of a Utica Headlight Works lamp in the *Railway Review,* June 11, 1881, p. 330, and such lettering is occasionally readable on some photographs of the period. Other lamp cases are so distinctive that they can be identified with a little practice. John Crerar of Chicago manufactured a recognizable lamp case with two finials on either side of the top. Many of the Union Pacific engines of the 1860's had Crerar lamps. Some of them are pictured in G. M. Best's *Iron Horses to Promontory* (1969). James Radley, a railway supply merchant of New York City, had a distinctive style of domed chimney on his lamps; several examples are shown in this book (see Figs. 171, 221, and 226).

Page 218

The human craving for ornamentation led to the creation of refulgent locomotives. A few railroads, notably the Lehigh Valley and the Philadelphia, Wilmington, and Baltimore, refused to give up their bright colors even into the late 1880's.[19] Angus Sinclair appreciated these late-blooming beauties and said, "Our heart warms up at the sight of the red wheels" as the engine rolls into the station.

There was a cost attached to fanciful locomotive decor. In 1854/1855, the Amoskeag Manufacturing Company calculated that painting an engine cost $100. Added to this was $34 for Russia iron jacketing and $75 for spun brass casing around the steam dome.[20] Some twenty years later, the Pennsylvania Railroad, which after 1865 was not much inclined to indulge in highly decorated locomotives, used twenty-seven books of gold leaf for lettering and striping its passenger engines.[21]

Red wheels on a locomotive are understandable. They are currently in favor in China. But what about red smokestacks? Here was a true way to get attention. It was also an easy way to tell your engines from those of another railroad when more than one line used a terminal. In an 1860 New Jersey Railroad and Transportation Company rule book, "red smokepipes" are mentioned.[22] Both New Jersey and Erie trains operated into the Jersey City terminal. Switches were manually thrown at ground level by switchmen. It must have been difficult for these men to identify a train coming head on, so as to guide it into the proper track, but if the chimney of one railroad's engine was red and others were black, the task was a simple one. On dark or foggy days, the switchmen needed all the help they could receive.

To the novice eye, a red smokestack might be alarming. A Midwestern farm boy was frightened half to death on first seeing an Indianapolis, Decatur, and Springfield Railroad engine with a bright red smokestack. He ran home sounding the alarm: "She is going to bust sure, its red hot clean to the top of the stove pipe."[23]

Page 222, RC

On the subject of locomotive cabs, there is a revealing series of letters between H. R. Campbell and M. W. Baldwin in the Baldwin letters at the Historical Society of Pennsylvania. This correspondence dates between September and December of 1850. At the time, Campbell was supervising the construction of the Vermont Central. Being out in the field much of the time, he was getting a firsthand taste of New England winters, and this convinced him that engine cabs were necessary to protect the operating crews from the bitter cold. Campbell admonished Baldwin to pay more attention to the design of his cabs, noting that Baldwin cabs were "by no means tasty and give your engines a clumsy, clodhopper appearance, which strikes people unfavorably." Campbell urged the Philadelphia builder to make the sideboards narrower and add ornamental moldings under the roof eaves. Baldwin should copy Amoskeag's attractive plan, and show up the Yankee builders.

---

[19] *Railway Age,* May 13, 1887, p. 329; E. P. Alexander, *Collector's Book of the Locomotive* (New York, 1966), p. 39.
[20] *Railway and Locomotive Historical Society Bulletin,* No. 53, October, 1940, p. 66.
[21] James Dredge, *The Pennsylvania Railroad* (London, 1879), p. 136.

[22] New Jersey Railroad and Transportation Company rule book (1860), New Jersey Historical Society; data from Edward T. Francis of Livingston, N.J.
[23] *Railroad Gazette,* April 23, 1880, p. 223.

Page 223

Our scant store of knowledge about early tender design is hardly bolstered by the preserved examples housed in various museums around the nation. The provenance of most of these pieces is uncertain. A case in point is the tender of the 0–4–0 *Lion* produced by Hinkley in 1846 and now exhibited at the Maine State Museum in Augusta.[24] The who, when, and why of the tender's construction cannot be determined with any certainty. The where of its fabrication is most likely the railroad's shop at Machiasport, but even this is more tradition than fact. It is not even certain whether the tender is from the *Lion* or its somewhat earlier sister the *Tiger*, another Hinkley product of 1842. The gas-pipe spoke wheels add another mystery to the pedigree of the *Lion*'s tender. Were they from an earlier English locomotive that reportedly ran on the Machiasport line before the *Tiger* appeared in 1842? The remainder of the tank is so anonymous that even the most careful examination reveals few clues as to its history. Was it built in the 1840's, or at some later date? Such questions probably cannot be answered.

Sketches taken from the *Lion*'s tender led to the preparation of the drawing shown here as Fig. 277. I suppose that this tender is about as small a standard-gauge tender as was ever built, but the Palmer and Machiasport Railroad was a rather modest affair in itself. The line was just under 8 miles long, built with strap rail track and never intended for speeds above 10 miles per hour. It was a lumber road, and as such was a pioneer in this country.

Page 226, RC

The *Daniel Nason* is now exhibited at the National Museum of Transport, St. Louis, Missouri.

Page 232, RC

One of the early advocates of iron tender frames was Benjamin W. Healey, one-time superintendent of the Rhode Island Locomotive Works. He obtained a U.S. patent for such a frame on June 7, 1870. An example of a Healey fabricated iron truss system is illustrated in Fig. 8.20 of my *American Railroad Freight Car* (1993).

Page 233, RC

An argument could be made that the *Best Friend* was the first American tank locomotive. The water tank was in an unusual position, however, being at the front and under the main frame.

Page 233

I shall conclude my tender notes with the addition of drawings for a large, eight-wheel iron-frame tender built in 1884 for the Cleveland, Columbus, Cincinnati, and Indianapolis engine No. 636, a Mogul.[25] Since mid-century, tender size seemed pretty well frozen at 2,000 gallons and 2 tons of fuel (Fig. 278). The Southern Pacific and the Lehigh Valley built a number of high-capacity twelve-wheel tenders, but other designers found the standard eight-wheel plan readily expandable. The tender shown in Fig. 278 carried 3,000 gallons of water and 5 tons of coal, and weighed 61,630 pounds. The wheelbase was an even 15 feet. By the late 1890's, 4,000-gallon tenders were common for large engines such as 2–8–0's, and a few railroads were trying 6,000-gallon tanks.[26]

Page 248

More information on the *John Bull*, its sister locomotives, and the Camden and Amboy Railroad is given in a small volume, my *John Bull: 150 Years a Locomotive* (1981). Included is the story of how, as part of its 150th anniversary celebration, the engine was removed from the museum and operated. The museum staff gained a new respect for the ancient little steamer during its public trials in September, 1981. It proved to be a very responsive and easy-to-operate machine, which may explain its long service life. It took a few hours of practice to

---

[24] Paul E. Rivard, *Lion: the History of an 1846 Locomotive Engine in Maine* (Augusta, Me., 1987).

[25] *National Car Builder*, November, 1884, p. 141.
[26] *Railroad Gazette*, August, 12, 1898, p. 580; *Locomotive Engineering*, March, 1899, p. 119.

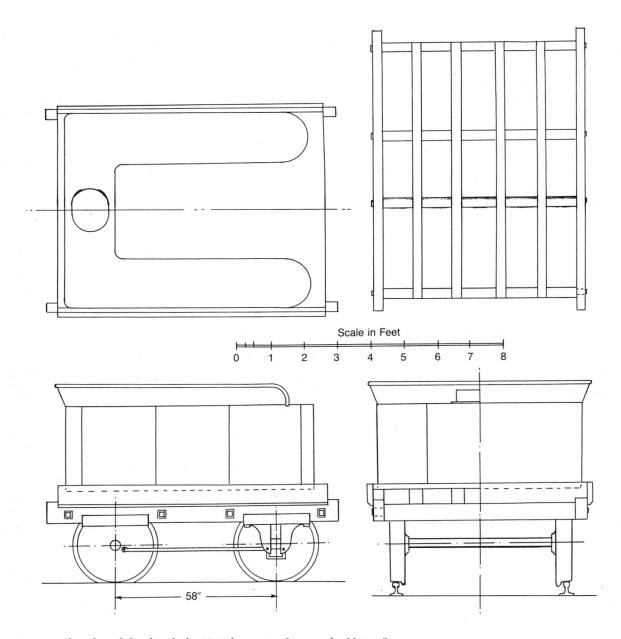

Fig. 277. A preserved tender exhibited with the 1846 locomotive Lion at the Maine State Museum, Augusta, Maine.

handle the valve gear, but once this was mastered, the *John Bull* proved surprisingly easy to manage. We all learned how to put the engine in forward and reverse "by feel," and this could be done quickly and easily. There was no owner's manual available to guide us, so we learned all of the *John Bull*'s operating idiosyncrasies by direct experience. About the only complaint we could make concerned the small size of the firebox and the very low position of the firebox door. This made the engine very difficult to fire. To keep up steam, the fireman had to maintain a position on his knees to one side of the fire door. Continuous firing was necessary, and each log had to be placed in just the right opening within the roaring-hot furnace.

Page 252

One other lesson was learned as a result of the risky operation of the venerable *John Bull*. Previously, I had questioned why the throttle valve was placed outside the steam dome. From a thermodynamic viewpoint, it seems so entirely wrong. Yet there appears to be a good reason for this exposed position, for it made the valve accessible. We experienced a stuck or frozen throttle valve toward the end of our operating adventure. The valve gradually tightened and then simply froze solid: it would not budge. The only remedy was to drop the fire, blow off steam, and disassemble the recalcitrant plug-style valve. A thin layer of scale was enough to cement the tapered plug to the valve body. Once cleaned and reassembled, the throttle worked as if new. Yet the steaming down and back up required about 2 hours. If the valve had been inside the steam dome, this procedure would have taken even longer. Hence it seems reasonable to assume that stuck throttles were a regular occurrence on the Camden and Amboy and that the outside placement of the valve was chosen to get engines back into service more quickly. The more fundamental remedy was, of course, to adopt a better style of throttle valve, and it is likely that the Camden and Amboy abandoned the cock design for the slide type of valve described on p. 145, above.

Page 311, RC

New information now suggests that the first railroad in continental South America was the Demerara Railway in British Guiana (now Guyana). This standard-gauge line, built with British capital, opened in November, 1848. The railroad was abandoned in two segments in 1972 and 1974. Refer to *Railroad History,* No. 166, Spring 1992, p. 126.

Page 330, RC

In the table of dimensions, "12½″ base" should read "12½″ bore."

Page 416, RC

The Erie Railway's 250 series Moguls had Graham's spring balances to hold down the rear safety valves. This device was patented on December 18, 1860 (Patent No. 30,964) and appears to have enjoyed a moderate popularity during that decade, for it is also shown in drawings for the Rogers 4–4–0, as depicted in Figs. 221 to 223. In conventional spring balances, a coil spring inside a telescoping brass tube provided the tension to hold the safety valves closed. Graham used a leaf spring. It was placed inside a round brass case mounted on top of the boiler inside the cab. A hand lever projecting out of the rear of the case allowed the crew to remove tension on the spring and release the safeties. One Graham valve took the place of two standard-style balances.

Page 538, RC

William Buchanan (1830–1910) was master mechanic on the Hudson River Railroad. Traffic moved slowly on most early U.S. railroads, as noted on pp. 320–46, but the Hudson River Railroad was an exception to this rule. Its tracks and alignments were made to sustain high-speed trains. Buchanan, as the road's master mechanic, had a unique opportunity to design express locomotives, and this he did in a most successful manner during his long career in the railway mechanical field. After rebuilding and improving most of the existing fleet, he turned to new construction in 1865. This machine, No. 79, was very large for its time, weighing 40 tons and having a 70-inch driver. Like all good locomotive men of his age, Buchanan remained loyal to the 4–4–0 wheel arrangement. If you needed a larger and faster express engine, you would simply produce a

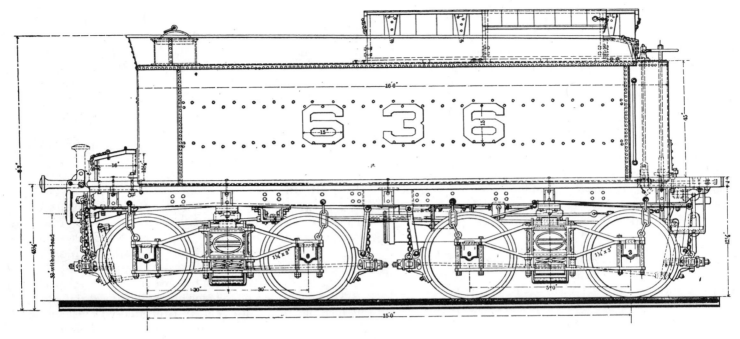

Side Elevation.

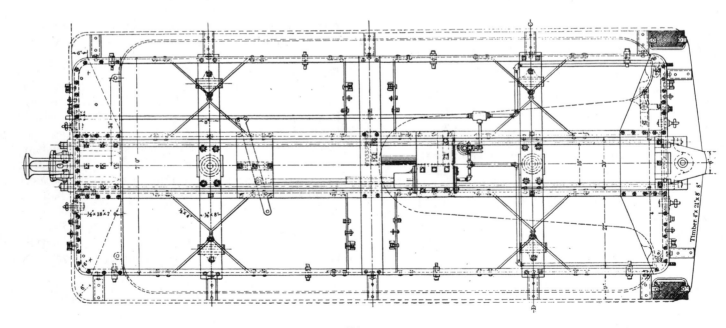

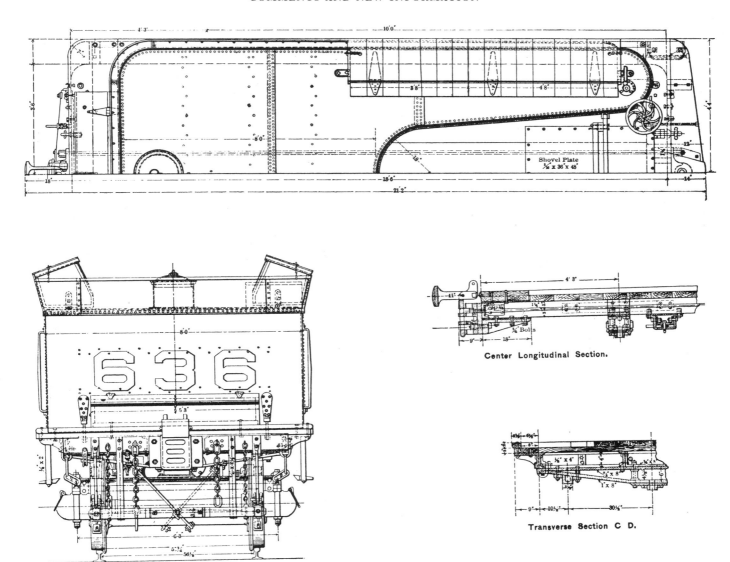

Fig. 278. A large, iron-framed tender built in 1884, representative of the larger, 3,000-gallon-capacity tenders coming into favor because of the use of larger freight locomotives.

larger 4–4–0. This straight line development culminated in Buchanan's celebrated No. 999 of 1893. It is claimed that this engine traveled at more than 112 miles per hour, a world record at the time. By today's standards, the record would surely be an unofficial one, but no one would deny that the engine was a fast runner.

Buchanan was born in Scotland, the son of a blacksmith. He came to the United States as a boy and became an apprentice machinist in 1847. He quickly moved on to become a locomotive engineer, then a shop foreman and a division master mechanic. In 1859 he was named master mechanic of the Hudson River Railroad. In 1881 he was elevated to the position of superintendent of motive power of the New York Central and Hudson River Railroad—a post he kept until retiring in 1899.

Page 538, RC
John Brandt's birth date is 1791.

Page 538, RC
Details on the life of Henry R. Campbell, uncovered by the research of Edward T. Francis of Livingston, New Jersey, were published in *Railroad History,* No. 157, Autumn, 1987, pp. 111–13. Campbell was born on September 9, 1807. His father, Amos, a carpenter and bridge builder, taught his son the basics of construction. In addition to the railroad projects outlined on p. 538, Campbell, a civil engineer, built the Camden and Woodbury Railroad in 1838. He remained active in the Philadelphia area for another twelve years and then worked in railroad building in New York and New England. Some twenty years later he returned to Philadelphia. He remained active until around 1875, and he died at his home in Woodbury, New Jersey, in February, 1879.

Page 545, RC
Andrew J. Stevens lived from 1833 to 1888. Like so many pioneer western railroad men, Stevens was a transplanted Yankee. Born in Vermont, he began his railroad career at age seventeen on the New England railway system. Four years later, he joined the Burlington as an engineer and machinist. A friend persuaded him to move to California just as the Civil War began. There was more opportunity for a young man in the far West, and Stevens advanced from a mechanic to a master mechanic on small railroads running out of San Francisco. In 1870, he joined the Central Pacific to govern the sprawling Sacramento repair shops, the largest industrial complex in the far West.

Once settled into his new position, Stevens began building new locomotives in Sacramento. The production was never great, for the costs were higher than those charged by the eastern builders. Stevens showed an independence and willingness to try new ideas. He experimented with outside valve gears and rotary valves. The scarcity of coal in the Pacific states prompted him to experiment with oil-fired boilers. Stevens also tended to think big. He favored large tenders. In 1882, he designed and built a 61½-ton 4–8–0, the largest engine built west of the Rockies and surely one of the heaviest to travel on any U.S. railroad at this time. This machine proved very successful, but the production models were built in the East, again because of the price. During the next year, Stevens produced an even larger locomotive, a monstrous 4–10–0 that weighed 73 tons and had 21-inch by 36-inch cylinders. The valve gear was so heavy that a power reverse was needed. For all of its size and its innovative features, this engine was a failure and was never duplicated. Yet, one failure hardly diminished the reputation of this pioneer who was willing to push locomotive development to its limits.

Page 575, LC
After the Forney locomotive was patented in 1866, few (not "none") were built until about 1878.

Page 580, LC
J. G. A. Meyer's *Modern Locomotive Construction* was reprinted by Lindsay Publications, Bradley, Illinois, in 1995.

Page 580, RC
Angus Sinclair's *Development of the Locomotive Engine* was reprinted in an annotated edition by MIT Press, Cambridge, Massachusetts, in 1970.

APPENDIX SECTION

APPENDIX A

BIOGRAPHICAL SKETCHES OF LOCOMOTIVE DESIGNERS AND BUILDERS

The following series of sketches is meant to identify the more prominent figures associated with American locomotive engineering in the nineteenth century. Nearly all of these individuals are mentioned in this work. In many cases little could be found on the life, much less the contributions, of these men. The fragments that have survived, usually obituaries, are memorial in nature and tend to credit sweeping improvements and reforms to the late, lamented subject of each particular notice. It is not uncommon to find the same invention credited to several mechanics.

If any general comment can be made on this group of men, it might concern the similarity of their histories. Their origins, almost to a man, were humble. Most started as blacksmiths, pattern makers, or draftsmen and rose to the top by luck and energy. This idea is currently unfashionable and is dismissed as a "Horatio Alger" fantasy, yet for the mechanic of the *early* nineteenth century it appears to have been a common phenomenon.

MATTHEW BAIRD (1817–1877) was for many years a principal partner at the Baldwin Locomotive Works after its founder's death. A native of Ireland, he followed his father as a coppersmith. He was apprenticed to the New Castle Manufacturing Company (Delaware) and there learned the locomotive business (1834–36). For a brief time he was superintendent of the Newcastle and Frenchtown Railroad's shops. In 1838 he was engaged by Baldwin as foreman of the boiler and sheet-metal shop. In 1854 he became a partner with Baldwin and continued as a major figure in the firm until his retirement in 1873.

Baird has been credited with using the firebrick arch in 1854, some three years before Griggs's patent. He was also the principal designer of the celebrated French and Baird spark arrester (1842).

MATTHIAS W. BALDWIN (1795–1866), possibly the most celebrated American locomotive builder, was born in Elizabethtown,

New Jersey, as the son of a prosperous carriage-builder. The early death of his father ended the family's prosperity, and young Baldwin was apprenticed to the jewelry trade when sixteen years of age. In 1825 he opened a small machine shop in Philadelphia in partnership with David Mason. He was successful in this venture and soon was recognized as one of the most able mechanics in the city. His reputation grew to such heights that in 1832 he was called upon by the Newcastle and Frenchtown Railroad to assemble its first locomotive imported from England, the *Delaware*. The year before, he built a model locomotive for the Peale Museum.

Encouraged by these ventures and his experience in stationary engine building, Baldwin built his first full-size locomotive in 1832, *Old Ironsides*. This machine was a direct copy of a Stephenson Planet type, the design apparently taken by Baldwin during the assembly of the *Delaware*. His next engine, and for several years thereafter his favorite design, was a 4–2–0 copied from Jervis' *Experiment*.

Within a few years Baldwin greatly expanded his locomotive business and became one of the largest builders in the United States. From what is known of his career, Baldwin's success was founded on the careful manufacture of locomotives on a conservative design. Except for the flexible-beam locomotive and a few other exceptions, Baldwin seems to have resisted the adoption of innovations in design. His notably late adoption of the 4–4–0 and 4–6–0 wheel arrangements and the link-motion valve gear is an example of conservative design policy. Such a position was admirably suited to the master mechanics of the time, who opposed radical and unorthodox designs in general.

ARETAS BLOOD (1816–1897), distinguished superintendent of the Manchester Locomotive Works, was a native of Weathersfield, Vermont. Blood was apprenticed as a blacksmith when seventeen years old. He learned the machinist trade and went to Lowell, Massachusetts, where he worked for the next seven years in the Locks and Canals machine shop. In about 1849 he became a "job hand" at the Essex Machine Shop, Lawrence, Massachusetts, where he produced small parts for locomotives. He acquired sufficient capital from this subcontracting venture to purchase a share in the newly founded Manchester Locomotive Works in 1853. For the first several years the shops were supervised by O. W. Bayley, formerly with the Amoskeag Manufacturing Company, who built on too light a pattern. Blood succeeded Bayley in 1857 and began the production of a more substantial machine that brought credit to the struggling firm. During succeeding years Blood became principal owner of the Works. In 1872 he added the fire engine business of the Amoskeag Manufacturing Company to his shop. Nearly 1,800 locomotives were built at the Manchester shops between its opening and 1901.

JOHN BRANDT (1785?–1860?), a native of Lancaster, Pennsylvania, trained as a blacksmith, became one of the most respected locomotive builders in the United States. His first railroad experience was acquired at the Parkesburg repair shops of the Philadelphia and Columbia Railroad where he was foreman from about 1833 to 1838. During the next two years he served as superintendent of motive power for the Georgia Railroad. In 1841 he held the same position on the Cumberland Valley Railroad but transferred to the New York and Erie Railway the next year. During his nine years on this road Brandt was responsible for the procurement of some of the largest locomotives in service in the United States. He was an early advocate, and possibly the originator, of the ten-wheel locomotive. In 1851 or 1852 Brandt became superintendent of the New Jersey Locomotive and Machine Company in Paterson, New Jersey. After a brief stay he returned to his home and joined the Lancaster Locomotive Works. Brandt's history is uncertain after the failure of the Lancaster Works in 1857. According to one report he retired with his son to Oregon.

HENRY R. CAMPBELL (1810?–1870?) is remembered for his 1836 patent for the 4–4–0 locomotive. Campbell's history is uncertain but this much is known: he was chief engineer of the Philadelphia, Germantown and Norristown Railroad from 1832 to 1839 and had formerly been an assistant engineer with the Philadelphia and Columbia Railroad; he was chief engineer on the Vermont Central from about 1848 to 1855. Although little is known of this engineer, his introduction of the 4-4-0 was a basic contribution to locomotive development.

ZERAH COLBURN (1832–1870) was a leading authority on locomotive engineering and one of the most gifted technical writers of the nineteenth century. Colburn was born in Saratoga Springs, New York, and spent most of his boyhood on a New Hampshire farm.

In 1847 he was employed as a clerk in a Lowell, Massachusetts, textile mill; later that year he was engaged by the Concord Railroad's mechanical department in a similar capacity. Between 1848 and 1851 he was employed as a draftsman by John Souther of Boston, from whom he learned the locomotive business. During this period he wrote his first book, *The Locomotive Engine* (1851), a small but valuable manual on American locomotive engineering. After a brief stay with Souther in Richmond, at the Tredegar Iron Works, Colburn was made mechanical editor of the *American Railroad Journal*. In 1854 he founded the *Railroad Advocate*.

Colburn served as consulting engineer for the New Jersey Locomotive and Machine Company from 1854 to 1858 and prepared designs for a remarkable wide firebox built by that firm. In later years he became editor of *The Engineer* (1858) and founded the journal *Engineering* (1866).

Colburn's most valuable contributions to railroad technology were the three folio-size publications (listed in the bibliography of this volume) that so clearly synthesized the theory and practice of the locomotive engine as developed up to that time. The last of these great works, *Locomotive Engineering and the Mechanism of Railways*, was published a year after Colburn's unfortunate suicide.

JOHN COOKE (1824–1882), born in Montreal, Canada, was apprenticed to Thomas Rogers of Paterson, New Jersey, when fifteen years of age. In about 1843 he was made superintendent of Rogers' shop and held that position until 1852 when he joined Charles Danforth to form a new locomotive-building firm in Paterson. This works, Danforth, Cooke and Company, was named after the principal partners and soon became a serious competitor of the Rogers shops. Cooke was directly in charge of locomotive production, Danforth being an expert in the area of textile machinery.

ISAAC DRIPPS (1810–1892) was born in Belfast, Ireland, and was brought to this country by his parents while still an infant. A machinist apprentice by the age of sixteen, he was hired by the Camden and Amboy Railroad in 1831 to assemble their first locomotive, the *John Bull*. He quickly rose in the mechanical department of this road and supervised the construction of locomotives, cars, and steamboat machinery. Dripps worked closely with the railroad's president, Robert L. Stevens, and it is difficult to assign credit for certain improvements to either man with absolute certainty. Among the important innovations produced were the cowcatcher (1832–33), the eight-wheel freight locomotive (1834–38), and the bonnet spark arrester (1833). In 1847 Dripps designed a high-wheel Crampton engine at the request of R. L. Stevens. This machine had a slope-backed firebox similar to that used by Winans and Millholland in later years.

In 1854 Dripps left the Camden and Amboy Railroad to become a partner in the Trenton Locomotive Works. The firm built a small number of engines before its failure in 1858. A year before its closing, Dripps designed and built the first pair of arch-bar trucks at these shops for the Lehigh Valley Railroad.

The remainder of Dripps's engineering career was spent with the Pennsylvania Railroad. He was superintendent of motive power from 1870 to 1872 but was forced to give up this position because of ill health. He retired in 1878.

WILSON EDDY (1813–1898) was born in Chelsea, Vermont. He learned the machinist trade at the Locks and Canals machine shop, starting there when nineteen years old. In 1840 he was made foreman of Western Railroad's locomotive repair shops at Springfield, Massachusetts. Ten years later he was appointed master mechanic of the road. The next year he completed his first locomotive, the *Addison Gilmore*, a fast passenger engine with an extraordinarily large heating surface. Eddy was a strong advocate of large fireboxes and developed a peculiar form of plate frame to permit wide, between-the-frame grates. While Eddy's engines were built on the conventional eight-wheel pattern, the design of various details was original. Among these was the use of the perforated dry pipe, a sliding connection at the cylinder saddle, and the large-diameter straight boiler without a steam dome. About 135 engines were built on this standard plan; the last was completed in 1881, a year after its designer's retirement.

For many years Eddy's "Clocks," so-called because of their smooth working, were regarded as first-rate examples of American locomotive engineering. After about 1870 Eddy's design became obsolete and his intemperate criticism of the new Mogul locomotives and steel boilers illustrated an unfortunate resistance to reform exhibited by many old-line mechanics.

MATTHIAS N. FORNEY (1835–1908), author of the standard text *Catechism of the Locomotive* (1874) was a technical journalist and mechanic of considerable skill. He was born in Hanover, Pennsylvania, and was apprenticed with Ross Winans to learn the machinist trade. A skillful draftsman, Forney worked in this position for the Baltimore and Ohio (1855–58), the Illinois Central (ca. 1861–64), and the Hinkley Locomotive Works (1865–70).

In 1870 Forney became associate editor of the *Railroad Gazette* and from this time forward rapidly gained the attention and respect of the railroad industry. He was one of the founders of the American Society of Mechanical Engineers and was an active member in many other engineering clubs, particularly the Master Car Builders Association.

The 1866 tank locomotive is the best known of Forney's thirty-three patented inventions. His attempt to supersede the established 4–4–0 with his new design failed, but the Forney locomotive performed well on the New York elevated railways for many years.

GEORGE S. GRIGGS (1805–1870) was one of the most influential railroad master mechanics of New England, and his designs were copied by nearly every commercial builder in that area. Griggs worked as a millwright and learned the machinist trade at the Locks and Canals machine shop before being appointed master mechanic of the Boston and Providence Railroad in 1834. His entire career was spent with this road.

Griggs built his first locomotive, the *Norfolk,* in 1845 at the Roxbury shops of the Boston and Providence. This machine was inside connected and had a riveted frame, a Stephenson boiler, and other features associated with typical New England 4–4–0's for the next ten years. Another twenty to thirty locomotives of this pattern were built under Griggs's direction. He obtained patents for car brakes, wooden cushion driving wheels, and the firebrick arch and was an early investigator of steam brakes and coal-burning fireboxes.

While Griggs has been mentioned as an innovator in connection with many mechanical improvements, he did not abandon a basic design even after it was long out of date. The Roxbury shops continued to build antique inside-connected locomotives three years after Griggs's death.

JOSEPH HARRISON, JR. (1810–1874), was born in Philadelphia and became a machinist apprentice when only fifteen years old. Five years later he was foreman at James Flint's machine shop. After working at several other Philadelphia machine shops, including William Norris's small locomotive establishment, he went to work for Garrett and Eastwick in the summer of 1835. Harrison applied the experience he had gained in locomotive design when working for Norris to the first engine built by his new employers. The success of this machine encouraged Garrett and Eastwick to more fully enter locomotive manufacture. Harrison's valuable contributions to this venture were rewarded by his being made a junior partner; in 1839 he became a full partner. The new firm of Eastwick and Harrison rapidly acquired a good reputation for building well-designed locomotives. They were the first builders to promote the 4–4–0, a wheel arrangement made successful by their development of the equalizing lever. In 1843 Harrison went to St. Petersburg, Russia, and there, in partnership with Andrew M. Eastwick and Thomas Winans, contracted for three million dollars' worth of rolling stock. Harrison returned to Philadelphia as a wealthy man in 1852. He continued his work as an engineer but produced no practical reforms in locomotive design after that time. His major contribution remained the equalizing lever.

HOLMES HINKLEY (1793–1866) was born in Hallowell, Maine, and worked as a carpenter until 1823 when he settled in Boston and entered the machinist trade. Eight years later he opened a small machine shop in partnership with Gardner P. Drury and Daniel F. Child. In 1840 Hinkley produced his first locomotive, a small 4-2-0 of ordinary design. Within a few years locomotives became Hinkley's most important product and his works became the largest locomotive manufacturer in New England. During the mid-1850's Hinkley was one of the major locomotive producers of the United States, but this position was short-lived. Production fell off rapidly before the works was closed in 1889.

Holmes Hinkley presumably had little to do with locomotive design; John Souther is credited with the design for his first engine. The products of his works were not notable in terms of advanced designs or innovations. Conventional designs were followed, and, as with most New England builders, inside connection was favored long after its rejection by other builders. In a biographical sketch entitled *Holmes Hinkley an Industrial Pioneer* (1913), Hinkley has been credited with the major contribution of using an arch rather than an English-dome firebox. In fact, however, he was copying the Stephenson boiler, which had already been used several years earlier by Locks and Canals machine shop.

WILLIAM S. HUDSON (1810–1881), born in Derby, England, learned the machinist trade at Stephenson's locomotive works at Newcastle. When twenty-four years of age he came to the United States and worked as an engineer on several northern New York railroads. In about 1838 he became engineer of the Auburn State Prison at Auburn, New York. He remained in the employ of this institution for eleven years but found time in 1842 to serve as chief engineer for Dennis, Wood and Russell, a small firm of Auburn which built three or four locomotives. It is possible that the locomotives were actually built in the prison's machine shop under Hudson's direction.

In 1849 Hudson became master mechanic of the Attica and Buffalo Railroad. Three years later he succeeded John Cooke as superintendent of the progressive Rogers, Ketchum and Grosvenor firm of Paterson, New Jersey, one of the largest locomotive builders in this country. Hudson continued in this position until his death. He obtained a number of patents for locomotive improvements, the most famous being the 1864 pony-truck equalizer patent.

WILLIAM T. JAMES (1786–1865) has been credited with such basic inventions as link motion and the spark arrester. James was thought to be a native of Rhode Island. He received a patent for making files in 1812. Between about 1820 and 1839 he operated a stove manufacturing business in New York City. James's personal interest appears to have been steam locomotion, for between 1828 and 1838 he is reported to have built four road steam carriages and three locomotives. Several of these machines were distinguished by the use of two-cylinder compound engines. In 1829 he built a large model of a locomotive. Two years later he built a similar full-size machine for the Baltimore and Ohio Railroad. It was not a success and was rebuilt or succeeded by a second machine in 1832. It was this engine, the *American,* that reportedly was first fitted with the link motion. It ran on both the Baltimore and Ohio and New York and Harlem railroads but was destroyed in 1834 when its boiler exploded.

James's final full-size locomotive, a 4-2-0 named the *Brother Jonathan,* was said to have been built in 1838 for the New York and Harlem Railroad. The existence of this machine might be questioned, as are most facts connected with James's life and work. James moved to Minnesota and died there in 1865.

JOHN B. JERVIS (1795–1885) was primarily a civil engineer who learned his trade in the field. His single contribution to locomotive design, the leading truck (1832), will forever grant him a place of honor among notable steam locomotive designers. Aside from his pioneering interest in coal-burning locomotives while chief engineer of the Mohawk and Hudson Railroad, he had little or nothing to do with locomotive construction. In later years he was involved with the construction and management of several major American railroads; among these were the Hudson River, the Michigan Southern and Northern Indiana, and the Chicago and Rock Island railroads. His book, *Railway Property* (1859), was long a standard text on railroad construction and management.

JOHN P. LAIRD (1826–1882) came to the United States from Scotland at the age of nineteen. He first worked for Rogers and then for the Ballardvale, Massachusetts, machine shop. In 1853 he became superintendent of the Latham Machine Shop of White River Junction, Vermont, where a few locomotives were constructed.

While serving as master mechanic of the Marietta and Cincinnati Railroad he devised a two-wheel equalizer leading truck (1857) for which he later received a patent. In 1862 he was made superintendent of motive power for the Pennsylvania Railroad and rebuilt much of the road's old power in succeeding years. Laird left the Pennsylvania in 1866 and worked in a similar capacity for several smaller lines in the years that followed.

STEPHEN H. LONG (1784–1864) was a native of New Hampshire and was one of the few engineers associated with early locomotives to have a college education. After his graduation from Dartmouth in 1809, Long became a lieutenant in the United States Army Corps of Engineers and taught for a time at West Point. In 1823–24 he led an expedition to the Rocky Mountains; the highest summit in that range was named in his honor.

Long next became interested in railroad engineering and secured several patents for locomotives. The first of these, granted in 1826, is significant because of its advocacy of coal-burning locomotives. Between 1827 and 1830 Long and several other prominent army engineers were engaged in the planning and construction of the Baltimore and Ohio Railroad. During this time Long published his *Railroad Manual* (1829), the first American work on the subject.

After leaving the Baltimore and Ohio he renewed his study of locomotives, obtained several more patents, and in 1832 formed the American Steam Carriage Company with William Norris and several other partners. Long handled the engineering; Norris promoted the business. However, Long's unorthodox designs, particularly the hard-coal-burning firebox, were not successful. After the building of about six machines the partnership was dissolved (1834) and Long quit the business of locomotive design. His most important contribution was the 1833 patent for four eccentric valve gears, although William T. James reportedly used the same plan several years earlier.

WALTER MCQUEEN (1817–1893) was born in Scotland and in 1830 emigrated to the United States, settling with his parents in upstate New York. He began in the machinist trade but was soon attracted to railroad work. Between 1840 and 1845 he worked as both a machinist and locomotive engineer for the Hudson and Berkshire and the Utica and Schenectady railroads. In 1840 McQueen built a small, 7-ton 4–2–0 locomotive at Albany for the Ithaca and Oswego Railroad. Five years later he was appointed master mechanic of the Albany and Schenectady Railroad. During the next three years he rebuilt several old engines and constructed a new machine, the *Mechanic*. One of the rebuilt engines, the *Mohawk*, is illustrated in the present work and is thought to be the earliest engine with a cylinder saddle. McQueen was hired as master mechanic of the Hudson River Railroad in about 1850 and designed most of the high-speed locomotives used by that road. In 1852 he left the Hudson River Railroad and became superintendent of the Schenectady Locomotive Works. He retired from this position in 1876 but remained a vice president of the works until his death.

McQueen cannot be credited specifically with any major locomotive improvement, although he does appear to have been the earliest user of the cylinder saddle and was the first to apply the air dome to feed pumps. However, his locomotives were regarded as first-class machines and his reputation was founded upon careful construction along conventional lines. Like several other old-line mechanics, McQueen was a loyal champion of the 4-4-0 and his refusal to accept new wheel arrangements was reportedly the cause of his retirement from active supervision of the Schenectady Works.

WILLIAM MASON (1808–1883), born in Mystic, Connecticut, spent his early years in the textile machinery trade. In 1835 he went to Taunton, Massachusetts. Ten years later, after devising many improvements in textile machinery and building up a good trade in that business, Mason built a new factory. In 1852 he decided to enter the locomotive business and completed his first engine the next year.

Mason's contribution to locomotive design has been greatly exaggerated, but there is no question that he was a master stylist and produced a first-class machine.

DAVID MATTHEW (1810?–1890?), an obscure mechanic whose name appears in nearly every account of early American locomotive building, was apprenticed at the West Point Foundry in about 1826. He assisted in the building of early locomotives at this works and was hired by John B. Jervis to assemble and run the *De Witt Clinton* on its delivery to the Mohawk and Hudson Railroad in 1831. In 1836 he was appointed as locomotive superintendent by the Utica and Schenectady Railroad. Matthew claims to have built the first roundhouse, snow plow, geared turntable, locomotive cab, spark arrester, a steam heater to warm feed water in the tender, and other basic railroad mechanisms. While there is little question that Matthew was a pioneer in the employment of these devices, it would be difficult to firmly establish his claims to their invention.

After leaving the Utica and Schenectady in 1842, Matthew quit railroad work to become manager of an iron foundry, but he continued to promote his smokestack and feed-water patents. His patent suit against Baldwin in 1860 for infringement of the smokestack patent produced the fine illustrative table on that subject reproduced earlier in this work. Matthew was last known to be living in San Francisco where he published a small illustrated folio of his sketches in 1887.

JAMES MILLHOLLAND (1812–1875) was born in Baltimore to a family of ship chandlers. At the age of eighteen he was an apprentice in George W. Johnson's machine shop and assisted in the construction of Peter Cooper's *Tom Thumb* as well as an engine built by his employer in 1831. Millholland's mechanical knowledge was broadened during the next several years while he worked at the Allaire Works, marine engine builders of New York City.

In 1838 he was appointed master mechanic of the Baltimore and Susquehanna Railroad. Millholland rebuilt several of this road's old engines and constructed two heavy 4-4-0 freight engines in the company shops. He also devised a cast-iron crank axle for inside-connected locomotives and in 1843 secured a patent for a six-wheel freight car.

These achievements, together with the construction of a remarkable iron-plate bridge in 1847, prompted the Philadelphia and Reading Railroad to hire the enterprising young mechanic in 1848. Millholland worked diligently to perfect a practical anthracite coal-burning firebox during the next several years. He perfected a successful design in about 1855 and succeeded in making the Reading one of the first all-coal-burning roads in America. He designed and built a large number of distinctive freight and passenger locomotives at the Reading, Pennsylvania, shops including a formidable 0-12-0 completed in 1863. Millholland was an early user of feed-water heaters, superheaters, and steel tires. His retirement from the Reading in 1866 ended his mechanical career.

ALEXANDER MITCHELL (1832–1908) was born in Nova Scotia. He entered railway service as a machinist in the Camden and Amboy Railroad repair shops. Between 1859 and 1861 he was assistant superintendent of the Trenton Locomotive Works. Next he began a life-long association with the Lehigh Valley Railroad which continued until his retirement in 1901.

Mitchell's chief contribution was the design of the 2-8-0

locomotive. First built in 1866, this wheel arrangement became one of the most important and successful freight locomotives ever developed. In 1867 Mitchell assisted in the design of the first 2–10–0's built in the United States. Two machines were built on this plan by the Lancaster Locomotive Works; one of them, the *Bee*, was rebuilt in 1883 as a 2–8–2 and was the first engine of this plan to be used in the United States.

RICHARD NORRIS (1807–1874) was a younger brother of William Norris who became active in the Norris Locomotive Works in 1839. Richard, like his brother, was not an engineer but apparently was a shrewd manager who made a great success of the business after forcing his brother out in about 1841. Under Richard's direction the firm prospered and by the mid-1850's had become the largest producer of locomotives in the United States during the mid-1850's. More than 1,000 locomotives were built by this firm between 1834 and 1867.

SEPTIMUS NORRIS (1818–1862) was the best-known engineer of the Norris brothers. While his invention of the ten-wheel locomotive is open to question, he was active in the development of coal-burning locomotives and was one of the first to employ the spread truck. His exaggerated praise of his brother William in the introduction of *Norris' Handbook for Locomotive Engineers and Machinests* (1852) and his activity in the Portland (1847) and Schenectady (1848) works indicate a lack of esteem for his more successful brother Richard. Septimus did, however, serve for a time as chief of the engineering department at the Philadelphia Works. He obtained patents for the ten-wheel locomotive, locomotive boilers, and with Jonathan Knight secured a patent for the locomotive running gear.

WILLIAM NORRIS (1802–1867) was probably the most overrated figure connected with locomotive development. Both Harrison and Whistler agree that he was not an engineer; rather, he was a zealous promoter who was attracted to locomotive building by the huge demand for such machines created by the rapidly expanding rail network.

Norris was a particularly ill-starred person; nearly every enterprise with which he was connected ended in failure. He had already failed in the dry goods business before organizing the American Steam Carriage Company in 1832. Stephen H. Long was the engineer of this firm, but his designs proved a failure and he retired from the enterprise in 1834. Norris next hired Fredrick de Sanno, who also proved unequal to producing a practical locomotive. It was not until Joseph Harrison entered Norris' employ in 1835 that a truly successful locomotive was produced under Norris' name.

After the remarkable performance of the *George Washington* in 1836, orders flowed to Norris' shop. This prosperity was quenched by the Panic of 1837. William's wealthy brother Richard was taken into the firm to buoy up the faltering enterprise but to no avail. The company failed in 1841 and William was forced out. In the next years William attempted to establish a locomotive works in Vienna (1844–48), but this too failed and he returned to the United States in 1848. Not being welcome back at the Philadelphia plant, which Richard had revived, William Norris went through a succession of failures which included an attempt to acquire the grant for the Panama Railroad (1848), open a gold mine, and construct a fast transatlantic steamer (1855). He apparently lived in retirement during his last years.

CHARLES T. PARRY (1821–1887), a native Philadelphian, was apprenticed to M. W. Baldwin when fifteen years old. Parry rose to the position of superintendent at the Baldwin Works in 1854 after first serving as pattern-maker and draftsman. He is credited with reforming the old-fashioned appearance of Baldwin's designs in the mid-1850's by adopting the designs of Rogers and Mason. He became a partner in the firm in 1867.

THOMAS ROGERS (1792–1856) was credited by Colburn as having done more than any one individual for the advancement of American locomotive design. Rogers was not an inventor but he was the first to recognize and incorporate into regular practice such basic reforms as spread trucks, wagon-top boilers, and link motion.

Rogers was born in Groton, Connecticut, and was schooled

in the carpenter and blacksmith trades before coming to Paterson, New Jersey, in 1812. After working as a pattern-maker and accumulating some capital in the manufacture of textile machinery, he established a new machine shop in 1832 under the title Rogers, Ketchum and Grosvenor. Several years later, after the introduction of railroads in the United States, he began making car wheels, axles, and other railroad fittings. In 1837 he completed his first locomotive, the *Sandusky*. Within a few years Rogers was operating one of the largest locomotive plants in this country. At first he seemed content to copy the leading designs of the nearby Philadelphia makers but in the early 1850's he adopted a progressive design that was soon copied by all other domestic builders. The reforms mentioned at the beginning of this sketch caused his business to prosper as new orders flowed in.

It should be noted that, while Rogers was a practical mechanic and undoubtedly had a direct hand in the design of products built at his shops, he was dependent upon William Swinburne, William S. Hudson, and others for many of the designs actually employed. It is therefore difficult to assign the several significant reforms originating at the Rogers plant exclusively to its proprietor.

JOHN SOUTHER (1818–1911) was born in Boston and worked first as a ship's carpenter. In 1840 he was engaged by Holmes Hinkley as a pattern-maker and in this capacity was said to be responsible for the design of Hinkley's first locomotive. In 1846 he started his own machine shop in south Boston and built locomotives, sugar-mill machinery, and steam excavators. Souther was fortunate to have Zerah Colburn on his staff for a short time. In 1852 he went to Richmond, Virginia, and managed the locomotive shop at the Tredegar Iron Works. Because of differences with the managers of the Tredegar works he returned to Boston in 1854 and opened a new factory. Souther was not a major locomotive builder, but his products were found in all parts of the United States before he ceased production in about 1864. No major improvements in locomotive design can be tracted to Souther; he built inside-connected engines almost exclusively until 1853 when he was forced to adopt more progresssive designs.

WILLIAM SWINBURNE (1805–1883) was born in Brooklyn, New York, and settled in Paterson, New Jersey, in 1833. About two years later he was employed by Thomas Rogers as a pattern-maker and assisted in the construction of Rogers' first locomotive, the *Sandusky*, in 1837. In 1848 he joined Samuel Smith of Paterson in forming a locomotive works. Three years later he formed an independent establishment for this purpose and is credited with producing some of the first spread-truck, level-cylinder engines built in this country. In about 1855 he built a number of 4-4-0's for the Chicago and Alton Railroad which were distinguished from the standard pattern by having their cylinders placed entirely behind the truck.

Swinburne's business was closed by the Panic of 1857 and his plant was purchased by the New York and Erie Railway the following year to serve as a repair shop. After this time Swinburne retired from the machine business and devoted his energies to the city of Paterson by holding several public offices.

GEORGE W. WHISTLER (1800–1849) was a graduate of West Point and was one of the few nineteenth-century figures connected with locomotive construction who had an academic rather than a "shop" education. Whistler was superintendent of the Locks and Canals machine shop from 1834 to 1837 and was engaged in the design of the earliest locomotive built in New England. Whistler copied Stephenson's Planet locomotives and appears to have introduced no original concepts to the arrangement of railway engines. Several years later when managing the Western Railroad of Massachusetts he recognized the need for heavier locomotives, but unfortunately he chose Ross Winans' unsuccessful 0-8-0 Crabs to meet this need. In 1842 Whistler went to Russia to supervise the construction of the Moscow to St. Petersburg Railway but died two years before its completion. Despite his failings as a locomotive designer he was considered a first-rate civil engineer.

APPENDIX A

SETH WILMARTH (1810–1886) was born in Brattleboro, Vermont, and learned the machinist trade in Pawtucket, Rhode Island. He gained further experience in the machine shop of Holmes Hinkley of Boston. In about 1836 he opened his own shop; by 1841 the enterprise had become so prosperous that a new factory, named the Union Works, was built in south Boston. Locomotive construction was soon begun but it never became Wilmarth's major line of work. His designs were generally orthodox and followed the ordinary New England style of inside-connected 4-4-0's. While engines were built for many railroads, total production did not exceed 200 machines. Wilmarth's shop was closed in 1854 after a disastrous contract with the New York and Erie Railway. The next year Wilmarth was appointed master mechanic of Boston's Charlestown Navy Yard and he continued in this capacity for twenty years.

ROSS WINANS (1796–1877), a strong, independent individual who started life as a farmer in his native state of New Jersey, was drawn to mechanics at an early age. First associated with the Baltimore and Ohio Railroad in 1829, he was appointed assistant engineer of machinery two years later. Soon he was main contractor for supplying the road's rolling stock and occupied the Mt. Clare Shops of the Baltimore and Ohio for this purpose. In about 1840 he built his own shop next to the Mt. Clare works and began a career as an independent locomotive builder. His original attempts at building coal-burning eight-wheel freight engines were not completely successful, but by 1848 he had perfected his ideas for the Camel locomotive. This powerful and original machine was well suited to slow-speed coal service but found few adherents outside of the Baltimore and Ohio and Reading. Winans produced a total of only about 300 locomotives. His refusal to adopt the reformed ideas of locomotive construction which developed so rapidly after 1850 caused his business to decline, and, after the Baltimore and Ohio's refusal to purchase more Camel engines, he was forced to close his shop in about 1860. Winans was far from a ruined man; he lived in comfortable circumstances and found expression for his strong-minded views by writing religious tracts and by joining his son Thomas in the construction of the Cigar boat.

Winans has been credited erroneously with the invention of the coned wheel, the chilled cast-iron car wheel, the eight-wheel car, and the leading truck. He was connected with these ideas at an early date, as he was with variable exhausts and coal-burning locomotives, but he cannot properly be given credit as their originator.

THOMAS WINANS (1820–1878), the eldest son of Ross Winans, was born in Vernon, New Jersey. A gifted mechanic with independent ideas much like his father, he was best known for his partnership with Harrison and Eastwick in the construction of rolling stock for the Moscow–St. Petersburg Railway. Many of the design features of these machines followed closely those of Ross Winans and were undoubtedly adopted through the agency of Thomas Winans.

On his return from Russia, Thomas assisted in the design and construction of two experimental passenger locomotives, the *Centipede* (1855), the first 4-8-0, and the *Celeste* (ca. 1854), a high-speed 4-4-0 tested on the Reading.

JOHN E. WOOTTEN (1822–1898), born in Philadelphia, was apprenticed to Baldwin in 1837 and worked in his shops for eight years. He then joined the Reading, becoming James Millholland's chief assistant. He succeeded Millholland in 1866 and was superintendent of motive power until 1871. Other promotions followed and he was made general manager of the entire railroad from 1877 until his retirement in 1886.

Wootten was a prolific inventor and held patents on smokestacks, steam gauge, car heater, and other devices. His most famous patent was for the wide firebox for burning waste anthracite coal, issued in 1877. This design, a development of earlier designs by Dripps, Winans, Millholland, and Colburn, was immensely successful; three to four thousand locomotives were equipped with Wootten fireboxes according to an estimate made in 1925.

APPENDIX B

CONTRACT FOR THE MOHAWK AND HUDSON RAILROAD'S EXPERIMENT, 1831*

Articles of Agreement made and concluded this sixteenth day of November in the year one thousand eight hundred and thirty one, Between the West Point Foundry association of the first part, and the Mohawk and Hudson Rail Road Company of the second part; Both of the state of New York: Witnesseth, that the said party of the first part, for the consideration hereinafter named, agree to furnish and deliver on board such vessel lying at their dock in the Harbour of New York as may be designated by the said party of the second part, a Locomotive steam engine; the said engine to be constructed agreeably to a plan of the same as furnished and explained by John B. Jervis Engineer of the party of the second part: the furnace part of the boiler to be rectangular; the flues to be made of copper, which are to pass through and be secured to the cylindrical part of the boiler: The dimensions of the several parts to be as represented on the plan aforesaid and a written statement with calculations furnished by the aforesaid Engineer. The engine to have two working cylinders each nine inches in diameter with a stroke of sixteen inches in length: each cylinder to have connected with it a force pump, with all necessary steam and water valves and pipes, eccentrics, rods, and other parts to make the working gear complete for the use intended. On the front end of the boiler a proper fixture to be made to receive a glass tube for a steam and water gauge so arranged as to admit of taking out a broken glass and conveniently substituting a new one; three extra glasses properly prepared to be furnished with the engine. The ends of the boiler to be secured by tie rods running longitudinally through the boiler. That part of the boiler which surrounds the furnace to be secured by tie bolts round the sides and on the top by iron ribs, of which there is to be two more than represented on the plan. The engine to be furnished with a good and convenient hand force pump, with copper pipes to connect with the water tank on the tender waggon. Oil cups to be provided for all parts that require oiling to be constructed with a cotton-wick syphon as in the Stevenson [sic] engine, belonging to the said party of the second part. The working wheels to be made of cast and

---

* From *Railway and Locomotive Historical Society Bulletin*, No. 55 (1941), pp. 21–22.

wrought iron on the plan of the wheels of the Locomotive engine De Witt Clinton; except that the wrought iron rim shall be made with a flange: The other wheels to have cast-iron naves with wooden spokes and felloes, and wrought iron flanged tire or rim. The frame to be made of good seasoned White Oak timber well ironed and braced where required. The axles, seats and all minor parts of the engine not represented on the drawing or description herein referred [sic] to, or explained shall be constructed agreeably to the direction of the aforesaid Engineer. A complete set of wrenches to work all the screw nuts to be furnished with the engine. The said party of the first part further agrees to furnish a Mechanik to superintend the transportation and to set up the engine on the Rail Road of the said party of the second part and to put the said engine in complete operation. (the said party of the second part to pay the charges of insurance and transportation after the engine is on board the vessel as aforesaid in the Harbour of New York).

The said party of the first part, further agree that all the materials and the workmanship shall be of the best character for the use intended and that the aforesaid engineer shall be inspector of the same.

The said party of the second part hereby agree to pay to the said party of the first part for completing this contract as aforesaid, to the satisfaction of the said engineer the sum of *Four thousand* and *Six hundred dollars* in the following manner to wit, eighty per cent of the contract price when the engine is ready to ship in New York, an addition of ten per cent on the certificate of the aforesaid Engineer that the engine is set up on the Mohawk and Hudson Rail Road agreeably to the provisions of this contract, and the balance (being ten per cent) on the further certificate of the said Engineer that the engine has run one month during which it has proved to be in all respects a compliance with this contract: Provided, however, that the said engine shall be ready to ship in the city of New York by the first day of April next; and should the engine not be ready by the first day of April next, it shall be at the option of the said party of the second part to receive it or not as they may prefer.

In witness the parties have hereunto set their hand.

APPENDIX C

SPECIFICATION FOR THE BALTIMORE & OHIO RAILROAD'S 0-8-0, 1847*

Proposals under seal will be received by the undersigned up to Saturday, the 6th of November, inclusive, for furnishing the Baltimore & Ohio Railroad Company with four locomotive engines, in conformity with the following specification:

1. The weight not to exceed 20 tons, of 2,240 pounds, and to come as near to that limit as possible.
2. The weight to be uniformly distributed upon all the wheels when the engine is drawing her heaviest load.
3. The number of wheels to be eight.
4. The diameter of the wheels to be 43 inches.
5. The four intermediate wheels to be without flanges.
6. The boiler to contain not less than 1,000 square feet of fire service, of which there shall be not less than one-fifteenth in the firebox.
7. The tubes of No. 11 flue iron, with not less than ¾ of an inch space between them in the tube sheets.
8. The firebox, with the exception of the tube and crown sheets to be of ⅖-inch copper.
9. The tube sheets to be ⅜ inch thick.
10. The boiler to be of No. 3 iron, of the best quality.
11. The firebox to be not less than 24 inches deep below the cylindrical part of the boiler.
12. The steam to be taken to the cylinder from a separate dome on the fore part of the boiler.
13. The frame, including the pedestals, to be entirely of wrought iron, and the boiler to be connected therewith, so as to allow of contraction and expansion without strain on either.
14. The cylinders to be 22 inches stroke, and not less than 17 inches diameter.
15. The cut-off to be effected by a double valve, worked by separate eccentrics.
16. The angle of the cylinder to be not greater than 13½ degrees with the horizontal line.
17. The frame and bearings to be inside the wheels, and the connection from the cylinder direct with the back pair of intermediate wheels.
18. The centers of the extreme wheels to be not more than 11½ feet [a]part.

* From *American Railroad Journal*, October 23, 1847.

APPENDIX C

19. The wheels to be of cast iron with chilled tire.

20. The means to be provided of varying the power of the exhaust in the blast pipe.

21. The engine to be warranted to do full work with Cumberland or other bituminous coal, in a raw state, as the fuel, and the furnace to be provided with an upper and lower fire door with that view.

22. The smoke-stack to be provided with a wire gauze covering.

23. Two safety valves to be placed upon the boiler, each containing not less than 5 square inches of surface, and one to be out of the reach of the engineman.

24. The tender to be upon eight wheels, and constructed upon such plan as shall be furnished by the company, and to carry not less than three cords of wood, or its equivalent in coal, and 1,500 gallons of water.

25. The materials and workmanship to be of the best quality, and the engine to be subjected to a trial of thirty days' steady work with freight upon the road, before acceptance by the company.

Payment to be made in cash on acceptance of the engine. The four engines to be delivered at the company's Mount Clare depot, in Baltimore; the first on the 1st of February, 1848, and the three others on the 1st of March, April and May ensuing.

The track is 4 feet 8½ inches gage, and the shortest curve of the road is 400 feet radius.

The company to be secured against all patent claims.

Further information will be communicated upon application to the undersigned, at the company's office, No. 23 Hanover Street, Baltimore, to which the proposals suitably endorsed will be addressed.

By order of the President and Directors.

BENJ. H. LATROBE,
Chief Engineer and Gen. Superintendent.

Baltimore, Sept. 18, 1847.

APPENDIX D

SPECIFICATION OF NEW YORK AND ERIE ENGINES, 1851*

As an example of the best constructed engines upon the Erie road, we present the following specification of the standard passenger engines, Nos. 100 to 105, inclusive. Nos. 106 to 111 are also the same, with the exception of the diameter of driving wheels and trucks, and the depth of furnace.

The engines of this class have not received as liberal allowances of heating surface, flue opening and steam room, as some of the earlier engines, the deficiency being evident in their performance, but for excellence of materials, and thorough, accurate and durable construction, we question if they are surpassed by any other engines upon the road or by any large number in the country. They have cast iron pumps where brass is now much used, and have cast iron instead of wrought iron rocker shafts, but with these exceptions they are constructed with every reference to the highest standard of workmanship.

These engines were built in the season of 1851, by Rogers, Ketchum and Grosvenor, of Paterson, New Jersey.

* From *American Railroad Journal*, July 23, 1853, pp. 467–68.

GENERAL ARRANGEMENT.

Horizontal cylinders, and double cranks; four driving wheels and four trucks, shifting link motion for working valves; steam chests outside of smoke box, framing inside of driving wheels.

BOILER.

Waist of boiler is of one-fourth inch Bowling plates; is 44½ inches in diameter, outside of main course, next firebox, and 43 inches at smoke-box end. Outside firebox is square, and of the "waggon-top" shape, and is of No. 3—or rather strong ¼ inch—iron. Furnace of one-fourth inch iron. Iron tube sheets at each end, three-eighths inch thick. The seams joining the firebox to the waist of boiler are double riveted.

Two round stay-rods pass from each stay bar, on crown sheet to outer shell of firebox. Stay-bars and longitudinal stay-rods, of usual number and strength. Grate 47 inches long, by 47½ inches wide, and for engines 100 to 105, inclusive, having 6 feet driving wheels, is 53 inches from crown sheet. Cast iron grates closed up for six inches of their length at their forward ends. 152, No. 13, copper tubes, 1 13-16 inches diameter outside of smoke-box ends, and 1 11-16 inches outside diameter, at fire-box ends of same. These are 11 feet 1 inch long between tube sheets, and are tightened at firebox ends by cast iron thimbles, of 1⅛ inch inside diameter.

Low wrought iron dome over fire-box, closed by a cast iron cap.

Lap welded iron steam pipe of six inches outside diameter, and one eighth inch thickness, (the pressure outside and inside of pipe being always equal) runs the entire length of boiler. Three longitudinal rows of slots or steam passages on upper side of pipe, commencing 24 inches from each end of pipe, being 3½ inches long each, and three eighths inch apart. One-third of their number towards fire-box are one-fourth inch wide, the middle portion are 5-16 inch, and those next the smoke-box are ⅜ inch wide. Throttle valve in a cast-iron chest within smoke-box. Joints of main steampipe, branch and blast pipes are ground turning joints, made with a composition ring on a cast-iron surface. Branch pipes 5½ inches diameter. Blast pipes have three sets of spare composition nozzles, 1⅞, 2, and 2⅛ inches diameter.

Cast iron boiler front, and hinged cast iron door.

Two safety valves, of two and a half inches diameter.

Whistle of six inches diameter, and bell of 120 pounds.

Pine lagging of three fourths inch thickness covered with Russia iron. Gauge cocks, frost and blow-off cocks, mud hole plugs, and Ashcroft's fusible safety-plug, in all the boilers.

CYLINDERS.

Seventeen inch bore, long enough for 20 inch stroke, which is 28¾ inch between ground faces. Cylinders are 7-8th inch thick; covers 1⅛ inch thick, The two cylinders are 37¼ inches from center to center. From back cylinder face to center of crank-axle is 9 feet 4½ inches. Cylinders are secured by a stout flange resting on the upper side of frame, and by a wide curved flange to under side of smoke box. They are also secured firmly together, while a stout flat brace comes from each forward driving axle jaw, and passes through a lug cast on cylinder, having a nut and flat key on each side of same. Valve faces are inclined out wards at an angle of 45°. Steam ports 14 inches, by 1⅝ inches. Exhaust port 14 inches, by two inches. Steam enters steam chest, through a port cast over induction port, and opening on face of cylinder. Composition main valve, with 11-16ths outside and no inside lap. Valve is encircled by a wrought iron hoop into which the valve stem is tapped, and into which is also tapped a guide spindle coming through forward end of chest and working in a brass sheath. Bolts to secure cover of chest pass down inside of chest so as to show a plain surface outside. Bridge between ports is 1 inch thick. Valve 8⅝ inches long, and has cavity of 4 inches in width.

Steam chests have each a double cock oil cup and there is also an oil cup in center of forward cover of each cylinder,

Pistons have one outside composition ring with two circumferential grooves, filled with Babbitt metal, and one inside ring of wrought iron. Outside ring is cut open obliquely at one place, and has a small wrought iron flap on each edge to prevent leakage of steam at the point of division.

Glands of piston rod and valve steam stuffing boxes are of cast iron lined with a tight composition bushing.

The casting for the truck pintal is attached to both cylinders by bolts, and forms the connection between the cylinders.

FRAME.

Is flat, 4 by 2 inches. The portion back of offset in forward jaw is 4½ inches wide. At back end and as far forward as offset the frame is 55½ inches apart in the clear; forward of offset it is 56½ inches apart. Top of back division of frame is 12 inches above the horizontal axis of cylinders; immediately in front of the forward jaws it drops 7½ inches, or to 4½ inches above horizontal axis of cylinders, and continues at that level to forward end. Connection of back end of frame is made by a heavy foot plate of forged iron. At forward end by an oak bunter beam. Pedestals for crank axles are forged at the Bowling works in England, and are welded to the bars forming the frame. Back pedestals are of cast iron. Both pedestals have keys on both sides of the bearings which they support, for the purpose of taking up the wear of the driving boxes against the pedestals. Cast iron thimbles and truss rods between back and forward pedestals. A brace of flat iron extends from forward jaw to cylinder, as already noticed. A diagonal brace extends from each back pedestal up to foot plate.

The boiler braces are riveted to boiler. The angle-iron braces,

APPENDIX D

which connect the fire-box with the frame, are held to the frame by long straps reaching over them and bolted down at the ends, and also at the middle, having oblong holes in the angle-iron where the bolts pass through.—An allowance is made, lengthwise, between the angle-irons and the straps which confine them, of 3-16 inch. The screws which secure the expansion braces to fire-box pass through the water space and are riveted inside of furnace. Before being entered in the inside sheet they have concave-faced nuts started upon them, and a packing of canvas and red lead, which are screwed tightly against the inner side of outer sheet. These stay bolts have square heads where they pass through the angle-iron braces, and have finished nuts and washers on their outer ends.

There are diagonal braces of round iron at each end of boiler, and four flat braces connecting the waist of the boiler with the frame. These, with the expansion braces, equalizing levers and spring straps are all finished to a smooth surface.

There is a deep cross-girt of wrought iron, supporting slides and rocker-shafts. It is ⅞ inch thick and is secured at each end to the frame, and at the center to the under side of waist of boiler. It is cut out to allow the connecting rods to work through it.

Outside rail of 3 in. angle-iron curved over driving wheels and continued to forward end, at the same level as inside frame.

WHEELS, AXLES AND SPRINGS.

The driving wheels are six feet in diameter on tread. Four Bowling tires, all with flanges, and 1⅞ inches thick; shrunk ¾ths inch in their inner circumference, and secured in addition by eight rivets passing through rim of wheel and tire.— Counterbalances for the driving wheels are bolted in, crank pins are forged in England from Bowling iron.

There are four truck wheels of 36 in. diameter. These are spoke wheels with chilled rims, and have their hubs banded with wrought iron. (These have since been changed for Bush and Lobdell's double plate wheels.) The trucks are secured each by one stout spline, and the drivers by two stout square keys.

Back and forward drivers are 6 ft. 9¾ inches between centers. Center of truck is 6 in. back of front face of cylinder, or 11 ft. 3¼ inches from center of crank, or eighteen feet one inch from center of back shaft. Truck wheels are 44 inches between centers, so that the entire length of rectangle covered by wheels of engine is 19 feet 11 inches.

Bowling crank axle, with bearings 6½ inches in diameter and 7 inches long. Crank wrist 7 inches diameter, and 4¼ inches long. 10 inches throw. Back shaft has same size of bearings as crank axle. Truck axle bearings 3⅞ inches diameter and 7 inches long. There is a collar on the truck axle, just inside of wheel, so that the box shall not work loose. Driving axle bearings, or boxes, are made with a composition lining, without Babbitt metal, and fitted tight to casting of driving box.

Driving springs, 40 inches long, 12 plates of ⅜ inch steel, 3½ inches wide, and have 11 plates of ⅜ inch iron in the center of spring between steel plates. Whole depth of spring in center 8⅝ inch. Flat spring straps bent over ends of springs.— Truck springs 46 inches long, have twenty-two plates of ⅜ steel, 3½ inches wide; 8¼ in. deep at center.

SLIDES, PUMPS AND CONNECTING RODS.

Slides are flat wrought iron bars, 3 inches by 1¼ inches, not case hardened. Cross head bearing of cast iron, without lined gibs, and is 9 inches long and 2 inches thick.

Pumps are of cast iron. Plungers 1⅞ inch diameter and are worked at full stroke from a pin on under side of cross head. Spindle pump valves, ground with flat faces and are 3 inches in diameter outside. Air chambers on forcing and on suction side of pump. Lap welded iron supply pipes 2¼ inches in diameter. Ball check valves, which are situated towards forward end of boiler, and opposite the center of its vertical diameter.

The main connecting rods have their straps secured by gib and key and one bolt. A portion of the flat face of the large end of the rod is cut out to reduce the weight. The boxes of the parallel rods are made to enclose the end of the crank pin to prevent the access of dirt to the bearing. The parallel rods are placed upon the crank pins when one axle stands at the same height in its jaw as the other, that is to say, when the engine stands on a level rail. Both halves of the box at forward end of the rod are made to fit close to the pin, as also does the inside box at back end while the inner surface of outside box is 1-16th inch from that side of the pin. This allows of the vertical movement of one pair of wheels in its jaw independent of the other without straining the rods, but produces a slight *slip* at every revolution.

VALVE MOTION.

There are four eccentrics of 5¼ inches throw, each eccentric being halved and secured by bolts. Wrought iron dowels are inserted, half in each of the two separate portions of the eccentric to prevent any side motion. The eccentrics are secured to the axle each by two square ended set screws pressing hardened steel *dies,* which are cut with sharp grooves, against the axle. The eccentric straps are of cast iron with oil cups cast on, and being grooved out inside so as to shut over eccentric and exclude dust. The eccentric rods are held by three square bolts passing sideways through a flat palm cast on eccentric strap. The back and front eccentric rods for each cylinder are attached to a curved wrought iron link, the curve of whose center is described with the radius of the eccentric rod, or three inches less than the distance from center of eccentric to center of link. The link is 17 inches long inside of slot, 2 inches thick, and the slot is 2¼ inches wide with 1½ inch thickness of iron around it. The link is raised and lowered upon a block on lower end of rocker arm. The studs to which the eccentric rods are attached are 1¼ inches in diameter outside of thimble which is ⅛ inch thick. The rod works loose on the stud the stud being secured tight to the link. The end of the rod has a bushing where it wears upon stud. The centers of studs are 3 inches from centers of links on some of the engines, as the No. 100. On others the distance is as little as 2⅜ inches. The rocker shaft is of cast iron and its boxes are secured to frame and to the cross girt or yoke sustaining the slides. Rocker arms 9⅛ inches long each way. The link motion is graduated to cut off at 17¾ inches, 15 inches, 12½ inches, 10 inches and 7½ inches respectively of each stroke. The lead under the greatest throw of the valve, (which is, 4¾ inches, or ½ inch less than the full throw of eccentrics), is 1-16th inch, and as the admission is reduced the lead increases, until when cutting off at 7½ inches of the stroke, it is 5-16 inch. The opening of the port with this admission is ⅜ inch or 5¼ square inches: cutting off at half stroke the opening of port is 7-16th inch, or 6⅛ square inches. The admissions for front and back strokes do not exceed ¼ inch for the difference between them.

The lifter shaft is 3 inches in diameter and carries a large counter weight to balance the weight of the eccentric rods and links. The arms of the lifter shaft are welded on. The links, blocks, studs, and the boxes of the valve stems are of wrought iron, case-hardened.

The reversing lever works in a graduated arc, and has a catch bolt to retain it in either one of the notches.

Radley & Hunter's, patent Spark Arrester, 16 feet high from rail.

Hand rail on boiler made of brass pipe.

Cow catcher of round rods, firmly braced.

Number plates have brass figures riveted upon a plate of Japanned iron, and enclosed within a plain brass border held by cup headed brass screws.

TENDER.

Tank of ⅛ inch iron in sides, well stayed, bottom ¼ inch thick, contains 1600 gallons, mounted on two trucks of four 33 inch wheels each. Trucks have iron frames, and inside bearings 3⅞ inch diameter, and 7 inch long. Tender frame of oak with center beam 20 inches by 5 inch; side sills 9 inches by 4 inches. Brakes for tender wheels are faced with wrought iron. Height of draw rod from rail 2 feet 8 inches.

APPENDIX E

SPECIFICATION FOR THE WESTERN AND ATLANTIC RAILROAD'S GENERAL, ROGERS LOCOMOTIVE WORKS, DECEMBER, 1855[*]

DESCRIPTION OF THE "GENERAL."

We are indebted to Mr. Louis L. Park, Chief Draughtsman of the Rogers Locomotive Works, Patterson, N. J., for the following information in regard to the "General," taken from the plans and specifications of that company:

Built by the Rogers Locomotive Works in December, 1855, for the Western & Atlantic R. R. An eight-wheel, wood-burning locomotive of type 440–50, weighing 50,300 pounds; gauge, 5 feet; cylinders, 15x22 inches; piston rod 2 1-4 inches in diameter; has four driving wheels, each 60 inches in diameter, made of cast iron, with journals 6 inches in diameter; driving wheel base, 7 feet; total wheel base of engine, about 20 feet, 6 inches; weight on drivers, 32,000 pounds; weight on truck, 18,000 pounds; heating surface: flues, 748.36 square feet; fire-box, 71.08 square feet; total heating surface, 819.44 square feet. Grate area, 12.46 square feet. Boiler of type known as Wagon Top, covered with felt and Russia iron; diameter inside first course, 40 inches; working pressure, about 140 pounds; thickness of barrel of boiler, 5-16 of an inch; thickness of dome course, 3-8 of an inch; fire-box: thickness of shell, 3-8 and 5-16 of an inch; thickness of crown, 3-8 of an inch; thickness of flue-sheet, 1-2 an inch; thickness of sides and back, 5-16 of an inch; length of grate, 46 inches; width 39 inches. Contains 130 flues, each 11 feet long by 2 inches in diameter. Steam-pipes 5 inches in diameter. Engine truck, 4-wheel, rigid center; tender trucks, 4-wheel, inside bearing. Diameter of wheels, 30 inches. Has 2 escape valves and 2 pumps. The smoke stack is of the old balloon type, and the cow-catcher is much longer and larger than those on modern engines.

[*] From *The Story of the "General,"* a booklet by the Nashville, Chattanooga, and St. Louis Railway, 1906.

APPENDIX F

PARTS AND WEIGHT LIST FOR THE HINKLEY LOCOMOTIVE WORKS'S 4-4-0, 1865[*]

THE WEIGHT AND NUMBER OF PARTS OF A LOCOMOTIVE AND TENDER IN DETAIL

We give, herewith, the weight of all the parts of a locomotive and tender, which were ascertained by carefully weighing each part when finished and ready to put on the engine. The number of pieces were counted and includes all excepting the nails in the wood work, and the pieces of wood in the lagging. The locomotives were built in 1865 for the Illinois Central Railroad Company by the Hinkley Locomotive Works of Boston, and were copies of locomotives built at the Rogers' Locomotive Works some years before that date. They were light engines compared with the practice of the present day. They were of the American type with 5 ft. driving-wheels and 15 x 22-in. cylinders. The fire-boxes or grates were 4½ ft. long, and the boilers had 165 tubes, 1¾ in. in diameter.

The weights of the brass parts, those of wrought-iron and steel, and those of cast-iron and also wood, etc., are given in separate columns. The number of parts was counted merely as a matter of curiosity. If some one would get the weight in detail of a standard engine of the present day in a similar way, it would be an extremely useful piece of work:

[*] From *Railroad Gazette,* December 8, 1882.

APPENDIX F

Number and Material of Parts of a Locomotive and Weights of Each.

	Brass lbs.	Wro't Iron lbs.	Cast Iron lbs.	Wood etc., lbs.	No. of p'c's
BOILER:					
Boiler, sheets, rivets and stay-bolts	......	6,299	......	......	......
Boiler-braces, crown-bars and stay-rods	......	1,635	......	......	1,010
Tubes and copper thimbles	......	3,034	......	......	128
1 brass ring for dry-pipe, riveted to tube sheet, 16 rivets for same	40	8	......	......	17
5 hand-hole plates, 5 holders, 5 bolts, nuts and washers, 5 pieces of rubber	......	15	17	......	25
1 furnace-door, hinges and chain	......	67	......	......	29
DOMES, ETC.:					
1 cast-iron dome	......	......	419	......	1
20 studs and nuts for same	......	12	......	......	40
1 top for front dome	......	......	171	......	1
14 studs and nuts for same	......	7	......	......	28
Whistle stud	......	5	......	......	1
1 set brass ornaments for back dome	35	......	......	......	4
1 set brass ornaments for front dome	39½	......	......	......	3
1 dome-base for back dome	......	......	112	......	1
1 dome-base for front dome	......	......	143	......	1
SAFETY-VALVES, ETC.:					
2 safety-valves	3	......	......	......	2
2 safety-valve seats	10	......	......	......	2
2 safety-valve levers and 4 fulcrums, 4 pins and 2 keys	......	18	......	......	14
2 spring balances and swivel	7½	8	......	......	32
2 cast-iron bushings in roof of cab	......	......	7	......	2
1 whistle	23	......	......	......	9
DRY PIPES:					
2 copper receiving-pipes	......	......	......	12	2
2 wrought-iron rings for same	......	12	......	......	2
4 wrought-iron yokes and 4 set-screws	......	32	......	......	8
2 cast-iron saddles	......	......	82	......	2
1 wrought-iron dry-pipe, 2 eye-bolts, and nuts for throttle stem	......	53	......	......	5
1 casting for back end of dry-pipe	......	......	11	......	1
1 wrought-iron ring for same	......	6	......	......	1
2 brass saddles	27	......	......	......	2
1 joint-ring at smoke-box	11	......	......	......	1
THROTTLE-VALVES, ETC.:					
Throttle-valve box and cover	......	......	193	......	3
8 studs and nuts and 8 bolts and nuts for do	......	18	......	......	32
1 brass stuffing-box for do	7½	......	......	......	1
1 brass gland for do	5½	......	......	......	1
2 bolts and 4 nuts for do	......	2	......	......	6
1 throttle-lever, back end of stem, screw and nut, 3 pins and keys, 2 links, 1 fulcrum	......	21	......	......	13
1 throttle stem	......	69	......	......	1
STEAM AND EXHAUST PIPES:					
2 steam-pipes	......	......	254	......	2
4 bolts and nuts for same	......	6	......	......	8
4 brass joint-rings for same	9½	......	......	......	4
2 exhaust-pipes	......	......	116	......	2
4 bolts and nuts, 2 sets screw for do	......	5	......	......	10
2 nozzles	......	......	7	......	2
PETTICOAT-PIPE, ETC.:					
1 Petticoat-pipe, 6 pieces, 3 braces	......	38	......	......	9
3 hangers, 6 bolts and nuts for same	......	14	......	......	15
BLOWER:					
1 blower-valve	16	......	......	......	3
1 stem and fulcrum, 3 nuts, 1 lever	......	6	......	......	5
1 rod, 1 guide, 1 handle	......	22	......	......	4
Pipe	......	10	......	......	1
SMOKE-BOX DOOR, ETC.:					
Smoke-box front, 2 pieces	......	......	228	......	2
29 studs and nuts for same	......	11	......	......	58
Smoke-box door	......	......	108	......	1
2 handles, 2 nuts, 2 pins and keys for hinges, 2 clips, number plate, 3 studs and nuts	......	34	......	......	17
1 brass ring for number-plate	9	......	......	......	1
SMOKE-STACK:					
Main pipe of smoke-stack, 11 pieces iron, 1 piece half-round iron, 5 bolts and nuts, 3 braces	......	325	......	......	25
Inside-pipe, 6 pieces iron, 3 rods, 6 nuts, 15 bolts and nuts, 3 brackets, bolts and nuts	......	117	......	......	54
1 deflector, 1 spark spout and cover	......	......	72	......	3
1 base for smoke-stack	......	......	128	......	1
6 bolts and nuts for same	......	10	......	......	12
GRATES, ASH-PAN, ETC.:					
6 grate-bars	......	......	504	......	6

APPENDIX F

Number and Material of Parts of a Locomotive and Weights of Each.

	Brass lbs.	Wro't Iron lbs.	Cast Iron lbs.	Wood etc., lbs.	No. of p'c's
2 side-plates			335		2
1 drop-door			155		1
2 rock-shafts			173		2
6 pins and keys, 1 pin and nuts		5			14
4 bearings for rock-shafts			80		4
8 bolts for same		10			8
2 bearings and bolts for side-plates		41			4
1 bearing-bar for drop-door			86		1
4 bolts for same		4			4
1 bar for shaking grate		37			1
1 lever for shaking grate, 2 bolts and nuts, and 2 pins and keys		32			9
1 fulcrum for shaking grate		24			1
1 handle for shaking grate		22			1
1 shaft for drop-door		54			1
1 lever for drop-door, handle and key		18			3
1 spring for drop-door, 1 bolt and nut		3			3
1 ash-pan, 4 sheets, 5 pieces angle iron, 2 doors, 4 slides, 2 handles		180			17
1 back door, hinges, 4 pieces		21			5
2 straps for ash-pan, 4 bolts and nuts		38			9

Boiler Braces:

	Brass lbs.	Wro't Iron lbs.	Cast Iron lbs.	Wood etc., lbs.	No. of p'c's
2 braces from bumper to smoke-box, 4 bolts and 4 bolts and nuts		99			14
2 braces from back end of frame to boiler, 12 bolts and 4 nuts		132			18
1 brace across frames in front of boiler, 3 studs and nuts		81			7
1 brace across frames behind boiler, 4 bolts		66			5
2 front braces from bed-casting to boiler, 8 bolts for same		59			10
1 angle-iron for back-brace on boiler		16			1
2 back-braces to boiler		92			2
2 front braces to boiler		79			2
4 bolts and nuts for same		4			8

Frames, Etc.:

	Brass lbs.	Wro't Iron lbs.	Cast Iron lbs.	Wood etc., lbs.	No. of p'c's
2 frames, 4 pieces, 4 braces for bottom of jaws, 8 bolts, 16 nuts for same, 10 bolts and nuts for bolting frame together, 4 keys		2,932			56
1 brace across frames behind smoke-box		41			1
1 brace across frames in front of fire-box, 4 bolts		43			5
1 brace across frames behind back jaws		40			1
1 back end-rail		198			1
2 centre lugs (on fire box) for expansion braces		82			2
2 outside lugs for expansion braces		58			2
4 end lugs for expansion braces		108			4
52 bolts and nuts for expansion braces		36			104

Bed-Casting, Etc.:

	Brass lbs.	Wro't Iron lbs.	Cast Iron lbs.	Wood etc., lbs.	No. of p'c's
1 bed-casting			908		1
1 piece angle-iron for same, 12 bolts and nuts		28			25
2 corner pieces, 2 bolts and nuts		1	4		2

Cylinders, Etc.:

	Brass lbs.	Wro't Iron lbs.	Cast Iron lbs.	Wood etc., lbs.	No. of p'c's
2 cylinders			2,393		2
2 front cylinder-heads			164		2
2 back cylinder-heads 2 glands			328		4
32 studs and nuts for same		29			64
2 brass bushings for glands and 2 for stuffing-boxes	5				4
4 studs and 8 nuts for stuffing-boxes		5			12
2 front cylinder-head covers and nuts	73				4
2 back cylinder-head covers and 8 screws	32				10
22 bolts and nuts for fastening cylinder		31			44
Wood-lagging				120	
2 pieces Russia-iron casing				16	2

Cylinder Cocks, Etc.:

	Brass lbs.	Wro't Iron lbs.	Cast Iron lbs.	Wood etc., lbs.	No. of p'c's
4 cylinder-cocks	10				16
4 pins, keys and washers, 2 rods		6			10
1 cock-shaft, 2 lower arms, 1 upper arm, 1 rod, 2 pins and nuts, 5 keys		29			14
2 stands for cock-shaft			12		2
1 rod for shaft, guide, handle		22			4

Steam-Chest, Etc.:

	Brass lbs.	Wro't Iron lbs.	Cast Iron lbs.	Wood etc., lbs.	No. of p'c's
2 steam-chests, 2 covers, 2 glands			623		6
2 bushings	1				2
6 studs, 12 nuts		3			18
24 studs and nuts for steam-chests		46			48
2 oil-cocks for steam-chests	15				8
2 outside covers for steam-chests			162		2
2 pieces Russia-iron for steam-chests		8			2

APPENDIX F

Number and Material of Parts of a Locomotive and Weights of Each.

	Brass lbs.	Wro't Iron lbs.	Cast Iron lbs.	Wood etc., lbs.	No. of p'c's
2 brass mouldings and 14 screws for steam-chests	20	1	......	......	16
2 name-plates, 16 screws	......	1	20	......	18
4 end plates, 20 screws	......	21	......	......	24
VALVES, ETC.:					
2 main slide-valves	......	......	132	......	2
2 main slide-valve yokes	......	80	......	......	2
2 brass guides for same	18	......	......	......	2
6 bolts and washers for same	......	2	......	......	12
PISTONS, ETC.					
2 piston-heads	......	......	138	......	2
4 packing-rings, 2 follower-plates	......	......	124	......	6
8 bolts and 2 pins	......	4	......	......	10
2 piston-rods	......	118	......	......	2
2 piston-rod keys and pins	......	5	......	......	4
CROSS-HEADS, GUIDES, ETC.:					
2 cross-heads	......	......	209	......	2
8 guide-bars, 4 lugs, 3 bolts, 4 nuts	......	509	......	......	19
8 cast-iron washers	......	......	31	......	8
2 brackets for guides, 4 bolts and nuts	......	80	......	......	10
2 bottom braces for same	......	20	......	......	2
2 braces to boiler, 4 studs and nuts	......	72	......	......	10
4 oilers for slides	12	......	......	......	4
CONNECTING-RODS, ETC.:					
2 main-rods, 4 straps, 4 gibs and keys, 6 bolts, 12 nuts, 4 set-screws, 4 bolts for oil-cellar	......	320	......	......	40
4 pairs brass boxes and 2 stoppers for oil-cellar	60	......	......	......	10
2 oil-cellars for same	......	......	14	......	2
2 parall[e]l rods, 4 straps, 16 bolts, 32 nuts, 4 keys, 4 set-screws, 4 steel plates	......	289	......	......	66
4 pairs brass boxes and 4 stoppers for oil-cellar	57	......	......	......	12
4 oil-cellars for same	......	......	28	......	4
CRANK-PINS, DRIVING-WHEELS AND AXLES:					
1 front driving-axle, 4 keys	......	540	......	......	5
1 back driving-axle, 4 keys, 4 sets screws	......	493	......	......	9
2 collars for driving-axle	......	......	107	......	2
4 driving-wheels	......	......	5,324	......	4
4 tires and 16 bolts and nuts	......	3,360	......	......	36
4 crank-pins and nuts {2 front, 80 / 2 back, 86}	......	166	......	......	8
Lead for counter-balances	......	......	......	640	4
DRIVING-SPRINGS, ETC.:					
4 driving-springs	......	416	......	......	44
4 straps, 4 rivets for same	......	30	......	......	8
2 long hangers, 2 pins and keys	......	62	......	......	6
2 short hangers, 2 pins and keys	......	42	......	......	6
4 short hangers, 4 keys	......	44	......	......	8
4 stirups and pins for driving-boxes	......	72	......	......	8
4 castings and rubber for hangers	......	50	......	10	8
2 equalizing-beams for engine	......	189	......	......	2
2 fulcrums for do	......	17	......	......	2
DRIVING-BOXES:					
4 driving-box housings and oil-cellars	......	......	430	......	8
4 driving-box brasses	112	......	......	......	4
8 bolts and keys for same	......	6	......	......	16
VALVE-GEARING:					
4 eccentrics	......	......	300	......	4
8 set-screws	......	3	......	......	8
4 pairs of eccentric-straps	......	......	316	......	8
13 bolts and 32 nuts for same	......	16	......	......	48
4 eccentric-rods	......	228	......	......	4
8 bolts for same	......	4	......	......	8
2 links (4 pieces), 2 saddles, nuts and washers, 2 hangers, 2 blocks, 4 bushings, 2 rocker-pins, nuts and washers; 4 eccentric-rod pins, nuts and washers; 2 hangers, pins and nuts; 12 bolts and nuts; 4 thimbles	......	152	......	......	70
1 lifting-shaft	......	185	......	......	1
2 hangers for same, 4 bolts and nuts, 6 screws	......	69	......	......	16
1 counter-balance spring-case	......	......	42	......	2
1 counter-balance spring	......	16	......	......	1
1 bolt for chain, 2 bolts and nuts for fastening to frame, 1 centre-bolt, nut and washer	......	5	......	......	8
1 counter-balancing chain, 1 bolt	......	9	......	......	2
2 rockers	......	......	136	......	2
2 top pins, nuts and washers	......	7	......	......	6
2 rocker-boxes	......	......	182	......	2

APPENDIX F

Number and Material of Parts of a Locomotive and Weights of Each.

Part	Brass lbs.	Wro't Iron lbs.	Cast Iron lbs.	Wood etc., lbs.	No. of p'c's
10 bolts and nuts, 2 bolts and washers		6			24
2 valve-stems, 2 straps, 2 keys, 2 gibs, 2 bolts and nuts, 2 set-screws, 2 knuckle-joints, 2 units, 2 pins and nuts		40			24
2 pair of boxes for same			8		4
Reverse-Lever, Etc.:					
Reverse-lever handle, 2 bolts and nuts, 1 catch, 1 spring, 1 spindle, 1 stop, 1 guide, 1 gib, 2 set-screws		54			14
1 brass thumb-piece	2½				1
Reverse-lever fulcrum			13		1
1 pin, 2 nuts, 2 bolts and nuts for do.		5			7
2 quadrants, 2 bolts and nuts		39			6
1 support for quadrant		26			1
1 reverse-rod, 2 pins and keys		94			5
1 casting for end of quadrant			6		1
3 bolts and 1 nut		8			4
Pumps:					
2 brass pumps and glands	243				4
4 studs and 8 nuts		2			12
4 pump valves and cages, and pipe for air-chamber	41				10
2 brass air chambers for pumps	80				2
6 bolts and nuts and 12 washers		4			24
2 lower air-chambers for pumps			144		2
6 bolts and nuts and 12 washers		4			24
2 feed-pipes from pump to check-valve, 4 brass nuts, 4 brass collars	50				10
2 pet-cocks	2				6
2 rods, etc.		18			8
2 pump-plungers, 4 nuts and 2 pins		61			8
1 right-hand feed-pipe, 1 brass thimble, 1 old hat, 1 gland	2	41	6		4
1 left-hand feed-pipe, 2 pieces, 3 brass thimbles, 1 gland, 1 old hat	6	37	6		7
1 T piece for same			7		1
1 lazy cock	12				4
1 handle for same, pin-plate, piece angle iron, 4 bolts, 2 washers, 2 nuts, 1 lower lever, 2 bolts, 1 set-screw, 1 bracket, 1 rod, 1 lever		56			19
2 hangers for feed-pipes, 2 bolts and nuts, 1 piece wood		30			12
Pump Check-Valves:					
2 check-chamber bases and tops			100		6
6 bolts and nuts, 8 studs and nuts for same		11			28
2 valve cages and seats for do	28				6
2 brass acorns, studs and brass casing	11	1			6
Injector, Etc.					
1 injector	12		16		17
1 casting for same			19		1
4 bolts and nuts for same, 1 small shaft, crank and handle, 2 nuts, 1 large shaft, 1 wheel, handle and pin		18			17
1 brass bushing	2				1
1 steam valve for injector	8	2			12
1 copper steam-feed-pipe for injector, 2 nuts and 1 thimble	16				4
Waste and water feed-pipes for injectors, 3 nuts and 3 thimbles	8				8
1 copper pipe from injector to check (2 pieces), 2 nuts, 2 collars	27				6
1 injector check-chamber, 4 studs and nuts		1	15		9
1 injector check-valve cage and top	5				3
Engine Truck:					
4 wheels			1,884		4
2 axles		604			2
4 cast-iron collars, 8 set-screws		3	66		12
4 housing-boxes and oil-cellars			223		8
4 brasses for same	72				4
8 bolts and keys		5			16
8 cast-iron jaws			346		8
1 truck frame		813			1
16 bolts and nuts for top of jaws		24			32
2 bottom braces		122			2
8 bolts and nuts for same		12			16
2 truck springs (17 plates)		337			34
2 straps, 2 pins, 2 rivets for do		26			6
2 equalizers		229			2
4 spring-hangers, 4 pins and keys, 4 stirrups, 4 pins and keys		126			24
4 castings for same			34		4
2 long braces for top		141			2
12 bolts and nuts for same		18			24
2 transverse braces		40			2

APPENDIX F

Number and Material of Parts of a Locomotive and Weights of Each.

	Brass lbs.	Wro't Iron lbs.	Cast Iron lbs.	Wood etc., lbs.	No. of p'c's
1 transverse brace	......	80	......	......	1
2 U bolts, 4 nuts for same	......	17	......	......	10
1 centre-pin and key	......	17	......	......	2
1 lower-centre casting	......	......	165	......	1
1 top-centre casting	......	......	200	......	1
3 bolts and nuts for same	......	8	......	......	6
2 I castings	......	......	24	......	2
Check-chains:					
4 check-chains and eyes	......	68	......	......	8
eyes, 4 bolts and nuts on truck frame for chains	......	41	......	......	10
2 hooks, bolts and nuts for do	......	22	......	......	6
Sand-box:					
1 base, 2 links, 2 valves, 2 cast-iron levers	......	......	218	......	7
1 wrought-iron lever, 10 bolts, 1 nut, 2 rods, 1 pin	......	12	......	......	15
1 sheet-iron cylinder for sand-box, 2 pieces Russia-iron, 1 welt	......	84	......	......	4
1 cover	......	......	65	......	1
2 brass moldings (top and bottom)	25	......	......	......	2
1 brass top	18	......	......	......	1
2 sand-pipes, 4 bolts and washers	......	48	......	......	10
2 brass lugs	2	......	......	......	2
Bell and Stand:					
1 bell	80	......	......	......	1
1 bell clapper	......	7	......	......	1
1 nut and eye for same	......	2	......	......	2
2 bell posts and yoke	68	......	......	......	3
1 lever, 2 bushings, 2 pins	......	4	......	......	5
1 base for bell-stand	......	......	52	......	1
2 pieces wrought-iron, 2 bolts	......	4	......	......	4
Hand-rail:					
4 pieces brass pipe, 4 acorns, 12 pins, 4 splices	63	......	......	......	24
8 supports and studs	......	44	......	......	16
Running-board:					
46 bolts and nuts, 2 plates, 6 washers	......	27	......	......	100
2 pieces plank for back end, 2 pieces plank for front end	......	......	......	334	4
4 pieces brass for same	60	......	......	......	4
8 brackets, 16 bolts	......	74	......	......	24
Wooden molding	......	......	......	......	8

	Brass lbs.	Wro't Iron lbs.	Cast Iron lbs.	Wood etc., lbs.	No. of p'c's
Cab, Foot-board, Etc.:					
1 cab, 22 bolts, nuts and washers	......	10	......	725	833
1 back foot-plate, 2 bolts	......	156	......	......	3
4 pieces flooring for same	......	......	......	90	4
2 side plates, 6 pieces angle-iron, 30 bolts, 12 nuts	......	88	......	......	50
2 cab-brackets	......	......	256	......	2
2 posts for foot-steps, 1 nut, 3 bolts and nuts	......	20	......	......	9
2 foot-steps and set screws	......	1	26	......	4
2 hand-holes for cab, 6 pieces, 4 bolts	4	10	......	......	10
6 sheets iron for front of cab, 10 bolts and nuts for do	......	38	......	......	26
Spring boxes	......	67	......	......	34
1 pulling-casting	......	......	164	......	1
1 wrought-iron piece, 4 bolts and nuts	......	24	......	......	9
1 bumper-timber	......	......	......	268	1
Tin roof and nails	......	20	......	......	
2 pieces pipe for roof	......	4	......	......	2
2 coupling-pins, handles and keys	......	31	......	......	6
1 draw-bar	......	96	......	......	1
Wheel-Covers:					
4 pieces sheet-iron for driving wheel covers, 8 brackets, 28 bolts and nuts, 4 pieces sheet-iron for truck wheels, 8 brackets, 26 bolts, 16 nuts	......	317	......	......	122
2 name-plates between drivers	......	......	120	......	2
6 brackets, 6 bolts and nuts	......	20	......	......	18
4 pieces brass-pipe for driving wheel covers, 8 acorns	115	......	......	......	12
4 pieces brass-pipe for truck, 6 wheel covers, 6 acorns	24	......	......	......	10
Cow-catcher, Etc.:					
Wooden parts of catcher	......	......	......	320	17
1 brace, 4 bolts, nuts and washers, 24 joint-bolts, nuts and washers, 1 shoe, 10 bolts, nuts and washers	......	170	......	......	116
1 large and two small braces	......	190	......	......	3
1 front bumper-casting	......	......	77	......	1
2 wrought-iron plates and 3 bolts and nuts	......	69	......	......	8
1 pin and key for pushing-bar	......	4	......	......	2
1 pushing-bar	......	63	......	......	1

APPENDIX F

Number and Material of Parts of a Locomotive and Weights of Each.

	Brass lbs.	Wro't Iron lbs.	Cast Iron lbs.	Wood etc., lbs.	No. of p'c's
1 front foot-plate		121			1
2 pieces sheet-iron for edge of front foot-plate, 2 pieces angle-iron, 18 bolts, 5 rivets		30			27
10 bolts, nuts and 4 washers		40			24
Head-light Brackets, Etc.:					
2 cast-iron brackets			68		2
6 bolts for same		3			6
1 bolt for lamp stand				49	1
3 pieces brass for same	7				3
6 brass posts for same	3				6
6 bolts and nuts for same		8			12
Steam-gauges, Cocks, Etc.:					
1 steam-gauge	18				46
4 gauge-cocks	8				16
2 heater-cocks	8				8
2 heater-cock pipes	12				2
1 back blow-off cock	10				4
1 side blow-off cock	9				4
1 surface-cock and nut	8				7
1 rod for opening side blow-off		10			2
1 rod for opening end blow-off		8			2
Pipe for surface-cock		8			1
Sundries:					
2 pieces rubber-hose, 4 brass thimbles and nuts, 2 pieces iron, 4 clamps, 4 bolts and nuts	16	18		17	22
2 number-plates, 2 pieces of wood, 4 pieces brass, 6 figures	18			17	14
1 brass bracket, set screw and rod	6				8
Sheet-iron casing		269			
6 Brass bands	48				6
12 knees for same, 6 bolts and nuts		32			24
2 brass flag-stands	26				8
Wood lagging on boiler				84	
Nails for wood lagging on boiler		10			
	1,948	29,627	19,785	2,718	4,904
Total weight of engine			54,078 lbs.		

TENDER, ETC.

	Brass.	Wr'ght. Iron.	Cast iron.	Wood. etc.	No. of pie's.
Tank		3,515			238
2 spouts			32		2
2 cast-iron rings			9		2
12 bolts and nuts for same		6			24
4 rubber gaskets				3	4
2 valves	6				2
2 strainers			36		2
2 stems, pins, keys and nuts		20			3
2 handles			8		2
2 top castings			18		2
4 bolts and nuts		2			8
2 handles, 4 bolts		9			6
4 brasses for same	5				4
1 cock	8				4
Tender-frame, Etc.:					
2 outside longitudinal timbers, 4 by 8 in				177	2
2 inside longitudinal timbers 4 by 8 in				166	2
3 inside longitudinal timbers 3 by 8 in				124	2
1 front piece, 5 in. thick by 15 in				170	1
1 front frame piece 5 by 8 in				90	1
2 top bolsters and truck pieces 4 by 15 in				379	2
1 back frame piece 7 by 8 in				137	1
1 piece for hanging brakes 7 by 3 in				51	1
1 front bumper, cast iron		67			1
4 bolts, nuts and washers for do		8			12
6 bolts, nuts and washers for front timber		10			18
4 corner castings			72		4
1 front pulling-casting			223		1
1 coupling-pin and key for do		11			2
8 bolts and nuts and 4 washers for do		16			20
2 rods, bolts, nuts and washers for holding down tank		46			8
8 bolts, nuts and washers in front corner casting		12			24
2 bolts and nuts for engine check-chain, 2 chains, 2 eyes		31			4
8 bolts, nuts and washers in truck timbers		24			24

APPENDIX F

Number and Material of Parts of a Locomotive and Weights of Each.

	Brass.	Wr'ght. Iron.	Cast iron.	Wood. etc.	No. of pie's.
2 top centre castings			304		2
8 bolts and nuts for same		25			16
4 bolster truss-rods, 8 nuts, 4 plates		122			16
4 bolts, nuts and washers for holding down tank		13			12
2 long longitudinal-rods, 4 nuts and washers		100			10
10 bolts, nuts and washers for brake-timber		25			30
2 eyes, nuts and washers for brake-hangers		8			6
12 bolts nuts and washers for end of bolster		30			36
4 bearing-plates for springs		78			4
8 bolts, nuts and washers for do		28			4
2 back end corner-plates		50			2
16 bolts and nuts for same		17			32
2 eyes for brake-hangers		8			2
4 bolts and nuts for same		4			8
2 long bolts nuts and washers for holding tank		22			6
2 back bumper-timbers, 2 keys				112	4
8 bolts, nuts and washers for do		30			24
2 end plates for do		12			2
4 castings for buffer			84		4
12 bolts, nuts and washers for do		33			36
1 buffer, 4 rivets		100			5
1 casting in buffer			7		1
1 piece wood in buffer				6	1
2 pieces wrought-iron		37			2
1 bolt and nut		4			2
1 rubber spring				7	1
1 coupling-pin, handle and key		15			3
2 back-end bearing pieces for buffer		49			2
4 bolts and nuts for same		14			8
2 thimbles		4			2
1 piece wood				2	1
1 piece sheet iron		2			1
2 centre-pins end keys		20			2
2 steps			32		2
4 hangers for same, 8 bolts, 8 nuts and 8 washers for do		40			24
18 pieces flooring				778	18
Nails for same		6			
13 pieces wood molding around tank				128	13
Nails for same		3			

	Brass.	Wr'ght. Iron.	Cast iron.	Wood. etc.	No. of pie's.
Tender Trucks:					
8 wheels			3,952		8
4 axles and collars		1,272			8
8 boxes			536		8
8 journal-bearings	64				8
8 leathers				6	8
8 leather holders, and bolts and nuts		20			24
8 box-covers			32		8
8 box-leathers				6	8
16 bolts for same		6			16
8 bolts for dirt-holes		3			8
4 springs, 16 plates		880			64
4 spring straps and rivets		108			8
4 bolts nuts and keys through strap		30			12
8 shoes			96		8
4 castings under springs			500		4
24 bolts and nuts through do		102			56
4 top pieces of truck frame		312			4
4 braces of truck frame		248			4
4 bottom pieces of truck frame		227			4
4 castings under truck frame			256		4
16 jaw castings			520		16
16 long bolts and 32 nuts for do		78			48
24 short bolts and nuts for jaws		40			48
8 check chains (5 links each)		68			40
2 bottom bolster timbers				670	2
4 truss-rods and 8 nuts		96			12
2 bottom centre-castings			150		2
8 bolts nuts and washers for do		21			24
2 truss castings			74		2
Brakes:					
2 brake-beams			70		2
4 brake blocks			86		4
8 castings for same			20		8
12 bolts and nuts and 4 washers		12			28
4 brake-hangers					4
4 bolts and nuts		18			8
4 brake-chains					4
8 eyes, nuts and washers		14			23
1 brake-lever		10			1
1 fulcrum, nut, pin, key and washer		3			5
1 lower rod, 2 pins and keys		13			5
2 straps, bolts and nuts		8			6
1 upper rod		21			1
1 brake-chain		5			1
1 brake shaft, 2 nuts		17			3

APPENDIX F

Number and Material of Parts of a Locomotive and Weights of Each.

	Brass.	Wr'ght. Iron.	Cast iron.	Wood. etc.	No. of pie's.
1 lower bracket for shaft, 2 bolts and nuts	......	14	......	......	5
1 top bracket, 3 bolts and nuts	......	8	......	......	7
1 ratchet-wheel, pin and pawl	......	......	11	......	3
1 brake-wheel	......	......	15	......	1
	83	8,330	6,987	3,118	1,366

Total weight of tender ....18,518 lbs.
    "   "   "   engine ....54,078 "
    Total weight of engine and tender ....72,596 lbs.
Total number of pieces in engine and tender............................ 6,270

APPENDIX G

DESCRIPTION OF A STANDARD GRANT LOCOMOTIVE WORKS 4-4-0, 1871*

The following description of a locomotive is taken from a volume published by the Grant Locomotive Works, as an excellent specimen of the detailed manner in which specifications should be made:

BOILERS.

The outside shell of the boiler is made with either a straight or wagon-top, as desired. Each course of the barrel of boiler is formed of a single sheet. The side sheets of outside shell of fire-box join the crown sheet of shell 1½ inches above the crown of fire-box.

Extra plates of iron are riveted to the inside of the side sheets, where the expansion braces are attached to give double thickness of metal for the studs. All the horizontal seams in the shell of boiler, and the seam which joins the barrel of the boiler to the fire-box shell, are double riveted. All sheets ⅜ inch thick are riveted with ¾ inch rivets, spaced two inches from center to center; 5-16 and ¼ inch sheets have ⅝ inch rivets spaced 1⅞ inches.

* From *Railroad Gazette,* May 27, 1871.

FIRE-BOX

at the bottom is as wide as possible, allowing sufficient water space, but is swelled out after passing the frames.

The flue and back sheets have flanges two and one-half inches wide turned on the sides and top; the crown sheet has flanges on the sides; the side sheets have no flanges. The space between inside and outside sheets at the bottom of fire-box is filled by a solid wrought-iron ring, two inches thick, and riveted to both sheets by long rivets.

The stay bolts are made of Low Moor iron ⅞ inch diameter, tapped through both sheets and riveted on both ends. They are spaced as near four inches from center to center as practicable. All fire-boxes are made of homogeneous steel plates unless otherwise ordered.

The fire-box door is made by flanging the back sheet of fire-box into the water space, and riveting a flanged sheet on outside of shell; the two flanges are connected by a welded ring of plate iron. The door is of cast iron and has an inside lining of cast iron perforated to admit air, the supply of which is regulated by a register on the outside.

APPENDIX G

CROWN BARS, BRACES, ETC.

The crown sheets are supported by bars formed of two pieces 4x¾ inches welded at ends. They are placed across, and ¾ inch above the crown sheet, each end having a lip turned down and resting on the edge of the side sheets. The bars are placed 5 inches from center to center, and are fastened to the crown sheet by T head bolts riveted on the under side of crown sheet. Each crown bar is connected to the outside shell by two braces bolted to crow feet.

The domes are also braced to the crown bars, by four braces. The back end and front tube sheet are braced by longitudinal braces. In wagon-top boilers the sides of throat are braced by angle irons riveted to them and connected by suitable braces.

SMOKE-BOXES

are circular, and have a solid wrought iron ring riveted to the front end to receive a cast iron front. The smoke-box door is circular, hinged on the side and fastened by four hook bolts held in position by lock nuts.

TUBES

are put in vertical rows. The two middle rows are arranged so that the tubes in them come opposite to each other. Iron tubes are set with copper ferrules at the fire-box end.

MUD HOLES.

An elliptically shaped mud hole, 2½x3¾ inches, is placed on each outside corner of the fire-box near the bottom ring. Each hole is covered with two cast iron plates, one inside and the other outside the boiler, and fastened with a bolt which passes through both.

BLOW-OFF COCKS.

A brass blow-off cock is placed in the back leg of the boiler, underneath the foot board, and arranged so it can be opened from the cab.

LAGGING.

The boilers are lagged with ⅞ inch pine, and sheathed with Russia iron. The latter is held on with brass bands, drawn tight with bolts and nuts. The domes are also lagged with ⅞ inch pine, covered with Russia iron. Each dome has a cast iron base, and an ornamental molding made of sheet iron or brass at the top and bottom.

FRAMES.

The frames and jaws are forged solid, of hammered scrap. The jaws, top of pedestal and feet in the jaws, are all forged out of one shape, which is bent so as to form each part. The top bar of the frame is then welded on at the point where the jaw and pedestal unite, and the bottom braces are welded to the feet.

Both sides of the frames are planed to a gauge, so as to make the thickness the same through their whole length. The jaws are all made tapered to receive the wedges. The form of the jaws, etc., is laid off from a template. Each pair of frames are bolted together and finished on a slotting machine, to the form and size of the template. The distance from the back face of the cylinder pocket to the center of the main driving-box is accurately gauged by a steel rod. All the holes are drilled from a cast iron gauge or template which is bolted to the frame.

A brace extends across the mouth of the jaws, from one foot to the other, and is held by a lug on each foot, which is let into a corresponding slot in the brace. The lugs and the slots to receive them, are both planed to gauges which fit to each other. The braces are bolted to the feet, and each bolt is secured with lock nuts.

The frames of all engines which have a four-wheeled truck are spliced in front of the front driving-wheel. For this purpose the frame is made with two braces, one welded to the top bar, and the other to the foot of the front jaw, and inclined towards each other, and left open to receive the front bar. The braces are bolted to the bar with bolts which pass through all three. The front bar has a ⊢ shaped lug on the end, which is bolted to the front jaw. The frames of engines without four-wheeled trucks are made solid, i.e., the front bar is welded to the back end of frame.

A recess or pocket is made in the front ends of the frames to receive the cylinders, which are keyed in with a key driven in in front of the cylinder.

WEDGES.

The forward jaw of each pedestal has a cast iron dead wedge fitted snugly between the top of pedestal and bottom brace. This wedge is bolted securely to the jaw with two bolts. The

APPENDIX G

back jaw of each pedestal has a cast iron movable wedge, which can be adjusted by a bolt tapped into the bottom brace, and secured with a lock nut underneath. The wedge is held in position by a movable bolt, which passes through a slot in the jaw. The faces of each pair of wedges are laid off on the frame by a gauge, so as to be square with the top of the frame and parallel with each other.

EXPANSION BRACES.

A wrought-iron expansion plate of an L shaped section is bolted on each side of the fire-box. The bottom flange of this plate rests upon the frame. On top of the flange of the expansion plate is another wrought iron plate, with a lug at each end, which rests on the frame and is bolted to it. The latter plate has round holes, and the expansion plate has oblong holes to receive the expansion bolts which pass through the frame. Each bolt has a washer on it of the same thickness as the flange, and received by the oblong holes. The top plate is then bolted down hard on these washers, and is stationary on the frame, while the expansion plate moves with the boiler. When a pair of drivers is behind the fire-box, a wrought iron clamp is bolted outside both the expansion plate and the frame. This clamp holds the fulcrum for the equalizing lever, and runs down with a foot and a bracket. The foot is bolted to the fire-box, while the bracket rests on the bottom bar of the frame, and has oblong holes the same as the expansion plate, with washers, and a top plate bolted down hard on the washers. The top bar of frame is fastened to the fire-box by additional wrought iron clamps, and the bottom bar by wrought iron brackets where they are necessary.

FRAME BRACES.

A tail brace made of hammered iron extends across, and is let into the back end of each frame. It has a ⌂ shaped piece forged on it, to which the draw casting is bolted. All engines, with the exception of some anthracite coal burners, with long fire boxes, have two braces made of rounded rolled iron bolted to the tail braces next the frames, and extend upward at an angle of about forty-five degrees to the shell of the fire box, to which they are bolted with a suitable foot.

A wrought iron cross brace is bolted to each frame in front of the main driving axle. An angle iron, made of boiler plate, is bolted to this brace, and riveted to the boiler. The angle iron has a projection or lug to hold the counterbalance spring.

The frames which are spliced have a brace, the foot of which is held by two of the splice bolts, and the other end is riveted to the shell of the boiler.

The front ends of the frames are braced to the smoke-box by a brace made of round rolled iron, and bolted to the top of the bumper timber and the frames by a foot with a lip, which locks against a corresponding lug on the frame. The foot has two bolts which pass through the timber and frame. The upper ends of the braces are bolted to the smoke-box by a suitable foot.

DRIVING SPRINGS

are made of cast steel. Each plate is punched in the center so as to make a cavity on one side and a projection on the other, which prevents the plates from slipping. A band is then shrunk on the center of the spring. The spring hangers are made of hammered iron. The ends which are attached to the frame are hung on rubber seats. The saddles for driving springs are made of cast iron, and rest in sockets cast for the purpose in the driving boxes.

EQUALIZING LEVERS.

All engines, with the exception of those with only four wheels, have suitable equalizing levers between the driving springs.

CYLINDERS

are all horizontal, cast of Lake Superior charcoal iron: one-half the saddle, the steam and exhaust pipes are cast on each cylinder. Both cylinders are alike, and are bolted together at the center of the engine. The smoke-box is round, the cylinders are fitted and securely bolted to it. The cylinders are bored, and the flanges turned and faced to gauges, then planed parallel with the bore, and the face of one where they join each other, drilled through a template. They are placed together, leveled by the frame seat, and the second cylinder marked off and drilled, then clamped together and the holes reamed, and the cylinders bolted together. The center casting for the truck and its seat are planed and drilled through a template and bolted in position.

The ports in the valve seat are milled to size by a cutter from a gauge which is bolted on the valve seat. The cutter works in a block which slides in slots in the gauge, corresponding to the ports. All the holes for steam-chest studs, cylinder-head, and frame bolt are drilled through templates. The templates are made of cast iron, and have the holes in them bushed with hardened steel.

CYLINDER-HEADS

are the same diameter as the flanges of the cylinders, and are fitted thereto with scraped joints. The front heads have a grooved turned in them next to the counter-bore, to protect the cylinders in case of accident. The flanges of the stuffing-boxes on the back heads are circular to receive the castings. Both the stuffing-box and gland have brass bushings fitted to the piston rods. The gland studs are case-hardened and provided with lock nuts.

STEAM-CHESTS

are made with pockets on the inside for the bolts, the covers have ribs both on the inside and the outside to stiffen them. The joints between the chest cover and cylinder are made with copper gaskets.

All the holes in the chest and covers are drilled through the same template that is used for the cylinder.

VALVES

are of cast iron, and have their faces scraped to the valve-seat. The stem is attached to the valve by a yoke which embraces the valve. The stem is connected to the valve rod by a socket joint.

PISTONS.

The heads and followers are made of cast iron, are turned to gauges, and fitted with brass packing rings. The piston rods are made of cast steel and are fitted taper into the pistons and cross-heads and keyed to them.

GUIDES

are made of cast steel, four to each cross-head, and are placed central on the cylinder head. Each pair of guides is bolted at each end to a block, one of which is fastened to the cylinder-head and the other to the guide yoke by a stud and nut. The guide blocks are faced off in a special chuck (which receives a full set) to an exact thickness. The guides and blocks are bolted together and planed and finished to a gauge.

The holes in cylinder head and guide yoke which receive the slide block studs, are drilled through a template.

GUIDE YOKES

are made of plate iron, planed and finished to a template. They are bolted to the frames by a lug and to the rocker box. They are fastened to the boiler by an angle iron, and form a bracket for the running board.

CROSS-HEADS

are made of hard cast iron, and are all fitted with glass bearings to prevent wear.

The neck of cross-head is first bored taper to fit the piston rod, the cross-head is then placed on a mandril with a gauge attached to it and planed. The journal for the connecting rod is cast in the cross-head, as is also the lug for driving the pump.

ROCKERS

are made of wrought iron, forged solid, and are finished all over, to gauges. Each arm has a boss on the end, and is furnished with case-hardened taper pins for valve rod, and link.

The rocker boxes are made in two pieces, bolted together, and also bolted to the frame and yoke brace.

VALVE GEARING

is the shifting link motion. The links are forged of Low Moor iron and case-hardened, and are made either solid or skeleton. They are hung in the center vertically, and back of the center horizontally.

The suspension pin is forged on the saddle, and the latter is bolted to the link.

CONNECTING RODS

are made of best hammered iron. The body of the rod is tapered from the front to the back end. The corners are chamfered off, and the rod accurately planed, and finished all over. The front end has a strap and two brasses. The lost motion is taken up with a key placed vertically in the stub end, and secured by double nuts. The key bears against a wrought iron plate. The straps are held by two bolts.*

The back stub end has two brasses, which are held by a strap bolted to the rod with two bolts. The lost motion is taken up with a key, which is secured with lock nuts, and bears against a wrought iron plate.

COUPLING RODS

are made of best hammered iron, planed and finished all over, with the corners of the body of the rod chamfered off. Each of the crank pin journals has two brasses, held with straps bolted to the rods with two bolts. The lost motion is taken up with suitable keys, secured with lock nuts.

LIFTING SHAFTS

are made of wrought iron forged solid. The arms have wide taper bearings for the pins of suspension links. The vertical arm to which the reach rod is attached is curved so as to clear the boiler. The ends are supported by cast iron stands, bolted to the frame.

In the centre of the shaft is a short arm to which is attached a rod which passes through two volute springs, which serve as a counter-balance for the links.

REVERSE LEVER.

The fulcrum is at the frame. The quadrant is made in two parts, case-hardened, and notched to hold the lever in the desired position. The lever is connected with the lifting shaft by the reach rod, which is supported by a bracket, fastened to the running board.

ECCENTRICS

are cast with a boss on one side; they are bored and turned to special gauges. Each eccentric is fastened to the axle with two steel set screws, cupped on the end.

ECCENTRIC STRAPS

are cast iron. They are bored out with a recess to receive the eccentrics; all parts accurately made to standard gauges. The two parts are joined at an angle of forty-five degrees with the center of the eccentric rod. An oil cup is cast on the top of the back half, and an oil cellar on the bottom of the front half.

ECCENTRIC RODS

are bolted with three bolts to the straps, and have a jaw on the front end to take the links. A pin with a steel thimble, which turns in the link, is fitted into the jaw.

DRIVING-WHEELS

The driving-wheel centers are made of cast iron, with hollow spokes and hollow rim. The section of the spokes is elliptical. The hubs for axles and crank pins are cast solid, and are flush with each other on the outside. The wheels are each keyed on the axles with a key, an inch square, and are placed with the right hand crank ahead. They are pressed on by a hydraulic press and the holes for the crank pins bored in a quartering machine. The outsides of the wheels are turned, the hubs bored and faced to gauges for each.

TIRES

are made of steel of an approved manufacture, and are bored out and shrunk on the centers, and secured with 1¼ inch bolts. The bolts are tapped into the rim of the wheel, and their ends are turned down to ⅞ inch diameter, and are fitted into a hole drilled into the tire to receive it. The hole is drilled deeper than the length of the bolts, so that the latter do not bear on the bottom of the hole.

DRIVING AXLES

are made of hammered iron. The main axles are turned their whole length to receive the eccentrics, the others are left rough between the inside collars. The collars are made of cast iron, shrunk on the axles, and form an inside bearing for the driving-boxes. The axles are all finished to gauges for their diameter and length.

CRANK PINS

are made of steel, fitted into the wheel with a straight bearing, and pressed in by a hydraulic press. The diameters and length are all turned to standard gauges. The main pins have a collar between the bearings for main and parallel rods, and all the pins have collars on the outer end.

---

* The rods which are used with cross-heads having two guides have solid ends in front, with two brasses, and a horizontal key.

All the brasses of the connecting rods are babitted, and each journal has one of Ricker's patent oilers.

DRIVING BOXES

are made of cast iron, with brass bearings babbitted. The top of the brass is round, and is turned where it bears against the box. The seat for the brass in the box is first laid off from the template, and then slotted out to a gauge. The sides and faces of the box are all planed to gauges for the length of bearings and thickness of flanges. The oil cellars and recesses to receive them are also planed, and the cellars held in position with two ½ inch bolts, which pass through the flanges of the box.

COUNTER-BALANCE WEIGHTS

are made of cast iron, and are bolted in pairs between the spokes of the wheels. Each pair is held in position by three bolts, with countersunk heads on the outside, and nuts which are let into a recess cast in the weights on the inside.

WHEEL-COVERS.

All the wheels have sheet iron covers, arranged to prevent the wheels from throwing mud over the engine.

ENGINE TRUCKS.

The frames for four-wheeled trucks are forged in one piece, and are planed on the outside edges and where the pedestals are bolted on. All the holes are drilled from a template, and the frame planed to a gauge. The pedestal jaws are made of wrought iron; the faces, sides, and top and bottom are planed to gauges. They are bolted to the frame with two bolts at the top, and with one bolt to the brace at the bottom. The boxes are planed and bored to gauges in the same way as the driving-boxes. Each box has a babbitted brass bearing, and an oil cellar. The axles are made of hammered iron or steel. The wheels are double plate or spoke cast iron, and are pressed on the axles by hydraulic pressure.

The trucks have a center bearing, with Smith's swing motion. A cast iron center casting is bolted to the bottom of the cylinder casting, and rests on the truck center. A center pin passes through both.

The truck springs are made of cast steel, in the same way as the driving-springs. They are placed underneath the truck frame, and are hung between two curved equalizing beams, which rest on the top of the truck boxes. Check chains are attached to the engine and truck frames, at each end of the latter. Suitable sheet-iron covers are arranged, so as to prevent the truck wheels from throwing mud on the engine.

PUMPS

are made of brass or cast-iron, are full stroke, driven from the cross-heads. They have cage-cup valves, and top and bottom air-chambers. The top chamber is connected to the check by a copper injection pipe, fitted with coupling nuts. The bottom chamber is connected with the suction pipe by a stuffing-box. The suction pipes are brass and are furnished with adjustable feed cocks, under control of the engineer. Each pump is supplied with a pet cock, worked from the cab, and also with frost plugs.

CHECK-VALVES

are of brass, and are attached to the boiler with ball joints; they have the same valves as the pumps.

THROTTLE-VALVES

are double seat poppet valves, with the spindles standing vertical. They are placed in the top of the dome. The seat is cast on the upper end of the dry pipe. The valve and seat are both made of cast iron. The spindle of valve extends down below the bottom of the dome, and is worked by a bell crank attached to the dry pipe. The crank is attached by a rod to the throttle lever.

THROTTLE-LEVERS

are made of the most approved pattern, and located in a convenient position, usually at the back end of the furnace.

DRY PIPES

consist of a vertical cast-iron pipe in the dome, which is connected to the smoke-box by a wrought iron pipe. The latter has a brass neck riveted to it, and is fastened with a strap bolt to two lugs on the cast iron pipe. To the front end of the dry pipe a brass sleeve is riveted, which makes a steam-tight joint, with another brass casting riveted to the inside of the front tube sheet.

STEAM PIPES.

A cast iron T pipe is bolted to the tube sheet in the smoke-box, and joins the brass sleeve in the dry pipe with a ball joint.

APPENDIX G

Two curved cast iron steam pipes connect the T pipe with the cylinders. The steam pipes are bolted at each end with two studs.

EXHAUST NOZZLES.

Double exhaust nozzles are bolted to the exhaust pipes. Three sizes are furnished with each engine, to be used to suit the condition under which the engine is worked.

PETTICOAT PIPES

are made with a flared mouth at the bottom. The pipe is made telescopic, the lower half sliding into the upper, and fastened to the latter with two bolts. The upper part is attached to the smoke box with three bolts, and the height is regulated with jam nuts on each of the bolts.

CYLINDER AND STEAM-CHEST CASINGS.

The cylinder heads are covered with either a cast iron or brass casing. The body of the cylinders are lagged with wood, and sheathed with brass or Russia iron. The casings for the covers of the steam-chests are made of cast iron, and that for the sides of Russia iron or sheet brass.

CYLINDER COCKS.

The cylinders each have two cocks, so arranged that they can be opened from the cab.

OIL COCKS.

Two oil cocks are located inside the cab, and are connected to the steam-chest by solid brass pipes, so that the slide-valves can be oiled from the cab. The cocks are so arranged that steam can be admitted behind the oil to force it into the steam-chest, or the pipes be cleaned when necessary. At the point where the pipe joins the steam-chest, a valve is provided which is closed by the pressure of the steam in the chest.

CAB.

The cabs are made of ash, with walnut moldings, all of good quality and well seasoned. They are well framed, and bolted together with ½ inch joint bolts. Two pieces of sheet iron ⅛ inch thick, are bolted, one inside and one outside the cab, where it rests on the boiler. The rafters are curved, and the roof covered with tin. All the windows and doors are fastened with convenient and substantial fastenings. The inside and the outside of cab are varnished. Suitable handles, and cast-iron steps to get on and off the engine are attached to the engine in convenient positions. A gong for the bell-rope is attached to the under side of roof or ceiling.

RUNNING-BOARDS

are made of ash 2½ inches thick, supported on wrought-iron brackets bolted to the boiler. The outside edges of the running-boards are bound with brass.

DRAW CASTINGS.

A suitable casting to receive the coupling and pin is bolted to each end of the engines.

FOOT-BOARDS

are made of ¼ inch sheet iron, covered with oak planking.

GRATES, ASH PANS, SMOKE STACKS AND PILOTS

will be made on the most approved plan, and adapted to the fuel and service for which the engines are to be used.

SAND-BOXES.

The top and bottom of the sand-boxes are made of cast iron, with ornamental moldings. The body is made of heavy sheet iron. Two iron pipes extend from the sand-box to within two inches of the rails forward of the front driving wheels. Suitable valves and a lever are attached to the cast iron base, to let the sand into the pipes.

The valves are worked by a rod attached to the lever, and extending to the cab.

BELLS AND STANDS.

The bells are hung by a cast iron yoke between two ornamental cast iron columns, fastened to a base which is bolted to the top of the boiler. The bell cord is attached to a brass arm fastened to the bell yoke.

HAND-RAILS

are made of brass pipes, supported by cast iron arms, which are screwed on to a stud tapped into the boiler.

LAMP BRACKETS.

Two cast iron brackets are bolted to the outside of the smoke box, and have a board, bound with brass, bolted to them to receive the head light.

NUMBER PLATES.

A circular cast iron plate with the number of the engine painted on it is put in the center of each smoke-box door.

FLAG-STANDS,

with ornamental brass bases are placed on each end of the front bumper timber.

MISCELLANEOUS.

A steam-gauge, whistle, two safety valves, gauge cocks, and two heater cocks are attached to each engine. Oil cans, jackscrews, pinch bars, wrenches, chains, tool boxes, extra bolts, and pokers are furnished with all engines.

GAUGES.

A complete system of standard gauges, male and female, is used for finishing all the work, so as to make it practically interchangeable. A special department is devoted to the manufacture and care of the gauges and tools.

BOLTS.

The heads and nuts of bolts are made of the same size, so that the same wrench will take either, and the size of finished heads and nuts are made the same as those which are rough. The system of threads used is the Franklin Institute U. S. Standard. The holes into which bolts are fitted are invariably reamed with reamers made and kept to the standard size.

TENDERS

have two four-wheeled trucks, either iron frames or wood.*

TANK

made of charcoal iron, riveted with ⅜ inch rivets, and strongly braced and securely fastened to the tender frame. The sides of the tank are made of No. 8, and the top and bottom of No. 6 iron. The sheets are secured together at the corners with angle iron. The legs of the tank are rounded at the front end, and taper back to the body of the tank. A suitable manhole is put on the top of the tank at the back end. The front end of each leg of the tank has a valve for letting the water into the feed-pipes, which are connected to the tank by rubber hose attached underneath the tender valves. A cock is put into the tank near the bottom, and furnished with a piece of rubber hose for wetting down the coal and foot-board.

TENDER FRAME

is made of three longitudinal, two end, two bolster, and one center transverse timber of well-seasoned oak. The outside timbers are framed and fastened together at the corners with strong castings. The bolsters are also attached to the outside timbers with castings, and the whole bolted together with 1 inch transverse rods, running through the frame from one side to the other. Cast iron brackets on the outside of the frames carry a timber on which the tank rests.

The flooring is 1½ inch thick, made of pine, and securely spiked down. A strong draw casting is attached to the frame at each end. The front end is coupled to the engine with a heavy bar. On each side of the bar are two heavy safety chains connecting the engine and tender together.

TENDER TRUCKS

each have four cast iron plate wheels. The axles have outside bearings. The front truck has a center, and the back truck side bearings. The frames are made of an approved plan, of the best material and workmanship, and have check chains on each corner of both trucks. The back trucks have brakes, with a wheel and shaft in a convenient position for setting the brakes.

PAINT.

The engine and tender each have one coat of priming, one of filling, and one of body color, and then ornamented, and finished with three coats of varnish.

---

* Four-wheeled switching engines sometimes have four-wheeled tenders.

APPENDIX H

CHRONOLOGY OF THE AMERICAN LOCOMOTIVE, 1795–1875

ca. 1795	Fitch locomotive model; some question as to whether it was a railway model.
1825	Colonel Stevens' experimental locomotive at Hoboken.
1825	Model of Stephenson locomotive brought to Philadelphia by William Strickland.
1828	Barlow and Bruen model locomotive at Lexington, Kentucky.
1829	Counterweight driving wheels—*Stourbridge Lion*.
1829	First locomotive imported—Delaware and Hudson's *Pride of Newcastle*.
1829	First locomotive operated—*Stourbridge Lion*.
1829	Feed-water heater used on *Stourbridge Lion*.
1830	First locomotive built in the United States by West Point Foundry—South Carolina Railroad's *Best Friend*.
1832	Gauge glass used on the *Experiment* of the Mohawk and Hudson Railroad.
1832	Jervis designs first 4–2–0, the *Experiment*, built for Mohawk and Hudson by West Point Foundry; incorporated first leading truck.
1832	Use of link motion (?) and four eccentrics by W. T. James.
1832	Baldwin's first locomotive, *Old Ironsides*.
1833	British build 4–2–0 for United States lines.
1833	Introduction of cowcatcher and pilot wheels on Camden and Amboy Railroad.
1833	Poppet-valve throttle—E. A. G. Young of the Newcastle Manufacturing Company.
1833	Bonnet stack invented by W. T. James or Isaac Dripps.
1834	Cast-iron driving-wheel centers adopted.
1835	Iron frame—Tuscumbia, Courtland and Decatur Railway, West Point Foundry.
1835	Center-bearing leading truck—G. E. Sellers.
1835	Bar frame—South Carolina Railroad's *Cincinnati*.
1836–38	First U.S. locomotive exported (?), Winans' *Columbus*.

APPENDIX H

1836	4–4–0 patented by H. R. Campbell.
1838	First 0–6–0 built in U.S.—Beaver Meadow Railroad's *Nonpareil*.
1838	Three hundred and forty-five locomotives in the U.S.
1838	Eight-wheel tender—New Jersey Railroad and Transportation Company's *Uncle Sam*.
1838	Equalizer patented by J. Harrison.
1838	Radial valve gear used on locomotive by Seth Boyden.
ca. 1838	First 0–8–0 built by Camden and Amboy—*Monster*.
1839	Wooden frame—last built by Baldwin.
1840	Only about twenty 4–4–0's in the U.S.
1840	About 590 locomotives in the U.S.
1840	Headlights introduced.
1841	Last British locomotive imported—Pennsylvania and Reading Railroad's *Gem*.
1842	French and Baird stack introduced.
1842	Flexible-beam locomotive patented by Baldwin.
ca. 1842	Jackets used on cylinders by Norris and Rodgers.
1842	Winans' first 0–8–0.
1842	Link motion introduced by Robert Stephenson and Company (England).
1845	The 4–4–0 becomes standard locomotive.
1845	Typical locomotive: 4–4–0, 13″ x 18″ cylinders, 54″ wheels, 13 tons, 80 pounds steam.
1845	Air dome first applied to feed pump by Walter McQueen.
1845	Inside connection revived by New England builders.
1847	First 4–6–0 built by S. Norris and Hinkley.
1847	Metal cab—Pennsylvania and Reading's *Novelty*.
1847	First link motion on U.S. locomotive—Eastern Railroad.
1848	First Winans' Camel built.
ca. 1849	Cylinder saddle designed by Walter McQueen for Albany and Schenectady's *Mohawk*.
1849	Rogers applies link motion to Hudson River Railroad's *Pacific*; soon becomes advocate of link motion.
1850	Typical locomotive: 4–4–0, 15″ x 20″ cylinders, 60″ wheels, 15 tons, 100 pounds steam.
1850	Variable cutoff used by Ethan Rogers at Cuyahoga Works.
1850	Wagon-top boiler—T. Rogers.
1851	Split frame—Wilson Eddy.
1851	Bourdon steam gauge introduced in U.S.
1852	First 4–6–0 built by Baldwin.
1852	*Pawnee*, a rigid 2–6–0 built by James Millholland.
1852	Level cylinders and spread truck introduced by Swinburne, Mason, Eddy, and Norris.
1852–53	Rogers, the most advanced builder in U.S., introduces "modern" 4–4–0.
1853	Center-bearing truck most popular.
1854	Wide firebox designed by Z. Colburn for Delaware, Lackawanna and Western.
1854	Fluted side rods used by Rogers.
1855	Typical locomotive: 4–4–0, 15″ x 20″ cylinders, 60″ wheels, 20 tons, 100 pounds steam.
1855	Inside connection out of favor.
1855	Approximately 6,000 locomotives in the U.S.
1855	Conversion from wood to coal begins in earnest.
1855	Link motion adopted as standard valve gear.
1856	Firebrick arch—G. S. Griggs (patented in December, 1857).
1857	Bissell four-wheel safety truck patented.
1857	Diamond stack design—G. S. Griggs.
1857	Half saddle—Baldwin.
1857	Bury boiler abandoned by most U.S. builders.
1859	Superheater used on *Hiawatha* by Millholland.
1859	First major road to abandon wood as fuel—Reading.
1860	Typical locomotive: 4–4–0, 16″ x 22″ cylinders, 60″ wheels, 25 tons, 110 pounds steam.
1860	Norris built 1000th locomotive.
1860	Steel piston rods used by Rogers.
1860	Steel firebox—Baldwin and Taunton.

APPENDIX H

1860	An estimated 9,000 locomotives in the U.S.
1860	Injector introduced on U.S. locomotives.
1860	Extended smokebox patented by J. Thompson.
1860	Outside frame rails abandoned.
1860	Steel tubes used by Baldwin.
1861	Steel boiler—Great Western Railway of Canada.
1865	First true 2-6-0—Hudson-Bissell truck.
1866	Richardson "pop" safety valve introduced.
1866	First 2-8-0 introduced—Lehigh Valley's *Consolidation*.
1866	Forney locomotive patented (none built until about 1878).
1867–68	First compound locomotive built in the U.S.—Erie Railroad's No. 122.
1870	The 2-6-0 and 2-8-0 begin to replace the 4-4-0 for freight.
1870	The 4-4-0 still most popular locomotive, about 85 percent of all U.S. power.
1870	Typical locomotive: 4-4-0, 17" x 24" cylinders, 63" wheels, 30 tons, 125 pounds steam.
1875	15,000 locomotives in the U.S.

BIBLIOGRAPHY

MANUSCRIPT SOURCES

Baldwin, M. W. Papers, Letters, and Account Books. *ca.* 1836–66. Historical Society of Pennsylvania, Philadelphia, Pa.

Baldwin Locomotive Works. Specifications. De Golyer Foundation Library, Southern Methodist University, Dallas, Texas.

Hinkley Locomotive Works. Register. 1841–56. Boston Public Library.

Patten Papers. Letters of G. W. Whistler and Joseph G. Swift. New York Public Library.

Robert Stephenson & Co. Specifications. 1828–40. English Electric Co., Ltd., Newton-le-Willows, England.

Rogers Locomotive Works. Specifications. 1845–46 (scattered copies). Smithsonian Institution, Washington, D.C.

BOOKS

Ahrons, E. L. *The British Steam Railway Locomotive, 1825–1925*. London: The Locomotive Publishing Co., 1927.
   An excellent historical survey of British locomotive engineering with line and halftone illustrations.

Alexander, Edwin P. *Iron Horses*. New York: W. W. Norton & Co., 1941.
   A pictorial work valued for the many contemporary lithographs reproduced. Also included are line drawings reprinted from Weissenborn's *American Locomotive Engineering and Railway Mechanism* and *Railroad Gazette*.

Bell, J. Snowden. *The Early Motive Power of the Baltimore and Ohio Railroad*. New York: Angus Sinclair Co., 1912.
   One of the few dependable technical histories on American locomotives. The author was a draftsman with the Baltimore and Ohio as a young man and had first-hand knowledge of his subject. Line and halftone illustrations.

BIBLIOGRAPHY

Bendel, A. *Aufsätze Eisenbahnwesen in Nord-Amerika* (text and atlas). Berlin, 1862.

    An official government report on U.S. railroad operations and equipment similar to Galton's 1857 report. Line drawings of Winans' Camel and Tyson's Ten Wheeler.

Brown, William H. *The History of the First Locomotive in America*. Revised ed. New York: D. Appleton & Co., 1874.

    A chatty account based largely on recollections. Concerned mainly with the *Stourbridge Lion* and locomotive "firsts." Little technical information. Line cuts.

Bruce, Alfred W. *The Steam Locomotive in America*. New York: W. W. Norton & Co., 1952.

    A comprehensive technical history of American locomotives *after* 1900. Bruce was chief engineer of the American Locomotive Company; his comments on modern practice are knowledgeable and represent an outstanding contribution to locomotive history. As a historian, however, Bruce failed badly; his comments on early practice are scanty and too often incorrect. Regrettably, no documentation was offered. Line drawings and photographs illustrate this valuable work.

Century of Reading Company Motive Power, A. Philadelphia: The Reading Company, 1941.

    An illustrated summary of locomotive development on Reading lines. Convenient for quick reference.

Clark, Daniel K. and Colburn, Zerah. *Recent Practice in the Locomotive Engine*. London, 1860.

    Offered as a supplement to Clark's early work on railway machinery, this volume was devoted almost entirely to British locomotives. The two American engines illustrated and Colburn's history of American practice made this work invaluable to any study of the subject.

Colburn, Zerah. *Locomotive Engineering and the Mechanism of Railways*. 2 vols. London and Glasgow, 1872.

    This work was nearly complete in 1864 but was not published until after the author's death. Only one American locomotive, a Rogers 4-4-0, is illustrated by a full set of drawings in this large work, but the detailed history offsets this omission in large part. It is an invaluable reference.

———. *The Locomotive Engine*. Boston, 1851.

    A small book intended for the practical mechanic. While largely nontheoretical, it treats the elements of steam and combustion. Specific locomotives are described, together with notes on their performance. The work is New England oriented. It was in print long after its content had become obsolete.

Colburn, Zerah, and Holley, Alexander L. *The Permanent Way and Coal-Burning Locomotives*. New York, 1858.

    Primarily European in content, it is useful for its illustrations of specialized fireboxes developed to burn coal.

De Pambour, C. F. M. G. *A Practical Treatise on Locomotive Engines upon Railways*. 1st English ed. London, 1836.

    The first technical book devoted entirely to locomotives. It established formulas, proportions, and general theories that were followed for many years in locomotive construction. A widely read and influential work. Illustrated with drawings for a Stephenson Planet engine.

Ferguson, Eugene S. (ed.). *Early Engineering Reminiscences (1815–1840) of George Escol Sellers*. Washington, D.C.: Government Printing Office, 1965.

    Recollections by an early locomotive builder who was associated with Baldwin, Norris, Brandt, and other pioneers in the field. Illustrations and annotations added by the editor.

Forney, Matthias N. *Catechism of the Locomotive*. 1st ed. New York: *Railroad Gazette*, 1874.

    Prepared as an instruction book for engineers and plain mechanics, this volume explains the workings and general construction of American locomotives. Properties of steam and the theory of the cutoff are explained. This and other early editions contain a set of assembly drawings for a Grant 4-4-0. The work remained a standard reference for many years and went through several editions.

———. *Locomotives and Locomotive Building*. Originally published New York: W. S. Gottsberger, 1886. 2nd ed., Berkeley, Calif.: Howell-North Books, 1963.

A combination illustrated history and catalog of the Rogers Locomotive and Machine Works. Many line cuts of early components and general designs for early Rogers locomotives as well as the history of the works and its founder are included.

Galton, Douglas. *Report to the Lords of the Committee of the Privy Council for Trade and Foreign Plantations, on the Railways of the United States.* 2 vols. London: Eyre and Spottiswoode, 1857–58. Text and atlas.

A general survey of operations, construction, and equipment with line drawings of cars, locomotives, and rails. A small but valuable report.

Gerstner, F. A. Ritter von. *Die innern communication der vereinigten Staaten von North-America.* Vienna, 1842–43.

A classic two-volume study of American railroads by a recognized Austrian engineer. Much technical information and detailed notes on operations are offered. Many engravings of early track are given, but only one plate on locomotives, a detail on spark arrestors, is included. Data gathered 1839–41.

Ghega, Karl von. *Die Baltimore-Ohio Eisenbahn. . . .* Vienna, 1844. Text and atlas of plates.

A more general treatise than the title implies, this work contains comments on Norris, Baldwin, and Eastwick and Harrison, as well as several fine engravings of locomotives. The data for the book were gathered from about 1840 or 1841 through 1842.

Harrison, Joseph, Jr. *The Iron Worker and King Solomon.* Philadelphia, 1869.

Contains a fine essay on Harrison's career as a locomotive builder both in Philadelphia and in St. Petersburg, Russia.

———. *The Locomotive and Philadelphia's Share in its Early Improvements.* Philadelphia: Gebbie, 1872.

The recollections of a central figure in the development of the American locomotive. Mention is made of Norris and Baldwin as well as of the more minor Philadelphia builders. Illustrated with line cuts.

History of the Baldwin Locomotive Works, The. Philadelphia: Baldwin Locomotive Works, 1923.

This is an updated version of the business history that first appeared in the 1873 Baldwin catalog. It presents a convenient, if incomplete, history of the firm. Illustrated with tables on production.

Hodge, Paul R. *The Steam Engine. . . .* New York: D. Appleton & Co., 1840. Text and atlas of plates.

The earliest work to contain complete drawings of an American locomotive. The author was a draftsman for Dunham and Rogers. Stationary and marine engines are also treated.

Holley, Alexander L. *American and European Railway Practice.* New York: Van Nostrand, 1861.

An expansion and revision of Colburn's and Holley's *The Permanent Way and Coal-Burning Locomotives.* It treats mainly coal-burning locomotives and track and is largely European in content.

Jervis, John B. *Railway Property.* New York, 1861.

A general manual treating construction and management of railroads, with historical references to Jervis' early work with truck locomotives.

Lardner, Dionysius. *Railway Economy.* New York and London, 1850.

A general work, mainly European in emphasis, but a useful firsthand account of early American locomotive operations.

Lucas, Walter A. *From the Hills to the Hudson.* New York, 1944.

Mainly a corporate history of the Paterson and Hudson River Railroad and associated lines. It contains some detailed data on the early locomotives of this line.

Marshall, C. F. Dendy. *A History of the Railway Locomotive Engine down to the End of the Year 1831.* London, 1953.

The best of several excellent works by this author on early locomotive and railway engineering. Marshall was an engineer who understood the historian's craft and gathered together much obscure information on the beginnings of steam locomotives.

Matthias (1st name unknown). *Darstellung, einer zum Trans-*

port-Betrieb auf der Berlin-Potsdamer Eisenbahn im Gebrauch befindlichen Locomotive . . . Norris in Philadelphia. Berlin, 1841.

Folio work devoted entirely to the description of the Norris 4–2–0. Plates are reproduced in the present work.

Meyer, J. G. A. *Modern Locomotive Construction.* New York, 1904.

Based on a series of articles which appeared in the *American Machinist* more than ten years earlier, the contents of this work treat locomotive designs of the 1880's. The text explains the rationale of the standard designs employed by American builders. Few general assemblies but over a thousand drawings of component parts are included. The author was chief draftsman at the Grant Locomotive Works.

Modern Locomotives. New York: *Railroad Gazette,* 1897.

A folio-size collection of drawings with a good introductory summary on locomotive developments since 1890. Reprinted in 1901.

Moné, Fredrick. *An Outline of Mechanical Engineering.* New York(?), 1851.

A complete set of working drawings and a fine mechanical description of the locomotive *Croton* are included in this work.

———. *A Treatise on American Engineering.* New York, 1854.

Devoted mainly to stationary and marine engines but containing drawings for two Hudson River Railroad locomotives, the *Columbia* and the *Superior.* Text is general and of minor importance compared to the drawings.

Norris, Septimus. *Norris' Handbook for Locomotive Engineers and Machinists.* Philadelphia, 1853.

Tables and very general comments copied from earlier British works.

Pangborn, J. G. *The World's Railway.* New York, 1894 and 1896.

A rambling, inaccurate history of locomotive development compiled by the manager of the Baltimore and Ohio exhibit at the Columbian Exposition. The author was a journalist and had little understanding of engineering. The work is valuable, however, for its record of the recollections of S. B. Dougherty, an early associate of W. T. James. The curious wash drawings illustrating the text are of questionable value.

Recent Locomotives. New York, 1883.

A collection of engravings and articles which appeared in the *Railroad Gazette* between 1871 and the time of publication. Some mechanical drawings included. Expanded and reprinted in 1886. Modern reprint of the first edition, by Graham Hardy (1950).

Reuter, Emil. *American Locomotives.* Philadelphia, 1849. Text and atlas of plates.

The first attempt to describe American locomotives, in which only about a third of the projected forty-two plates were printed. The text also appears to be incomplete. The drawings are of great value and include complete plans for the *Philadelphia,* an 0–6–0, the *Delaware,* an 0–8–0, and the *John Stevens,* a 6–2–0. The author was a draftsman employed by both Millholland and Winans.

Ringwalt, James L. *Development of the Transportation Systems of the United States.* Philadelphia, 1888.

A useful digest of transport history to that date. Little is offered on locomotive construction, but some good material on railroad operations is included.

Sinclair, Angus. *Development of the Locomotive Engine.* New York, 1907.

An ambitious attempt to chronicle the American locomotive which has been a standard reference on the subject since its publication. Its value is greatly impaired by numerous errors, a lack of documentation, and too great a reliance on recollections. Should be cautiously consulted.

Stevens, Frank W. *The Beginnings of the New York Central Railroad.* New York, 1926.

Primarily a corporate history of the early roads that later combined to form the New York Central. The author has assembled much good data on the early locomotives of the Mohawk and Hudson Railroad.

Stevenson, David. *Sketch of the Civil Engineering of North America.* London, 1838.

BIBLIOGRAPHY

A general travel account that includes a chapter on American railroads. Notable for its description of the cowcatcher.

Tanner, Henry S. *A Description of the Canals and Railroads of the United States.* New York, 1840.

Little material on railroad equipment but several detailed descriptions of operations and track construction.

Vose, George L. *Handbook of Railroad Construction.* Boston, 1857.

Brief description of locomotives including a few outline drawings. Valuable for its references on fuels.

Warren, J. G. H. *A Century of Locomotive Building by Robert Stephenson & Co., 1823–1923.* Newcastle, England, 1923.

An excellent history of the world's first commercial locomotive builder, with outstanding technical descriptions of design and development illustrated by original drawings. Many early locomotives exported to the United States are described. In the present author's opinion, it is the finest locomotive history printed in the English language.

Weissenborn, Gustavus. *American Engineering.* New York, 1861 (text) and 1857 (atlas).

A collection of handsome line engravings described by a very general text. The engravings of the *Talisman* are the only locomotive drawings in this work.

———. *American Locomotive Engineering and Railway Mechanism.* New York, 1871.

A large folio work with magnificent line drawings for eleven American locomotives, most of which date from the late 1860's. Baldwin, Cooke, Hinkley, and the products of several railroad shops are represented. One of the best works ever published on American locomotives. The plates were reprinted by the Glenwood Publishers (1967).

White, John H., Jr. *Cincinnati Locomotive Builders, 1845–1868,* Washington, D.C.: Government Printing Office, 1965.

A business history of the largest center of early Midwest builders. Some notes on design and construction.

Wood, Nicholas. *A Practical Treatise on Rail-Roads.* 1st ed. London, 1825.

One of the earliest works to discuss locomotive construction. The 1832 edition has an appendix on American railways.

Yoder, Jacob H., and Wharen, George B. *Locomotive Valves and Valve Gears.* New York, 1917.

Excellent technical work. Except for link motion, only modern valve gears are treated.

REPORTS

American Railway Master Mechanics Association Annual Report. 1868 to the present.

Detailed reports on mechanical problems and developments. Some line drawings after about 1875.

American v. English Locomotives, Correspondence, Criticism and Commentary Respecting their Relative Merits. New York, 1880.

Baltimore and Ohio Annual Report. 1827–75.

Cohen, Mendes. *Report on Coke and Coal Used with Passenger Trains, on the Baltimore and Ohio Railroad.* Baltimore, Md., 1854.

Eastern Railroad Association—Annual Reports of Executive Committee. Boston, Springfield, and New York, 1867–83.

The main function of this association was to report on patent claims made against member railroads.

Knight, J., and Latrobe, B. H. *Report upon the Locomotive Engines . . . Several of the Principal Rail Roads in the Northern and Middle States.* Baltimore, Md., 1838.

New England Association of Railway Superintendents—*Report of the Trial of Locomotive Engines.* Lowell, Mass., 1852.

Palmer, William J. *Report of Experiments with Coal Burning Locomotives Made on the Pennsylvania Railroad.* Philadelphia, 1860.

Philadelphia and Reading Annual Report. 1839–70.
    Lists locomotives, with notes on rebuilding.

Report of the Canal Commissioners of Pennsylvania. Harrisburg, Pa., 1825–ca. 1857.
    Reports for 1835 to 1840, valuable for locomotive data on the Philadelphia and Columbia and Allegheny Portage railroads.

Report of the Railroad Commissioners of New York. Albany, N.Y., 1856 only.
    Operating statistics and list of locomotives for all New York railroads.

South Carolina Canal and Railroad Company. *Report of the Committee on Cars. . . .* Charleston, S.C., 1833.
    Mainly concerned with locomotives, with special reference to Allen's articulated locomotives.

Strickland, William. *Reports on Canals, Railways, Road, and Other Subjects Made to Pennsylvania Society for the Promotion of Internal Improvements.* Philadelphia, 1826.
    American report on British transport.

Whistler, George W., Jr. *Report upon the Use of Anthracite Coal in Locomotive Engines on the Reading Rail Road.* Baltimore, Md., 1849.

PERIODICALS

The various railroad and technical journals listed below were by far the most valuable sources of information found for specific technical details. The journals were far more pertinent to engineering history and development than were any of the books available on the subject.

American Railroad Journal. New York, 1832–86. Little on engineering after about 1860.

American Railway Review. New York, 1859–61.

American Railway Times. Boston, 1849–72.

Engineer (London). 1855 to the present.

Engineer (Philadelphia). 1860 only. Zerah Colburn was the editor. The journal was similar to the *Railroad Advocate*.

Engineering. London, 1866 to the present.

Journal of the Franklin Institute. Philadelphia, 1826 to the present.

Locomotive Engineering. New York, 1888–1928. Angus Sinclair was the editor. See *Development of the Locomotive Engine.*

Railroad Advocate. New York, 1854–56. The only early American railroad journal devoted to technical matters.

Railroad Gazette. New York, 1870–1908.

Railway Age. Chicago and New York, 1876 to the present.

Railway and Locomotive Historical Society Bulletin. Boston, 1922 to the present. A rich source and a rare hobby journal that is not devoted to club news.

Railway Master Mechanic Magazine. 1886–1916.

Scientific American. New York, 1845 to the present.

INDEX

A

A. and W. Denmead and Sons, 172, 336, 370, 371
Active, 187*n*, 272
Adams and Westlake, 527
Adams, Isaac, 189
Addison Gilmore, 110, 172, 205, 498–501, 539–40
Adhesion, 74–75
Adirondack, 57
Advance, 201–2
Agenoria, 240, 242, 244
Air brakes, 524
Albany, 328*n*
Albany and Schenectady Railroad, 235, 542
Albany and Schenectady Railroad shops, 51
Albany Iron and Nail Company, 156
Allegheny, 55
Allegheny Portage Railroad, 172
Allen, Horatio, 3, 5, 167, 168, 190, 215, 239, 240, 242, 509
Allen, Samuel F., 138
Amenia, 164, 166, 172, 233, 234
America, 39, 40, 41, 42, 53, 172, 220, 240
American, 188, 196, 541
American Locomotive Company, 14, 33*n*
American Steam Carriage Company, 25, 233*n*, 542, 544
American Steam Gauge Company, 523
American type. *See* Wheel arrangements, 4–4–0
Ames Iron Company, 180
Amoskeag Manufacturing Company, 120*n*, 163, 194, 526, 528, 538
Anderson, Richard K., Jr., 500
Anthracite, 235
Arabian, 71
Archimedes Works, 280
Articulated locomotives, 509–10
Ashcroft, E. H., 134, 136
Atalanta, 297
Atlantic, 71, 190
Atlantic and Great Western Railway, 204, 484
Atlas, 204, 398
Atlas Works, 162
Attica and Buffalo Railroad, 541
Auburn and Syracuse Railroad, 215
Aurora Shops (CB&Q), 495
Axle boxes, 58, 62
Axles, 31, 32, 51, 62, 152, 153, 178–79, 235, 262, 271, 396, 397, 408, 417, 471, 520, 569

B

Babbitt, Isaac, 186, 198, 207
Baden State Railway, 197
Baird, Matthew, 108, 121, 428, 537
Baldwin Locomotive Works, 13, 14, 16, 17, 20, 21, 22, 23, 24, 25, 27–28, 45, 46, 52, 57, 59, 62, 64, 65, 66, 67, 69, 75, 80, 94, 98, 101, 104, 105, 108, 111, 116, 121, 122, 131, 137, 141, 145, 157, 158, 160–61, 164, 170, 171, 175, 177, 178, 179, 180, 184, 187, 188, 190, 194, 197, 198, 201, 206, 207, 208, 212, 213, 219, 223, 226, 228, 231, 232, 235, 269, 270, 271, 272, 274, 275, 289, 314, 396–99, 400, 406, 427–36, 437–42, 471, 474–75, 487, 513, 514–15, 517, 528, 537–38, 543, 544
Baldwin, Matthias W., 5, 8, 12, 23, 25, 34, 45, 46, 48, 57, 59, 69, 101, 138, 156, 168, 221, 249, 270, 271, 281, 396–99, 400, 406, 528, 537–38
Baldwin Ten Wheelers, 437–42, 471
Baltimore, 8, 9, 175, 213, 224, 510
Baltimore and Ohio Railroad, 3, 4, 12, 19, 45, 46, 59, 64, 65, 66, 67, 69, 71, 73, 77, 78, 79, 80, 84, 85, 86, 87, 89, 95, 100, 104, 112, 113, 118, 124, 126, 134, 136,

137, 144, 152, 161, 163, 167, 171, 181, 187, 188, 196, 198, 207, 217, 232, 242, 297, 298, 347, 348, 349, 350, 351, 352, 353, 366, 368, 369, 385, 391, 490, 494, 540, 541, 542, 546, 549–50
Baltimore and Ohio Railroad, numbered engines: *No. 7*, 224n; *No. 16*, 46, 47; *No. 17*, 224n; *No. 199*, 351; *No. 201*, 170, 171; *No. 222*, 366; *No. 223*, 366; *No. 224*, 366; *No. 225*, 366, 371; *No. 226*, 366; *No. 227*, 366, 370; *No. 228*, 366; *No. 229*, 366; *No. 230*, 366, 369; *No. 232*, 386; *No. 292*, 370; Mogul *No. 600*, 64
Baltimore and Susquehanna Railroad, 7, 8, 51, 146, 168, 175, 181, 188, 212, 510, 518, 543
Baltimore Locomotive Works, 352
Bangor and Piscataquis Canal and Railroad Company, 224, 225
Barnes, 502
Barnes, J. B., 232
Barnes, James, 231
Batteries, and traction, 235
Bayley, O. W., 450
Beaver, 226, 233, 312, 316
Beaver Meadow Railroad, 4, 48, 66, 87, 104, 152
Bedford, 65
Bee, 544
Belgium State Railway, 201
Bell cranks, 34
Bell, J. Snowden, 207, 297, 385, 391
Bell stands, 30
Bells, 30, 212–14, 218
Belmont Incline, 72
Bendel, A., 353, 354
Bessemer process, 156
Best Friend, 13, 21, 175, 512
Best, Gerald M., 528
Beyer, Charles, 162, 163
Birmingham and Gloucester Railway, 133
Bissell, Levi, 62, 156, 173, 174, 175
Black Hawk, 270
Blakslee, James I., 428, 429
Blast pipes, 7, 111–14
Blood, Aretas, 538
Boardman, Horace, 105
Boilers, 5, 7, 14, 16, 30–34, 48, 50, 52, 59, 62, 65, 67, 74, 88, 93–149, 242, 252, 253, 258, 271, 286, 287, 288, 312, 322–23, 329, 337, 341–42, 347, 349, 350, 351, 352–53, 358, 367, 372, 384, 397, 400, 416, 422, 429, 434, 437, 438, 449, 498, 512, 551–52
Boreas, 383
Boston, 192
Boston and Albany Railroad, 26, 57, 126, 385, 498–99
Boston and Albany Railroad engine *No. 39*, 163, 498–502
Boston and Albany Railroad shops, 163, 498–99
Boston and Lowell Railroad, 8, 73, 77, 114, 213, 223
Boston and Maine Railroad, 59, 348–49, 350
Boston and Providence Railroad, 7, 78, 81, 84, 85, 86, 88, 90, 178, 184, 208, 209, 226, 233, 322, 495, 540
Boston and Providence Railroad shops, 208, 235
Boston and Worcester Railroad, 78, 84, 204, 208, 212–13, 215, 224, 225
Boston, Clinton and Fitchburg Railroad, 201
Boston Locomotive Works, 14, 19, 20, 21, 26. *See also* Hinkley Locomotive Works
Boston, Lowell and Nashua Railroad, 204
Bourdon, Eugene, 134, 136, 523
Boyden, Seth, 181, 200
Braithwaite, John, 8, 512
Brakes, 184, 450, 504, 524
Brandt, John, 57, 58, 59, 358, 534, 538
Brattleboro, 460
Breese, Kneeland and Company, 24, 330, 341, 342, 343
Briscoe, Benjamin, 122
British industry and American industry, 3–4, 34, 73, 93, 96, 99, 102, 103, 158, 444. *See also British manufacturers by name*
Brook, James, 46
Brooklyn Elevated Railway, 503–5
Brosnan, William D., 521
Brother Jonathan, 33–34, 35, 45, 117, 125, 158, 541
Brown, 57
Browning, William, 280
Buchanan, William, 520, 531
Buffalo and Corning Railroad, 172
Buffalo and Erie Railroad, 66, 98, 187, 203, 212
Buffalo and State Line Railroad, 66, 185, 198, 199
Builders. *See* Locomotive manufacturers
Bullock, William, 120n, 173n–74n
Burnham, George, 428

Bury, Edward, 8, 94, 97, 151
Bush and Lobdell, 182, 367

C

C. E. Detmold, 349
C. P. Huntington, 233
Cab-in-front, 512
Cabs, 30, 31, 221–23, 305, 315, 347, 370, 377, 385, 477, 528, 531
Camden and Amboy Railroad, 7, 8, 66, 86–87, 100, 103, 105, 107–8, 118, 124n, 158, 175, 176, 181, 185, 189, 211, 217, 232, 248, 249, 251, 252, 253, 254, 256, 280, 408, 539, 543
Camden and Amboy Railroad engines: *No. 6*, 189; *No. 9*, 189, 519, 529
Camden and Woodbury Railroad, 8, 12
Camel, 348
Camels, 22, 69, 110, 113, 136, 163, 186, 187, 190, 221, 297, 347–57, 366, 369, 398, 513, 546
Campbell, Alexander, 311–12
Campbell, Allan, 311–12
Campbell, Henry R., 46, 48, 528, 534, 538–39
Carbon, 398
Carey Manufacturing Company, 100n
Carroll of Carrollton, 5, 156, 175, 176n, 235, 337
Caruthers, C. H., 304n
Cattle guards. *See* Cowcatchers
Celeste, 546
Centipede, 126, 203, 208, 350, 546
Central Ohio Railroad, 181
Central Pacific Railroad, 199, 201, 517
Central Railroad (Georgia), 197
Central Railroad of New Jersey, 62, 64, 78, 85, 173, 203, 233, 235, 359, 407
Champlain, 94, 101n, 190, 209, 222, 320, 322, 324, 325–27
Chanute, Octave, 519
Chapman, William, 151, 167
Charles Tayleur and Company, 7, 8, 12, 159, 176
Chesapeake, 57, 58, 59
Cheshire Railroad, 8, 525
Chicago, Alton and St. Louis Railroad, 24
Chicago and Alton Railroad, 457
Chicago and North Western Railroad, 184, 202, 221

Chicago and North Western Railway, 140, 160, 206, 513
Chicago and Rock Island Railroad, 73, 138, 141, 542
Chicago, Burlington and Quincy Railroad, 89, 141, 144, 495, 497
Chicago Feed Water Heater Company, 138
Chichester, 48n
Child, Daniel F., 444, 541
Chilean Railways, 312
Cincinnati, 11, 12, 177, 187
Cincinnati, Hamilton and Dayton Railroad, 521
Cincinnati, New Orleans and Texas Pacific Railroad, 114
Clark, David K., 138, 194, 199
Clement, W. H., 101
Cleveland, 192
Cleveland and Pittsburgh Railroad, 123, 124, 192, 219n, 231, 350
Cleveland and Toledo Railroad, 524
Cleveland, Columbus, Cincinnati and Indianapolis Railroad, 529, 532–33
Cleveland, Columbus, Cincinnati and St. Louis Railroad, 26, 192, 470
Coal. *See* Fuel
Coal chutes, 352, 353
Cocks, 30, 132–33, 342
Colburn, Zerah, 14–15, 22, 52, 72n, 104, 106, 110, 112, 133, 161, 167, 174, 178, 195, 196, 197, 198, 209, 218, 220, 223, 281, 347, 348–49, 358, 422, 443, 463, 519, 539, 544, 545, 546
Columbia, 99, 103, 173, 192, 207, 321, 330, 337, 342
Columbus, 27
Colvin, F. H., 202
Combination valve gear, 198–99
Comet, 159
Compound locomotives, 209–10
Concord Railroad, 539
Congdon, Isaac H., 521
Connecticut, 50, 51
Connecting rods. *See* Rods
Consolidation, 64, 65, 126, 161, 174, 213, 232, 348, 427–36
Consolidations, 104, 174, 427–36, 519. *See also* Wheel arrangements, 2–8–0
Consuelo, 476–77, 480, 482
Cooke, John, 19, 539, 541. *See also* Danforth, Cooke and Company
Cooper, Peter, 543

Copiapo, 186, 189, 192, 213, 222, 233, 234, 311–19
Copiapo Railroad, 311
Copper firebox, 30, 103–4, 367, 521, 549
Corliss, George H., 201, 202
Corry, 164
Costs: boiler, 98, 103; firebox, 103; fuel, 77, 78, 80, 81, 83, 84, 85, 87, 88, 354; labor, 77–78, 79; locomotive, initial, 4, 12, 13, 14–15, 21–23, 24, 46, 240, 249, 269, 351, 392, 428, 443, 450, 484, 487; lubricating, 79–80; per mile of railroad, 3–4; operating, 73, 77–78, 80–81, 520; repair, 79; and speed, 73, 74; tire, 183; tube, 100
Courier, 197
Cowcatchers, 30, 59, 211–12, 265, 313, 524
Crabs, 67, 137, 221, 545
Crampton, 107, 158
Cramptons, 105, 176n, 185. 235
Crank axles, 31, 208, 209, 281, 326
Crawford, 514–15
Crerar, John, 528
Crossheads, 125, 126, 186–87, 242, 272, 281, 288, 353, 400, 408
Croton, 121, 188, 192, 209, 212, 226, 232, 321, 328–30, 331, 332–36, 337
Cumberland, 62, 63, 190
Cumberland and Pennsylvania Division of Pennsylvania Railroad, 233, 349, 352
Cumberland Valley Railroad, 112, 126, 220, 233, 538
Cunningham, James, 462
Cushing, George W., 184, 514
Cushion wheels, 177–78
Cutting, James A., 304
Cuyahoga Steam Furnace Company, 192, 193, 199, 213, 484, 494
Cyclops, 192
Cylinder cocks, 342
Cylinder lagging, 30
Cylinders, 5, 30, 31, 32, 45, 46, 50, 51, 52, 58, 59, 62, 66, 67, 74, 75, 76, 205–10, 242, 253, 260, 270, 285, 287–88, 305, 313, 329, 337, 349, 350, 351, 353, 359, 368, 383, 398, 400, 416, 422, 429, 552

D

Danforth, Charles, 113, 539
Danforth, Cooke and Company, 66, 220, 233, 392, 393, 407, 416, 418–21, 539

Daniel Nason, 178, 226, 529
Danville, 476, 480, 482
Danville, Hazleton and Wilkes Barre Railroad, 437, 438, 441, 442
Davis, J. C., 64, 513
Davis, Thomas S., 204
Davy Crockett, 12, 168, 169, 512
De Pambour, Francois M. G., 75, 113, 133, 134, 188
De Sanno, Fredrick, 219n, 544
De Sanno, Walter, 218–19
De Witt Clinton, 21, 66, 86, 125, 176, 221, 543, 548
Decoration, of locomotives, 218–21, 504, 528
Dedham, 235
DeGolyer Library, 509
Delaware (1832), 8, 12, 145, 249, 538
Delaware (1847), 88, 297
Delaware and Hudson Canal Company, 3, 7, 86, 239, 240, 241, 242, 243
Delaware and Hudson Railroad, 86
Delaware, Lackawanna and Western Railroad, 62, 106, 110, 219, 350, 392, 393, 483, 512
Demerara Railway, 531
Denmead. *See* A. and W. Denmead and Sons
Dennis, C. C., 198
Dennis, Wood and Russell, 541
Design of Locomotives, 454–55, 477, 487, 498–99, 513–16
Design, standardization of, 25–27, 443–45, 513–16
Detmold, Christian E., 513
Detroit Locomotive Works, 16, 163
Detroit Lubricator Company, 205
Dickson, 75
Dimensions. *See* Specifications
Dimpfel, F. P., 93, 102, 105
Dr. Ordway, 8, 10, 156
Don Pedro II Railway, 398
Dorsey, Edward B., 520
Dotterer, Samuel H., 241
Dougherty, Samuel B., 196
Dover, 175, 176n
Dripps, Isaac, 103, 105, 106, 118, 124n, 211, 233, 249, 253, 254, 539, 546
Drivers, 31, 34, 45, 46, 48, 50, 51, 52, 58, 59, 62, 64, 65, 66, 67, 69, 168, 172, 175–79, 184, 234, 235, 243, 254, 263, 264, 271–72, 287, 313, 326, 337, 342, 379, 396, 397, 398, 483–84

Drury, Gardner P., 541
Dunham, 112, 134, 186, 188, 205, 207
Dunham, Henry R., 145, 280, 282–86. *See also* H. R. Dunham and Company
Dutch Wagons, 171
Duty, on imported locomotives, 12

E

E. L. Miller, 34, 270
Eames, Lovett, 235
Early, Patrick, 241
Eastern Counties Railway, 174, 183
Eastern Counties Railway engine *No. 248,* 174
Eastern Railroad, 197
Eastwick and Harrison, 27, 48, 108, 159–60, 161, 170–71, 287, 288, 296, 546. *See also* Garrett and Eastwick
Eastwick, Andrew M., 48, 152, 188, 288, 289, 540–41, 546
Ebbert, Peter S., 140, 141
Eccentric-driven pumps, 126
Eddy Clocks, 57, 163, 498–99, 540
Eddy, Wilson, 16, 19, 26, 57, 96–97, 101, 104–5, 106, 110, 126, 161, 163, 172, 203, 204, 205, 234, 385, 498–99, 524, 539–40
Edson, William, 329
Edwin R. Bennett and Company, 121
Eggerton, Albert, 521
El Paso and South Western Railroad engine *No. 1,* 162, 163
Elephant, 462–69
Engineering Societies Library, 513
Equalizers, 64, 152–53, 154, 156, 171, 172, 174, 288, 296, 417, 519
Erie and Kalamazoo Railroad, 397
Erie Railroad, 22, 26, 62, 97, 100, 183, 222, 348, 350, 407
Erie Railroad engine *No. 254,* 112, 146, 416–21
Erie Railroad Mogul, 147
Erie Railway, 24, 52, 57, 58, 59, 65, 79, 104, 126*n*, 146, 214, 416–21
Erie Railway engines: *No. 112,* 198; *No. 122,* 209, 210
Essex, 88, 200, 216, 331
Essex Machine Shop, 538
Esslinger Works, 397
Evans, Oliver, 242

Evansville, Hendersonville and Nashville Railroad, 440
Exhaust pipes. *See* Blast pipes
Exhaust steam heaters, 137–38, 139–40
Experiment, 21, 33–34, 35, 73, 86, 132, 168, 176, 538, 547–48
Exports, 27–28

F

Fairlie, R. F., 5
Falcon, 141
Fans, draft, 112
Farley, H. W., 197
Feed-water heaters, 4, 137–42, 242, 512
Feed-water pumps, 30, 31, 59, 124–27, 128, 232, 242, 253, 272, 288, 303–4, 313, 353, 367–68, 400, 416–17, 437, 456, 491, 498, 553
Financing, 22, 23–24
Fire Fly, 8, 176
Fireboxes, 7, 30, 32, 34, 52, 57, 58, 59, 93, 95, 102–8, 110, 271, 287, 288, 296, 312, 323, 329, 337, 341, 347, 348, 349, 351, 352, 358, 367, 416, 422, 430, 437, 438, 521–22
Fisher, M., 201
Fishlow, Albert, 509, 520
Fitchburg Railroad, 102, 194, 199, 226, 235, 460
Flexible-beam-truck engines, 396–406, 428
Flint, James, 540
Fogel, Robert W., 509
Force, E., 213
Forest State, 477, 481
Forney, Matthias N., 144, 172, 233, 503–4, 534, 540
Fort Erie, 160
Foster, Rastwick and Company, 8, 21, 240, 244
Foster, William A., 8, 199
Frames, 57, 158–66, 261, 271, 313, 318, 323, 326, 341, 342, 348, 353, 358, 369, 373, 374, 381, 383, 385, 434, 456, 470, 552–53
Freedom Iron Works, 179, 180
Freemantle, William, 242
French and Baird, 121, 304*n*
French, R., 121
Fuel: and boilers, 94, 102; coal, 78, 83, 84–85, 86–90, 102; coke, 89; and economy,

443–44, 499, 523; general, 34, 57, 58, 73, 76; oil, 83, 90, 534; and water consumption, 223, 504, 521; wood, 78, 83–86, 89, 102, 520–21. *See also* Costs, fuel; Tenders

G

Galena and Chicago Union Railroad, 89, 141, 181
Galton, Douglas, 4, 369
Garrett and Eastwick, 48, 152, 540. *See also* Eastwick and Harrison
Garrison, C. K., 462–63
Gasconade, 194, 324
Gauges, 30, 132–36, 314
Gem, 7
General, 101, 467
General Dix, 213
General Stark, 120*n*
George Salter and Company, 147
George Washington, 71–72, 544
Georgia Railroad, 538
Georgia Railroad and Banking Company, 397
Ghega, Karl von, 73, 159, 287–88, 289
Giffard, Henri J., 128, 130
Gleason, John, 203
Globe, 159*n*
Globe Locomotive Works. *See* Souther, John
Gooch, Daniel, 13, 198, 514
Goss, William F. M., 499
Gould, E. P., 341
Gowan and Marx, 24, 48, 72, 88, 95, 159, 161, 171, 188, 213, 224, 287–89, 291–95
Graham spring balance, 531
Grand Junction Railway, 203
Grand Trunk Railroad of Canada, 143, 144, 476–77
Grant, D. B., 217, 358, 428
Grant Locomotive Works, 25–26, 135, 494, 565
Grasshoppers, 66, 71, 95, 112, 137, 242, 494
Grates, 76, 106, 108–10, 348, 352, 430
Gray, Horace, 192, 194
Gray, John, 192
Grazi Tsaritzin Railway, 90
Great Western Railway, 198, 331, 514
Great Western Railway shops, 98
Greyhound, 120*n*
Griggs, George S., 4, 19, 34, 81, 88, 107,

108, 120, 178, 184, 208, 209, 226, 233, 235, 322, 540
Grimes, William C., 116, 121
"Gunboats," 161, 513
Gurney, G., 137

H

H. & F. Blandy, 158
H. R. Dunham and Company, 13, 111, 280, 281, 282
Hackworth, Timothy, 7, 147, 151, 152, 199
Half-stroke pumps, 126
Hall, Adam, 513
Hampshire, 8
Hand pumps, 125
Hannibal and St. Joseph Railroad, 123
Harris, D. M., 162, 181
Harrison, Joseph, Jr., 48, 152, 153, 272, 287, 288, 289, 540–41, 544. *See also* Eastwick and Harrison
Hayes, Samuel J., 90, 104
Hayes, Samuel L., 163, 351–52, 369
Hayward-Bartlett company, 352. *See also* Baltimore Locomotive Works
Hazleton Railroad, 87
Hazleton Shops (of Lehigh Valley Railroad), 199
Headlights, 215–17, 218, 477, 495, 520, 527
Healey, Benjamin W., 529
Hedley, William, 7, 242
Henszey, William P., 428
Herald, 7, 8, 168
Hercules, 48, 152
Hero, 68, 69
Hiawatha, 109, 143, 223, 416
Hichens and Harrison, 72
Hicksville, 215
Hill, S., 217
Hindle, Brooke, 520
Hinkley, Holmes, 14, 19, 20, 21, 26, 59, 185, 541, 545, 546
Hinkley Locomotive Works, 101*n*, 120*n*, 126, 159*n*, 165, 175, 192, 194, 198, 210, 213, 222, 226, 454–61, 483–85, 540, 541, 556
Hodge, Paul R., 280–81
Holley, Alexander L., 97, 143, 198, 201
Hollins, John S., 168
Hoosac Tunnel, 235

Horsepower, 76, 281, 287. *See also* Locomotive power
Housatonic Railroad, 175
Hovey, Jacob, 123–24
Howe, William, 196, 197
Hudson and Berkshire Railroad, 542
Hudson River Railroad, 73–74, 77, 100, 108, 141, 143, 197, 271, 320, 321, 322, 325, 328, 329, 333, 337, 341, 343, 520, 531, 542
Hudson, William S., 53, 62, 64, 65, 137, 138, 139, 144, 174, 175, 206, 408, 417, 428, 523, 541, 545
Huffman, Wendell W., 462*n*, 517
Hunter, John W., 121
Huntington and Broad Top Mountain Railroad, 350, 352*n*, 353
Huntington, C. P., 517
Huntsville, 470
Huskisson, William, 73

I

Illinois, 185
Illinois Central Railroad, 24, 77, 78, 84, 89–90, 104, 141, 183, 540
Imports, 7–12, 512
Independent cutoffs, 52, 189–92, 198–99
Indianapolis, Decatur and Springfield Railroad, 528
Injectors, 124, 126, 128–32, 417, 430, 437, 523
Inside-connected cylinders, 34, 252, 332
Inside-connected engines, 4, 14, 15, 16, 208–9
Insulation, 207, 208
Iron Mountain Railroad, 359, 398
Ithaca and Oswego Railroad, 542

J

J. and C. Carmichael, 188, 272
J. H. Devereux, 360
J. J. Chetwood, 199
Jackets. *See* Lagging
James Brook Company, 46
James S. Corry, 59, 60
James, William T., 118, 121, 187–88, 196, 297, 541, 542
Jersey City Locomotive Company, 163*n*, 204. *See also* Breese, Kneeland and Company
Jervis, John B., 12, 33, 34, 73, 77, 86, 167–68, 169, 221, 239–40, 241, 270, 320, 538, 541–42, 543, 547
Job hands, 14–15
Johann, Jacob, 516
John Bull, 7, 8, 21, 34, 37, 45, 99, 103, 116, 125, 145, 175, 176, 188, 211, 232, 248, 249, 251–54, 255–56, 257, 258–68, 529, 531, 539
John C. Calhoun, 397, 400
John Cockerill company, 143
Johns River Railroad, 8
Johnson, George W., 543
Journals, 31, 59
Joy, David, 199
Juniata, 137, 146
Juno, 134, 224*n*, 281, 297, 298

K

Kasson Locomotive Express, 516–17
Kennedy, James, 159
Ketchum, Morris, 24
Kinderhook, 328*n*
Kite, Joseph S., 173, 392
Kittatinny, 219
Knight, Jonathan, 3, 544
Knight, William, 154, 167
Krupp Works, 156, 179, 182, 183

L

L. B. Tyng and Company, 138, 182
La Junta, 449–53
Lackawanna and Bloomsburg Railroad, 219, 520
"Lackawanna Coal," 86
Lackawanna Railroad. *See* Delaware, Lackawanna and Western Railroad
Lady Elgin, 476–79
Lafayette, 160
Lagging, 30, 100–2, 207, 219, 521
Laird, John P., 65, 174, 187
Lancashire Witch, 156, 189
Lancaster, 21, 103, 111, 112, 125, 134, 177, 188, 205, 269–70, 271, 272, 273, 277–79
Lancaster Locomotive Works, 538, 544
Lapped valves, 187–88
Lardner, Dionysius, 302, 303

Latham Machine Shop, 542
Latta, A. B., 93
Lawrence Manufacturing Company, 14, 22, 24, 26, 194
Lay, J. L., 209, 210
Leading trucks, 5, 8, 16, 34, 46, 52, 62, 151, 152, 167–75, 254, 262, 313, 323, 342, 385, 392, 397, 408, 417, 422, 427, 429, 523
Leaf springs, 152–53, 156, 235, 271
Lebanon, 235
Lehigh, 106, 110
Lehigh and Mahanoy Railroad, 427, 428, 431–36
Lehigh Valley Railroad, 199, 201, 428, 429, 520, 539, 543
Lehigh Valley Railroad engine *No. 120*, 199
Leipzig and Dresden Railroad, 27
Leominster, 194
Level cylinders, 5, 14, 172
Lewis, Enoch, 130, 288, 289
Licking, 45, 226, 314
Lightning, 176n, 216
Lima Locomotive Works, 15, 120n, 123
Link motion, 14, 52, 59, 192, 194–99, 201, 297, 342, 369, 422
Linsay, William, 241
Lion, 529–30
Lisle, 483–85
Little Miami Railroad, 100, 101, 181, 517, 524
Little Schuylkill Railroad, 398
Littleton, 194n
Liverpool, 159
Liverpool and Manchester Railway, 73, 111, 113, 133, 192, 200, 203–4, 224n, 249
Liverpool Line, 133
Locks and Canals company, 8, 14, 24, 48, 213, 215, 455, 538, 539, 540, 541, 545
Locomotion, 253
Locomotive Engine Safety Truck Company, 175
Locomotive fleet management, 520
Locomotive manufacturers, general, 4–5, 8, 12, 13–28, 513. *See also by name*
Locomotive power, 71, 74–76. *See also* Horsepower
London and North Western Railway, 205
Long Island Railroad, 48n, 78, 158, 215, 271
Long, Stephen H., 12, 86, 88, 188, 189, 444, 542, 544

Louisville and Nashville Railroad, 62, 490–93
Lowell Machine Shop, 8, 232, 328, 331, 333, 337, 338, 510
Lubrication, 31, 74, 79–80, 204–6, 499, 510, 524
Luttgens, H. A., 143, 199, 358–59
Lycoming, 172
Lyon, 487

M

M. W. Baldwin and Company. *See* Baldwin Locomotive Works
McCarroll, W. J., 430
McDaniel and Horner Company, 156
McMahon, William, 182
McNeill, 8, 113, 213
McNeill, William G., 167
McQueen, Walter, 45, 51, 52, 125, 161, 163, 172–73, 192, 207, 235, 314, 321, 328, 329, 337, 542–43
Mad River and Lake Erie Railroad, 470–72
Madison, 95
Madison and Indianapolis Railroad, 95
Main rods. *See* Rods
Maine, 59
Maine Historical Society, 509
Maine State Museum, 529
Major Whistler, 297
Mallets, 66
Manchester Locomotive Works, 450
Manhattan Elevated Railroad, 223
Mann, Z. H., 138
Manometers, 134
Mansfield and Sandusky Railroad, 45
Marietta and Cincinnati Railroad, 174, 542
Marks, Joseph, 204
Marmora, 499
Mars, 224n
Martin, James, 143, 144
Martin Van Buren, 275
Maryland, 67, 257
Mason Machine Works, 194, 213, 214, 219
Mason, N. W., 215
Mason, William, 14, 94, 97, 120, 136, 172, 198, 201, 283–91, 538, 543, 544
Massachusetts, 50, 51, 101n, 120n, 126n, 159n, 164, 165, 192
Matanzas Railroad, 449

Materials: for boilers, 97–99, 103, 104; general, 22–23, 29–32, 51, 59. *See also individual components*
Matteawan, 328n
Matthew, David, 45–46, 73, 84, 116, 164, 189–90, 221, 543
Maynard, George C., 239n
Mechanic, 542
Medford, 204
Melling, John, 200
Memphis, 194, 195
Mercury, 152, 160, 171, 204
Meteor, 224, 225
Mexican Central Railroad engine *No. 3*, 223
Meyer, J. G. A., 100, 492
Michigan, 185
Michigan Central Railroad, 16, 176, 208, 280, 454
Michigan Southern and Northern Indiana Railroad, 542
Midland Railway, 205
Miles per year, or mileage, 77, 512, 520
Milford, 101
Mill Dam Foundry, 14
Miller, Ezra L., 34, 269, 270, 512–13, 518
Millholland, James, 5, 16, 19, 51, 58–59, 62, 88, 103, 104, 105–6, 108, 109, 110, 130, 137, 143, 146, 181, 183, 185, 221, 223, 304, 305, 306, 352, 353, 407, 416, 429, 539, 543, 546
Milwaukee and Mississippi Railroad, 163
Milwaukee and St. Paul Railroad, 204
Mine Hill and Schuylkill Haven Railroad, 138
Mine Hill Railroad, 101, 138, 184, 398
Mine Hill Railroad engine *No. 30*, 138, 141, 194
Minnehaha, 177n
Minnesota, 136, 353
Minnie, 184
Mississippi, 280
Mississippi and Alabama Railroad, 86
Missouri, 514–15
Mitchell, Alexander, 65, 427–28, 543–44
Mogul, 64, 407
Moguls, 104, 147, 174, 213, 232, 407–15, 416–21, 486, 540. *See also* Wheel arrangements, 2-6-0
Mohawk, 51–52, 53, 161, 173, 207, 212, 329, 542
Mohawk and Hudson Railroad, 8, 12, 33,

34, 45, 73, 86, 125, 132, 175, 176, 221, 270, 542, 543, 547–48
Monster, 66–67, 87, 106, 107, 519
Montcheuil, Jean de, 144
Montereau-Troyes Railway, 144
Morris and Essex Railroad, 126*n,* 175, 190, 200
Moscow and St. Petersburg Railway, 27, 289, 457, 546
Moses and Sons, 220*n*
Moses Starr and Sons, 271
Mott, Jordan L., 234
Mt. Clare Shops (of B & O), 19, 64, 187, 297, 357, 366, 369, 513, 546
Moving locomotives, 486
Mud Diggers, 67, 161
Mullen, James, 183

N

Nashua Iron Company, 176–77, 180
National Museum of Transport, 499, 529
Naylor and Company (Philadelphia), 208
Naylor, Vickers and Company (England), 208
Necker, 197, 226, 228
Neversink, 133
New Hampshire, 59, 519
New Hampshire Central Railroad, 222
New Jersey, 53, 56
New Jersey Locomotive and Machine Company, 62, 66, 199, 358–59, 360, 361, 407, 538, 539
New Jersey Railroad and Transportation Company, 62, 64, 72, 198, 199, 213, 231, 232, 280, 407, 408
New Jersey Railroad and Transportation Company engines: *No. 8,* 423; *No. 35,* 407, 408; *No. 36,* 407, 408–15; *No. 39,* 64, 407, 408
New Orleans, 194
New Orleans, Jacksonville and Great Northern Railroad, 194
New York, 337
New York and Erie Railroad, 52, 194, 519, 538, 545, 546
New York and Erie Railroad engines: *No. 13,* 54; *Nos. 100–105,* 551–54
New York and Harlem Railroad, 166, 280, 322, 541

New York Central Railroad, 33*n,* 45, 74, 77, 85, 98, 181, 320, 329, 520, 534
New York Central Railroad engine *No. 10,* 329
New York Elevated Railroad, 503–6, 523–24
New York Locomotive Works, 162–63, 341, 342. *See also* Breese, Kneeland and Company
New York, New Haven and Hartford Railroad, 124
Newcastle and Frenchtown Railroad, 8, 12, 86, 114, 145, 176, 249, 252, 257, 538
Newcastle and Frenchtown Railroad shops, 449
Newcastle Manufacturing Company, 48, 146, 158, 290, 537
Newport Iron Works, 181
Nicolls, G. A., 5, 88, 287, 288
Niles and Company, 161, 163, 174, 201, 204, 206*n*
Nonpareil, 66
Norfolk, 208, 540
Norris and Long, 208, 233. *See also* American Steam Carriage Company
Norris, Edward S., 226, 230
Norris Locomotive Works, 13, 17, 18, 19, 20, 21, 23–24, 48, 62, 66, 94, 102–3, 105, 111, 125, 133, 158, 160, 169, 170, 173, 179, 180, 185, 186, 188, 190, 191, 197, 212, 213, 219, 220, 222, 226, 229, 302, 303, 310, 311, 312, 313, 315, 316, 544
Norris, Octavius J., 58*n*
Norris, Richard, 24, 456
Norris, Septimus, 16, 57–58, 59, 154, 172, 544
Norris, William, 25, 27, 28, 188, 189, 215, 540, 542, 544
Norris-Portland. *See* Portland Locomotive Works
Norris-Schenectady. *See* Schenectady Locomotive Works
North America, 256
North Midland Railway engine *No. 71,* 197
North Pennsylvania Railroad, 398
North Star, 172
Northern Central Railroad, 297, 347
Northern Central Railroad engine *No. 47,* 213
Northern Railway, 476

Northumberland, 187, 398, 405
Northumbrian, 73, 224*n*
Norwich, 213
Norwick and Worcester Railroad, 213
Novelty, 5, 88, 223

O

Ohio and Mississippi Railroad, 359, 392
Oil cups, 204–5
Oiling. *See* Lubrication
Old Colony Railroad, 175
Old Ironsides, 8, 24, 249, 538
Oneida, 48*n*
Ontalaunee, 234, 289, 290
Ontario, 72
Ontario, Simcoe and Huron Railroad, 476
Opelousas, 517
Orange, 181
Orray Taft, 202
Ottawa, 524
Outside cranks, 52
Outside frames, 52
Outside-cylinder engines, 16

P

Pacific, 197, 321
Pacific Railroad (Missouri), 324
Palmer and Machiasport Railroad, 529
Panama Railroad, 200, 518, 544
Papin, Denis, 146
Paris-Orleans Railway, 202
Park, Louis L., 555
Parry, Charles T., 25*n,* 544
Parts of locomotive, 521, 564
Passaic, 22, 62
Paterson and Hudson River Railroad, 8, 22, 95*n,* 113, 213
Paterson Iron Company, 180
Patrick, 213
Patt, Ornam L., 202
Pawnee, 62
Peekskill, 328*n*
Pennsylvania, 145*n,* 197, 233
Pennsylvania Canal Commission, 8, 241
Pennsylvania Central Railroad, 212
Pennsylvania Public Works Line, 241
Pennsylvania Railroad, 19, 26–27, 59, 62, 65, 72, 74, 77, 78, 104, 122, 130, 137, 149, 174, 175, 179, 187, 216, 217, 232,

233, 252, 254, 321, 398, 405, 407, 520, 528, 539, 542
Pennsylvania Railroad engines: *No. 44*, 398; *No. 45*, 398; *No. 118*, 471, 474; *No. 128*, 398; *No. 129*, 398; *No. 214*, 130, 131
Pennsylvania Railroad shops, 252
Peoples Railway engine *No. 3*, 95, 228, 296, 311
Performance: general, 71–81; of locomotives, 33, 45, 46, 48, 57, 58, 59, 62, 65–66, 281, 287, 322, 354, 366, 429, 510
Perkins, Jacob, 142
Perkins, Thatcher, 69, 137, 182, 471, 490–93
Perrin, P. I., 15, 181
Petersburg Railroad, 113, 159
Petticoat pipe, 112
Phantom, 97, 99, 161, 177, 198, 207, 219, 233, 383–85, 387–90
Philadelphia (1838), 27, 160, 169, 219, 396
Philadelphia (1844), 66, 113, 125, 146, 147, 186, 189, 207, 223, 302, 304–9
Philadelphia and Columbia Railroad, 4, 34, 72, 77, 85, 86, 95n, 114, 121, 122, 170, 176, 178, 215, 219, 229, 234, 269, 275, 538
Philadelphia and Reading Railroad, 3, 7, 48, 57, 58–59, 62, 84, 88, 106, 113, 130, 136, 143, 146, 165, 179, 181, 182, 186, 187n, 190, 212, 221, 222, 223, 235, 272, 287, 289, 290, 297, 302, 303, 304, 347, 350, 352, 353, 354, 355, 398, 454, 512, 543
Philadelphia and Reading Railroad engine *No. 405*, 61
Philadelphia and Reading Railroad shops, 183
Philadelphia and Trenton Railroad, 269
Philadelphia, Germantown and Norristown Railroad, 24, 46, 160, 233, 538
Philadelphia, Wilmington and Baltimore Railroad, 85, 88, 181, 182
Phillip Thomas, 179
Phleger, Leonard, 105, 120, 219
Phoenix, 111, 175
Pilots. *See* Cowcatchers
Pioneer (1832), 113, 160, 224
Pioneer (1851), 126, 186n, 220, 233n
Pioneer (1869), 463
Piston pumps, 125
Piston rods, 207, 208, 272
Pistons, 204, 207–8

Pittsburgh, Fort Wayne and Columbus Railroad, 99, 316
Pittsburgh Locomotive Works, 504
Planet, 249
Planet class, 8, 186, 225, 323, 454–55, 538, 545
Ponchartrain Railroad, 114n
Pony, 494
Pony trucks, 174, 408
Poppet throttles, 5, 145–46, 304, 416, 430, 437
Portland Locomotive Works, 8, 14, 16, 19, 120n, 176–77, 470–72, 476–82, 509, 544
Pride of Newcastle, 21, 240, 241
Production statistics, 5, 7, 8, 13, 15, 16, 17, 19–23, 26, 27, 28, 45, 57, 62, 64, 65, 66, 69, 270, 347, 358, 396, 397, 398
Proprietors of Locks and Canals on the Merrimack River, 8. *See also* Locks and Canals company
Providence and Worcester Railroad, 202
Pyle, George C., 217

Q

Quillacq, M., 144

R

R. and W. Hawthorn, 97, 142, 152n, 162
Radley and Hunter, 120n, 121, 122, 123, 329
Radley, James, 121, 528
Radley, McAlister and Company, 217
Rahway, 213
Rainhill Trials, 133, 134, 240
Ramapo, 101n
Ramsbottom, John, 205
Rastrick, John U., 8, 240
Rauch, E. J., 183
Reading Railroad, 5, 24, 69, 72, 74, 78, 95n, 104, 105, 114, 133, 136, 137, 161, 176, 223, 349, 546
Reading Railroad shops, 104, 109
Rebuilding of locomotives, 449–50, 463, 511–12
Red Rover, 8
Reindeer, 222, 297
Rensselaer, 321, 337
Renwick, James, 241, 243, 281

Repair of locomotives, 79, 516, 519–20, 531
Return-crank pumps, 126
Reuter, Emil, 304, 305
Rhode Island Locomotive Works, 202, 495–96, 499, 504, 529
Richards, George, 184
Richardson, George W., 147, 148, 149, 203
Richmond, 103, 302–4, 305
Roanoke, 159n, 463, 466
Robert Fulton, 8, 34, 36, 37, 94, 112, 145, 156, 175, 186, 253–54
Robert Stephenson and Company, 7, 8, 12, 21, 34, 66, 94, 98, 99, 102, 103, 124n, 125, 145, 147, 158, 162, 169, 175–76, 196, 197, 198, 201, 204, 225, 232, 240, 249, 251, 253, 255, 257, 271, 342. *See also* Stephenson, Robert
Rochester, 45, 125
Rochester, Lockport and Niagara Falls Railroad, 83–84
Rocket, 102, 103, 133, 159, 192, 444, 512
Rockport, 45
Rods: connecting, 31, 62, 66, 553–54; general, 184–87; main, 48, 50; piston, 207, 208, 272; solid end, 353
Rogers, Ethan, 192, 193
Rogers, Ketchum and Grosvenor, 541, 545, 551
Rogers Locomotive Works, 13, 14, 16, 20, 21, 22, 23, 24, 28, 45, 48, 50, 53, 59, 62, 64, 66, 72, 75, 77, 94, 101n, 105, 111, 112, 123, 125–26, 137, 138, 147, 158, 161, 164, 166, 172, 173, 174, 175, 177, 185, 187, 198, 199, 203, 206, 207, 208, 213, 216, 219, 226, 231, 232, 271, 281, 297, 358, 359, 383, 407, 408–14, 422–26, 428, 449, 519, 542, 545, 555
Rogers, Nathan, 215, 216
Rogers, Thomas, 8, 28, 52–53, 95, 97, 101, 172, 179, 190, 197, 215, 280, 281, 321, 366, 539, 544–45. *See also* Rogers Locomotive Works
Rothwell and Hick, 8, 113
Rough and Ready, 8, 297
Roxbury, 235
Royal George, 137, 152
Royal Württemberg Railway, 27
Running gear, 48, 58, 151–210, 278, 288, 317
Russell and Howells, 100
Russia Iron, 101, 521, 528

S

Sacramento Valley Railroad, 462–63
Safety valves, 146–49, 342, 378, 416, 498, 531
St. Helens Railway, 128
Sam Gutherie, 136
Sam Sloan, 392
Sampson, 249
Sampson class, 34, 66, 168, 252
Sandboxes, 30, 34, 58, 234, 305, 313, 353, 449, 456, 495, 504
Sandusky (1837), 177, 215, 280–81, 545
Sandusky (1851), 470–72
Santiago De Cuba, 398, 406
Saratoga and Schenectady Railroad, 12, 168, 169, 212
Saratoga and Whitehall Railroad, 359
Schenectady Locomotive Works, 20, 21, 98, 163, 216, 226, 230, 542, 543, 544
Schmidt, William, 144
Schultz, William, 121–22
Schuylkill, 398
Scotia, 98
Scott, Irving M., 487
Sellers, Charles, 34, 235
Sellers, Coleman, 206
Sellers, George E., 34, 57–58, 159, 170, 178, 215, 235
Sellers, William, 128. *See also* William Sellers and Company
Seminole, 219
Seneca, 322
Serrell, Edward W., 235
Sharp, Stewart and Company, 128
Shaw and Justice, 134
Shepard Iron Works, 209, 210
Shipping of locomotives, 486, 516–18
Short-stroke pumps, 125–26
Shutt End Railway, 240
Silsby Manufacturing Company, 127
Simpson, John, 241
Sinclair, Angus, 383
Sing Sing, 328n
Smith, Alba F., 108, 112, 143, 173–74, 175
Smith and Perkins, 62, 126, 471, 473
Smith, James, 122
Smith, Samuel, 545
Smokeboxes, 31, 59, 111, 206–7, 208, 253, 271, 312–13, 358, 370, 422
Smokestack heaters, 138, 140–41

Smokestacks, 30, 31, 114, 116–22, 253, 297, 313, 353, 370, 410, 521–22, 528
Snook, T., 217
South Carolina, 167
South Carolina Railroad, 12, 13, 59, 73, 159, 167, 215, 223, 270, 297, 509–10, 521
Souther and Anderson, 159n
Souther, John, 14, 22, 24, 219, 222, 232, 451, 453, 455, 462–64, 545
Southern Pacific Railroad, 233, 529, 534
Southern Railroad of Chile, 138
Southport, 99, 112, 161, 392–95
Spark arrestors, 111, 114–24, 504
Specifications: *Atlantic*, 467; Baldwin 4–4–0, 46; Baldwin Ten Wheeler, 437, 438; B & O engine *No.* 600, 64–65; B & O 0–8–0, 461–62; *Champlain*, 323; *Chesapeake*, 58; *Columbia*, 337; *Consolidation*, 429, 430; *Copiapo*, 314; *Croton*, 330; Erie Mogul, 417; flexible-beam-truck engine, 399; Grant 4–4–0, 565–72; *Gowan and Marx*, 287, Hinkley 4–4–0 (1865), 556–64; *James S. Corry*, 59; *John Bull*, 248–49; *Lancaster*, 269, 272; New York and Erie 1851 engines, 551–54; *Phantom*, 385; *Philadelphia*, 305; *Richmond*, 302–3; Rogers 4–4–0, 422; Rogers Mogul, 408; *Southport*, 392; *Susquehanna*, 347, 349, 352, 354; *Talisman*, 359; Tyson Ten Wheeler, 367, 369; *Winans* 4–4–0, 297
Speeds, 71, 72, 73–74, 77, 240, 251, 281, 287, 320, 322, 328, 337, 349, 351, 354, 366
Speedwell Iron Works, 180
Spitfire, 512
Spuyten Duyvil, 328n
State of Maine, 120n
Steam pressure, 5, 7, 45, 46, 72n, 75, 76, 96, 281, 288
Stephenson, George, 156, 159, 187, 240
Stephenson, Robert, 7, 8, 12, 73, 111, 112, 156, 159, 167, 168, 176, 184, 186, 188, 189, 194, 198, 224, 240, 248, 252, 253, 322, 323. *See also* Robert Stephenson and Company
Stevens, A. J., 199, 201, 534
Stevens, John, 93
Stevens, Robert L., 189, 248, 249, 252, 253, 254, 539
Stevenson, David, 211

Stimpson, James, 121
Stockbridge, 175
Stonington Line, 510
Storey, 486–89
Storm, W. M., 144
Stourbridge Lion, 21, 86, 137, 159, 175, 239, 240–41, 242, 243, 245–47
Strait, Hiram, 203
Strickland, William, 3
Stuart, John, 241
Stubbs and Gryll, 108
Stub key, 59
Subtreasury, 120
Suffolk, 495–96
Sullivan, 126
Superheaters, 5, 142–44
Superior, 163, 173, 176n, 198, 207, 330, 341–46
Suspension, 46, 48, 62, 64, 151–57, 235, 271, 353, 359, 369, 385, 408, 417, 553
Susquehanna, 58, 59, 98, 110, 163, 190, 214, 215, 271, 321, 349, 352–55
Sussex, 126n, 190
Sweet, Allen S., 212, 215
Swinburne, William, 24, 53, 172, 271, 280n, 358, 545
Switchers, 484, 494–97, 512
Syracuse and Utica Railroad, 172
Syracuse, Binghamton and New York Railroad, 483

T

Taber, Thomas T., 213
Taber, Thomas T., III, 483, 513
Talisman, 186, 199, 232, 358–59, 361–65
Tank locomotives, 233, 494, 503–4, 512
Taunton Locomotive Manufacturing Company, 14, 15, 26, 64, 88, 101n, 104, 112, 179, 192, 194, 220, 322, 324, 325
Tayleur, Charles. *See* Charles Tayleur and Company
Telegraph, 204
Tenders, 30, 34, 58, 59, 76, 223–33, 235, 247, 254, 266, 267, 288, 295, 297, 315, 330, 336, 337, 354, 359, 365, 369, 380, 381, 385, 390, 391, 404, 408, 415, 436, 450, 456, 460–61, 469, 477, 482, 495, 529, 554
Tennessee, 397
Terre Haute and Indianapolis Railroad, 233

INDEX

Thomas Prosser and Son, 183
Thomas Rogers, 219
Thomas, William, 117
Thompson, John, 123
Throttles, 52, 59, 145–46, 260, 271, 304, 341, 353, 358, 378, 392, 398–99, 531
Tiger (1842), 529
Tiger (1856), 219
Tioga, 52, 54, 161, 186, 190, 226, 229, 232, 312
Tip Top, 187n
Tires, 31, 32, 59, 179–83, 470
Toledo and Illinois Railroad, 383, 387
Tolles, Elisha, 34, 234
Tom Thumb, 86, 95, 543
Tracks, 62, 71, 74, 77, 93
Traction, 34, 46, 71, 72n, 75, 76, 174, 234–35, 321, 397. *See also* Sandboxes
Train size, 71, 72, 487, 503, 510, 519
Tredegar Iron Works, 15, 451, 463, 545
Trenton Locomotive Works, 66, 249, 539, 543
Trevithick and Vivian, 137
Trevithick 1808 engine, 240
Trevithick, Richard, 111, 142
Troy and Boston Railroad, 148, 149, 158, 271
Truck frames, 48, 51, 167, 168, 169–70, 171, 172–73, 314, 330, 337
Trucks, 5, 14, 52, 58, 62, 64, 65, 66, 69, 151, 167–75, 170–71, 271, 318, 330, 337, 354, 359, 369, 370, 375, 382, 385, 389, 396, 397, 408, 422, 430, 435, 523. *See also* Leading trucks; Wheels
Tuscarora, 350, 353
Tuscumbia, Courtland and Decatur Railroad, 159, 234
Tyng, L. B., 138, 182
Tyson, Henry, 99, 103, 113, 124, 144, 351–52, 366, 367–69
Tyson Ten Wheelers, 163, 177, 182, 187, 207, 232, 233, 366–82, 389

U

Uhry, H., 143, 199, 358–59
Uncle Sam, 72, 231
Underhill, Arthur B., 499
Union Iron Works, 15, 487
Union Pacific Railway, 219, 521, 523, 524, 528

Union Works, 546
United of Yucatan Railways, 57
United Railroads of New Jersey, 407
United States Military Railroads, 23, 213, 360, 390
Urquhart, Thomas, 90
Useful, 495
Utah and Northern Railroad, 523
Utica and Schenectady Railroad, 45, 84, 116n, 177–78, 189–90, 211, 221, 542, 543
Utica and Schenectady Railroad engines: *No. 1,* 178, 271, 274; *No. 11,* 190
Utica and Syracuse Railway, 48, 518
Utica Head Light Works, 528

V

Vail, Stephen, 180
Valve gears, 30, 31, 34, 52, 74, 187–202, 242, 253, 260, 268, 272, 288, 289, 304–5, 313, 326, 329, 354, 358–59, 369, 376, 384–85, 398, 399, 450, 456, 531, 534
Valve ports, 203, 204, 288
Variable cutoffs, 190, 192–94, 195, 463
Variable exhausts, 113–14, 304, 348, 367, 374, 398, 430
Vermont Central Railroad, 86, 321, 539
Viaduct, 209
Vickers steel, 514
Victoria, 224n
Victory, 216
Virginia, 117, 126, 145n, 154, 155, 186
Virginia and Tennessee Railroad, 195, 466
Virginia and Truckee Railroad, 486–89
Volcano, 66, 187
Volga-Don Railway, 288
Von Gerstner, F. A. R., 73
Vose, George L., 85
Vulcan, 98, 222
Vulcan Foundry, 7, 187. *See also* Charles Tayleur and Company
Vulcan Iron Works, 15

W

W. C. Alderson, 199
W. E. Dodge, 392
Wabash Railroad, 232
Wakeley, William, 215, 216
Walker, Herbert L., 392n

Walloonsac, 149
Walschaert, Eglide, 199, 201
Ward, Henry, 216–17
Ward, Joseph, 303
Washington, 44, 219
Water: consumption of, 76, 223, 519; effect of impurities in, 96
Water tanks, 65, 223–26, 232
Waterman, Henry, 116, 153
Watkins, J. E., 249, 252
Watson, 161, 164, 518
Watt, James, 75
Weights, of components, 556–64
Weights, of locomotives, 6, 21–22, 25, 26, 46, 48, 51, 52, 57, 58, 64, 66, 71, 72, 75, 76, 287, 288, 428–29, 438
Weissenborn, Gustavus, 220, 358, 359, 408
Wellington, Arthur M., 494, 509, 519–20
Wells, Reuben, 136
West Point, 66, 223, 328n
West Point Foundry Association of New York City, 13, 21, 23, 33, 35, 145, 158, 159, 167, 175, 176, 208, 223, 240, 510, 513, 524, 543, 547–48
Western and Atlantic Railroad, 85, 194, 555
Western Railroad of Massachusetts, 8, 67, 95, 120n, 190, 192, 203, 204, 222, 226, 231, 498–99, 539, 545
Western Railroad shops, 205, 498–99
Wheel arrangements: general, 33–69; *0–4–0,* 8, 34, 66–69, 95n, 159, 168, 249; *0–6–0,* 25, 33, 62, 66–69, 167, 174, 187, 302, 396, 397, 398; *0–8–0,* 22, 62, 66–69, 95, 163, 174, 184, 187, 398, 427, 429, 545, 549–50; *0–12–0,* 429; *2–2–0,* 175, 249; *2–2–2,* 184; *2–4–0,* 197, 397; *2–6–0,* 22, 57, 62–65, 72, 75, 174, 196, 398, 407, 416, 427, 444; *2–8–0,* 22, 57, 64, 65–66, 72, 75, 196, 444, 543–44; *2–8–2,* 544; *2–10–0,* 544; *4–2–0,* 8, 12, 21, 25, 33–46, 48, 59, 72, 110, 145, 151, 159, 160, 168, 171n, 175, 186, 189, 235, 270, 271, 276, 302, 397, 444, 538, 541; *4–2–2,* 175, 235, 321; *4–2–4,* 175; *4–4–0,* 21, 22, 25, 26, 33, 45, 46–57, 59, 62, 64, 65, 66, 72, 75, 96, 102, 110, 112, 145, 147, 152, 154, 161, 164, 171, 174, 176, 186, 190, 197, 198, 199, 202, 207, 208, 213, 231, 235, 287, 288, 297, 299, 311, 314, 321, 329, 422, 444, 445, 538, 539, 540, 543, 545, 546, 556–64, 565–72; *4–4–4,* 175; *4–6–*

0, 33, 57–62, 186, 199, 398, 427, 438, 538; 4–8–0, 208, 350, 546. *See also* Baldwin Ten Wheelers; Tyson Ten Wheelers
Wheels, 7, 16, 31, 76, 168–69, 175–77, 242–43, 253–54, 263, 271, 288–89, 327, 353, 398, 400, 435, 553. *See also* Drivers
Wheelwright, William, 311
Whetstone, John L., 174, 200, 201
Whistler, 7
Whistler, George Washington, 3, 27, 167, 544, 545
Whistles, 30, 213, 214–15, 524
Whiting, E. C., 269–70
Whitney, Asa, 34, 327, 330
Whyte, Fredric M., 33*n*
Wilder, R. A., 101, 102, 138
Wilke, Jim, 523

Wm. H. Watson, 51, 52
Wm. Mason, 201
William Penn, 484
William Sellers and Company, 4, 13, 128, 129, 130, 132, 200, 206
Williams, William, 196–97
Wilmarth, Seth, 24, 233, 546
Wilmington and Raleigh Railroad, 397, 400
Wilmington and Susquehanna Railroad, 214
Wilmore, 471, 473
Winans, Ross, 3, 5, 19, 22, 27, 34, 66, 67, 69, 88, 95, 101, 102, 103, 105, 106, 108, 110, 112, 113, 117, 125, 136, 137, 145, 147, 152, 156, 161, 163, 170, 175, 177, 181, 182, 186, 187, 190, 208, 221, 226, 231–32, 235, 288, 298, 299, 300, 301, 304, 337, 347–54, 366, 367, 369, 398, 539, 540, 545, 546
Winans, Thomas, 27, 203, 540–41, 546
Winchester and Potomac Railroad, 154, 155
Windale, H., 249
Wood, Nicholas, 73, 167, 240
Wootten, John E., 106, 136, 137, 546
Worthington, Henry R., 127
Wurts, Charles S., 241
Württemberg State Railway, 197, 226, 228, 397
Wyoming, 134, 219, 220

Y

Yates, 519
Young, E. A. G., 5, 145–46
Young, O. W., 202

Library of Congress Cataloging-in-Publication Data

White, John H., 1933–
    American locomotives : an engineering history, 1830–1880 / by
John H. White, Jr.—Rev. and expanded ed.
      p.   cm.
    Includes bibliographical references and index.
    ISBN 0-8018-5714-7 (alk. paper)
    1. Steam locomotives—United States—History.   I. Title.
TJ603.2.W48   1997
625.26′1′0973—DC21   97-18800
                                                                         CIP

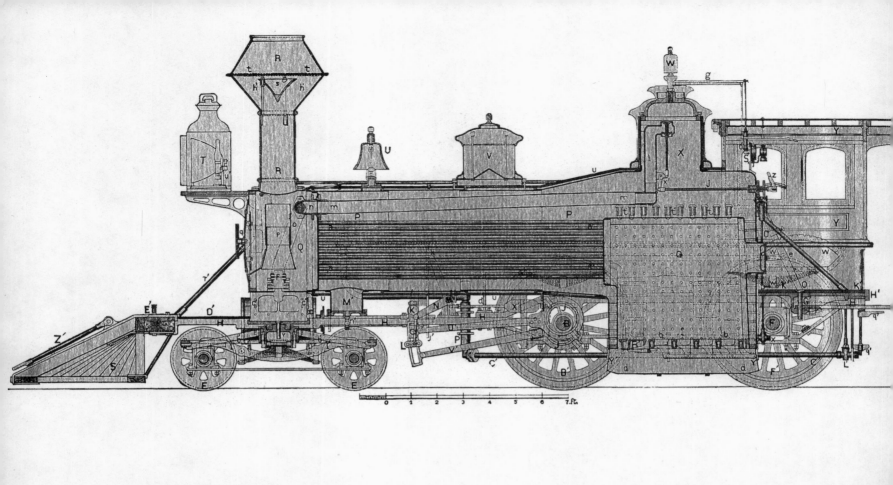
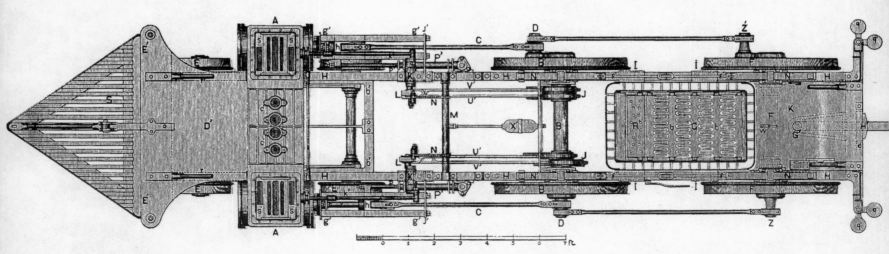